Undergraduate Lecture Notes in Physics

Undergraduate Lecture Notes in Physics (ULNP) publishes authoritative texts covering topics throughout pure and applied physics. Each title in the series is suitable as a basis for undergraduate instruction, typically containing practice problems, worked examples, chapter summaries, and suggestions for further reading.

ULNP titles must provide at least one of the following:

- An exceptionally clear and concise treatment of a standard undergraduate subject.
- A solid undergraduate-level introduction to a graduate, advanced, or non-standard subject.
- A novel perspective or an unusual approach to teaching a subject.

ULNP especially encourages new, original, and idiosyncratic approaches to physics teaching at the undergraduate level.

The purpose of ULNP is to provide intriguing, absorbing books that will continue to be the reader's preferred reference throughout their academic career.

Series editors

Neil Ashby
Professor Emeritus, University of Colorado, Boulder, CO, USA

William Brantley
Professor, Furman University, Greenville, SC, USA

Michael Fowler
Professor, University of Virginia, Charlottesville, VA, USA

Morten Hjorth-Jensen
Professor, University of Oslo, Oslo, Norway

Michael Inglis
Professor, SUNY Suffolk County Community College, Long Island, NY, USA

Heinz Klose
Professor Emeritus, Humboldt University Berlin, Germany

Helmy Sherif
Professor, University of Alberta, Edmonton, AB, Canada

More information about this series at http://www.springer.com/series/8917

Siegfried Hess

Tensors for Physics

 Springer

Siegfried Hess
Institute for Theoretical Physics
Technical University Berlin
Berlin
Germany

ISSN 2192-4791   ISSN 2192-4805   (electronic)
Undergraduate Lecture Notes in Physics
ISBN 978-3-319-12786-6   ISBN 978-3-319-12787-3   (eBook)
DOI 10.1007/978-3-319-12787-3

Library of Congress Control Number: 2015936466

Springer Cham Heidelberg New York Dordrecht London

Printed on acid-free paper

Springer International Publishing AG Switzerland is part of Springer Science+Business Media (www.springer.com)

Preface

Tensors are needed in Physics to describe anisotropies and orientational behavior. While every physics student knows what a vector is, there is often an uneasiness about the notion *tensor*. In lectures, I used to tell students: "you can be a good physicist without knowing much about tensors, but when you learn how to handle tensors and what they are good for, you will have a considerable advantage. And here is your chance to learn about tensors as a mathematical tool and to get familiar with their applications to physics."

This book is, up to Chap. 14, largely based on the two books:

Siegfried Hess, Vektor- und Tensor-Rechnung, which, in turn, was based on lectures for first-year physics students, and
Siegfried Hess and Walter Köhler, Formeln zur Tensor-Rechnung, a collection of computational rules and formulas needed in more advanced theory.

Both books were published by *Palm and Enke*, Erlangen, Germany in 1980, reprinted in 1982, but are out of print since many years.

Here, the emphasis is on Cartesian tensors in 3D. The applications of tensors to be presented are strongly influenced by my presentations of the standard four courses in Theoretical Physics: Mechanics, Quantum Mechanics, Electrodynamics and Optics, Thermodynamics and Statistical Physics, and by my research experience in the kinetic theory of gases of particles with spin and of rotating molecules, in transport, orientational and optical phenomena of molecular fluids, liquid crystals and colloidal dispersions, in hydrodynamics and rheology, as well as in the elastic and plastic properties of solids. The original publications cited, in particular in the second part of the book, show a wide range of applications of tensors. An outlook to 4D is provided in Chap. 18, where the Maxwell equations of electrodynamics are formulated in the appropriate four-dimensional form.

While learning the mathematics, first- and second-year students may skip the applications involving physics they are not yet familiar with, however, brief introductions to basic physics are given at many places in the book. Exercises are found throughout the book, answers and solutions are given at the end.

Here, I wish to express my gratitude to Prof. Ludwig Waldmann (1913–1980), who introduced me to Cartesian Tensors, quite some time ago, when I was a student. I thank my master- and PhD-students, postdocs, co-workers, and colleagues for fruitful cooperation on research projects, where tensors played a key role. I am grateful to Springer for publishing this Tensor book in the series *Undergraduate Lecture Notes in Physics*, and I thank Adelheid Duhm, Project Coordinator at Production Physics Books of Springer in Heidelberg for her diligent editorial work.

Berlin   Siegfried Hess

Contents

Part I   A Primer on Vectors and Tensors

1   Introduction   3
   1.1   Preliminary Remarks on Vectors   3
       1.1.1   Vector Space   3
       1.1.2   Norm and Distance   5
       1.1.3   Vectors for Classical Physics   6
       1.1.4   Vectors for Special Relativity   7
   1.2   Preliminary Remarks on Tensors   7
   1.3   Remarks on History and Literature   8
   1.4   Scope of the Book   9

2   Basics   11
   2.1   Coordinate System and Position Vector   11
       2.1.1   Cartesian Components   11
       2.1.2   Length of the Position Vector, Unit Vector   12
       2.1.3   Scalar Product   13
       2.1.4   Spherical Polar Coordinates   14
   2.2   Vector as Linear Combination of Basis Vectors   14
       2.2.1   Orthogonal Basis   14
       2.2.2   Non-orthogonal Basis   15
   2.3   Linear Transformations of the Coordinate System   16
       2.3.1   Translation   16
       2.3.2   Affine Transformation   17
   2.4   Rotation of the Coordinate System   19
       2.4.1   Orthogonal Transformation   19
       2.4.2   Proper Rotation   21
   2.5   Definitions of Vectors and Tensors in Physics   22
       2.5.1   Vectors   22
       2.5.2   What is a Tensor?   23

2.5.3   Multiplication by Numbers and Addition
        of Tensors   23
2.5.4   Remarks on Notation.   24
2.5.5   Why the Emphasis on Tensors?   24
2.6   Parity   25
2.6.1   Parity Operation   25
2.6.2   Parity of Vectors and Tensors.   26
2.6.3   Consequences for Linear Relations   27
2.6.4   Application: Linear and Nonlinear Susceptibility
        Tensors   27
2.7   Differentiation of Vectors and Tensors with Respect
      to a Parameter   28
2.7.1   Time Derivatives   28
2.7.2   Trajectory and Velocity   29
2.7.3   Radial and Azimuthal Components of the Velocity . . .   30
2.8   Time Reversal   30

3   **Symmetry of Second Rank Tensors, Cross Product**   33
3.1   Symmetry   33
3.1.1   Symmetric and Antisymmetric Parts   33
3.1.2   Isotropic, Antisymmetric and Symmetric
        Traceless Parts   34
3.1.3   Trace of a Tensor   34
3.1.4   Multiplication and Total Contraction of Tensors,
        Norm   35
3.1.5   Fourth Rank Projections Tensors   36
3.1.6   Preliminary Remarks on "Antisymmetric Part
        and Vector"   37
3.1.7   Preliminary Remarks on the Symmetric
        Traceless Part.   37
3.2   Dyadics   37
3.2.1   Definition of a Dyadic Tensor   37
3.2.2   Products of Symmetric Traceless Dyadics   38
3.3   Antisymmetric Part, Vector Product   40
3.3.1   Dual Relation   40
3.3.2   Vector Product   41
3.4   Applications of the Vector Product   43
3.4.1   Orbital Angular Momentum   43
3.4.2   Torque   43
3.4.3   Motion on a Circle   44
3.4.4   Lorentz Force.   45
3.4.5   Screw Curve   45

4 **Epsilon-Tensor** 47
4.1 Definition, Properties. 47
    4.1.1 Link with Determinants 47
    4.1.2 Product of Two Epsilon-Tensors. 48
    4.1.3 Antisymmetric Tensor Linked with a Vector 50
4.2 Multiple Vector Products 50
    4.2.1 Scalar Product of Two Vector Products 50
    4.2.2 Double Vector Products. 50
4.3 Applications. 51
    4.3.1 Angular Momentum for the Motion on a Circle 51
    4.3.2 Moment of Inertia Tensor 52
4.4 Dual Relation and Epsilon-Tensor in 2D 53
    4.4.1 Definitions and Matrix Notation 53

5 **Symmetric Second Rank Tensors.** 55
5.1 Isotropic and Symmetric Traceless Parts 55
5.2 Principal Values 56
    5.2.1 Principal Axes Representation 56
    5.2.2 Isotropic Tensors 56
    5.2.3 Uniaxial Tensors. 57
    5.2.4 Biaxial Tensors. 58
    5.2.5 Symmetric Dyadic Tensors 59
5.3 Applications. 60
    5.3.1 Moment of Inertia Tensor of Molecules. 60
    5.3.2 Radius of Gyration Tensor. 62
    5.3.3 Molecular Polarizability Tensor 63
    5.3.4 Dielectric Tensor, Birefringence 63
    5.3.5 Electric and Magnetic Torques 64
5.4 Geometric Interpretation of Symmetric Tensors. 65
    5.4.1 Bilinear Form. 65
    5.4.2 Linear Mapping 66
    5.4.3 Volume and Surface of an Ellipsoid 67
5.5 Scalar Invariants of a Symmetric Tensor 69
    5.5.1 Definitions. 69
    5.5.2 Biaxiality of a Symmetric Traceless Tensor 69
5.6 Hamilton-Cayley Theorem and Consequences. 71
    5.6.1 Hamilton-Cayley Theorem. 71
    5.6.2 Quadruple Products of Tensors. 72
5.7 Volume Conserving Affine Transformation 73
    5.7.1 Mapping of a Sphere onto an Ellipsoid 73
    5.7.2 Uniaxial Ellipsoid. 73

6   Summary: Decomposition of Second Rank Tensors   75

7   Fields, Spatial Differential Operators.   77
     7.1   Scalar Fields, Gradient.   78
            7.1.1   Graphical Representation of Potentials.   78
            7.1.2   Differential Change of a Potential, Nabla Operator . . .   79
            7.1.3   Gradient Field, Force.   79
            7.1.4   Newton's Equation of Motion, One and More
                     Particles.   80
            7.1.5   Special Force Fields   81
     7.2   Vector Fields, Divergence and Curl or Rotation   84
            7.2.1   Examples for Vector Fields   84
            7.2.2   Differential Change of a Vector Fields.   88
     7.3   Special Types of Vector Fields.   90
            7.3.1   Vorticity Free Vector Fields, Scalar Potential   90
            7.3.2   Poisson Equation, Laplace Operator   91
            7.3.3   Divergence Free Vector Fields, Vector Potential.   91
            7.3.4   Vorticity Free and Divergence Free Vector Fields,
                     Laplace Fields   93
            7.3.5   Conventional Classification of Vector Fields   94
            7.3.6   Second Spatial Derivatives of Spherically
                     Symmetric Scalar Fields   94
     7.4   Tensor Fields   95
            7.4.1   Graphical Representations of Symmetric Second
                     Rank Tensor Fields.   95
            7.4.2   Spatial Derivatives of Tensor Fields   96
            7.4.3   Local Mass and Momentum Conservation,
                     Pressure Tensor   97
     7.5   Maxwell Equations in Differential Form   98
            7.5.1   Four-Field Formulation   98
            7.5.2   Special Cases.   100
            7.5.3   Electromagnetic Waves in Vacuum.   101
            7.5.4   Scalar and Vector Potentials.   102
            7.5.5   Magnetic Field Tensors   103
     7.6   Rules for Nabla and Laplace Operators   105
            7.6.1   Nabla   105
            7.6.2   Application: Orbital Angular Momentum Operator . . .   106
            7.6.3   Radial and Angular Parts of the Laplace Operator. . . .   108
            7.6.4   Application: Kinetic Energy Operator in Wave
                     Mechanics   108

8   Integration of Fields   111
     8.1   Line Integrals.   111
            8.1.1   Definition, Parameter Representation.   111
            8.1.2   Closed Line Integrals   112

8.1.3   Line Integrals for Scalar and Vector Fields 113
8.1.4   Potential of a Vector Field 114
8.1.5   Computation of the Potential for a Vector Field 115
8.2   Surface Integrals, Stokes 117
8.2.1   Parameter Representation of Surfaces 117
8.2.2   Examples for Parameter Representations
of Surfaces 118
8.2.3   Surface Integrals as Integrals Over Two
Parameters 120
8.2.4   Examples for Surface Integrals 121
8.2.5   Flux of a Vector Field 123
8.2.6   Generalized Stokes Law 124
8.2.7   Application: Magnetic Field Around an Electric
Wire 127
8.2.8   Application: Faraday Induction 128
8.3   Volume Integrals, Gauss 129
8.3.1   Volume Integrals in R^3 129
8.3.2   Application: Mass Density, Center of Mass 131
8.3.3   Application: Moment of Inertia Tensor 134
8.3.4   Generalized Gauss Theorem 136
8.3.5   Application: Gauss Theorem in Electrodynamics,
Coulomb Force 138
8.3.6   Integration by Parts 140
8.4   Further Applications of Volume Integrals 140
8.4.1   Continuity Equation, Flow Through a Pipe 140
8.4.2   Momentum Balance, Force on a Solid Body 142
8.4.3   The Archimedes Principle 143
8.4.4   Torque on a Rotating Solid Body 144
8.5   Further Applications in Electrodynamics 145
8.5.1   Energy and Energy Density in Electrostatics 145
8.5.2   Force and Maxwell Stress in Electrostatics 146
8.5.3   Energy Balance for the Electromagnetic Field 147
8.5.4   Momentum Balance for the Electromagnetic Field,
Maxwell Stress Tensor 149
8.5.5   Angular Momentum in Electrodynamics 151

Part II   Advanced Topics

9   Irreducible Tensors 155
9.1   Definition and Examples 155
9.2   Products of Irreducible Tensors 157
9.3   Contractions, Legendre Polynomials 157
9.4   Cartesian and Spherical Tensors 158

9.4.1   Spherical Components of a Vector   158
           9.4.2   Spherical Components of Tensors   159
   9.5   Cubic Harmonics   161
           9.5.1   Cubic Tensors   161
           9.5.2   Cubic Harmonics with Full Cubic Symmetry   161

10   Multipole Potentials   163
   10.1   Descending Multipoles   163
          10.1.1   Definition of the Multipole Potential Functions.   163
          10.1.2   Dipole, Quadrupole and Octupole Potentials.   164
          10.1.3   Source Term for the Quadrupole Potential   164
          10.1.4   General Properties of Multipole Potentials   165
   10.2   Ascending Multipoles   166
   10.3   Multipole Expansion and Multipole Moments
         in Electrostatics   167
          10.3.1   Coulomb Force and Electrostatic Potential   167
          10.3.2   Expansion of the Electrostatic Potential   168
          10.3.3   Electric Field of Multipole Moments   170
          10.3.4   Multipole Moments for Discrete Charge
                  Distributions   171
          10.3.5   Connection with Legendre Polynomials   172
   10.4   Further Applications in Electrodynamics   172
          10.4.1   Induced Dipole Moment of a Metal Sphere   172
          10.4.2   Electric Polarization as Dipole Density   173
          10.4.3   Energy of Multipole Moments in an External Field. . .   174
          10.4.4   Force and Torque on Multipole Moments
                  in an External Field   175
          10.4.5   Multipole–Multipole Interaction   176
   10.5   Applications in Hydrodynamics   177
          10.5.1   Stationary and Creeping Flow Equations   177
          10.5.2   Stokes Force on a Sphere   178

11   Isotropic Tensors   183
   11.1   General Remarks on Isotropic Tensors.   183
   11.2 Δ-Tensors   184
          11.2.1   Definition and Examples   184
          11.2.2   General Properties of Δ-Tensors.   184
          11.2.3 Δ-Tensors as Derivatives of Multipole Potentials   186
   11.3   Generalized Cross Product, $\square$-Tensors.   186
          11.3.1   Cross Product via the $\square$-Tensor   186
          11.3.2   Properties of $\square$-Tensors.   188

	11.3.3	Action of the Differential Operator $\mathscr{L}$ on Irreducible Tensors	189
	11.3.4	Consequences for the Orbital Angular Momentum Operator	190
11.4		Isotropic Coupling Tensors	191
	11.4.1	Definition of $\Delta^{(\ell,2,\ell)}$-Tensors	191
	11.4.2	Tensor Product of Second Rank Tensors	192
11.5		Coupling of a Vector with Irreducible Tensors	193
11.6		Coupling of Second Rank Tensors with Irreducible Tensors	194
11.7		Scalar Product of Three Irreducible Tensors	195
	11.7.1	Scalar Invariants	195
	11.7.2	Interaction Potential for Uniaxial Particles	196

12 Integral Formulae and Distribution Functions **199**

12.1		Integrals Over Unit Sphere	199
	12.1.1	Integrals of Products of Two Irreducible Tensors	200
	12.1.2	Multiple Products of Irreducible Tensors	201
12.2		Orientational Distribution Function	202
	12.2.1	Orientational Averages	202
	12.2.2	Expansion with Respect to Irreducible Tensors	203
	12.2.3	Anisotropic Dielectric Tensor	204
	12.2.4	Field-Induced Orientation	205
	12.2.5	Kerr Effect, Cotton-Mouton Effect, Non-linear Susceptibility	208
	12.2.6	Orientational Entropy	209
	12.2.7	Fokker-Planck Equation for the Orientational Distribution	210
12.3		Averages Over Velocity Distributions	212
	12.3.1	Integrals Over the Maxwell Distribution	213
	12.3.2	Expansion About an Absolute Maxwell Distribution	214
	12.3.3	Kinetic Equations, Flow Term	216
	12.3.4	Expansion About a Local Maxwell Distribution	218
12.4		Anisotropic Pair Correlation Function and Static Structure Factor	221
	12.4.1	Two-Particle Density, Two-Particle Averages	222
	12.4.2	Potential Contributions to the Energy and to the Pressure Tensor	223
	12.4.3	Static Structure Factor	224
	12.4.4	Expansion of $g(\mathbf{r})$	225
	12.4.5	Shear-Flow Induced Distortion of the Pair Correlation	227

        12.4.6 Plane Couette Flow Symmetry   229
        12.4.7 Cubic Symmetry.   231
        12.4.8 Anisotropic Structure Factor.   232
   12.5 Selection Rules for Electromagnetic Radiation   234
        12.5.1 Expansion of the Wave Function   234
        12.5.2 Electric Dipole Transitions.   236
        12.5.3 Electric Quadrupole Transitions   237

13 Spin Operators.   239
   13.1 Spin Commutation Relations   239
        13.1.1 Spin Operators and Spin Matrices.   239
        13.1.2 Spin 1/2 and Spin 1 Matrices   240
   13.2 Magnetic Sub-states   241
        13.2.1 Magnetic Quantum Numbers and Hamilton Cayley. . . .   241
        13.2.2 Projection Operators into Magnetic Sub-states   241
   13.3 Irreducible Spin Tensors   242
        13.3.1 Defintions and Examples.   242
        13.3.2 Commutation Relation for Spin Tensors   243
        13.3.3 Scalar Products.   244
   13.4 Spin Traces   245
        13.4.1 Traces of Products of Spin Tensors.   245
        13.4.2 Triple Products of Spin Tensors   246
        13.4.3 Multiple Products of Spin Tensors   247
   13.5 Density Operator   247
        13.5.1 Spin Averages   247
        13.5.2 Expansion of the Spin Density Operator   248
        13.5.3 Density Operator for Spin 1/2 and Spin 1   249
   13.6 Rotational Angular Momentum of Linear Molecules,
         Tensor Operators   250
        13.6.1 Basics and Notation   250
        13.6.2 Projection into Rotational Eigenstates, Traces.   251
        13.6.3 Diagonal Operators   252
        13.6.4 Diagonal Density Operator, Averages   253
        13.6.5 Anisotropic Dielectric Tensor of a Gas of Rotating
                Molecules   255
        13.6.6 Non-diagonal Tensor Operators   255

14 Rotation of Tensors   259
   14.1 Rotation of Vectors.   259
        14.1.1 Infinitesimal and Finite Rotation.   259
        14.1.2 Hamilton Cayley and Projection Tensors   260
        14.1.3 Rotation Tensor for Vectors   261
        14.1.4 Connection with Spherical Components.   262

14.2   Rotation of Second Rank Tensors.   262
        14.2.1   Infinitesimal Rotation   262
        14.2.2   Fourth Rank Projection Tensors   263
        14.2.3   Fourth Rank Rotation Tensor   264
14.3   Rotation of Tensors of Rank ℓ   265
14.4   Solution of Tensor Equations   266
        14.4.1   Inversion of Linear Equations.   266
        14.4.2   Effect of a Magnetic Field on the Electrical
                  Conductivity   267
14.5   Additional Formulas Involving Projectors   268

15 Liquid Crystals and Other Anisotropic Fluids   273
15.1   Remarks on Nomenclature and Notations.   274
        15.1.1   Nematic and Cholesteric Phases, Blue Phases.   274
        15.1.2   Smectic Phases.   276
15.2   Isotropic $\leftrightarrow$ Nematic Phase Transition.   277
        15.2.1   Order Parameter Tensor.   277
        15.2.2   Landau-de Gennes Theory   279
        15.2.3   Maier-Saupe Mean Field Theory.   283
15.3   Elastic Behavior of Nematics   284
        15.3.1   Director Elasticity, Frank Coefficients   285
        15.3.2   The Cholesteric Helix   287
        15.3.3   Alignment Tensor Elasticity   288
15.4   Cubatics and Tetradics.   290
        15.4.1   Cubic Order Parameter   291
        15.4.2   Landau Theory for the Isotropic-Cubatic
                  Phase Transition   292
        15.4.3   Order Parameter Tensor for Regular Tetrahedra   293
15.5   Energetic Coupling of Order Parameter Tensors   294
        15.5.1   Two Second Rank Tensors.   294
        15.5.2   Second-Rank Tensor and Vector.   296
        15.5.3   Second- and Third-Rank Tensors   297

16 Constitutive Relations   299
16.1   General Principles.   300
        16.1.1   Curie Principle   300
        16.1.2   Energy Principle.   302
        16.1.3   Irreversible Thermodynamics, Onsager Symmetry
                  Principle   302
16.2   Elasticity   304
        16.2.1   Elastic Deformation of a Solid, Stress Tensor.   304
        16.2.2   Voigt Coefficients.   305
        16.2.3   Isotropic Systems   306

16.2.4   Cubic System.   307
16.2.5   Microscopic Expressions for Elasticity Coefficients. . .   309
16.3   Viscosity and Non-equilibrium Alignment Phenomena.   312
16.3.1   General Remarks, Simple Fluids.   312
16.3.2   Influence of Magnetic and Electric Fields   314
16.3.3   Plane Couette and Plane Poiseuille Flow   315
16.3.4   Senftleben-Beenakker Effect of the Viscosity   318
16.3.5   Angular Momentum Conservation, Antisymmetric
         Pressure and Angular Velocity   320
16.3.6   Flow Birefringence   322
16.3.7   Heat-Flow Birefringence   326
16.3.8   Visco-Elasticity   326
16.3.9   Nonlinear Viscosity.   328
16.3.10 Vorticity Free Flow.   330
16.4   Viscosity and Alignment in Nematics   332
16.4.1   Well Aligned Nematic Liquid Crystals
         and Ferro Fluids.   332
16.4.2   Perfectly Oriented Ellipsoidal Particles   335
16.4.3   Free Flow of Nematics, Flow Alignment
         and Tumbling.   337
16.4.4   Fokker-Planck Equation Applied to Flow
         Alignment   338
16.4.5   Unified Theory for Isotropic and Nematic Phases   343
16.4.6   Limiting Cases: Isotropic Phase, Weak Flow
         in the Nematic Phase.   345
16.4.7   Scaled Variables, Model Parameters   347
16.4.8   Spatially Inhomogeneous Alignment   349

17   Tensor Dynamics   351
17.1   Time-Correlation Functions and Spectral Functions.   351
17.1.1   Definitions.   351
17.1.2   Depolarized Rayleigh Scattering   353
17.1.3   Collisional and Diffusional Line Broadening   356
17.2   Nonlinear Relaxation, Component Notation   357
17.2.1   Second-Rank Basis Tensors   357
17.2.2   Third-Order Scalar Invariant and Biaxiality
         Parameter.   359
17.2.3   Component Equations   359
17.2.4   Stability of Stationary Solutions   360
17.3   Alignment Tensor Subjected to a Shear Flow   362
17.3.1   Dynamic Equations for the Components   362
17.3.2   Types of Dynamic States.   362
17.3.3   Flow Properties   365

17.4 Nonlinear Maxwell Model............................ 365
   17.4.1 Formulation of the Model 366
   17.4.2 Special Cases.............................. 366

18 From 3D to 4D: Lorentz Transformation, Maxwell Equations ... 369
18.1 Lorentz Transformation 369
   18.1.1 Invariance Condition........................ 369
   18.1.2 4-Vectors................................. 370
   18.1.3 Lorentz Transformation Matrix................ 372
   18.1.4 A Special Lorentz Transformation. 372
   18.1.5 General Lorentz Transformations 373
18.2 Lorentz-Vectors and Lorentz-Tensors 373
   18.2.1 Lorentz-Tensors 373
   18.2.2 Proper Time, 4-Velocity and 4-Acceleration....... 374
   18.2.3 Differential Operators, Plane Waves 376
   18.2.4 Some Historical Remarks..................... 377
18.3 The 4D-Epsilon Tensor 378
   18.3.1 Levi-Civita Tensor 378
   18.3.2 Products of Two Epsilon Tensors 379
   18.3.3 Dual Tensor, Determinant 379
18.4 Maxwell Equations in 4D-Formulation 381
   18.4.1 Electric Flux Density and Continuity Equation 381
   18.4.2 Electric 4-Potential and Lorentz Scaling. 381
   18.4.3 Field Tensor Derived from the 4-Potential 382
   18.4.4 The Homogeneous Maxwell Equations 383
   18.4.5 The Inhomogeneous Maxwell Equations 383
   18.4.6 Inhomogeneous Wave Equation 384
   18.4.7 Transformation Behavior of the Electromagnetic
        Fields 384
   18.4.8 Lagrange Density and Variational Principle 385
18.5 Force Density and Stress Tensor....................... 386
   18.5.1 4D Force Density 386
   18.5.2 Maxwell Stress Tensor 387

Appendix: Exercises: Answers and Solutions. 389

References...................................... 423

Index 433

Part I
A Primer on Vectors and Tensors

Chapter 1
Introduction

Abstract In this chapter, preliminary remarks are made on vectors and tensors. The axioms of a vector space and the norm of a vector are introduced, the role of vectors for classical physics and for Special Relativity is discussed. The scope of the book as well as a brief overview of the history and literature devoted to vectors and tensors are presented. Before tensors and their properties are introduced here, it is appropriate to discuss the question: what is a vector? As we shall see, mathematicians and physicists give somewhat different answers.

1.1 Preliminary Remarks on Vectors

Some physical quantities like the *mass, energy* or *temperature* are quantified by a single numerical value. Such a quantity is referred to as *scalar*. For other physical quantities, like the *velocity* or the *force* not only their magnitude but also their direction has to be specified. Such a quantity is a *vector*. In the three-dimensional space we live in, three numerical values are needed to quantify a vector. These numbers are, e.g. the three components in a rectangular, Cartesian coordinate system.

In general terms, a vector is an element of a vector space. The axioms obeyed by these elements are patterned after the rules for the addition of arrows and for their multiplication by real numbers.

1.1.1 Vector Space

Consider special vectors, represented by arrows, which have a length and a direction. The rules for computations with vectors can be visualized by manipulations with arrows. Multiplication of a vector by a number means: the length of the arrow is multiplied by this number. The relation most typical for vectors is the addition of two vectors **a** and **b** as indicated in Fig. 1.1.

The operation **a** + **b** means: attach the tail of **b** to the arrowhead of **a**. The sum is the arrow pointing from the tail of **a** to the arrowhead of **b**. The sum **b** + **a**, indicated by dashed arrows, yields the same result, thus

© Springer International Publishing Switzerland 2015
S. Hess, *Tensors for Physics*, Undergraduate Lecture Notes in Physics,
DOI 10.1007/978-3-319-12787-3_1

Fig. 1.1 Vector addition

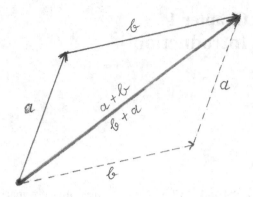

$$\mathbf{a} + \mathbf{b} = \mathbf{b} + \mathbf{a}. \qquad (1.1)$$

As a side remark, one may ask: how was the rule for the vector addition conceived? A vector **a** can be associated with the displacement or shift along a straight line of an object, from point 0 to point A. This is the origin for the word *vector*: it *carries* an object over a straight and directed distance. The vector **b** corresponds to a shift from 0 to point B. The vector addition $\mathbf{a} + \mathbf{b}$ means: make first the shift from 0 to point A and then the additional shift corresponding to vector **b**, which has to start from point A. For this reason, the tail of the second vector is attached to the head of the first vector in the vector addition operation.

When three vectors **a**, **b** and **c** are added, it makes no difference when first the vector sum of **a** and **b** is computed and then the vector **c** is added or when **a** is added to the sum of **b** and **c**:

$$(\mathbf{a} + \mathbf{b}) + \mathbf{c} = \mathbf{a} + (\mathbf{b} + \mathbf{c}). \qquad (1.2)$$

The vector sum is also used to define the difference between two vectors according to

$$\mathbf{a} + \mathbf{x} = \mathbf{b} \quad \rightarrow \quad \mathbf{x} = \mathbf{b} - \mathbf{a}. \qquad (1.3)$$

When the vector **b** in (1.3) is equal to zero, then one has

$$\mathbf{x} = -\mathbf{a}. \qquad (1.4)$$

This vector has the same length as **a** but the opposite direction, i.e. arrowhead and tail are exchanged.

For real numbers k and ℓ, the following rules hold true for any vector **a**:

$$(k + \ell)\mathbf{a} = k\mathbf{a} + \ell\mathbf{a},$$
$$k(\ell\mathbf{a}) = (k\ell)\mathbf{a}, \qquad (1.5)$$
$$1\mathbf{a} = \mathbf{a}.$$

Furthermore, for any real number k and two vectors $\mathbf{a}$ and $\mathbf{b}$ one has:

$$k(\mathbf{a} + \mathbf{b}) = k\mathbf{a} + k\mathbf{b}. \tag{1.6}$$

Mathematical objects which obey the rules or axioms (1.1)–(1.6) are elements of a *vector space*. For mathematician, the answer to the question "what is a vector?" is: "it is an element of a vector space". In addition to the arrows we discussed, there are many other types of vector spaces. Examples are,

1. real numbers or complex numbers,
2. polynomials of order n,
3. quadratic matrices,
4. ordered n-tuples $(a_1, a_2, \ldots, a_n)$ with real numbers a_1, a_2 to a_n.

In physics, the notion *vector* is used in a more special sense. Before this is discussed, a brief remark on the *norm of a vector* is in order.

1.1.2 Norm and Distance

It is obvious that an arrow has a length. For an element $\mathbf{a}$ of an abstract vector space the *norm* $||\mathbf{a}|| \geq 0$ corresponding to the length of a vector has to be defined by rules. Computation of the norm requires a *metric*. Without going into details, the general properties of a norm are listed here.

1. When the norm of a vector equals zero, the vector must be the *zero-vector*. Likewise, the norm of the zero-vector is equal to zero, thus,

$$||\mathbf{a}|| = 0 \leftrightarrow \mathbf{a} = 0. \tag{1.7}$$

2. For any real number r with the absolute magnitude $|r|$, one has:

$$||r\mathbf{a}|| = |r|\, ||\mathbf{a}||. \tag{1.8}$$

3. The norm of the sum of two vectors $\mathbf{a}$ and $\mathbf{b}$ cannot be larger than the sum of the norm of the two vectors:

$$||\mathbf{a} + \mathbf{b}|| \leq ||\mathbf{a}|| + ||\mathbf{b}||. \tag{1.9}$$

The relation (1.9) is obvious for the addition of he arrows as shown in Fig. 1.1.

The distance $d(\mathbf{a}, \mathbf{b})$ between two vectors $\mathbf{a}$ and $\mathbf{b}$ is defined as the norm of the difference vector $\mathbf{a} - \mathbf{b}$:

$$d(\mathbf{a}, \mathbf{b}) := ||\mathbf{a} - \mathbf{b}||. \tag{1.10}$$

For vectors represented as arrows with their tails located at the same point, this corresponds to the length of the vector joining the arrowheads.

The distance is translationally invariant. This means: addition of the same vector $\mathbf{x}$ to both $\mathbf{a}$ and $\mathbf{b}$ does not change their distance:

$$d(\mathbf{a} + \mathbf{x}, \mathbf{b} + \mathbf{x}) = d(\mathbf{a}, \mathbf{b}). \tag{1.11}$$

Furthermore, the distance is homogeneous. This means: multiplication of both vectors $\mathbf{a}$ and $\mathbf{b}$ by the same real number k implies the multiplication of the distance by the absolute value $|k|$:

$$d(k\mathbf{a}, k\mathbf{b}) = |k| d(\mathbf{a}, \mathbf{b}). \tag{1.12}$$

As stressed before, in many applications in physics, the notion *vector* refers to a more special mathematical object. Before details are discussed in the following section, here a short answer is given to the question: what is special about vectors in physics? Vectors in two and three dimensions, as used in classical physics, have to be distinguished from the four-dimensional vectors of special relativity theory.

1.1 Exercise: Complex Numbers as 2D Vectors
Convince yourself that the complex numbers $z = x + iy$ are elements of a vector space, i.e. that they obey the rules (1.1)–(1.6). Make a sketch to demonstrate that $z_1 + z_2 = z_2 + z_1$, with $z_1 = 3 + 4i$ and $z_2 = 4 + 3i$, in accord with the vector addition in 2D.

1.1.3 Vectors for Classical Physics

The *position* of a physical object is represented by an arrow pointing from the origin of a coordinate system to the center of mass of this object. This *position vector* is specified by the coordinates of the arrowhead. This ordered set of two or three numbers, in two-dimensional (2D) or three-dimensional (3D) space R^3, is referred to as *the components* of the position vector. It is convenient to use a *Cartesian coordinate system* which has rectangular axes. Then the length (norm or magnitude) of a vector is just the square root of the sum of the components squared.

In a coordinate system rotated with respect to the original one, the same position vector has different components. There are well defined rules to compute the components in the rotated system from the original components. This is referred to as *transformation of the components upon rotation of the coordinate system*.

Now we are in the position to state what is special about vectors in physics:
A vector is a quantity with two or three components which transform like those of the position vector, upon a rotation of the coordinate system.

The vectors used in classical physics like the velocity or the force are elements of a vector space, do have a norm, and they possess an additional property, viz. a specific transformation behavior of their components.

A *scalar* is a quantity which does not change upon a rotation of the coordinate system. Examples for scalars are the mass or the length of the position vector.

1.1.4 Vectors for Special Relativity

Vectors with four components are used in special relativity theory. The basic vector is composed of the three components of the position vector; the fourth one is the time, multiplied by the speed of light. The components of this 4-*vector* change according to the *Lorentz transformation* when the original coordinate system is replaced by a coordinate system moving with constant velocity with respect to the original one. The time is also changed in this transformation. Physical quantities with four components, which transform with the same rule, are referred to as *Lorentz vectors*. Properties of Lorentz vectors and their application in physics, in particular in electrodynamics, are discussed in the last chapter of this book.

1.2 Preliminary Remarks on Tensors

For the first time, students hear about tensors in connection with the *moment of inertia tensor* linking the rotational angular momentum with the rotational velocity of a rotating solid body. In such a linear relation, two vectors are not just parallel to each other. Tensors also describe certain orientational dependencies in anisotropic media. Examples are electric and magnetic susceptibility tensors, mobility and diffusion tensors. To be more precise, these are tensors of rank 2. Vectors are also referred to as tenors of rank 1. There are second rank tensors which are physical variables of their own, like the stress tensor and the strain tensor. The linear relation between two second rank tensors is described by a tensor of rank 4. An example is the elasticity tensor linking the stress tensor with the strain tensor. Tensors of different ranks are used to characterize orientational distributions.

Definitions, properties and applications of tensors represented by their components in a 3D coordinate system are discussed in detail in the following sections. Here just a brief, preliminary answer is given to the question: *What is a tensor?*

The rule which links the components of the position vector in a rotated coordinate system with the components of the original one involves a transformation matrix, referred to as *rotation matrix*. The product of ℓ components of the position vector needs the product of ℓ rotation matrices for their interrelation between the rotated system and the original one.

A tensor of rank ℓ is a quantity whose components are transformed upon a rotation of the coordinate system with the ℓ-fold product of the rotation matrix.

In this sense, scalars and vectors are tensors of rank $\ell = 0$ and $\ell = 1$. Often, tensors of rank $\ell = 2$ are just referred to as tensors. Properties and applications of second rank tensors, as well as of higher rank tensors, are discussed in the following sections.

1.3 Remarks on History and Literature

Cartesian *tensors*, as they are used here for the description of material properties in anisotropic media, were first introduced by the German physicist *Woldemar Voigt* around 1890. This is documented in his books on the physics of crystals [1]. In his lectures on theoretical physics, published around 1895, Voigt worked with tensors, without using the word. In the books on crystal physics, the term *vector* is used as if everybody is familiar with it. Tensors, in particular of rank two, are discussed in detail. Tensors of rank three and four are applied for the description of the relevant physical phenomena. Voigt refers to Pierre Curie [2] as having very similar ideas about tensors and symmetries.

In connection with differential geometry, the notion *tensor*, however, not the word, was already invented by *Carl Friedrich Gauss*, who had lived and worked in Göttingen, more than half a century before Voigt. At about the same time as Voigt, the Italian mathematicians Tullio Levi-Civita and Gregorio Ricci Curbastro formulated *tensor calculus*, as it is used since then in connection with differential geometry [3]. Their work provided the mathematical foundation for Einstein's *General Relativity Theory* [4]. This topic, however, is not treated here. The necessary mathematical tool for General Relativity are found in the text books devoted to this subject, e.g. in [5–7].

As stated before, the emphasis of this book is on Cartesian tensors in three dimensions and applications to physics, in particular for the description of anisotropic properties of matter. Classic books on the subject were published between 1930 and 1960, by Jeffrey [8], Brillouin [9], Dusschek and Hochrainer [10], and Temple [11]. In the Kinetic Theory of gases, tensors and the importance of the use of irreducible tensors was stressed in the book of Chapman and Cowling [12] and in the *Handbuch* article by Waldmann [13]. I was introduced to Cartesian tensors in lectures on electrodynamics by Ludwig Waldmann in 1963. Applications to the kinetic theory of molecular gases, in the presence of external fields, as well as to optics and transport properties of liquid crystals required efficient use of tensor algebra and tensor calculus. This strongly influenced a book for an introductory course to vectors and tensors [14], for first year students of physics, and led to a collection of computational rules and formulas needed in more advanced theory [15]. For the application of tensors in the kinetic theory of molecular gases, see also [16, 17].

In the following, no references will be given to the physics which is standard in undergraduate and graduate courses, the reader may consult her favorite text book or internet source. In the second part of the book, devoted to more specialized subjects, references to original articles will be presented. The particular choice of these topics largely reflects my own research experience.

A remark on vectors is in order. The notion of *vector*, i.e. that some physical quantities, like velocity and force, have both a magnitude or strength and a direction and that the combined effect of two vectors follow a rule which we call the *addition of vectors*, was known long before the word *vector* was used. Among others, Isaac Newton was well aware, how vectorial quantities should be handled. Gauss and others used a geometric representation of complex numbers in a way which was essentially equivalent to dealing with two-dimensional vectors. William Rowan Hamilton tried to extend this concept to three dimensions. He did not succeed but, in 1843, he invented the four-dimensional *quaternions*. James Clark Maxwell did not encourage the application of quaternions to the theory of electrodynamics, but rather favored a vectorial description. Vector analysis, as it is used nowadays, was strongly promoted, around 1880, by Willard Gibbs [18].

1.4 Scope of the Book

The first part of the book, Chaps. 2–8, is a primer on vectors and tensors, it provides definitions, rules for calculations and applications every student of physics should become familiar with at an undergraduate level. The symmetry of second rank tensors, viz. their decomposition into isotropic, antisymmetric and symmetric traceless parts, the connection of the antisymmetric part with a dual vector, in 3D, as well as the differentiation and integration of vector and tensor fields, including generalizations of the laws of Stokes and Gauss play a central role.

Part II, Chaps. 9–18, deals with more advanced topics. In particular, Chaps. 9–11 are devoted to irreducible tensors of rank ℓ, multipole potentials and multipole moments, isotropic tensors. Integral formulae and distribution functions, spin operators and the active rotation of tensors are presented in Chaps. 12–14. The properties of liquid crystals intimately linked with tensors, constitutive relations for elasticity, viscosity and flow birefringence, as well as the dynamics of tensors obeying nonlinear differential equations are treated in Chaps. 15–17. Whereas this book is mostly devoted to tensors in 3D, Chap. 18 provides an outlook to the 4D formulation of electrodynamics. Answers and solutions to the exercises are given at the end of the book.

The examples presented are meant to show the applications of tensors in a variety of physical properties and phenomena, occurring in different branches of physics. The examples are far from exhaustive. They are closely linked with the author's experience in teaching and research. Applications to *Mechanics* and to *Electrodynamics and Optics* are, e.g., found in Chaps. 2–10 and in Chaps. 7–14, as well as in Chap. 18. Applications to *Quantum Mechanics* and properties of *Atoms and Molecules* are discussed in Chaps. 5, 7, and 10–13. *Elasticity, Hydrodynamics* and *Rheology* are treated in Chaps. 7–10 and 16. Problems of *Statistical Physics*, of the *Physics of Condensed Matter* and *Material Properties* are addressed in Chaps. 12–16. Applications to *Non-equilibrium Phenomena* like *Transport and Relaxation Processes* and *Irreversible Thermodynamics* are presented in Chaps. 12, 14, 16, and 17. The physics underlying the various applications of tensors is discussed to an extend considered appropriate, without the intention to replace any textbook or monograph on the topics considered.

Chapter 2
Basics

Abstract This chapter is devoted to the basic features needed for Cartesian tensors: the components of a position vector with respect to a coordinate system, the scalar product of two vectors, the transformation of the components upon a change of the coordinate system. Special emphasis is put on the orthogonal transformation associated with a rotation of the coordinate system. Then tensors of rank $\ell \geq 0$ are defined via the transformation behavior of their components upon a rotation of the coordinate system, scalars and vectors correspond to the special cases $\ell = 0$ and $\ell = 1$. The importance of tensors of rank $\ell \geq 2$ for physics is pointed out. The parity and time reversal behavior of vectors and tensors are discussed. The differentiation of vectors and tensors with respect to a parameter, in particular the time, is treated.

2.1 Coordinate System and Position Vector

2.1.1 Cartesian Components

Given the origin of a coordinate system, the position of a particle or the center of mass of an extended object is specified by the position vector $\mathbf{r}$, as indicated in Fig. 2.1. In the three-dimensional space we live in, this vector has three components, often referred to as the x-, y- and z-components. We use a (space-fixed) right-handed rectangular coordinate system, also called *Cartesian coordinate system*. It is convenient to label the axes by 1, 2 and 3 and to denote the components of the position vector by r_1, r_2, and r_3. Sometimes, the vector is written as an ordered triple of the form (r_1, r_2, r_3).

For these Cartesian components of the position vector the notation r_μ is preferred, where it is understood that μ, or any other Greek letter used for the subscript, also called indices, can have the value 1, 2 or 3. Of course, the mathematical content is unaffected, when Latin letters are used as subscripts instead of the Greek ones. Here, Latin letters are reserved for the components of two- and four-dimensional vectors or for components in a coordinate system with axes which are not orthogonal.

The components of the sum $\mathbf{S} = \mathbf{r} + \mathbf{s}$ of two vectors $\mathbf{r}$ and $\mathbf{s}$, with the Cartesian components r_1, r_2, r_3 and s_1, s_2, s_3, are given by $r_1+s_1, r_2+s_2, r_3+s_3$. This standard

© Springer International Publishing Switzerland 2015   11
S. Hess, *Tensors for Physics*, Undergraduate Lecture Notes in Physics,
DOI 10.1007/978-3-319-12787-3_2

Fig. 2.1 Position vector in a
Cartesian coordinate system.
The *dashed lines* are guides
for the eye

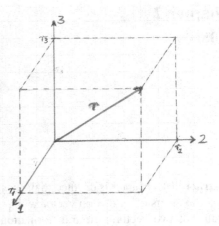

rule for the addition of two vectors can also be written as

$$S_\mu = r_\mu + s_\mu, \tag{2.1}$$

with $\mu = 1, 2, 3$. The multiplication of the vector **r** with a real number k, i.e. $\mathbf{R} = k\mathbf{r}$
means, that each component is multiplied by this number, viz.,

$$R_\mu = k\, r_\mu. \tag{2.2}$$

We are still dealing with the same vectors when other Greek letters, like v, λ, ... or
$\alpha, \beta, \ldots$ are used as subscripts.

2.1.2 Length of the Position Vector, Unit Vector

For the rectangular coordinate system, the length r of the vector **r** is given by the
Euclidian norm:

$$r^2 = \mathbf{r} \cdot \mathbf{r} = r_1^2 + r_2^2 + r_3^2 := r_\mu r_\mu. \tag{2.3}$$

Thus one has

$$r = \sqrt{r_\mu r_\mu}. \tag{2.4}$$

The length of the vector is also referred to as its *magnitude* or its *norm*.

Here and in the following, the *summation convention* is used: Greek subscripts
which occur twice are summed over. This implies that on one side of an equation,
each Greek letter can only show up once or twice as a Cartesian index. Einstein
introduced such a summation convention for the components of four-dimensional
vectors. For this reason, also the term *Einstein summation convention*, is used.

The vector $\mathbf{r}$, divided by its length r, is the dimensionless *unit vector* $\widehat{\mathbf{r}}$:

$$\widehat{\mathbf{r}} = r^{-1}\mathbf{r}, \tag{2.5}$$

or, in component notation:

$$\widehat{r}_\mu = r^{-1}r_\mu. \tag{2.6}$$

The unit vector has magnitude 1:

$$\widehat{r}_\mu \widehat{r}_\mu = 1. \tag{2.7}$$

2.1.3 Scalar Product

The scalar product of two position vectors $\mathbf{r}$ and $\mathbf{s}$ with components r_μ and s_μ is

$$\mathbf{r} \cdot \mathbf{s} = r_1 s_1 + r_2 s_2 + r_3 s_3 := r_\mu s_\mu. \tag{2.8}$$

Clearly, the length squared (2.3) of the vector $\mathbf{r}$ is its scalar product with itself. Just as in (2.3), the *center dot* "·" is essential to indicate the scalar product, when the vectors are written with bold face symbols. The summation convention is used for the component notation. Notice that the "name" of the summation index does not matter, i.e. $r_\mu s_\mu = r_\nu s_\nu = r_\lambda s_\lambda$. What really matters is: a Greek letter occurs twice (and only twice) in a product.

The scalar product has a simple geometric interpretation. In general, the two vectors $\mathbf{r}$ and $\mathbf{s}$ span a plane. We choose the coordinate system such that $\mathbf{r}$ is parallel to the 1-axis and $\mathbf{s}$ is in the 1–2-plane, see Fig. 2.2. Then the components of $\mathbf{r}$ are $(r_1, 0, 0)$ and those of $\mathbf{s}$ are $(s_1, s_2, 0)$. The scalar product yields $\mathbf{r} \cdot \mathbf{s} = r_1 s_1$. The lengths of the two vectors are given by $r = r_1$ and $s = \sqrt{s_1^2 + s_2^2}$. The angle between $\mathbf{r}$ and $\mathbf{s}$ is denoted by φ, see Fig. 2.2. One has $s_1 = r \cos\varphi$, and

$$\mathbf{r} \cdot \mathbf{s} = r s \cos\varphi \tag{2.9}$$

Fig. 2.2 For the geometric interpretation of the scalar product

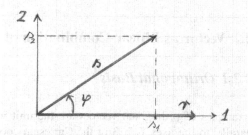

holds true. Or in words: the scalar product of two vectors is equal to the product of their lengths times the cosine of the angle between them. The scalar product of **s** with the unit vector $\hat{\mathbf{r}}$ is equal to $s \cos \varphi$. The vector $s \cos \varphi \hat{\mathbf{r}} = (\mathbf{s} \cdot \hat{\mathbf{r}})\hat{\mathbf{r}}$ is called the *the projection of* **s** *onto the direction of* **r**.

The value of the scalar product reaches its maximum and (negative) minimum when the vectors are parallel ($\varphi = 0$) and anti-parallel ($\varphi = \pi$). The scalar product vanishes for two vectors which are perpendicular to each other, i.e. for $\varphi = \pi/2$. Such vectors are also referred to as *orthogonal* vectors.

2.1 Exercise: Compute Scalar Product for Given Vectors

Compute the length, the scalar products and the angles between the vectors **a**, **b**, **c** which have the components $\{1, 0, 0\}$, $\{1, 1, 0\}$, and $\{1, 1, 1\}$.

2.1.4 Spherical Polar Coordinates

As stated before, the position vector **r** has a length, specified by its magnitude $r = \sqrt{\mathbf{r} \cdot \mathbf{r}}$, and a direction, determined by the unit vector $\hat{\mathbf{r}}$, cf. (2.5) and (2.6). These parts of the vector are often referred to as *radial part* and *angular part*. Indeed, the unit vector and thus the direction of **r** can be specified by the two *polar angles* ϑ and φ. Conventionally, a particular coordinate system is chosen, the Cartesian coordinates $\{r_1, r_2, r_3\}$ are denoted by $\{x, y, z\}$ which, in turn, are related to the *spherical polar coordinates* r, ϑ, φ by

$$x = r \sin \vartheta \cos \varphi, \quad y = r \sin \vartheta \sin \varphi, \quad z = r \cos \vartheta. \tag{2.10}$$

Notice, the three numbers for r, ϑ, φ are not components of a vector.

The information given by the Cartesian components of a unit vector corresponds to a point on the unit sphere, identified by the two angles, similar to positions on earth. Notice, however, that the standard choice made for the angle ϑ would correspond to associate $\vartheta = 0$ and $\vartheta = 180°$ with the North Pole and the South Pole, respectively, whereas the equator would be at $\vartheta = 90°$. For positions on earth, one starts counting ϑ from zero on the equator and has to distinguish between North and South, or plus and minus. In any case, the angle spans an interval of $180°$, or just π, whereas that of φ is $360°$, or 2π.

2.2 Vector as Linear Combination of Basis Vectors

2.2.1 Orthogonal Basis

Examples of orthogonal vectors are the unit vectors $\mathbf{e}^{(i)}$, $i = 1, 2, 3$, which are parallel to the axes 1, 2, 3 of the Cartesian coordinate system. These vectors have the properties $\mathbf{e}^{(1)} \cdot \mathbf{e}^{(1)} = 1$, $\mathbf{e}^{(1)} \cdot \mathbf{e}^{(2)} = 0, \ldots$, in more general terms,

$$\mathbf{e}^{(i)} \cdot \mathbf{e}^{(j)} = \delta_{ij}. \qquad (2.11)$$

Here δ_{ij} is the Kronecker symbol, i.e. $\delta_{ij} = 1$ for $i = j$ and $\delta_{ij} = 0$ for $i \neq j$.

The position vector $\mathbf{r}$ can be written as a linear combination of these unit vectors $\mathbf{e}^{(i)}$ according to

$$\mathbf{r} = r_1 \mathbf{e}^{(1)} + r_2 \mathbf{e}^{(2)} + r_3 \mathbf{e}^{(3)}. \qquad (2.12)$$

Since the *basis vectors* are not only orthogonal, but also normalized to 1, the Cartesian components are equal to the scalar product of $\mathbf{r}$ with the basis vectors, e.g. $r_1 = \mathbf{e}^{(1)} \cdot \mathbf{r}$.

2.2.2 Non-orthogonal Basis

Three vectors $\mathbf{a}^{(i)}$, with $i = 1, 2, 3$, which are not within one plane, can be used as basis vectors. Then the vector $\mathbf{r}$ can be represented by the linear combination

$$\mathbf{r} = \xi^1 \mathbf{a}^{(1)} + \xi^2 \mathbf{a}^{(2)} + \xi^3 \mathbf{a}^{(3)}, \qquad (2.13)$$

with the coefficients ξ^i. Scalar multiplication of (2.13) with the basis vectors $\mathbf{a}^{(i)}$ yields

$$\xi_i = \mathbf{a}^{(i)} \cdot \mathbf{r} = \sum_{j=1}^{3} g_{ij} \xi^j. \qquad (2.14)$$

The coefficient matrix

$$g_{ij} = \mathbf{a}^{(i)} \cdot \mathbf{a}^{(j)} = g_{ji}, \qquad (2.15)$$

is determined by the scalar products of the basis vectors. The coefficients ξ^i and ξ_i are referred to as *contra-* and *co-variant* components of the vector in a coordinate system with axes specified by the basis vectors $\mathbf{a}^{(i)}$.

In this basis, the square of the length or of the magnitude of the vector is given by

$$\mathbf{r} \cdot \mathbf{r} = \sum_i \sum_j \xi^i \xi^j \mathbf{a}^{(i)} \cdot \mathbf{a}^{(j)} = \sum_i \sum_j \xi^i \xi^j g_{ji} = \sum_i \xi^i \xi_i. \qquad (2.16)$$

The coefficient matrix g_{ij} characterizes the connection between the co- and the contra-variant components and it is essential for the calculation of the norm. Thus it determines the *metric* of the coordinate system.

The geometric meaning of the two different types of components is demonstrated in Fig. 2.3 for the 2-dimensional case.

The intersection of the dashed line parallel to the 2-axis with the 1-axis marks the component ξ^1, similarly ξ^2 is found at the intersection of the 2-axis with the dashed line parallel to the 1-axis. The component ξ_1 and ξ_2 are found at the intersections

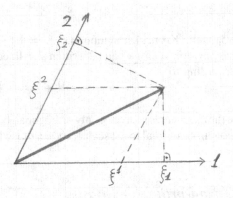

Fig. 2.3 Components of the position vector in a non-orthogonal coordinate system

of the dashed lines perpendicular to the axes. It is understood that the basis vectors along the axes, not shown in Fig. 2.3, are unit vectors.

For basis vectors which are mutually perpendicular and normalized to 1, the matrix g_{ij} reduces to unit matrix δ_{ij}. Consequently co- and the contra-variant components are equal. This is also obvious from Fig. 2.3. The two types of components coincide when the basis vectors are orthogonal. We do not have to distinguish between co- and the contra-variant components when we use the Cartesian coordinate system.

2.3 Linear Transformations of the Coordinate System

The laws of physics do not depend on the choice of a coordinate system. However, in many applications, a specific choice is made. Then it is important to know, how components have to be transformed such that the physics is not changed, when another coordinate system is chosen. Here, we are concerned with *linear transformations* where the coordinates in the new system are linked with those of the original coordinate system by a linear relation. The two types of linear transformations, *translations* and *affine transformations*, also referred to as *linear maps*, are discussed separately. The rotation of a coordinate system is a special case of an affine transformations. Due to its importance, an extra section is devoted to rotations.

2.3.1 Translation

Consider a new coordinate system, that is shifted with respect to the original one by a constant vector **a**. Such a shift is referred to as *translation of the coordinate system*. In Fig. 2.4, a translation within the 1,2-plane is depicted.

Fig. 2.4 Components of the
position vector **r** in shifted
coordinate system

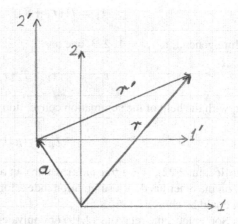

The position vector $\mathbf{r}'$ with respect the origin of the shifted coordinate system is related to the original $\mathbf{r}$ by

$$\mathbf{r}' = \mathbf{r} - \mathbf{a}, \tag{2.17}$$

or in component notation,

$$r'_\mu = r_\mu - a_\mu. \tag{2.18}$$

The inverse transformation, which brings the shifted coordinate system back to the original one, corresponds to a shift by the vector $-\mathbf{a}$.

Notice: the translation of the coordinate system is a passive transformation, which has to be distinguished from the *active translation* of the position of a particle or of an object from $\mathbf{r}$ to $\mathbf{r} + \mathbf{a}$.

2.3.2 Affine Transformation

For an affine transformation, the components r'_1, r'_2, r'_3 of the position vector $\mathbf{r}'$ in the new coordinate system are linear combinations of the components r_1, r_2, r_3 in the original system. When the components of the vectors are written in columns, the linear mapping can be expressed in the form

$$\begin{pmatrix} r'_1 \\ r'_2 \\ r'_3 \end{pmatrix} = \begin{pmatrix} T_{11} & T_{12} & T_{13} \\ T_{21} & T_{22} & T_{23} \\ T_{31} & T_{32} & T_{33} \end{pmatrix} \begin{pmatrix} r_1 \\ r_2 \\ r_3 \end{pmatrix}. \tag{2.19}$$

The elements $T_{11}, T_{12}, \ldots$ of the matrix $\mathbf{T}$ characterize the affine transformation. The determinant of $\mathbf{T}$ must not be zero, such that the reciprocal matrix $\mathbf{T}^{-1}$ exists. Standard matrix multiplication is assumed in (2.19). This means, e.g.

$$r_1' = T_{11} r_1 + T_{12} r_2 + T_{13} r_3. \tag{2.20}$$

More general, for $\mu = 1, 2, 3$, one has

$$r_\mu' = T_{\mu 1} r_1 + T_{\mu 2} r_2 + T_{\mu 3} r_3, \tag{2.21}$$

or, with the help of the summation convention

$$r_\mu' = T_{\mu\nu} r_\nu. \tag{2.22}$$

Notice: in (2.22), μ is a *free index* which can have any value 1, 2 or 3. The subscript ν, on the other hand, is a summation index, for which any other Greek letter, except μ, could be chosen here.

Sometimes, the relations (2.19) or equivalently (2.22) are expressed in the form

$$\mathbf{r}' = \mathbf{T} \cdot \mathbf{r}, \tag{2.23}$$

where the matrix-character of $\mathbf{T}$ is indicated by the *bold face sans serif* letter and the center dot "·" implies the summation of products of components.

Notice: in such a notation, the order of factors matters, in contradistinction to the component notation. The equation $\mathbf{r}' = \mathbf{r} \cdot \mathbf{T}$ corresponds to $r_\mu' = r_\nu T_{\nu\mu} = T_{\nu\mu} r_\nu$ which is different from (2.22), unless the transformation matrix $\mathbf{T}$ is symmetric, i.e. unless $T_{\nu\mu} = T_{\mu\nu}$ holds true.

The *inverse transformation*, also called *back-transformation*, links the components of $\mathbf{r}$ with those of $\mathbf{r}'$, according to

$$\mathbf{r} = \mathbf{T}^{-1} \cdot \mathbf{r}', \tag{2.24}$$

with the inverse transformation matrix $\mathbf{T}^{-1}$. Insertion of (2.23) into (2.24) leads to $\mathbf{r} = \mathbf{T}^{-1} \cdot \mathbf{r}' = \mathbf{T}^{-1} \cdot \mathbf{T} \cdot \mathbf{r}$ which implies

$$\mathbf{T}^{-1} \cdot \mathbf{T} = \delta, \tag{2.25}$$

or in component notation,

$$T_{\mu\lambda}^{-1} T_{\lambda\nu} = \delta_{\mu\nu}. \tag{2.26}$$

Notice: here μ and ν are free indices, λ is the summation index. The symbol δ indicates the unit matrix, viz.:

$$\delta := \begin{pmatrix} 1\,0\,0 \\ 0\,1\,0 \\ 0\,0\,1 \end{pmatrix}, \tag{2.27}$$

or equivalently, $\delta_{\mu\nu} = 1$ for $\mu = \nu$, and $\delta_{\mu\nu} = 0$ for $\mu \neq \nu$.

Similarly, insertion of (2.24) into (2.23) leads to

$$\mathbf{T} \cdot \mathbf{T}^{-1} = \delta, \tag{2.28}$$

or in component notation

$$T_{\mu\lambda}\, T_{\lambda\nu}^{-1} = \delta_{\mu\nu}. \tag{2.29}$$

For the affine transformation, the left-inverse and the right-inverse matrices are equal.

2.4 Rotation of the Coordinate System

2.4.1 Orthogonal Transformation

Affine transformations, which conserve the rule for the computation of the length or the norm of the position vector, and likewise the scalar product of two vectors, are of special importance. Coordinate transformations with this property are called *orthogonal transformations*. Proper rotations and rotations combined with a mirroring of the coordinate system are special cases to be discussed in detail.

The orthogonal transformations are defined by the requirement that

$$r'_\mu\, r'_\mu = r_\mu\, r_\mu, \tag{2.30}$$

where it is understood, that a relation of the form (2.23) holds true. This then is a condition on the properties of the transformation matrix $\mathbf{T}$. Here and in the following, the symbol $\mathbf{U}$ is used for the norm-conserving orthogonal transformation matrices. The letter "U" is reminiscent of "unitarian".

The property of the orthogonal matrix is inferred as follows. Use of $r'_\lambda = U_{\lambda\mu}r_\mu$ and $r'_\lambda = U_{\lambda\nu}r_\nu$ yields $r'_\lambda r'_\lambda = U_{\lambda\mu}U_{\lambda\nu}r_\mu r_\nu$. On the other hand (2.30) requires this expression to be equal to $r_\mu r_\mu = \delta_{\mu\nu}r_\mu r_\nu$. Thus one has

$$U_{\lambda\mu}U_{\lambda\nu} = \delta_{\mu\nu}. \tag{2.31}$$

Notice: here the summation index λ is the front index for both matrices $\mathbf{U}$. Reversal of the order of the subscripts yields the corresponding component of the *transposed matrix*, labelled with the superscript "T". Thus one has $U_{\lambda\mu} = U_{\mu\lambda}^{\mathrm{T}}$, and (2.31) is equivalent to

$$U_{\mu\lambda}^{\mathrm{T}}\, U_{\lambda\nu} = \delta_{\mu\nu}. \tag{2.32}$$

This orthogonality relation for the transformation matrix is equivalent to

$$\mathbf{U}^{\mathrm{T}} \cdot \mathbf{U} = 1, \tag{2.33}$$

where it is understood that the 1 on the right hand side stands for the unit matrix. Comparison of (2.32) and (2.33) with (2.26) and (2.25) reveals: the inverse $\mathbf{U}^{-1}$ of the orthogonal matrix $\mathbf{U}$ is just its transposed $\mathbf{U}^{\mathrm{T}}$:

$$\mathbf{U}^{-1} = \mathbf{U}^{\mathrm{T}}, \tag{2.34}$$

or

$$U_{\mu\nu}^{-1} = U_{\nu\mu}. \tag{2.35}$$

Use of the inverse transformation in considerations similar to those which lead to (2.31) and of (2.35) yield the orthogonality relation with the summation index at the back,

$$U_{\mu\lambda}U_{\nu\lambda} = \delta_{\mu\nu}, \tag{2.36}$$

or, equivalently,

$$\mathbf{U} \cdot \mathbf{U}^{\mathrm{T}} = 1. \tag{2.37}$$

Summary

The coordinate transformation

$$r'_{\mu} = U_{\mu\nu}\, r_{\nu}, \tag{2.38}$$

where the matrix $U_{\mu\nu}$ has the property

$$U_{\mu\lambda}\, U_{\nu\lambda} = U_{\lambda\mu}\, U_{\lambda\nu} = \delta_{\mu\nu} \tag{2.39}$$

guarantees that the scalar product of two vectors (2.8) and consequently, the expression (2.4) for the length of a vector are invariant under this transformation. Furthermore, the relation (2.39) means that the reciprocal $\mathbf{U}^{-1}$ of $\mathbf{U}$ is equal to the transposed matrix $\mathbf{U}^{\mathrm{T}}$ which, in turn is defined by $U_{\mu\nu}^{\mathrm{T}} = U_{\nu\mu}$.

Simple Examples

The simplest examples for transformation matrices which obey (2.39) are $U_{\mu\nu} = \delta_{\mu\nu}$ and $U_{\mu\nu} = -\delta_{\mu\nu}$, or in matrix notation:

$$\mathbf{U} = \delta := \begin{pmatrix} 1 & 0 & 0 \\ 0 & 1 & 0 \\ 0 & 0 & 1 \end{pmatrix}, \quad \mathbf{U} = -\delta := \begin{pmatrix} -1 & 0 & 0 \\ 0 & -1 & 0 \\ 0 & 0 & -1 \end{pmatrix}, \tag{2.40}$$

which, respectively, induce the identity transformation and a reversal of the directions of the coordinate axes. The latter case means a transformation to the 'mirrored' coordinate system.

Fig. 2.5 The components of
the position vector **r** in the
original coordinate system
and in one rotated about the
3-axis by the angle α are
given by the projections of **r**
on the coordinate axes 1, 2
and $1'$, $2'$, respectively

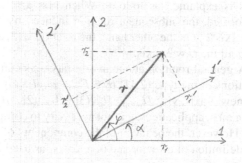

2.4.2 Proper Rotation

In general, an orthogonal transformation is either a proper rotation or a rotation combined with mirrored axes. The relation (2.39) implies $(\det \mathbf{U})^2 = 1$, thus $\det \mathbf{U} = \pm 1$. In the case of a *proper rotation*, the determinant "det" of the transformation matrix is equal to 1. Check the sign of the determinant for the simple matrices shown in (2.40).

An instructive nontrivial special case is the rotation of the coordinate system about one of its axes, e.g. the 3-axis as in Fig. 2.5 by an angle α. Let **r** be a vector located in the 1–2-plane, the angle between **r** and the 1-axis is denoted by φ. Then one has $r_1 = r \cos \varphi, r_2 = r \cos \varphi, r_3 = 0$, where r is the length of the vector. From the figure one infers: $r_1' = r \cos(\varphi - \alpha) = r(\cos \varphi \cos \alpha + \sin \varphi \sin \alpha) = r_1 \cos \alpha + r_2 \sin \alpha$ and $r_2' = r \sin(\varphi - \alpha) = r(\sin \varphi \cos \alpha - \cos \varphi \sin \alpha) = -r_1 \sin \alpha + r_2 \cos \alpha$; furthermore $r_3' = 0$. Thus the rotation matrix $U_{\mu\nu} = U_{\mu\nu}(3|\alpha)$, also denoted by $\mathbf{U}(3|\alpha)$, reads:

$$\mathbf{U} = \mathbf{U}(3|\alpha) := \begin{pmatrix} \cos \alpha & \sin \alpha & 0 \\ -\sin \alpha & \cos \alpha & 0 \\ 0 & 0 & 1 \end{pmatrix}. \tag{2.41}$$

A glance at (2.41) shows $\mathbf{U}(3|-\alpha) = \mathbf{U}^{\mathrm{T}}(3|\alpha)$. This is expected on account of (2.39) equivalent to $\mathbf{U}^{-1} = \mathbf{U}^{\mathrm{T}}$, since the rotation by the angle $-\alpha$ corresponds to the inverse transformation.

By analogy to (2.41), the transformation matrix for a rotation by the angle β about the 2-axis is

$$\mathbf{U} = \mathbf{U}(2|\beta) := \begin{pmatrix} \cos \beta & 0 & -\sin \beta \\ 0 & 1 & 0 \\ \sin \beta & 0 & \cos \beta \end{pmatrix}. \tag{2.42}$$

The two rotation matrices $\mathbf{U}(3|\alpha)$ and $\mathbf{U}(2|\beta)$ do not commute, i.e. one has

$$U(3|\alpha)_{\mu\lambda} U(2|\beta)_{\lambda\nu} \neq U(2|\beta)_{\mu\lambda} U(3|\alpha)_{\lambda\nu}.$$

This is explained as follows. When first a rotation $\mathbf{U}(3|\alpha)$ about the 3-axis is performed, the subsequent rotation induced by $\mathbf{U}(2|\beta)$ is about the new coordinate axis $2'$. On the other hand, the rotation $\mathbf{U}(3|\alpha)$, performed after the $2'$ rotation, is about the new $3'$-axis.

A general rotation about an arbitrary axis can be expressed by three successive rotations of the type (2.41) by 3 *Euler angles* about the 3-axis, the new 2-axis, and the new 3-axis, viz.: $U_{\mu\nu} = U_{\mu\lambda}(3|\gamma)U_{\lambda\kappa}(2|\beta)U_{\kappa\nu}(3|\alpha)$.

In most applications, it is not necessary to compute or to perform rotations explicitly. However, the behavior of the components of the position vector is essential for the definition of a vector and of a tensor, as used in physics.

2.5 Definitions of Vectors and Tensors in Physics

2.5.1 Vectors

A quantity **a** with Cartesian components a_μ, $\mu = 1, 2, 3$ is called a *vector* when, upon a rotation of the coordinate system, its components are transformed just like the components of the position vector, cf. (2.38). This means, the components a'_μ, in the rotated coordinate system, are linked with the components in the original system by

$$a'_\mu = U_{\mu\nu}\, a_\nu. \tag{2.43}$$

Here $U_{\mu\nu}$ are the elements of a transformation matrix for a proper rotation of the coordinate system.

Differentiation with respect to time t does not affect the vector character of a physical quantity. Thus the velocity $v_\mu = dr_\mu(t)/dt$ and the acceleration $dv_\mu(t)/dt$ are vectors. The linear momentum **p**, being equal to the mass of a particle times its velocity, and the force **F** are vectors. This guarantees that Newton's equation of motion $d\mathbf{p}/dt = \mathbf{F}$, or in components

$$\frac{dp_\mu}{dt} = F_\mu, \tag{2.44}$$

is form-invariant against a rotation of the coordinate system.

Warning

A rotated coordinate system must not be confused with a *rotating coordinate system*. A rotating coordinate system is an accelerated system where additional forces, like the Coriolis force and a centrifugal force, have to be taken into account in the equation of motion.

2.5.2 What is a Tensor?

Tensors are important "tools" for the characterization of anisotropies; but what is meant by the notion *tensor*? Here mainly Cartesian tensors of rank ℓ, $\ell = 0, 1, 2, \ldots$ are treated. These are quantities with ℓ indices which change in a specific way, when the coordinate system is rotated. More specifically: a Cartesian tensor of rank ℓ is a quantity with ℓ indices, e.g. $A_{\mu_1\mu_2\ldots\mu_\ell}$, whose Cartesian components $A'_{\mu_1\mu_2\ldots\mu_\ell}$ in a rotated coordinate system are obtained from the original ones by the application of ℓ rotation matrices $\mathbf{U}$ to each one of the indices, viz.:

$$A'_{\mu_1\mu_2\ldots\mu_\ell} = U_{\mu_1\nu_1} U_{\mu_2\nu_2} \ldots U_{\mu_\ell\nu_\ell} A_{\nu_1\nu_2\ldots\nu_\ell}. \tag{2.45}$$

In this sense, *scalars* and *vectors* are tensors of rank $\ell = 0$ and $\ell = 1$. Examples for vectors are the position vector $\mathbf{r}$ of a particle, its velocity $\mathbf{v}$, its linear momentum $\mathbf{p}$, as already mentioned before, but also its orbital angular momentum $\mathbf{L}$, its spin $\mathbf{s}$, as well as an electric field $\mathbf{E}$ and a magnetic field $\mathbf{B}$.

Tensors of rank $\ell = 2$ are frequently referred to as *tensors* without indicating their rank. Examples are the moment of inertia tensor, the pressure tensor or the stress tensor. Applications will be discussed later.

A second rank tensor can also be written as a matrix. However, it is distinguished from an arbitrary 3×3-matrix by the transformation properties of its components, just as not any 3-tuple is a vector in the sense described above. Of course, the matrix notation does not work for tensors of rank 3 or of higher rank.

2.5.3 Multiplication by Numbers and Addition of Tensors

The multiplication of a tensor by real number k means the multiplication of all its elements by this number, which is almost trivial in component notation:

$$k\,(\mathbf{A})_{\mu_1\mu_2\ldots\mu_\ell} = k\,A_{\mu_1\mu_2\ldots\mu_\ell}. \tag{2.46}$$

The addition of two tensors of the same rank implies that the corresponding components are added. When a tensor $\mathbf{C}$ is said to be the sum of the tensors $\mathbf{A}$ and $\mathbf{B}$, this means:

$$C_{\mu_1\mu_2\ldots\mu_\ell} = A_{\mu_1\mu_2\ldots\mu_\ell} + B_{\mu_1\mu_2\ldots\mu_\ell}. \tag{2.47}$$

The addition of two tensors makes sense only when both have the same rank ℓ. Of course, the rank of the resulting sum is also ℓ.

Notice, though it may sound somewhat confusing, tensors of a fixed rank ℓ (with $\ell = 0, 1, 2, \ldots$) are elements of a vector space.

2.5.4 Remarks on Notation

The Cartesian components of tensors are unambiguously specified by Greek subscripts, e.g. a_μ and $a_{\mu\nu}$. As practiced already above, it is sometimes more convenient
to use *boldface* and *boldface sans serif* letters, e.g. **a** and **a** to indicate that a quantity
is a vector or (second rank) tensor. An alternative "invariant" notation for tensors of
rank ℓ (which is preferred in hand writing) is to underline a letter ℓ times, e.g. $\underline{a}$ and
$\underline{\underline{a}}$ for a vector and a tensor of rank 2. When Cartesian components are not written
explicitly, a *center dot* $\cdot$ must be used to indicate a "contraction", i.e. a summation
over indices. The scalar product $\mathbf{a} \cdot \mathbf{b} = a_\mu b_\mu$ has to be distinguished from the *dyadic
product* $\mathbf{a}\,\mathbf{b}$, equivalent to $a_\mu b_\nu$, which is a second rank tensor. The scalar product
of a second rank tensor with a vector, e.g. $\mathbf{C} \cdot \mathbf{b}$, equivalent to $C_{\mu\nu} b_\nu$, is a vector
whose components are computed by analogy to the multiplication of a matrix with
a "column vector". The quantity $\mathbf{C}\,\mathbf{b}$, on the other hand, stands for the third rank
tensor $C_{\mu\nu} b_\lambda$.

The invariant notation appears to be "simpler" than the component notation. Here
both notations are used. The components of Cartesian tensors are specified explicitly
when new relations are introduced and when ambiguities in the order of subscripts
could arise as, e.g., in the products $a_{\mu\nu} b_{\nu\mu}$ and $a_{\mu\nu} b_{\mu\nu}$ of two tensors **a** and **b**.
The invariant notation is preferred only when it can be translated uniquely into the
component form.

2.5.5 Why the Emphasis on Tensors?

The physical content of equations must be invariant under a rotation of the coordinate
system. For the linear relation

$$b_\mu = C_{\mu\nu}\, a_\nu, \tag{2.48}$$

between two vectors **a** and **b**, this implies that the components of the coefficient
matrix **C** have to transform under a rotation like the components of a tensor of rank
2. In short, **C** *is* a second rank tensor. The proof is as follows. We assume that **a** and
b are vectors. This means, in the rotated coordinate system, the components of **b** are
related to the original ones by $b'_\mu = U_{\mu\lambda} b_\lambda$. Use of (2.48) leads to

$$b'_\mu = U_{\mu\lambda}\, C_{\lambda\kappa}\, a_\kappa.$$

The components of **a** are related to those of **a**$'$ by $a_\kappa = U^{-1}_{\kappa\nu} a'_\nu = U_{\nu\kappa} a'_\nu$. In the last
equality it has been used that the inverse and the transposed of the transformation
matrix **U**, cf. (2.35), are equal. Insertion into the previous equation leads to $b'_\mu =
U_{\mu\lambda} C_{\lambda\kappa} U_{\nu\kappa} a'_\nu$, which is equivalent to

$$b'_\mu = C'_{\mu\nu}\, a'_\nu, \tag{2.49}$$

with the quantity $\mathbf{C}'$ linked with $\mathbf{C}$ by

$$C'_{\mu\nu} = U_{\mu\lambda} U_{\nu\kappa} C_{\lambda\kappa}. \tag{2.50}$$

The relation (2.50) proves: $\mathbf{C}$ is a second rank tensor.

Examples for linear relations like (2.48) are those between the angular momentum and the angular velocity of a solid body, where the moment of inertia tensor occurs, and between the electric polarization and the electric field in a "linear medium". Here, the susceptibility tensor plays the role of $\mathbf{C}$.

Similarly, the linear relation $b_{\mu\nu} = C_{\mu\nu\lambda\kappa} a_{\lambda\kappa}$ between two second rank tensors $\mathbf{a}$ and $\mathbf{b}$ implies that, in this case, $\mathbf{C}$ is a tensor of rank 4. The elasticity and the viscosity tensors linking the stress tensor or the pressure tensor with the gradient of the displacement and of the velocity field, respectively, are of this type.

The generalization of (2.48) is a linear relation between a tensor $\mathbf{b}$ of rank ℓ with tensor $\mathbf{a}$ of rank k of the form

$$b_{\mu_1\mu_2...\mu_\ell} = C_{\mu_1\mu_2...\mu_\ell\, \nu_1\nu_2...\nu_k}\, a_{\nu_1\nu_2...\nu_k}. \tag{2.51}$$

Here $\mathbf{C}$ is a tensor of rank $\ell + k$.

In physics, examples for linear relations linking tensors of rank 1 with tensors of rank 1, 2, 3 and of tensors of 2 with tensors of rank 1, 2, 3 of tensors, and so on, were already discussed over hundred years ago in the book *Lehrbuch der Kristallphysik* where Woldemar Voigt introduced the notion *tensor*.

The relation (2.51) is a *linear mapping* of $\mathbf{a}$ on $\mathbf{b}$. Nevertheless, the physical content may describe non-linear effects, when the tensor $\mathbf{a}$ stands for a product of tensors. Examples occur in non-linear optics. For strong electric fields, the induced electric polarization contains not only the standard term linear in the field, but also contributions bilinear and of third order in the electric field. The material coefficient characterizing these effects, called higher order susceptibilities, are tensors of rank 3 and 4.

2.6 Parity

2.6.1 Parity Operation

In addition to their rank, tensors are classified by their *parity*. The parity is either equal to 1 or -1 when the physical quantity considered is an eigenfunction of the *parity operator* $\mathscr{P}$. The *parity operation* is an active transformation where the position vector $\mathbf{r}$ is replaced by $-\mathbf{r}$, cf. Fig. 2.6. This active 'mirroring' should not be confused with the mirroring of the coordinate system as described by the transformation matrix $U_{\mu\nu} = -\delta_{\mu\nu}$.

Fig. 2.6 Parity operation:
$\mathbf{r} \rightarrow -\mathbf{r}$

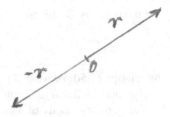

The parity operator $\mathscr{P}$, when applied on any function $f(\mathbf{r})$, yields $f(-\mathbf{r})$:

$$\mathscr{P} f(\mathbf{r}) = f(-\mathbf{r}). \tag{2.52}$$

Clearly, one has $\mathscr{P} f(-\mathbf{r}) = f(\mathbf{r})$ and consequently

$$\mathscr{P}^2 = 1, \tag{2.53}$$

or $(\mathscr{P} - 1)(\mathscr{P} + 1) = 0$. Thus the eigenvalues of the parity operator are

$$P = \pm 1. \tag{2.54}$$

Usually eigenfunctions are referred to as having *positive* or *negative* parity, when $P = 1$ and $P = -1$, respectively, applies.

2.6.2 Parity of Vectors and Tensors

In most applications tensors, and this includes vectors, are eigenfunction of the parity operator. Tensors of rank ℓ with

$$\mathscr{P} = (-1)^{\ell} \tag{2.55}$$

are called *proper tensors*, those with

$$\mathscr{P} = -(-1)^{\ell} = (-1)^{\ell+1} \tag{2.56}$$

are referred to as *pseudo tensors*.

For vectors ($\ell = 1$), also the terms *polar vector* and *axial vector* are used to distinguish between proper and pseudo vectors. Examples for polar vectors are the linear momentum $\mathbf{p}$ of a particle and the electric field, whereas the angular momentum and the magnetic field are axial vectors, as will be discussed later.

2.6.3 Consequences for Linear Relations

The electromagnetic interaction underlying all relevant interactions encountered in every days life, i.e. in gases, liquids and solids, is invariant under the parity operation. The equations governing physical properties and phenomena must not violate this parity invariance. This means, for example, when the vector $\mathbf{b}$ in the relation $b_\mu = C_{\mu\nu} a_\nu$ has the parity -1 (polar vector), the vector $\mathbf{a}$ and the tensor $\mathbf{C}$ must have the parities -1 and 1 (polar vector and proper tensor) or 1 and -1 (axial vector and pseudo tensor). More general, let P_a, P_b, P_C the values of the parities of the tensors $\mathbf{a}$, $\mathbf{b}$, $\mathbf{C}$ in the linear relation (2.51). Parity invariance requires

$$P_b = P_C \, P_a. \tag{2.57}$$

Likewise, when the parities of $\mathbf{a}$ and $\mathbf{b}$ are given by their physical meaning, the coefficient tensor $\mathbf{C}$ must have the parity

$$P_C = P_a \, P_b, \tag{2.58}$$

in order that the linear relation (2.51) does not violate parity.

2.6.4 Application: Linear and Nonlinear Susceptibility Tensors

The electric field $\mathbf{E}$, the electric displacement field $\mathbf{D}$ and the electric polarization $\mathbf{P}$ used in *electrodynamics* are polar vectors. They are linked by the general relation

$$\mathbf{D} = \varepsilon_0 \, \mathbf{E} + \mathbf{P},$$

where ε_0 is the electric permeability coefficient of the vacuum. In a material, called *linear medium*, the electric polarization is linearly related to the electric field, according to

$$P_\mu = \varepsilon_0 \, \chi_{\mu\nu} \, E_\nu,$$

where $\chi_{\mu\nu}$ is the *linear susceptibility tensor*. In the special case of a linear medium, one has

$$D_\mu = \varepsilon_0 \, \varepsilon_{\mu\nu} \, E_\nu, \quad \varepsilon_{\mu\nu} = \varepsilon_0 \, (\delta_{\mu\nu} + \chi_{\mu\nu}),$$

with the dimensionless dielectric tensor $\varepsilon_{\mu\nu}$. In general, in particular for strong electric fields as, e.g. encountered in a (focussed) laser beam, terms nonlinear in the electric field give significant contributions to the electric polarization. Up to third order in the electric field, the electric polarization is given by

$$P_\mu = \varepsilon_0 \, (\chi_{\mu\nu}^{(1)} \, E_\nu + \chi_{\mu\nu\lambda}^{(2)} \, E_\nu E_\lambda + \chi_{\mu\nu\lambda\kappa}^{(3)} \, E_\nu E_\lambda E_\kappa + \ldots). \qquad (2.59)$$

Here $\chi_{\mu\nu}^{(1)} \equiv \chi_{\mu\nu}$ is the linear susceptibility tensor. The third and fourth rank tensors $\chi_{\mu\nu\lambda}^{(2)}$ and $\chi_{\mu\nu\lambda\kappa}^{(3)}$ characterize the second and third order susceptibilities. In optics, these terms are responsible for the second and third harmonics generation, where a part of the incident light with frequency ω is converted into light with the frequencies 2ω and 3ω, respectively.

Both the electric field and the electric polarization have negative parity. Conservation of parity enforces that the linear and the third order susceptibility tensors must have positive parity, i.e. they are proper tensors of rank 2 and 4, respectively. In the simple case of an *isotropic medium*, these tensors reduce to $\chi_{\mu\nu}^{(1)} = \chi^1 \delta_{\mu\nu}$ and $\chi_{\mu\nu\lambda\kappa}^{(3)} = \chi^3 \delta_{\mu\nu} \delta_{\lambda\kappa}$, with (proper) scalar coefficients χ^1 and χ^3. The second order susceptibility, underlying the second harmonic generation (and also the generation of a zero frequency field), must have negative parity. This can be provided by a polar vector $\mathbf{d}$ in the medium, such as dipole moment or internal electric field, or even by the vector normal to a surface. Then the second order susceptibility tensor $\chi_{\mu\nu\lambda}^{(2)}$ will contain contributions proportional to $d_\mu \delta_{\nu\lambda}$ and to $\delta_{\mu\nu} d_\lambda$.

Notice: as far as the tensor algebra is concerned, the terms nonlinear in the electric field in (2.59) still are "linear relations" between P_μ and the tensors $E_\nu E_\lambda$ and $E_\nu E_\lambda E_\kappa$, which are of second and third order in the components of the electric field vector.

2.7 Differentiation of Vectors and Tensors with Respect to a Parameter

2.7.1 Time Derivatives

Just like scalars, vectors and tensors can dependent on parameters. In most applications in physics, one deals with functions of the time t. The time derivative of a tensor $\mathbf{A}$ is a tensor again. It is defined as the time derivatives of all its components, viz.,

$$\left(\frac{\mathrm{d}}{\mathrm{d}t} \mathbf{A} \right)_{\mu\nu\ldots} \equiv (\dot{\mathbf{A}})_{\mu\nu\ldots} = \frac{\mathrm{d}}{\mathrm{d}t} A_{\mu\nu\ldots}. \qquad (2.60)$$

It is recalled that the tensor character of a quantity is intimately linked with the transformation behavior of its components under a rotation of the coordinate system, cf. (2.45). Since the transformation matrix $\mathbf{U}$ is "timeless", the differentiation with respect to time and the rotation of the coordinate system commute. Thus the time derivative of a tensor of rank ℓ obeys the same transformation rules, it is also a tensor of rank ℓ.

The parity operation is also timeless. Thus it commutes with the differentiation with respect to time. Consequently, the time derivative $\dot{\mathbf{A}}$ of a tensor has the same parity as the original tensor $\mathbf{A}$.

In short: neither the property of a physical quantity being a tensor, nor its parity behavior are affected by differentiating it with respect to time.

2.7.2 Trajectory and Velocity

The trajectory of a mass point or of the center of mass of any solid object is described by the time dependence of its position vector $\mathbf{r} = \mathbf{r}(t)$, or equivalently, $r_\mu = r_\mu(t)$, $\mu = 1, 2, 3$. The velocity $\mathbf{v}$ is defined by

$$v_\mu = \frac{d}{dt} r_\mu \equiv \dot{r}_\mu. \tag{2.61}$$

The velocity is a polar vector, just as the position vector.

The unit vector

$$\widehat{v}_\mu = v^{-1} v_\mu = (\dot{r}_\nu \dot{r}_\nu)^{-1/2} \dot{r}_\mu, \tag{2.62}$$

points in the direction of the tangent of the curve describing the trajectory. It is referred to as *tangential vector*.

Two simple types of motion are considered next.

1. **Motion along a straight line**. The trajectory is determined by

$$r_\mu(t) = r_\mu^0 + f(t) e_\mu,$$

where r_μ^0 and the unit vector e_μ are constant. The differentiable function $f(t)$ is assumed to be equal to zero for $t = 0$, then $r_\mu(0) = r_\mu^0$. For $r_\mu^0 = 0$, the line runs through the origin. The resulting velocity is

$$v_\mu(t) = \frac{df}{dt} e_\mu.$$

Here, one has $\widehat{v}_\mu = e_\mu = $ const. and $v = \dot{f}$. For a *straight uniform motion*, not only the direction of the velocity, but also the speed v is constant. Then $f(t) = vt$ hold true.

2. **Motion on a circle**. The motion on a circle with the radius R and the angular velocity w is described by

$$r_1 = R\cos(wt), \quad r_2 = R\sin(wt), \quad r_3 = 0,$$

where, obviously, the circle lies in the 1–2-plane. The origin of the coordinate system is the center of the circle. At time $t = 0$, the position vector points in the

1-direction. For $w > 0$, the motion runs counterclockwise, i.e. in the mathematically positive sense. Assuming $R = \text{const.}$ and $w = \text{const.}$, the components of the velocity are

$$v_1 = -Rw \sin(wt), \quad v_2 = Rw \cos(wt), \quad v_3 = 0.$$

In this case, the velocity is perpendicular to the position vector, it is purely tangential.

2.7.3 Radial and Azimuthal Components of the Velocity

The position vector $r_\mu(t)$ can be written as a product of its magnitude $r = (r_\nu r_\nu)^{1/2}$ and the unit vector $\widehat{r}_\mu(t)$, according to $r_\mu = r\widehat{r}_\mu$. Then one has

$$\frac{d}{dt}r_\mu = \frac{dr}{dt}\widehat{r}_\mu + r\frac{d}{dt}\widehat{r}_\mu. \tag{2.63}$$

The *radial* component of the velocity is the first term on the right hand side of (2.63), which is parallel to the position vector $\mathbf{r}$. It describes the change of the length of $\mathbf{r}$. The second term, associated with the change of the direction of $\mathbf{r}$, is called the *azimuthal component*, sometimes also the *tangential component* of the velocity, because it is perpendicular to $\mathbf{r}$. This can be seen quickly as follows. Notice that $\widehat{r}_\nu \widehat{r}_\nu = 1$. The time derivative of this equation yields $2\widehat{r}_\nu \frac{d}{dt}\widehat{r}_\nu = 0$, which implies that the derivative of the radial unit vector $\widehat{\mathbf{r}}$ is perpendicular to $\widehat{\mathbf{r}}$. Alternatively, the definition of the unit vector, viz., $\widehat{r}_\mu = r_\mu r^{-1} = r_\mu (r_\nu r_\nu)^{-1/2}$ and the chain rule can be used to obtain

$$\frac{d}{dt}\widehat{r}_\mu = r^{-1}\frac{d}{dt}r_\mu - r^{-3}r_\mu r_\nu \frac{d}{dt}r_\nu = r^{-1}\left(\delta_{\mu\nu} - \widehat{r}_\mu \widehat{r}_\nu\right)v_\nu. \tag{2.64}$$

The projection tensor $\delta_{\mu\nu} - \widehat{r}_\mu \widehat{r}_\nu$ guarantees that $\frac{d}{dt}\widehat{r}_\mu$ is perpendicular to $\widehat{r}_\mu$.

Notice that the word *tangential* is used with two slightly different meanings, which only coincide for the motion on a circle. In one case it refers to the tangent of a trajectory which points in the direction of the velocity. In the second case, just discussed here, where the word "azimuthal" is more appropriate, it means the direction perpendicular to the position vector.

2.8 Time Reversal

The trajectory of a particle or of the center of mass of an extended object is described by the time dependence of the position vector $\mathbf{r} = \mathbf{r}(t)$. One may ask the question: does the trajectory $\mathbf{r}(-t)$ also describe a physically possible motion? In other words, does physics allow the backward motion just as well as the forward motion. If the answer to this question is "yes", the motion is called *reversible*, otherwise it is referred to as *irreversible*. In movies and in computer simulations, one can let the time run backwards. In real physics, just as in real life, this is not possible. On the other hand, physics deals both with reversible processes, like the celestial motion of a planet around the sun and with irreversible processes, like an earthly motion, damped by friction. It is desirable to know, whether the equations governing the dynamics, describe a reversible or an irreversible behavior, even before these equations are solved. This can be found out by inspecting the *time reversal behavior* of all terms in the relevant equations.

The time reversal behavior of a physical quantity is called *even* or *odd*, or also denoted by plus $+$ or minus $-$, depending on whether the *time reversal operator*, applied on this quantity, leaves it unchanged or changes its sign. The time reversal operator does not change the position vector $\mathbf{r}$. Application to the velocity $\mathbf{v} = \frac{d}{dt}\mathbf{r}$ yields $-\mathbf{v}$. More generally, the first derivative of a physical variable has a time reversible behavior, which is just opposite to that of the original variable. Clearly, the acceleration $\mathbf{a} = \frac{d}{dt}\mathbf{v} = \frac{d^2}{dt^2}\mathbf{r}$ is even under time reversal.

The idea behind these considerations is as follows: observe a process, e.g. the trajectory of a particle, from time $t = 0$ to the time t_{obs}, then change the sign of the velocity and of all relevant variables, which are odd under the time reversal operation and let the time run forward till $2t_{obs}$. When the process comes back to the original state, e.g. a particle runs back to its initial position, the process is called *reversible*. If the process does not return to its original state, it is called *irreversible*. When all physical variables in an equation governing the dynamics of a process have the same time reversal behavior, *time reversal invariance* is obeyed, otherwise the time reversal invariance is violated. A simple example is Newton's equation of motion for a single particle. Mass times acceleration is even under time reversal. When the force is just a function of the position vector, it is also even and, as a consequence, the equation describes a reversible dynamics. When, on the other hand, the force has a frictional contribution proportional to the velocity, the equation of motion involves terms with different time reversal behavior, the motion is irreversible. The motion is damped provided that the friction coefficient has the correct sign.

To distinguish in the theoretical description between reversible and irreversible phenomena, it is important to know the time reversal behavior of vectors and tensors used in physics. As already mentioned, the position vector $\mathbf{r}$ is not affected by the time reversal operator, the velocity $\mathbf{v} = \frac{d}{dt}\mathbf{r}$, however, changes sign, when t is replaced by $-t$. Likewise, the linear momentum $\mathbf{p} = m\mathbf{v}$, and also the orbital angular momentum, as discussed later, are odd under time reversal. The acceleration, being the second derivative of $\mathbf{r}$ with respect to time, is even under time reversal.

In Table 2.1, parity and the time reversible behavior of some vectors are indicated by plus or minus. The parity of all these vectors is uniquely determined. This is also true for the time reversal behavior of $\mathbf{r}$, $\mathbf{v}$, $\mathbf{p}$ and of the acceleration $\mathbf{a}$, of the angular velocity $\mathbf{w}$, and of the orbital angular momentum $\mathbf{L}$. As will be discussed later, this also applies for the electric and magnetic fields $\mathbf{E}$ and $\mathbf{B}$. When the time

Table 2.1 The parity and time-reversal behavior of some vectors

Physical quantity	r	v	p	a	F	w	L	T	E	B
Parity	−	−	−	−	−	+	+	+	−	+
Time reversal	+	−	−	+	±	−	−	±	+	−

reversal behavior of the force $\mathbf{F}$ and of the torque $\mathbf{T}$ are positive, the dynamics is reversible. Forces and torques, however, contain terms with the other time reversal behavior, when friction plays a role. Then the dynamics is irreversible. The time reversal behavior of tensors occurring in applications will be discussed later.

Chapter 3
Symmetry of Second Rank Tensors, Cross Product

Abstract This chapter deals with the symmetry of second rank tensors and the definition of the cross product of two vectors. In general, a second rank tensor contains a part which is symmetric and a part which is antisymmetric with respect to the interchange of its indices. For 3D, there exists a dual relation between the antisymmetric part of the second rank tensor and a vector. The symmetric part of the tensor is further decomposed into its isotropic part involving the trace of the tensor and the symmetric traceless part. Fourth rank projection tensors are defined which, when applied on an arbitrary second rank tensor, project onto its isotropic, antisymmetric and symmetric traceless parts. The properties of dyadics, viz. second rank tensors composed of the components of two vectors, are discussed. The dual relation between its antisymmetric part and a vector corresponds to the definition of the cross product or vector product, various physical applications are presented.

3.1 Symmetry

3.1.1 Symmetric and Antisymmetric Parts

An arbitrary tensor **A** of rank 2 can be decomposed into its *symmetric* and *antisymmetric* parts $\mathbf{A}^{\text{sym}}$ and $\mathbf{A}^{\text{asy}}$ according to

$$A_{\mu\nu}^{\text{sym}} = \frac{1}{2}\left(A_{\mu\nu} + A_{\nu\mu}\right), \qquad A_{\mu\nu}^{\text{asy}} = \frac{1}{2}\left(A_{\mu\nu} - A_{\nu\mu}\right). \tag{3.1}$$

Clearly, the interchange of subscripts implies

$$A_{\mu\nu}^{\text{sym}} = A_{\nu\mu}^{\text{sym}}, \qquad A_{\mu\nu}^{\text{asy}} = -A_{\nu\mu}^{\text{asy}}. \tag{3.2}$$

In three dimensions, $\mathbf{A}^{\text{sym}}$ and $\mathbf{A}^{\text{asy}}$ have 6 and 3 independent components.

© Springer International Publishing Switzerland 2015
S. Hess, *Tensors for Physics*, Undergraduate Lecture Notes in Physics,
DOI 10.1007/978-3-319-12787-3_3

3.1.2 Isotropic, Antisymmetric and Symmetric Traceless Parts

The symmetric part of a second rank tensor **A** can be decomposed further into an *isotropic part* proportional to the product of the isotropic tensor δ and the *trace* tr **A** $= A_{\lambda\lambda}$ and a symmetric traceless part $\overline{\mathbf{A}}$ defined by

$$\overline{A_{\mu\nu}} = \frac{1}{2}\left(A_{\mu\nu} + A_{\nu\mu}\right) - \frac{1}{3}A_{\lambda\lambda}\delta_{\mu\nu}. \tag{3.3}$$

Thus the tensor **A** is decomposed into its isotropic, antisymmetric and symmetric traceless parts according to

$$A_{\mu\nu} = \frac{1}{3}A_{\lambda\lambda}\delta_{\mu\nu} + A_{\mu\nu}^{\text{asy}} + \overline{A_{\mu\nu}}. \tag{3.4}$$

This decomposition is invariant under a rotation of the coordinate system.

The symbol $\overline{\cdots}$ used to indicate the symmetric traceless part of a tensor, was introduced by Ludwig Waldmann around 1960. Compared with the double arrow $\longleftrightarrow$, which also occurs in printing, the $\overline{\cdots}$ has the advantage that it can be drawn in one stroke. For second rank tensors, $\overline{\cdots}$ first appeared in print in [20], and in [21], it was applied for irreducible tensors of any rank. Alternative notations used in the literature for symmetric traceless tensors are mentioned in Sect. 3.1.7.

3.1.3 Trace of a Tensor

The isotropic part involves the *trace of the tensor*

$$\text{tr}(\mathbf{A}) = A_{\lambda\lambda} = A_{11} + A_{22} + A_{33}. \tag{3.5}$$

It is a scalar (tensor of rank $\ell = 0$), i.e., it is invariant under a rotation of the coordinate system. The proof is: the tensor property $A'_{\mu\nu} = U_{\mu\kappa}U_{\nu\lambda}A_{\kappa\lambda}$ implies $A'_{\mu\mu} = U_{\mu\kappa}U_{\mu\lambda}A_{\kappa\lambda}$, and due to the orthogonality (2.31) of the transformation matrix, one has $A'_{\mu\mu} = \delta_{\kappa\lambda}A_{\kappa\lambda} = A_{\kappa\kappa}$.

The term *isotropic* is used since the unit tensor $\delta_{\mu\nu}$ has no directional properties, it is not affected by a rotation of the coordinate system. Here the other orthogonality (2.36) is used for the proof: $\delta'_{\mu\nu} = U_{\mu\kappa}U_{\nu\lambda}\delta_{\kappa\lambda} = U_{\mu\lambda}U_{\nu\lambda} = \delta_{\mu\nu}$.

Notice: the antisymmetric part of the tensor does not contribute to the trace:

$$\text{tr}(\mathbf{A}) = A_{\lambda\lambda} = A_{\lambda\lambda}^{\text{sym}} = \text{tr}(\mathbf{A}^{\text{sym}}). \tag{3.6}$$

The trace of a second rank tensor is also given by the total contraction of this tensor with the unit tensor:

$$\text{tr}(\mathbf{A}) = \delta_{\mu\nu}A_{\mu\nu} = \delta_{\mu\nu}A_{\nu\mu} = A_{\nu\nu}. \tag{3.7}$$

Again, notice that summation indices can have different names, as long as no index appears more than twice.

The trace of the unit tensor is equal to the dimension D, here $D = 3$. Thus

$$\delta_{\nu\nu} = 3. \tag{3.8}$$

This is the reason why the fraction $\frac{1}{3}$ occurs in (3.3) and (3.4).

3.1.4 Multiplication and Total Contraction of Tensors, Norm

The multiplication of a tensor $A_{\mu\nu}$ with a tensor $B_{\lambda\kappa}$ yields a fourth rank tensor. The contraction with $\nu = \lambda$, corresponding to a "dot-product" $\mathbf{A} \cdot \mathbf{B}$, gives a second rank tensor. The total contraction or "double dot-product"

$$\mathbf{A} : \mathbf{B} = A_{\mu\nu} B_{\nu\mu} \tag{3.9}$$

is a scalar. The order of the indices is such that it corresponds to the trace of the matrix product of $\mathbf{A}$ with $\mathbf{B}$.

In such a total contraction, the symmetry of one tensor is imposed on the other one. This means, e.g. when $\mathbf{A}$ is symmetric, the symmetric part of $\mathbf{B}$ only contributes in the product $A_{\mu\nu} B_{\nu\mu}$. Likewise, when $\mathbf{A}$ is antisymmetric, the antisymmetric part of $\mathbf{B}$ only contributes in the product $A_{\mu\nu} B_{\nu\mu}$. Furthermore, when $\mathbf{A}$ is isotropic, i.e. proportional to the unit tensor, then the trace of $\mathbf{B}$ only contributes to the product. When $\mathbf{A}$ is symmetric traceless, then only the symmetric traceless part of $\mathbf{B}$ gives a contribution. With both tensors decomposed according to (3.4), one obtains

$$A_{\mu\nu} B_{\nu\mu} = \frac{1}{3} A_{\lambda\lambda} B_{\kappa\kappa} + A_{\mu\nu}^{\text{asy}} B_{\nu\mu}^{\text{asy}} + \overline{A_{\mu\nu}}\, \overline{B_{\nu\mu}}. \tag{3.10}$$

Notice that one has $A_{\mu\nu} B_{\nu\mu} = A_{\mu\nu} B_{\mu\nu}$ only when at least one of the two tensors is symmetric.

The square of the norm or magnitude $|\mathbf{A}|$ of a second rank tensor is determined by the total contraction of $\mathbf{A}$ with its transposed $\mathbf{A}^{\mathsf{T}}$, viz.:

$$|\mathbf{A}|^2 = A_{\mu\nu} A_{\nu\mu}^{T} = A_{\mu\nu} A_{\mu\nu}. \tag{3.11}$$

Of course, the order of the subscript does not matter when $\mathbf{A}$ is symmetric.

3.1.5 Fourth Rank Projections Tensors

The decomposition (3.4) of a second rank tensor $\mathbf{A}$ into its isotropic ($i = 0$), its antisymmetric ($i = 1$), and its symmetric traceless ($i = 2$) parts $\mathbf{A}^{(i)}$ can also be accomplished by application of fourth rank projection tensors $P^{(i)}_{\mu\nu\mu'\nu'}$ on the components $A_{\mu'\nu'}$ according to

$$A^{(i)}_{\mu\nu} = P^{(i)}_{\mu\nu\mu'\nu'} A_{\mu'\nu'}. \tag{3.12}$$

Here pairs of subscripts are used like one index. Furthermore, notice that $A^{(1)}_{\mu\nu} \equiv A^{\text{asy}}_{\mu\nu}$ and $A^{(2)}_{\mu\nu} \equiv \overline{A_{\mu\nu}}$. The projection tensors are defined by

$$P^{(0)}_{\mu\nu,\mu'\nu'} := \frac{1}{3} \delta_{\mu\nu}\delta_{\mu'\nu'}, \quad P^{(1)}_{\mu\nu\mu'\nu'} := \frac{1}{2}(\delta_{\mu\mu'}\delta_{\nu\nu'} - \delta_{\mu\nu'}\delta_{\nu\mu'}), \tag{3.13}$$

and

$$P^{(2)}_{\mu\nu\mu'\nu'} \equiv \Delta_{\mu\nu,\mu'\nu'} := \frac{1}{2}(\delta_{\mu\mu'}\delta_{\nu\nu'} + \delta_{\mu\nu'}\delta_{\nu\mu'}) - \frac{1}{3}\delta_{\mu\nu}\delta_{\mu'\nu'}. \tag{3.14}$$

In the applications presented later, the symbol $\Delta_{...}$ is preferred over $P^{(2)}_{...}$.

The projection tensors have the properties

$$P^{(i)}_{\mu\nu\alpha\beta} P^{(j)}_{\alpha\beta\mu'\nu'} = \delta^{ij} P^{(i)}_{\mu\nu\mu'\nu'}, \tag{3.15}$$

where δ^{ij} is the Kronecker symbol, being equal to 1, when $i = j$ and 0 when $i \neq j$, and they obey the 'sum rule' or 'completeness relation'

$$P^{(0)}_{\mu\nu,\mu'\nu'} + P^{(1)}_{\mu\nu,\mu'\nu'} + P^{(2)}_{\mu\nu,\mu'\nu'} = \delta_{\mu\mu'}\delta_{\nu\nu'}. \tag{3.16}$$

The contraction $\nu' = \nu$ of the projectors yields

$$P^{(0)}_{\mu\nu,\mu'\nu} = \frac{1}{3}\delta_{\mu\mu'}, \quad P^{(1)}_{\mu\nu,\mu'\nu} = \delta_{\mu\mu'}, \quad P^{(2)}_{\mu\nu\mu'\nu} \equiv \Delta_{\mu\nu,\mu'\nu} = \frac{5}{3}\delta_{\mu\mu'}. \tag{3.17}$$

The subsequent complete contraction, corresponding to $\mu' = \mu$, gives the numbers of the independent components of the isotropic, antisymmetric and symmetric traceless parts of second rank tensor in 3D, viz.:

$$P^{(0)}_{\mu\nu,\mu\nu} = 1, \quad P^{(1)}_{\mu\nu,\mu\nu} = 3, \quad P^{(2)}_{\mu\nu\mu\nu} \equiv \Delta_{\mu\nu,\mu\nu} = 5. \tag{3.18}$$

Generalized Delta-tensors of rank 2ℓ which, when applied to tensors of rank ℓ, project out the symmetric traceless part of that tensor, will be introduced later.

3.1.6 Preliminary Remarks on "Antisymmetric Part and Vector"

The three independent components of the antisymmetric part, in 3D, can be linked with a vector (tensor of rank $\ell = 1$). This property is specific for 3D, whereas most relations formulated here, apply also to Cartesian components in 2, 4 and higher dimensions. This is seen as follows. In n dimensions, the number of elements of a second rank tensor is n^2. There are n elements in the diagonal, consequently one has $n^2 - n$ off-diagonal elements. The number of independent elements of the symmetric part is

$$N^{\text{sym}} = n + \frac{1}{2}n(n-1) = \frac{1}{2}n(n+1), \tag{3.19}$$

that of the antisymmetric part is

$$N^{\text{asy}} = \frac{1}{2}n(n-1). \tag{3.20}$$

The number of elements of a vector is n. For $n > 0$, the relation $n = N^{\text{asy}}$ has just the solution $n = 3$.

3.1.7 Preliminary Remarks on the Symmetric Traceless Part

The symmetric traceless part cannot be expressed in terms of lower rank tensors. For this reason, it is also referred to as the *irreducible* part of the tensor. In 3D, it has 5 independent components.

The symbol $\overline{\cdots}$ is also used for tensors of rank $\ell \geq 2$ in order to indicate the symmetric traceless (irreducible) part which, in general, has $2\ell + 1$ independent components, details later.

Different notations for the symmetric traceless part of tensor are found in the literature. Sometimes the double arrow $\overleftrightarrow{\cdots}$, in same cases the brackets $[\ldots]_0$ or double brackets $[[\ldots]]_0$, where the subscript 0 indicates that the trace is zero, are used instead of $\overline{\cdots}$.

3.2 Dyadics

3.2.1 Definition of a Dyadic Tensor

A second rank tensor constructed from the components of two vectors, e.g. **a** and **b** is called a *dyadic tensor*, sometimes also just *dyadic* or *dyad*. The quantity

$$\mathbf{a}\,\mathbf{b} := \begin{pmatrix} a_1b_1 & a_1b_2 & a_1b_3 \\ a_2b_1 & a_2b_2 & a_2b_3 \\ a_3b_1 & a_3b_2 & a_3b_3 \end{pmatrix} \tag{3.21}$$

is a tensor. It must not be confused with the scalar product $\mathbf{a}\cdot\mathbf{b} = a_1b_1+a_2b_2+a_3b_3$.
With $A_{\mu\nu} = a_\mu b_\nu$, the decomposition (3.4) reads:

$$a_\mu b_\nu = \frac{1}{3}\,a_\lambda b_\lambda \delta_{\mu\nu} + \frac{1}{2}\,\left(a_\mu b_\nu - a_\nu b_\mu\right) + \overline{a_\mu b_\nu}\,. \tag{3.22}$$

The symmetric traceless part, in accord with (3.3), given by

$$\overline{a_\mu b_\nu} = \frac{1}{2}\,\left(a_\mu b_\nu + a_\nu b_\mu\right) - \frac{1}{3}\,a_\lambda b_\lambda \delta_{\mu\nu}. \tag{3.23}$$

The trace of the dyadic $\mathbf{a}\,\mathbf{b}$ is the scalar product $\mathbf{a}\cdot\mathbf{b}$. In 3D, the antisymmetric part of the dyadic is linked with the cross product $\mathbf{a}\times\mathbf{b}$, details later.

The product of dyadic tensors, with total contraction, can be inferred from (3.10). The case where both dyadics are symmetric traceless is discussed in Sect. 3.2.2.

3.1 Exercise:
Symmetric and Antisymmetric Parts of a Dyadic in Matrix Notation
Write the symmetric traceless and the antisymmetric parts of the dyadic tensor $A_{\mu\nu} = a_\mu b_\nu$ in matrix form for the vectors $\mathbf{a}$: $\{1, 0, 0\}$ and $\mathbf{b}$: $\{0, 1, 0\}$. Compute the norm of the symmetric and the antisymmetric parts and compare with $A_{\mu\nu}A_{\mu\nu}$ and $A_{\mu\nu}A_{\nu\mu}$.

3.2.2 Products of Symmetric Traceless Dyadics

Consider two dyadics, formed by the pairs of vectors $\mathbf{a}$, $\mathbf{b}$ and $\mathbf{c}$, $\mathbf{d}$, respectively. As discussed above for the tensor multiplication with contraction, the expression $\mathbf{a}\,\mathbf{b}\cdot\mathbf{c}\,\mathbf{d}$ stands for the dyadic $\mathbf{a}\,\mathbf{d}$, multiplied by the scalar product $\mathbf{b}\cdot\mathbf{c}$. The double dot product then yields $\mathbf{a}\cdot\mathbf{d}\,\mathbf{b}\cdot\mathbf{c}$.

Of particular interest is the case, where the symmetric traceless parts of these dyadics are multiplied. Here one has

$$\begin{aligned}
(\overline{\mathbf{a}\,\mathbf{b}} \cdot \overline{\mathbf{c}\,\mathbf{d}})_{\mu\nu} &= \overline{a_\mu b_\lambda}\,\overline{c_\lambda d_\nu} \\
&= \frac{1}{4}\,(\mathbf{b}\cdot\mathbf{c}\,a_\mu d_\nu + \mathbf{b}\cdot\mathbf{d}\,a_\mu c_\nu + \mathbf{a}\cdot\mathbf{c}\,b_\mu d_\nu + \mathbf{a}\cdot\mathbf{d}\,b_\mu c_\nu) \\
&\quad - \frac{1}{6}\,\mathbf{c}\cdot\mathbf{d}\,(a_\mu b_\nu + a_\nu b_\mu) - \frac{1}{6}\,\mathbf{a}\cdot\mathbf{b}\,(c_\mu d_\nu + c_\nu d_\mu) \\
&\quad + \frac{1}{9}\,\mathbf{a}\cdot\mathbf{b}\,\mathbf{c}\cdot\mathbf{d}\,\delta_{\mu\nu}.
\end{aligned} \tag{3.24}$$

The further contraction $\mu = \nu$ leads to

$$\overline{\mathbf{ab}} : \overline{\mathbf{cd}} = \overline{a_\mu b_\lambda}\; \overline{c_\lambda d_\mu} = \frac{1}{2}\left(\mathbf{a}\cdot\mathbf{c}\,\mathbf{b}\cdot\mathbf{d} + \mathbf{a}\cdot\mathbf{d}\,\mathbf{b}\cdot\mathbf{c}\right) - \frac{1}{3}\mathbf{a}\cdot\mathbf{b}\,\mathbf{c}\cdot\mathbf{d}. \quad (3.25)$$

For the special case $\mathbf{a} = \mathbf{b}$ and $\mathbf{c} = \mathbf{d}$, relation (3.25) reduces to

$$\overline{\mathbf{aa}} : \overline{\mathbf{cc}} = \overline{a_\mu a_\lambda}\; \overline{c_\lambda c_\mu} = (\mathbf{a}\cdot\mathbf{c})^2 - \frac{1}{3}a^2 c^2$$

$$= a^2 c^2 \left((\hat{\mathbf{a}}\cdot\hat{\mathbf{c}})^2 - \frac{1}{3}\right) = a^2 c^2 \left(\cos^2\varphi - \frac{1}{3}\right), \quad (3.26)$$

where $\hat{\mathbf{a}}$ and $\hat{\mathbf{c}}$ are unit vectors, and φ is the angle between the vectors $\mathbf{a}$ and $\mathbf{c}$. Clearly, for $\mathbf{a} = \mathbf{c}$, corresponding to $\varphi = 0$, one finds

$$\overline{\mathbf{aa}} : \overline{\mathbf{aa}} = \overline{a_\mu a_\lambda}\; \overline{a_\lambda a_\mu} = \frac{2}{3} a^4. \quad (3.27)$$

Notice, the double dot product of two symmetric traceless dyadic tensors constructed from orthogonal vectors is not zero. For $\mathbf{c}$ perpendicular to $\mathbf{a}$, corresponding to $\varphi = 90°$, relation (3.26) implies

$$\mathbf{a}\perp\mathbf{c} \implies \overline{\mathbf{aa}} : \overline{\mathbf{cc}} = \overline{a_\mu a_\lambda}\; \overline{c_\lambda c_\mu} = -\frac{1}{3}a^2 c^2. \quad (3.28)$$

On the other hand, the two dyadic tensors $\overline{\mathbf{aa}}$ and $\overline{\mathbf{cc}}$ are "orthogonal", in the sense that their double dot product vanishes, when the angle between the two vectors is given by the "magic angle" $\varphi = \arccos(1/\sqrt{3}) \approx 54.7°$. Applications of these relations for the double dot product of dyadics are discussed later.

3.2 Exercise: Symmetric Traceless Dyadics in Matrix Notation

Write the symmetric traceless parts of the dyadic tensor $C_{\mu\nu} = C_{\mu\nu}(\alpha) = 2a_\mu b_\nu$ in matrix form for the vectors $\mathbf{a} = \mathbf{a}(\alpha) : \{c, -s, 0\}$ and $\mathbf{b} = \mathbf{b}(\alpha) : \{s, c, 0\}$, where c and s are the abbreviations $c = \cos\alpha$ and $s = \sin\alpha$. Discuss the special cases $\alpha = 0$ and $\alpha = \pi/4$. Compute the product $B_{\mu\nu}(\alpha) = \overline{C_{\mu\lambda}}\,(0)\,\overline{C_{\lambda\nu}}\,(\alpha)$, determine the trace and the symmetric traceless part of this product. Determine the angle α, for which one has $B_{\mu\mu} = 0$.

3.3 Antisymmetric Part, Vector Product

3.3.1 Dual Relation

As already mentioned before, the antisymmetric part $(1/2)(A_{\mu\nu} - A_{\nu\mu})$ of a second rank tensor $A_{\mu\nu}$, in three dimensions, can be linked with the three components of a vector $\mathbf{a}$. This link, referred to as *dual relation*, is:

$$a_1 = A_{23} - A_{32} = 2A_{23}^{\text{asy}},$$
$$a_2 = A_{31} - A_{13} = 2A_{31}^{\text{asy}},$$
$$a_3 = A_{12} - A_{21} = 2A_{12}^{\text{asy}}. \tag{3.29}$$

Clearly, the order of the subscript $1, 2, 3$ in the second line is a cyclic permutation of that one in the first line, and so on.

The relation (3.29) can be inverted in the sense that, in matrix notation, the antisymmetric tensor is given by

$$\mathbf{A}^{\text{asy}} = \begin{pmatrix} 0 & \dfrac{1}{2}a_3 & -\dfrac{1}{2}a_2 \\ -\dfrac{1}{2}a_3 & 0 & \dfrac{1}{2}a_1 \\ \dfrac{1}{2}a_2 & -\dfrac{1}{2}a_1 & 0 \end{pmatrix}. \tag{3.30}$$

The 'proof' that $\mathbf{a}$, defined by (3.29) indeed transforms like a vector, when $\mathbf{A}$ is a tensor, is presented next. This is not self-evident since a vector is transformed with one rotation matrix, whereas the second rank tensor is transformed with a product of two transformation matrices.

The tensor property of $\mathbf{A}$ implies, that the first component of $\mathbf{a}$, in the rotated system, is given by

$$a_1' = A_{23}' - A_{32}' = U_{2\mu}U_{3\nu}A_{\mu\nu} - U_{3\mu}U_{2\nu}A_{\mu\nu} = U_{2\mu}U_{3\nu}(A_{\mu\nu} - A_{\nu\mu}). \tag{3.31}$$

Note: in the last term before the last equality sign, the summation indices μ, ν, have been interchanged. When a_1', given by this relation is the component of a vector, the quantity b_μ defined by

$$b_\mu = U_{\mu\lambda}^{-1}a_\lambda' = U_{\lambda\mu}a_\lambda', \tag{3.32}$$

must be equal to a_μ. The first component of (3.32) reads

$$b_1 = U_{11}a_1' + U_{21}a_2' + U_{31}a_3'. \tag{3.33}$$

Use of (3.31) for a_1' and of the corresponding expressions for a_2', a_3' (obtained by the cyclic permutations of 1, 2, 3) leads to

$$b_1 = (U_{11}U_{2\mu}U_{3\nu} + U_{21}U_{3\mu}U_{2\nu} + U_{31}U_{1\mu}U_{2\nu})(A_{\mu\nu} - A_{\nu\mu}). \qquad (3.34)$$

Now we consider the special case $a_1 = A_{23} - A_{32} \neq 0$ and $a_2 = A_{31} - A_{13} = 0$, $a_3 = A_{12} - A_{21} = 0$. Then (3.34) yields, with the summation over double indices written explicitly

$$b_1 = (U_{11}(U_{22}U_{33} - U_{23}U_{32}) + U_{21}(U_{32}U_{13} - U_{33}U_{12})$$
$$+ U_{31}(U_{12}U_{23} - U_{13}U_{22}))a_1 = \det(\mathbf{U})a_1, \qquad (3.35)$$

where $\det(\mathbf{U})$ is the determinant of the transformation matrix. For a proper rotation $\det(\mathbf{U}) = 1$ holds true, and consequently $b_1 = a_1$. The proof for the equality of the other components $\mathbf{b}$ and $\mathbf{a}$ requires just the cyclic permutation of 1, 2, 3. This then completes the proof that the three-component quantity $\mathbf{a}$ linked with the antisymmetric part of a tensor by the duality relation (3.29) is a vector in the sense used here.

3.3.2 Vector Product

In the case of a dyadic $\mathbf{A} = \mathbf{ab}$ and with $\mathbf{a}$, in (3.29), replaced by $\mathbf{c}$, the relation (3.29) corresponds to the usual *cross product* or *vector product* $\mathbf{c} = \mathbf{a} \times \mathbf{b}$ of the two vectors $\mathbf{a}$ and $\mathbf{b}$. More specifically one has

$$c_1 = (\mathbf{a} \times \mathbf{b})_1 = a_2b_3 - a_3b_2,$$
$$c_2 = (\mathbf{a} \times \mathbf{b})_2 = a_3b_1 - a_1b_3,$$
$$c_3 = (\mathbf{a} \times \mathbf{b})_3 = a_1b_2 - a_2b_1. \qquad (3.36)$$

By analogy to (3.30), an antisymmetric tensor is linked with the cross product. In particular, one has for the 12-component:

$$2\,(\mathbf{a}\,\mathbf{b})_{12}^{\text{asy}} := a_1b_2 - a_2b_1 = (\mathbf{a} \times \mathbf{b})_3. \qquad (3.37)$$

The other components are obtained by a cyclic interchange of 1, 2, 3.

The antisymmetric part of the dyadic $\mathbf{a}\,\mathbf{b}$, as well as the cross product $\mathbf{c} = \mathbf{a} \times \mathbf{b}$ vanish, when the vectors $\mathbf{a}$ and $\mathbf{b}$ are parallel to each other.

As an alternative to (3.36), the components of the cross product can be expressed with the help of a determinant according to

$$c_\mu := \begin{vmatrix} \delta_{1\mu} & a_1 & b_1 \\ \delta_{2\mu} & a_2 & b_2 \\ \delta_{3\mu} & a_3 & b_3 \end{vmatrix} = \begin{vmatrix} \delta_{1\mu} & \delta_{2\mu} & \delta_{3\mu} \\ a_1 & a_2 & a_3 \\ b_1 & b_2 & b_3 \end{vmatrix}. \qquad (3.38)$$

To verify this expression, e.g. use $\mu = 1$ and note that $\delta_{11} = 1$, $\delta_{21} = \delta_{31} = 0$. Then one obtains the first line of (3.36). Similarly, the second and third line are recovered with $\mu = 2$ and $\mu = 3$.

The cross product of the two vectors is antisymmetric with respect their exchange:

$$(\mathbf{b} \times \mathbf{a}) = -(\mathbf{a} \times \mathbf{b}). \tag{3.39}$$

When the two vectors are parallel, i.e. when one has $\mathbf{b} = k\mathbf{a}$ with some numerical factor k, the cross product is equal to zero, thus

$$\mathbf{a} \times \mathbf{b} = 0 \Longleftrightarrow \mathbf{a} \parallel \mathbf{b}. \tag{3.40}$$

These properties follow from the definition of the vector product. Likewise, with the help of (3.38), the scalar product of a vector $\mathbf{d}$ with the vector $\mathbf{c}$ which, in turn, is the vector product of vectors $\mathbf{a}$ and $\mathbf{b}$ is given by the *spate product*:

$$\mathbf{d} \cdot \mathbf{c} = \mathbf{d} \cdot (\mathbf{a} \times \mathbf{b}) = \begin{vmatrix} d_1 & a_1 & b_1 \\ d_2 & a_2 & b_2 \\ d_3 & a_3 & b_3 \end{vmatrix} = \begin{vmatrix} d_1 & d_2 & d_3 \\ a_1 & a_2 & a_3 \\ b_1 & b_2 & b_3 \end{vmatrix}. \tag{3.41}$$

Two vectors $\mathbf{d}$ and $\mathbf{c}$ are orthogonal, when the scalar product $\mathbf{d} \cdot \mathbf{c}$ vanishes. From (3.41) follows that the spate products $\mathbf{a} \cdot (\mathbf{a} \times \mathbf{b})$ and $\mathbf{b} \cdot (\mathbf{a} \times \mathbf{b})$ are zero, since a determinant with two equal columns or rows is zero. Thus the vector product $\mathbf{a} \times \mathbf{b}$ of two vectors $\mathbf{a}$ and $\mathbf{b}$ is perpendicular to both $\mathbf{a}$ and $\mathbf{b}$. Of course, $\mathbf{a}$ and $\mathbf{b}$ are assumed not to be parallel to each other.

In summary, the vector product of two vectors which span a plane is defined such that $\mathbf{a} \times \mathbf{b}$ is perpendicular to this plane. The direction of this vector is parallel to the middle finger of the *right hand* when $\mathbf{a}$ points along the thumb and $\mathbf{b}$ is parallel to the pointing finger. The magnitude of the vector product is given by the magnitude of $\mathbf{a}$ times the magnitude of $\mathbf{b}$ times the magnitude of the sine of the angle φ between $\mathbf{a}$ and $\mathbf{b}$, viz.:

$$|\mathbf{a} \times \mathbf{b}| = |\mathbf{a}|\,|\mathbf{b}|\,|\sin \varphi|. \tag{3.42}$$

For the proof of (3.42), see Fig. 3.1. The coordinate system is chosen such that $\mathbf{a}$ is parallel to the 1-axis and $\mathbf{b}$ is in the 1–2-plane, their components then are $(a_1, 0, 0)$ and $(b_1, b_2, 0)$, with $a_1 = a$, $b_1 = b \cos \varphi$, $b_2 = b \sin \varphi$, $a = |\mathbf{a}|$ and $b = |\mathbf{b}|$.

Due to (3.36), the components of $\mathbf{c} = \mathbf{a} \times \mathbf{b}$ are $(0, 0, c_3)$, with $c_3 = a_1 b_2 = ab \sin \varphi$. The magnitude of c_3 then is given by (3.42). The magnitude of the vector product assumes it maximum value when the two vectors are orthogonal.

Fig. 3.1 Vector product

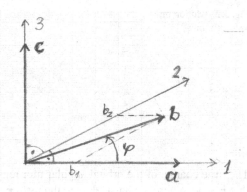

3.4 Applications of the Vector Product

3.4.1 Orbital Angular Momentum

The *orbital angular momentum* $\mathbf{L}$ of a mass point at the position $\mathbf{r}$ with the *linear momentum* $\mathbf{p}$ is defined by

$$\mathbf{L} = \mathbf{r} \times \mathbf{p}. \tag{3.43}$$

When the linear momentum is just mass m times the velocity $\mathbf{v}$, (3.43) is equivalent to $\mathbf{L} = m\mathbf{r} \times \mathbf{v}$.

Notice: the orbital angular momentum depends on the choice of the origin of the coordinate system where $\mathbf{r} = 0$. Furthermore, $\mathbf{L}$ is non-zero even for a motion along a straight line, as long as the line does not go through the point $\mathbf{r} = 0$. For constant speed v, the magnitude L of $\mathbf{L}$ is determined by $r^0 mv$ where r^0 is the shortest distance of the line from $\mathbf{r} = 0$, cf. Fig. 3.2. The angular momentum is perpendicular to the plane, pointing downward.

3.3 Exercise: Angular Momentum in Terms of Spherical Components
Compute the z-component of the angular momentum in terms of the spherical components (2.10).

3.4.2 Torque

According to Newton, the time change of the linear momentum a particle, subjected to a force $\mathbf{F}$, is determined by

$$\frac{d\mathbf{p}}{dt} := \dot{\mathbf{p}} = \mathbf{F}. \tag{3.44}$$

Fig. 3.2 Motion on
a straight line

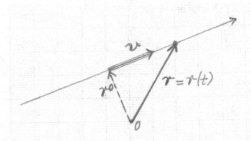

The time change of the orbital angular momentum $\mathbf{L}$ is $\dot{\mathbf{L}} = \dot{\mathbf{r}} \times \mathbf{p} + \mathbf{r} \times \dot{\mathbf{p}}$. Notice
that $\dot{\mathbf{r}} = \mathbf{v}$. When one has $\mathbf{p} = m\,\mathbf{v}$, the term $\dot{\mathbf{r}} \times \mathbf{p}$ is zero. Then the time change of
the orbital angular momentum is given by

$$\frac{d\mathbf{L}}{dt} := \dot{\mathbf{L}} = \mathbf{r} \times \dot{\mathbf{p}} = \mathbf{r} \times \mathbf{F}. \tag{3.45}$$

The quantity $\mathbf{r} \times \mathbf{F}$ is called *torque*.
Notice: there are two cases where $\dot{\mathbf{L}} = 0$ and where, as a consequence, the angular
momentum $\mathbf{L}$ is constant:

1. no force is acting, $\mathbf{F} = 0$,
2. the force $\mathbf{F}$ is parallel to the position vector $\mathbf{r}$.

A force with this property is called *central force*.

Furthermore, notice: the orbital angular momentum and the torque are axial vectors. These quantities do not change sign upon the parity operation. This follows from the fact that they are bilinear functions of polar vectors. By definition, the angular momentum changes sign under time reversal, its time derivative does not change sign. Thus (3.45) describes a reversible dynamics provided that the torque $\mathbf{r} \times \mathbf{F}$ does not change sign under time reversal. This, in turn, is the case when there is no rotational friction proportional to the rotational velocity.

3.4 Exercise: Torque Acting on an Anisotropic Harmonic Oscillator
Determine the torque for the force

$$\mathbf{F} = -k\,\mathbf{r} \cdot \mathbf{e}\,\mathbf{e} - (\mathbf{r} - \mathbf{e}\,\mathbf{r} \cdot \mathbf{e}),$$

where the parameter k and unit vector $\mathbf{e}$ are constant. Which component of the angular momentum is constant, even for $k \neq 1$?

3.4.3 Motion on a Circle

The velocity $\mathbf{v}$ of a mass point on a circular orbit can be expressed as

$$\mathbf{v} = \mathbf{w} \times \mathbf{r}, \tag{3.46}$$

where **w** is an axial vector which is perpendicular to the plane of motion. The magnitude of this vector is the angular velocity w. The point $\mathbf{r} = 0$ is located on the *axis of rotation*. For a circle with radius R, the magnitude of the velocity is $v = R\,w$.

3.4.4 Lorentz Force

The force **F** acting on a particle with charge e, moving with velocity **v**, in the presence of an electric field **E** and a magnetic field (flux density) **B** is

$$\mathbf{F} = e\,\mathbf{E} + e\,\mathbf{v} \times \mathbf{B}. \tag{3.47}$$

This expression is called *Lorentz force*. Notice that **F**, **E** and **v** are polar vectors, with negative parity, whereas **B** is an axial vector with positive parity. Thus parity is conserved in (3.47).

What about time reversal invariance? The electric field **E** is even under time reversal, the **B**-field and the velocity are odd, i.e. they do change sign under time reversal. Thus both terms on the right hand side of (3.47) do not change sign and the same is true for the resulting force. Consequently, the Lorentz force conserves parity and is time reversal invariant.

3.4.5 Screw Curve

The position vector $\mathbf{s} = \mathbf{s}(\alpha)$ describing a screw-like curve, as function of the angle α (Fig. 3.3), is given by

$$\mathbf{s} = \rho\,(\mathbf{e}\cos\alpha + \mathbf{u}\sin\alpha) + \chi\,\frac{\alpha}{2\pi}\,\mathbf{e} \times \mathbf{u}. \tag{3.48}$$

Fig. 3.3 A screw curve for $\rho = 1$ and $\chi = 1/3$. The unit vectors **u** and **v** are pointing in the x- and y-directions. The vertical line indicates the axis of the screw

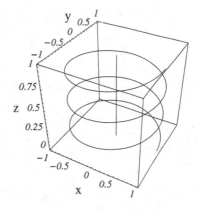

Here **e** and **u**, with $\mathbf{e} \cdot \mathbf{u} = 0$, are orthogonal unit vectors. The parameter ρ is the radius of the screw, projected into the in the **e**–**u**-plane. The sign of the chirality parameter

$$\chi = \mathbf{s}(2\pi) \cdot (\mathbf{e} \times \mathbf{u}) \tag{3.49}$$

determines, whether the curve describes a right-handed or a left-handed screw. The magnitude of χ is the pitch of the screw. The chirality, being the spate product of three polar vectors, is a pseudo vector, which changes sign under the parity operation.

3.5 Exercise: Velocity of a Particle Moving on a Screw Curve

Hint: Use $\alpha = \omega t$ for the parameter occurring in the screw curve (3.48), ω is a frequency.

Chapter 4
Epsilon-Tensor

Abstract The third rank epsilon-tensor is used to formulate the dual relation between an antisymmetric second rank tensor and a vector or vice versa, in three-dimensional space. In this chapter, the properties of this isotropic tensor are presented. From the rules for the multiplication of two of these tensors follow relations for the scalar product of two vector products and the double vector product. Some applications are presented, involving the orbital angular momentum, the torque, the motion on a circle and on a screw curve, as well as the Lorentz force. The dual relation in two-dimensional space is discussed.

4.1 Definition, Properties

4.1.1 Link with Determinants

The dual relation between an antisymmetric second rank tensor and a vector, as well as the properties of the vector product can be formulated more efficiently with the help of the third rank *epsilon-tensor*, which is also called *Levi-Civita tensor*. It is defined by

$$\varepsilon_{\mu\nu\lambda} := \begin{vmatrix} \delta_{1\mu} & \delta_{1\nu} & \delta_{1\lambda} \\ \delta_{2\mu} & \delta_{2\nu} & \delta_{2\lambda} \\ \delta_{3\mu} & \delta_{3\nu} & \delta_{3\lambda} \end{vmatrix}. \tag{4.1}$$

This implies

$$\varepsilon_{\mu\nu\lambda} = \begin{matrix} 1, \ \mu\,\nu\,\lambda = 123, 231, 312 \\ -1, \ \mu\,\nu\,\lambda = 213, 132, 321 \\ 0, \ \mu\,\nu\,\lambda = \text{else}, \end{matrix} \tag{4.2}$$

or, equivalently, $\varepsilon_{123} = \varepsilon_{231} = \varepsilon_{312} = 1$, $\varepsilon_{213} = \varepsilon_{132} = \varepsilon_{321} = -1$, and $\varepsilon_{...} = 0$ for all other combinations of subscripts.

© Springer International Publishing Switzerland 2015
S. Hess, *Tensors for Physics*, Undergraduate Lecture Notes in Physics,
DOI 10.1007/978-3-319-12787-3_4

The epsilon-tensor is totally antisymmetric, i.e. it changes sign, when two indices are interchanged. It is equal to zero, when two indices are equal. Furthermore, the tensor $\varepsilon_{\mu\nu\lambda}$ is isotropic. This means, just like the unit tensor $\delta_{\mu\nu}$, it is form-invariant upon a rotation of the coordinate system.

The dual relation between a vector and the antisymmetric part of a tensor, as given by (3.29), is equivalent to

$$a_\mu = \varepsilon_{\mu\nu\lambda} A_{\nu\lambda}. \tag{4.3}$$

To verify this relation, consider the 1-component. Then one has $a_1 = \varepsilon_{1\nu\lambda} A_{\nu\lambda} = \varepsilon_{123} A_{23} + \varepsilon_{132} A_{32} = A_{23} - A_{32}$. Notice that the positions of the summation indices matter, not their names. The double contraction of the epsilon-tensor with a second rank tensor, as in (4.3), projects out the antisymmetric part of the tensor, i.e.

$$\varepsilon_{\mu\nu\lambda} A_{\nu\lambda} = \varepsilon_{\mu\nu\lambda} A_{\nu\lambda}^{\mathrm{asy}}. \tag{4.4}$$

This can be seen as follows. Trivially, since $1 = \frac{1}{2} + \frac{1}{2}$, one has $\varepsilon_{\mu\nu\lambda} A_{\nu\lambda} = \frac{1}{2}\varepsilon_{\mu\nu\lambda} A_{\nu\lambda} + \frac{1}{2}\varepsilon_{\mu\nu\lambda} A_{\nu\lambda}$. The renaming $\nu, \lambda \to \lambda\nu$ of the summation indices in the second term on the right hand side, leads to $\varepsilon_{\mu\nu\lambda} A_{\nu\lambda} = \frac{1}{2}\varepsilon_{\mu\nu\lambda} A_{\nu\lambda} + \frac{1}{2}\varepsilon_{\mu\lambda\nu} A_{\lambda\nu}$. Next, $\varepsilon_{\mu\lambda\nu} = -\varepsilon_{\mu\nu\lambda}$ is used. Then $\varepsilon_{\mu\nu\lambda} A_{\nu\lambda} = \frac{1}{2}\varepsilon_{\mu\nu\lambda}(A_{\nu\lambda} - A_{\lambda\nu})$ is obtained, which corresponds to (4.4).

By analogy to (4.3), the vector product of two vectors $\mathbf{a}$ and $\mathbf{b}$, defined by (3.38), can be written as

$$c_\mu = \varepsilon_{\mu\nu\lambda} a_\nu b_\lambda. \tag{4.5}$$

The properties of the vector product discussed above follow from the properties of the epsilon-tensor.

The spate product $\mathbf{d} \cdot (\mathbf{a} \times \mathbf{b})$ corresponds to $\varepsilon_{\mu\nu\lambda} d_\mu a_\nu b_\lambda$. A cyclic renaming of the summation indices and the use of $\varepsilon_{\mu\nu\lambda} = \varepsilon_{\nu\lambda\mu} = \varepsilon_{\lambda\mu\nu}$ implies

$$\varepsilon_{\mu\nu\lambda} a_\mu b_\nu d_\lambda = \varepsilon_{\mu\nu\lambda} b_\mu d_\nu a_\lambda, \tag{4.6}$$

or, equivalently

$$\mathbf{d} \cdot (\mathbf{a} \times \mathbf{b}) = \mathbf{a} \cdot (\mathbf{b} \times \mathbf{d}) = \mathbf{b} \cdot (\mathbf{d} \times \mathbf{a}). \tag{4.7}$$

Of course, the symmetry of the spate product can also be inferred from the symmetry properties of the determinant shown in (3.41).

4.1.2 Product of Two Epsilon-Tensors

The product of two epsilon-tensors is a tensor of rank 6 which can be expressed in terms of triple products of the unit second rank tensor, in particular

$$\varepsilon_{\mu\nu\lambda}\,\varepsilon_{\mu'\nu'\lambda'} = \begin{vmatrix} \delta_{\mu\mu'} & \delta_{\mu\nu'} & \delta_{\mu\lambda'} \\ \delta_{\nu\mu'} & \delta_{\nu\nu'} & \delta_{\nu\lambda'} \\ \delta_{\lambda\mu'} & \delta_{\lambda\nu'} & \delta_{\lambda\lambda'} \end{vmatrix}. \tag{4.8}$$

The rows and columns in the determinant can be interchanged and one has $\varepsilon_{\mu\nu\lambda}\,\varepsilon_{\mu'\nu'\lambda'} = \varepsilon_{\mu'\nu'\lambda'}\varepsilon_{\mu\nu\lambda}$. Written explicitly, (4.8) is equivalent to

$$\begin{aligned} \varepsilon_{\mu\nu\lambda}\,\varepsilon_{\mu'\nu'\lambda'} = {} & \delta_{\mu\mu'}\delta_{\nu\nu'}\delta_{\lambda\lambda'} + \delta_{\mu\nu'}\delta_{\nu\lambda'}\delta_{\lambda\mu'} + \delta_{\mu\lambda'}\delta_{\nu\mu'}\delta_{\lambda\nu'} \\ & - \delta_{\mu\mu'}\delta_{\nu\lambda'}\delta_{\lambda\nu'} - \delta_{\mu\nu'}\delta_{\nu\mu'}\delta_{\lambda\lambda'} - \delta_{\mu\lambda'}\delta_{\nu\nu'}\delta_{\lambda\mu'}. \end{aligned} \tag{4.9}$$

The relation (4.8) can be inferred from (4.1) as follows. First, notice that the determinant of the product of two matrices is equal to the product of the determinants of these matrices. Furthermore, in the defining relation (4.1) for $\varepsilon_{\mu'\nu'\lambda'}$, the rows and columns are interchanged in the determinant. Then $\varepsilon_{\mu'\nu'\lambda'}\varepsilon_{\mu\nu\lambda}$ is equal to the determinant of the matrix product

$$\begin{pmatrix} \delta_{1\mu'} & \delta_{2\mu'} & \delta_{3\mu'} \\ \delta_{1\nu'} & \delta_{2\nu'} & \delta_{3\nu'} \\ \delta_{1\lambda'} & \delta_{2\lambda'} & \delta_{3\lambda'} \end{pmatrix} \begin{pmatrix} \delta_{1\mu} & \delta_{1\nu} & \delta_{1\lambda} \\ \delta_{2\mu} & \delta_{2\nu} & \delta_{2\lambda} \\ \delta_{3\mu} & \delta_{3\nu} & \delta_{3\lambda} \end{pmatrix}.$$

Matrix multiplication, row times column, yields $\delta_{1\mu'}\delta_{1\mu} + \delta_{2\mu'}\delta_{2\mu} + \delta_{3\mu'}\delta_{3\mu} = \delta_{\kappa\mu'}\delta_{\kappa\mu} = \delta_{\mu'\mu} = \delta_{\mu\mu'}$ for the 11-element. Similarly, the other elements are obtained, as they appear in the determinant (4.8).

In many applications, a contracted version of the product of two epsilon-tensors is needed where two subscripts are equal and summed over. In particular, for $\lambda = \lambda'$, (4.8) and (4.9) reduce to

$$\varepsilon_{\mu\nu\lambda}\,\varepsilon_{\mu'\nu'\lambda} = \begin{vmatrix} \delta_{\mu\mu'} & \delta_{\mu\nu'} \\ \delta_{\nu\mu'} & \delta_{\nu\nu'} \end{vmatrix} = \delta_{\mu\mu'}\delta_{\nu\nu'} - \delta_{\mu\nu'}\delta_{\nu\mu'}. \tag{4.10}$$

Notice that $\varepsilon_{\mu\nu\lambda}\varepsilon_{\mu'\nu'\lambda} = \varepsilon_{\mu\nu\lambda}\varepsilon_{\lambda\mu'\nu'}$. Formulae known for the product of two vector products follow from (4.10).

The further contraction of (4.10), with $\nu = \nu'$, yields

$$\varepsilon_{\mu\nu\lambda}\,\varepsilon_{\mu'\nu\lambda} = 2\,\delta_{\mu\mu'}. \tag{4.11}$$

The total contraction of two epsilon-tensors is equal to 6, viz.:

$$\varepsilon_{\mu\nu\lambda}\,\varepsilon_{\mu\nu\lambda} = 6. \tag{4.12}$$

This value is equal to the number of non-zero elements of the epsilon-tensor in 3D. Side remark: by analogy, a totally antisymmetric isotropic tensor can be defined in D dimensions. Then the total contraction of this tensor with itself, corresponding to (4.12) is D!. Clearly, this value is 6 for D = 3.

4.1.3 Antisymmetric Tensor Linked with a Vector

With the help of the epsilon-tensor and its properties, the dual relation (4.3) which links a vector with an antisymmetric second rank tensor can be inverted. This is seen as follows. First, by renaming indices, (4.3) is rewritten as $a_\lambda = \varepsilon_{\lambda\mu'\nu'}A_{\mu'\nu'}$. Multiplication of this expression by $\varepsilon_{\mu\nu\lambda}$ and use of (4.10) leads to

$$\varepsilon_{\mu\nu\lambda}\, a_\lambda = (\delta_{\mu\mu'}\delta_{\nu\nu'} - \delta_{\mu\nu'}\delta_{\nu\mu'})A_{\mu'\nu'} = A_{\mu\nu} - A_{\nu\mu} = 2\, A_{\mu\nu}^{\text{asy}}, \tag{4.13}$$

or

$$A_{\mu\nu}^{\text{asy}} = \frac{1}{2}\,\varepsilon_{\mu\nu\lambda}\, a_\lambda. \tag{4.14}$$

4.2 Multiple Vector Products

4.2.1 Scalar Product of Two Vector Products

Let $\mathbf{a}$, $\mathbf{b}$, $\mathbf{c}$, $\mathbf{d}$ be four vectors. The scalar product of two vector product formed with these vectors can be expressed in terms of products of scalar products:

$$(\mathbf{a} \times \mathbf{b}) \cdot (\mathbf{c} \times \mathbf{d}) = \mathbf{a} \cdot \mathbf{c}\, \mathbf{b} \cdot \mathbf{d} - \mathbf{a} \cdot \mathbf{d}\, \mathbf{b} \cdot \mathbf{c}. \tag{4.15}$$

For the proof, notice that the left hand side of (4.15) is equivalent to

$$\varepsilon_{\lambda\mu\nu}a_\mu b_\nu\, \varepsilon_{\lambda\mu'\nu'}c_{\mu'}d_{\nu'}.$$

Use of the symmetry properties of the epsilon-tensor and of (4.10) allows one to rewrite this expression as $(\delta_{\mu\mu'}\delta_{\nu\nu'} - \delta_{\mu\nu'}\delta_{\nu\mu'})a_\mu b_\nu c_{\mu'}d_{\nu'} = a_\mu c_\mu b_\nu d_\nu - a_\mu d_\mu b_\nu c_\nu$, which corresponds to the right hand side of (4.15).

Now the special case $\mathbf{c} = \mathbf{a}$, $\mathbf{d} = \mathbf{b}$ is considered. Then (4.15) implies

$$|\mathbf{a} \times \mathbf{b}|^2 = a^2 b^2 - (\mathbf{a} \cdot \mathbf{b})^2 = a^2 b^2 \left(1 - (\widehat{\mathbf{a}} \cdot \widehat{\mathbf{b}})^2\right) = a^2 b^2 (1 - \cos^2 \varphi) = a^2 b^2 \sin^2 \varphi. \tag{4.16}$$

Here $\widehat{\mathbf{a}}$ and $\widehat{\mathbf{b}}$ are unit vectors and φ is the angle between the vectors $\mathbf{a}$ and $\mathbf{b}$. The last equality is equivalent to (3.42).

4.2.2 Double Vector Products

Let $\mathbf{a}$, $\mathbf{b}$, $\mathbf{c}$ be three vectors. The double vector product $\mathbf{a} \times (\mathbf{b} \times \mathbf{c})$ is a vector with contributions parallel to $\mathbf{b}$ and $\mathbf{c}$, in particular

$$\mathbf{a} \times (\mathbf{b} \times \mathbf{c}) = \mathbf{a} \cdot \mathbf{c}\, \mathbf{b} - \mathbf{a} \cdot \mathbf{b}\, \mathbf{c}. \qquad (4.17)$$

Notice that the position of the parenthesis (. . .) is essential in this case. The expression $\mathbf{a} \times \mathbf{b} \times \mathbf{c}$ is not well defined since $\mathbf{a} \times (\mathbf{b} \times \mathbf{c}) \neq (\mathbf{a} \times \mathbf{b}) \times \mathbf{c}$, in general. The latter double cross product yields a vector with component parallel to $\mathbf{a}$ and $\mathbf{b}$.

The proof of (4.17), is also based on the relation (4.10). In component notation, one has

$$[\mathbf{a} \times (\mathbf{b} \times \mathbf{c})]_\mu = \varepsilon_{\mu\nu\lambda} a_\nu (\mathbf{b} \times \mathbf{c})_\lambda = \varepsilon_{\mu\nu\lambda} \varepsilon_{\lambda\mu'\nu'} a_\nu b_{\mu'} c_{\nu'}.$$

Now use of (4.10), together with the symmetry properties of the epsilon-tensor yields

$$[\mathbf{a} \times (\mathbf{b} \times \mathbf{c})]_\mu = b_\mu\, a_\nu c_\nu - c_\mu\, a_\nu b_\nu, \qquad (4.18)$$

which is equivalent to (4.17).

For the special case $\mathbf{a} = \mathbf{c} = \mathbf{e}$, where $\mathbf{e}$ is a unit vector, (4.17) reduces to

$$\mathbf{e} \times (\mathbf{b} \times \mathbf{e}) = \mathbf{b} - \mathbf{e} \cdot \mathbf{b}\, \mathbf{e} = \mathbf{b} - \mathbf{b}^\| := \mathbf{b}^\perp. \qquad (4.19)$$

Here $\mathbf{b}^\| := \mathbf{e} \cdot \mathbf{b}\, \mathbf{e}$ and $\mathbf{b}^\perp$ are the parts of $\mathbf{b}$ which are parallel and perpendicular, respectively, to $\mathbf{e}$. Clearly, one has $\mathbf{b}^\| + \mathbf{b}^\perp = \mathbf{b}$.

4.3 Applications

4.3.1 Angular Momentum for the Motion on a Circle

The linear momentum $\mathbf{p} = m\mathbf{v}$ of a particle with mass m and velocity $\mathbf{v}$, moving on a circle, with the angular velocity $\mathbf{w}$, cf. (3.46), is given by

$$\mathbf{p} = m\, \mathbf{w} \times \mathbf{r}.$$

Here $\mathbf{r} = 0$ is a point on the axis of rotation which is parallel to $\mathbf{w}$. With the help of (4.18), the resulting orbital angular momentum $\mathbf{L} = \mathbf{r} \times \mathbf{p}$ is found to be

$$\mathbf{L} = m\mathbf{r} \times (\mathbf{w} \times \mathbf{r}) = m\, (r^2 \mathbf{w} - \mathbf{r} \cdot \mathbf{w}\mathbf{r}). \qquad (4.20)$$

In component notation, this equation linking the angular momentum with the angular velocity, is equivalent to

$$L_\mu = m\, (r^2 w_\mu - r_\nu w_\nu\, r_\mu) = m\, (r^2 \delta_{\mu\nu} - r_\mu r_\nu)\, w_\nu. \qquad (4.21)$$

When the center of the circle is chosen as the origin $\mathbf{r} = 0$, the position vector $\mathbf{r}$ is perpendicular to $\mathbf{w}$, thus $\mathbf{r} \cdot \mathbf{w} = 0$ and $\mathbf{L} = mr^2 \mathbf{w} = mR^2 \mathbf{w}$, where R is the radius of the circle. For a single particle, such a special choice of the origin, where $\mathbf{r}$ has only components perpendicular to the rotation axis, can always be made. This, however, is not possible for a rotating solid body which is composed of many mass points.

4.3.2 Moment of Inertia Tensor

A *solid body* is composed of N mass points, i.e. atoms, molecules or small parts of the body, with masses $m_1, m_2, \ldots, m_N$ located at positions $\mathbf{r}^{(1)}, \mathbf{r}^{(2)}, \ldots, \mathbf{r}^{(N)}$. The total mass is $M = \sum_{i=1}^{N} m_i$. *Solid* means: the distances between the constituent parts of the body do not change, the body moves as a whole. In the case of a rotation with the angular velocity $\mathbf{w}$, each mass point has the velocity $\mathbf{v}^{(i)} = \mathbf{w} \times \mathbf{r}^{(i)}$, the linear momentum $\mathbf{p}^{(i)} = m_i \mathbf{v}^{(i)}$, and the angular momentum $\mathbf{L}^{(i)} = \mathbf{r}^{(i)} \times \mathbf{p}^{(i)}$, for $i = 1, 2, \ldots, N$. Again, the origin of the position vectors is a point on the rotation axis, $\mathbf{w}$ is parallel to this axis. By analogy to (4.21), the total angular momentum is found to be

$$L_\mu = \sum_{i=1}^{N} L_\mu^{(i)} = \Theta_{\mu\nu} \, w_\nu, \tag{4.22}$$

with the *moment of inertia tensor* $\Theta_{\mu\nu}$ given by

$$\Theta_{\mu\nu} = \sum_{i=1}^{N} m_i \, [(r^{(i)})^2 \, \delta_{\mu\nu} - r_\mu^{(i)} r_\nu^{(i)}]. \tag{4.23}$$

By definition, the moment of inertia tensor is symmetric: $\Theta_{\mu\nu} = \Theta_{\nu\mu}$.

The equation (4.22) is an example for a linear relation between two vectors governed by a second rank tensor. Here both vectors $\mathbf{w}$ and $\mathbf{L}$ have positive parity, i.e. they are axial or pseudo vectors. The moment of inertia tensor has also positive parity, i.e. it is a proper tensor of rank 2.

The *moment of inertia* for a rotation about a fixed axis is defined via the linear relation between the component of the angular momentum parallel to this axis and the magnitude w of the angular velocity. Scalar multiplication of (4.22) with the unit vector $\widehat{w}_\mu = w_\mu / w$ leads to

$$\widehat{w}_\mu L_\mu = \widehat{w}_\mu \Theta_{\mu\nu} \, w_\nu = \Theta \, w,$$

with the moment of inertia

$$\Theta = \widehat{w}_\mu \Theta_{\mu\nu} \widehat{w}_\nu = \sum_{i=1}^{N} m_i \left(r_\perp^{(i)} \right)^2. \tag{4.24}$$

Here

$$\left(r_\perp^{(i)}\right)^2 = \left(r^{(i)}\right)^2 - \left(\widehat{w}_\nu r_\nu^{(i)}\right)^2$$

is the square of the shortest distance of mass point i from the rotation axis.

The moment of inertia tensor and the inertia moment, as defined by (4.23) and (4.24) depend on the choice of the origin. As mentioned before, the origin for the position vectors has to be on the axis of rotation. When the axis of rotation goes through the center of mass of the body, it is convenient to choose the center of mass as the origin. In this case one has $\sum_{i=1}^N m_i \mathbf{r}^{(i)} = 0$.

Due to its definition, the moment of inertia tensor does not change its sign both under the parity operation nor under time reversal.

When one talks about *the* moment of inertia tensor or *the* moment of inertia of a solid body, one means quantities, which are characteristic for the shape of the body: they are computed via (4.23) and (4.24) with $\mathbf{r}^{(i)} = 0$ corresponding to the center of mass of the solid body. Furthermore, expressions analogous to (4.23) and (4.24) are presented later for continuous mass distributions, where the sum over discrete masses is replaced by an integral over space.

Examples for the moment of inertia tensor are discussed in Sects. 5.3.1, 8.3.3 and computed in the Exercises 5.1 and 8.4.

4.4 Dual Relation and Epsilon-Tensor in 2D

4.4.1 Definitions and Matrix Notation

Let a_i and b_i, $i = 1, 2$, be the Cartesian components of two vectors in 2D. The antisymmetric part of the dyadic constructed from these components is related to a scalar c according to

$$c = a_1 b_2 - a_2 b_1. \tag{4.25}$$

This dual relation can be written as

$$c = \varepsilon_{ij} a_i b_j, \tag{4.26}$$

where the summation convention is used for the Latin subscripts. By analogy to the 3D case, ε_{ij} is defined by

$$\varepsilon_{ij} := \begin{vmatrix} \delta_{1i} & \delta_{1j} \\ \delta_{2i} & \delta_{2j} \end{vmatrix}. \tag{4.27}$$

This 2D ε-tensor can also be expressed in matrix notation:

$$\varepsilon_{ij} = \begin{pmatrix} 0 & 1 \\ -1 & 0 \end{pmatrix}. \tag{4.28}$$

In 3D, a corresponding notation requires a three dimensional '3 $\times$ 3 $\times$ 3-matrix'.

4.1 Exercise: 2D Dual Relation in Complex Number Notation

Let the two 2D vectors (x_1, y_1) and (x_2, y_2) be expressed in terms of the complex numbers $z_1 = x_1 + iy_1$ and $z_2 = x_2 + iy_2$. Write the dual relation corresponding to (4.25) in terms of the complex numbers z_1 and z_2. How about the scalar product of these 2D vectors?

Hint: the complex conjugate of $z = x + iy$ is $z^* = x - iy$.

Chapter 5
Symmetric Second Rank Tensors

Abstract This chapter deals with properties and applications of symmetric second rank tensors which are composed of isotropic and symmetric traceless parts. A principle axes representation is considered and the cases of isotropic, uniaxial and biaxial tensors are discussed. Applications comprise the moment of inertia tensor, the radius of gyration tensor, the molecular polarizability tensor, the dielectric tensor and birefringence, electric and magnetic torques. Geometric interpretations of symmetric tensors are possible via bilinear forms or via a linear mapping. The scalar invariants are discussed. The consequences of a Hamilton-Cayley theorem for triple and quadruple products of symmetric traceless tensors are presented. A volume conserving affine mapping of a sphere onto an ellipsoid is considered.

5.1 Isotropic and Symmetric Traceless Parts

Here the properties of symmetric tensors are discussed. For a tensor **S** this means

$$S_{\mu\nu} = S_{\nu\mu}.$$

As mentioned before, cf. (3.3) and (3.4), such a tensor is equal to the sum of its isotropic part, involving its trace $S_{\lambda\lambda}$ and its symmetric traceless part:

$$S_{\mu\nu} = \frac{1}{3} S_{\lambda\lambda} \delta_{\mu\nu} + \overline{S_{\mu\nu}}, \tag{5.1}$$

with

$$\overline{S_{\mu\nu}} = S_{\mu\nu} - \frac{1}{3} S_{\lambda\lambda} \delta_{\mu\nu}. \tag{5.2}$$

This decomposition is invariant under a rotation of the coordinate system. Notice that $\overline{S_{\nu\nu}} = 0$, that is why this part of the tensor is called *symmetric traceless*. Frequently, it is also referred to as *irreducible* part, because it can not be associated with a lower rank tensor. Here, it is also called *anisotropic* part, because the symmetric traceless part of tensors used in applications characterizes the anisotropy of physical properties.

© Springer International Publishing Switzerland 2015

S. Hess, *Tensors for Physics*, Undergraduate Lecture Notes in Physics,
DOI 10.1007/978-3-319-12787-3_5

In solid state mechanics, the symmetric traceless part is commonly referred to *deviatoric* part, because it indicates a deviation from isotropy. Notice, not only various names are used for the same item, also different notations are found in the literature.

5.2 Principal Values

5.2.1 Principal Axes Representation

A symmetric tensor **S** can be brought into a diagonal form with the help of an appropriate rotation of the coordinate system. Then, in this *principal axes system*, it is presented as

$$\mathbf{S} := \begin{pmatrix} S^{(1)} & 0 & 0 \\ 0 & S^{(2)} & 0 \\ 0 & 0 & S^{(3)} \end{pmatrix}, \tag{5.3}$$

with real *principal values*, also called *eigenvalues* $S^{(i)}$, $i = 1, 2, 3$. The axes of the particular coordinate system in which the tensor is diagonal are referred to as *principal axes*. Unit vectors parallel to these axes are denoted by $\mathbf{e}^{(i)}$, $i = 1, 2, 3$. The order 1, 2, 3 can be chosen conveniently. With the help of the dyadics $\mathbf{e}^{(i)}\mathbf{e}^{(i)}$, the symmetric tensor **S** can be written in the *eigen-representation*:

$$\mathbf{S} = \sum_{i=1}^{3} S^{(i)} \mathbf{e}^{(i)} \mathbf{e}^{(i)}. \tag{5.4}$$

The validity of the standard eigenvalue equation

$$\mathbf{S} \cdot \mathbf{e}^{(i)} = S^{(i)} \mathbf{e}^{(i)} \tag{5.5}$$

follows from (5.4) and the orthogonality $\mathbf{e}^{(i)} \cdot \mathbf{e}^{(j)} = \delta^{(ij)}$ of the eigen vectors $\mathbf{e}^{(i)}$.

Notice that a symmetric second rank tensor in 3D has 6 independent components. How come there are only 3 numbers specified by the 3 eigenvalues? The answer is: 3 additional numbers, e.g. Eulerian angles, are needed to determine the directions of the principal axis in relation to an arbitrary, space fixed, coordinate system.

5.2.2 Isotropic Tensors

In the special case where all three principal values are equal,

$$S^{(1)} = S^{(2)} = S^{(3)} = S,$$

the tensor **S** is *isotropic*, i.e. it is proportional to the unit tensor:

$$S_{\mu\nu} = S\delta_{\mu\nu}. \tag{5.6}$$

This follows from (5.4), due to the completeness relation for the orthogonal unit vectors $e^{(i)}$, viz.,

$$\sum_{i=1}^{3} e_{\mu}^{(i)} e_{\nu}^{(i)} = \delta_{\mu\nu}. \tag{5.7}$$

By definition, an isotropic tensor has no *anisotropic* part. Whenever the three eigenvalues are not equal to each other, the symmetric tensor possesses a non-zero traceless part, sometimes also called *anisotropic part*. Two cases can be distinguished: (i) two eigenvalues are equal, but different from the third one, and (ii) three different eigenvalues. These cases are referred to with the labels *uniaxial* and *biaxial*.

5.2.3 Uniaxial Tensors

When only two the principal values are equal but different from the third one, say $S^{(1)} = S^{(2)} \neq S^{(3)}$, the tensor is called *uniaxial*. It possesses a symmetry axis which is parallel to $e^{(3)}$ in the special case considered. It is convenient to use the notation $e^{(3)} = e$ for the unit vector parallel to the symmetry axis and to denote the eigenvalues associated with the directions parallel and perpendicular to this direction by $S_{\parallel}$ and $S_{\perp}$, respectively. This means:

$$S^{(1)} = S^{(2)} = S_{\perp}, \quad S^{(3)} = S_{\parallel}.$$

Thus the uniaxial tensor can be written as

$$S_{\mu\nu} = S_{\parallel} e_{\mu} e_{\nu} + S_{\perp} (\delta_{\mu\nu} - e_{\mu} e_{\nu}), \tag{5.8}$$

and

$$S_{\mu\nu} = \frac{1}{3}(S_{\parallel} + 2 S_{\perp}) \delta_{\mu\nu} + (S_{\parallel} - S_{\perp}) \overline{e_{\mu} e_{\nu}}. \tag{5.9}$$

In matrix notation, this expression is equivalent to

$$\mathbf{S} := \frac{1}{3}(S_{\parallel} + 2 S_{\perp}) \begin{pmatrix} 1 & 0 & 0 \\ 0 & 1 & 0 \\ 0 & 0 & 1 \end{pmatrix} + \frac{2}{3}(S_{\parallel} - S_{\perp}) \begin{pmatrix} -\frac{1}{2} & 0 & 0 \\ 0 & -\frac{1}{2} & 0 \\ 0 & 0 & 1 \end{pmatrix}. \tag{5.10}$$

The factor $\frac{2}{3}$ in the symmetric traceless part stems from $\overline{e_{\mu} e_{\nu}} \, \overline{e_{\mu} e_{\nu}} = \frac{2}{3}$.

5.2.4 Biaxial Tensors

The general symmetric second rank tensor with three different principal values is referred to as *biaxial* tensor. With the abbreviations

$$\bar{S} = \frac{1}{3}\left(S^{(1)} + S^{(2)} + S^{(3)}\right), \quad s = S^{(3)} - \frac{1}{2}\left(S^{(1)} + S^{(2)}\right), \quad q = \frac{1}{2}\left(S^{(1)} - S^{(2)}\right),$$

$$\text{(5.11)}$$

the decomposition of the tensor according to (5.1), can be written as

$$S_{\mu\nu} = \bar{S}\,\delta_{\mu\nu} + s\,\overline{e_\mu^{(3)} e_\nu^{(3)}} + q\left(\overline{e_\mu^{(1)} e_\nu^{(1)}} - \overline{e_\mu^{(2)} e_\nu^{(2)}}\right). \quad \text{(5.12)}$$

To check the validity of this relation, notice that $S^{(i)} = e_\mu^{(i)} S_{\mu\nu} e_\nu^{(i)}$ and $e_\mu^{(i)} \overline{e_\mu^{(j)} e_\nu^{(j)}} e_\mu^{(i)}$ is equal to 2/3, for $i = j$ and equal to $-1/3$, for $i \neq j$, cf. (3.27) and (3.28). Both, i and j, can be equal to 1, 2 or 3. The result is

$$S^{(1)} = \bar{S} - \frac{1}{3}s + q, \quad S^{(2)} = \bar{S} - \frac{1}{3}s - q, \quad S^{(3)} = \bar{S} + \frac{2}{3}s. \quad \text{(5.13)}$$

Furthermore, notice that $\overline{e_\mu^{(1)} e_\nu^{(1)}} - \overline{e_\mu^{(2)} e_\nu^{(2)}} = e_\mu^{(1)} e_\nu^{(1)} - e_\mu^{(2)} e_\nu^{(2)}$. Alternatively, to obtain (5.12), one may start from (5.4) and decompose the tensor into its isotropic and anisotropic parts according to

$$\mathbf{S} = \frac{1}{3}\sum_{i=1}^{3} S^{(i)}\delta_{\mu\nu} + \sum_{i=1}^{3} S^{(i)}\,\overline{e_\mu^{(i)} e_\nu^{(i)}}.$$

Since

$$S^{(1)}\,\overline{e_\mu^{(1)} e_\nu^{(1)}} + S^{(2)}\,\overline{e_\mu^{(2)} e_\nu^{(2)}}$$

is equal to

$$\frac{1}{2}\left(S^{(1)} + S^{(2)}\right)\left(\overline{e_\mu^{(1)} e_\nu^{(1)}} + \overline{e_\mu^{(2)} e_\nu^{(2)}}\right) + \frac{1}{2}\left(S^{(1)} - S^{(2)}\right)\left(\overline{e_\mu^{(1)} e_\nu^{(1)}} - \overline{e_\mu^{(2)} e_\nu^{(2)}}\right)$$

and

$$\overline{e_\mu^{(1)} e_\nu^{(1)}} + \overline{e_\mu^{(2)} e_\nu^{(2)}} = -\overline{e_\mu^{(3)} e_\nu^{(3)}},$$

equation (5.12) is recovered.

In matrix notation, (5.12) is equivalent to

$$
S_{\mu\nu} = \bar{S} \begin{pmatrix} 1 & 0 & 0 \\ 0 & 1 & 0 \\ 0 & 0 & 1 \end{pmatrix} + \frac{2}{3} s \begin{pmatrix} -\frac{1}{2} & 0 & 0 \\ 0 & -\frac{1}{2} & 0 \\ 0 & 0 & 1 \end{pmatrix} + q \begin{pmatrix} 1 & 0 & 0 \\ 0 & -1 & 0 \\ 0 & 0 & 0 \end{pmatrix}. \tag{5.14}
$$

For a symmetric traceless tensor one has $S^{(1)} + S^{(2)} + S^{(3)} = 0$, i.e. $\bar{S} = 0$. The anisotropic part of the tensor is characterized by the linear combinations s and q of its principal values. For $q = 0$ and $s \neq 0$, the uniaxial case, discussed above, is recovered. In the other special case $s = 0$, but $q \neq 0$, the symmetric traceless part of the tensor is referred to as being *planar biaxial*. In general, a symmetric tensor, in principle axis representation, is characterized by its three principal values or by its trace and two parameters specifying its symmetric traceless part, as introduced in (5.11). In principle, the labeling 1, 2, 3 of the principle axes can be interchanged. Preferentially, the 3-axis is associated either with the largest or the smallest eigenvalue.

5.2.5 Symmetric Dyadic Tensors

In general, the principal directions and values of a symmetric tensor can be found by the methods used in Linear Algebra for the diagonalization of matrices. In many problems of physics, the principal directions are obvious by symmetry considerations, and then the principal values can be inferred from the eigenvalue equation. As an example, the special case of a symmetric dyadic tensor constructed from two unit vectors **u** and **v** is considered. Thus we have

$$
S_{\mu\nu} = \frac{1}{2} (u_\mu v_\nu + u_\nu v_\mu). \tag{5.15}
$$

Let **h** be a vector parallel to a principal direction. Then the eigenvalue equation $S_{\mu\nu} h_\nu = S h_\mu$ implies that the principal value S is determined by $h_\mu S_{\mu\nu} h_\nu = S h_\mu h_\mu$ and consequently

$$
S = h_\mu u_\mu v_\nu h_\nu / h_\kappa h_\kappa. \tag{5.16}
$$

First, the case **v** parallel to **u** is considered. Then one principal direction is parallel to **u** and the directions of the two other ones are not uniquely determined, but they lie in the plane perpendicular to **u**. The principal values are $S = 1$, for $\mathbf{h} = \mathbf{u}$, and two principal values are 0, for two orthogonal directions which are perpendicular to **u**. In short, the principal values are $\{1, 0, 0\}$. Clearly, the tensor is uniaxial. The symmetry direction can be identified with any one of the coordinate axes, frequently either the 1- or the 3-axis is chosen.

Now the case is considered, where $\mathbf{v}$ is not parallel to $\mathbf{u}$. By symmetry, one of the principal directions is parallel to the vector product $\mathbf{u} \times \mathbf{v}$. The pertaining principal eigenvalue is 0. The two other principal directions are parallel to the bisectors between the vectors $\mathbf{u}$ and $\mathbf{v}$, i.e. they are parallel to $\mathbf{u} \pm \mathbf{v}$. With $\mathbf{h} = \mathbf{u} \pm \mathbf{v}$, one has $\mathbf{h} \cdot \mathbf{u} = 1 \pm c$, $\mathbf{h} \cdot \mathbf{v} = c \pm 1$, and $\mathbf{h} \cdot \mathbf{h} = 2(1 \pm c)$, with $c = \mathbf{u} \cdot \mathbf{v} = \cos \vartheta$, where ϑ is the angle between the vectors $\mathbf{u}$ and $\mathbf{v}$. Thus the two other principal values are found to be $(c \pm 1)/2$. In summary, the principle values are

$$\left\{ \frac{1}{2}(c+1), \frac{1}{2}(c-1), 0 \right\}.$$

Clearly, for $\vartheta = 0$, corresponding to $\mathbf{u} = \mathbf{v}$, one has $c = 1$, and the uniaxial case is recovered. For $\mathbf{u} \perp \mathbf{v}$, $c = 0$ applies and the principal values are $\{1/2, -1/2, 0\}$. This corresponds to a symmetric traceless planar biaxial tensor.

The trace of the tensor given by (5.15) is the scalar product $c = \mathbf{u} \cdot \mathbf{v}$. Thus the principal values of the symmetric traceless tensor $\overline{u_\mu v_\nu}$ are

$$\left\{ \frac{1}{6}c + \frac{1}{2}, \frac{1}{6}c - \frac{1}{2}, -\frac{1}{3}c \right\}, \quad c = \mathbf{u} \cdot \mathbf{v} = \cos \vartheta. \tag{5.17}$$

Comparison with (5.14) shows that the coefficients s and q, introduced in Sect. 5.2.4, are here given by $s = -c/2$ and $q = 1/2$.

5.3 Applications

5.3.1 Moment of Inertia Tensor of Molecules

The moment of inertia tensor $\Theta_{\mu\nu}$, as defined in (4.23), is symmetric. When the origin of the position vectors of the constituent parts of a solid body coincides with the center of mass of the body, the moment of inertia tensor reflects and characterizes the shape and symmetry of the body. Molecules in their vibrational ground state can be looked upon as "solid bodies". Some simple examples are considered next.

(i) Linear Molecules

A linear molecule is composed of two atoms, with masses m_1 and m_2, separated by the distance d. The unit vector parallel to the axis joining the nuclei of the two atoms is denoted by $\mathbf{u}$. Their position vectors, with respect to the center of mass of the molecule, are $\mathbf{r}^{(1)} = d_1 \mathbf{u}$ and $\mathbf{r}^{(2)} = -d_2 \mathbf{u}$, with the distances d_1 and d_2 determined by $d_1 + d_2 = 0$ and $m_1 d_1 = m_2 d_2$. Use of (4.23) leads to the moment of inertia tensor

$$\Theta_{\mu\nu} = \left(m_1 d_1^2 + m_2 d_2^2 \right) \left(\delta_{\mu\nu} - u_\mu u_\nu \right) = \Theta \left(\delta_{\mu\nu} - u_\mu u_\nu \right), \tag{5.18}$$

with the moment of inertia

$$\Theta = m_1 d_1^2 + m_2 d_2^2 = m_{12} d^2. \tag{5.19}$$

Here $m_{12} = m_1 m_2/(m_1 + m_2)$ is the reduced mass.

Notice that the tensor $\delta_{\mu\nu} - u_\mu u_\nu$, when multiplied with a angular velocity vector w_ν, projects onto the directions perpendicular to $\mathbf{u}$. Thus the angular momentum $L_\mu = \Theta_{\mu\nu} w_\nu$ is perpendicular to $\mathbf{u}$, and $\mathbf{L} \cdot \mathbf{u} = 0$. The rotational motion of a linear molecule, in 3D, has just 2 and not 3 degrees of freedom. This also holds true, when the rotational motion is treated quantum mechanically.

Examples for linear molecules are the homo-nuclear molecules ($m_1 = m_2$) of hydrogen and nitrogen, viz.: H_2 and N_2. Hetero-nuclear molecules ($m_1 \neq m_2$) are, e.g. hydrogen-deuterium HD and hydrogen chloride HCl. Also some tri-atomic molecules are linear, e.g. carbon-dioxide CO_2. In this case, the center of mass coincides with the central C-atom and the moment of inertia is determined by the masses and distances of the O-atoms.

(ii) Symmetric Top Molecules

The moment of inertia tensor of molecules with a symmetry axis parallel to the unit vector $\mathbf{u}$ is of the form

$$\Theta_{\mu\nu} = \Theta_\parallel u_\mu u_\nu + \Theta_\perp (\delta_{\mu\nu} - u_\mu u_\nu), \tag{5.20}$$

where $\Theta_\parallel$ and $\Theta_\perp$ are the moments of inertia for the angular velocity parallel and perpendicular to $\mathbf{u}$, respectively. Examples for symmetric top molecules are CH_3Cl and $CHCl_3$, or C_6H_6. For *prolate*, i.e. elongated, particles, one has $\Theta_\parallel > \Theta_\perp$. Particles with $\Theta_\parallel < \Theta_\perp$ are referred to as *oblate* or disc-like.

The linear molecules discussed above correspond to $\Theta_\parallel = 0$.

(iii) Spherical Top Molecules

The special case $\Theta_\parallel = \Theta_\perp$ applies to *spherical top* molecules which have an isotropic moment of inertia tensor

$$\Theta_{\mu\nu} = \Theta \delta_{\mu\nu}, \tag{5.21}$$

with the moment of inertia Θ. The regular tetrahedral molecules, CH_4 and CF_4, as well as the regular octahedral molecules CF_6 and SF_6, are spherical top molecules.

Notice that physical properties described by a second rank tensor, like the moment of inertia tensor, are isotropic not only for spheres, but also for regular tetrahedra, cubes and regular octahedra. Physical properties associated with higher rank tensors are needed to distinguish between the different symmetries.

(iv) Asymmetric Top Molecules

In general, the moment of inertia tensor is characterized by three different principal values $\Theta^{(1)}$, $\Theta^{(2)}$ and $\Theta^{(3)}$. By definition, one has $\Theta^{(i)} \geq 0$ for $i = 1, 2, 3$.

A molecule with three different moments of inertia is referred to as *asymmetric top* molecule. On the level of the second rank tensor, it has the same symmetry as a brick stone.

5.1 Exercise: Show that the Moment of Inertia Tensors for Regular Tetrahedra and Octahedra are Isotropic

Hint: Use the coordinates $(1, 1, 1)$, $(-1, -1, 1)$, $(1, -1, -1)$, $(-1, 1, -1)$ for the four corners of the tetrahedron and $(1, 0, 0)$, $(-1, 0, 0)$, $(0, 1, 0)$, $(0, -1, 0)$, $(0, 0, 1)$, $(0, 0, -1)$, for the six corners of the octahedron.

5.3.2 Radius of Gyration Tensor

Consider a cloud of N particles or N monomers of a polymer molecule located at positions $\mathbf{r}^{(1)}, \mathbf{r}^{(2)}, \ldots, \mathbf{r}^{(N)}$. The geometric center shall correspond to $\mathbf{r} = 0$, thus $\sum_{i=1}^{N} \mathbf{r}^{(i)} = 0$. The *radius of gyration tensor* is defined by

$$G_{\mu\nu} = \sum_{i=1}^{N} r_{\mu}^{(i)} r_{\nu}^{(i)}. \tag{5.22}$$

The trace of this tensor is the square of the average radius R of the group of particles considered:

$$R^2 = G_{\lambda\lambda} = \sum_{i=1}^{N} r_{\lambda}^{(i)} r_{\lambda}^{(i)}, \tag{5.23}$$

where R is a measure for the size of the group of particles. The full tensor

$$G_{\mu\nu} = \frac{1}{3} G_{\lambda\lambda} \delta_{\mu\nu} + \overline{G_{\mu\nu}},$$

characterizes the size and the shape of the group of particles under consideration. The symmetric traceless part, in particular, is a measure for the deviation from a spherical symmetry. When all particles considered have the same mass m, the moment of inertia tensor $\Theta_{\mu\nu}$ is related to $G_{\mu\nu}$ by

$$\Theta_{\mu\nu} = m \left(G_{\lambda\lambda} \delta_{\mu\nu} - G_{\mu\nu} \right). \tag{5.24}$$

In applications, the average of the right hand side of (5.22) is referred to as radius of gyration tensor, viz. $G_{\mu\nu} = \langle \sum_{i=1}^{N} r_{\mu}^{(i)} r_{\nu}^{(i)} \rangle$.

5.3.3 Molecular Polarizability Tensor

An electric field **E** causes a slight average shift of the electrons in an atom or a molecule. The center of charge of the electrons is displaced with respect to the center of charge of the nuclei. Thus the electric field induces an electric dipole moment $\mathbf{p}^{\text{ind}}$. It has to be distinguished from a permanent dipole moment $\mathbf{p}^{\text{perm}}$ which some molecules posses. The molecular polarizability tensor $\alpha_{\mu\nu}$ characterizes the size and direction of the induced dipole moment according to

$$p_\mu^{\text{ind}} = \varepsilon_0 \, \alpha_{\mu\nu} \, E_\nu, \tag{5.25}$$

where ε_0 is the dielectric permeability of the vacuum.

For molecules with a symmetry axis parallel to the unit vector **u**, the molecular polarizability tensor is of the form

$$\alpha_{\mu\nu} = \alpha_\| \, u_\mu u_\nu + \alpha_\perp \left(\delta_{\mu\nu} - u_\mu u_\nu \right), \tag{5.26}$$

where $\alpha_\|$ and $\alpha_\perp$ are the polarizability for an electric field parallel and perpendicular to **u**, respectively. The standard decomposition of the polarizability tensor into its isotropic and symmetric traceless parts is

$$\alpha_{\mu\nu} = \bar{\alpha} \, \delta_{\mu\nu} + \left(\alpha_\| - \alpha_\perp \right) \overline{u_\mu u_\nu}, \tag{5.27}$$

with the average polarizability $\bar{\alpha} = (\alpha_\| + 2\alpha_\perp)/3$. The polarizability has the dimension of a volume. For atoms and molecules, it is of the order of a molecular volume. For a metallic sphere of radius R, it is $4\pi R^3$, cf. Sect. 10.4.1.

5.3.4 Dielectric Tensor, Birefringence

In an anisotropic linear medium the electric displacement field **D** is linked with the electric field **E** via the linear relation

$$D_\mu = \varepsilon_0 \, \varepsilon_{\mu\nu} \, E_\nu = \varepsilon_0 \left(\bar{\varepsilon} \, E_\mu + \overline{\varepsilon_{\mu\nu}} \, E_\nu \right), \tag{5.28}$$

where ε_0 is the dielectric permeability of the vacuum, and $\bar{\varepsilon} = \varepsilon_{\lambda\lambda}/3$. The connection between the dielectric tensor and the molecular polarizability is discussed in Sects. 12.2.3 and 13.6.5.

The symmetric dielectric tensor $\varepsilon_{\mu\nu}$ depends on the frequency ω of the electric field. Its symmetric traceless part $\overline{\varepsilon_{\mu\nu}}$ characterizes the dielectric or optical anisotropy of the medium. *Double refraction*, which is also called *birefringence*, occurs only when $\overline{\varepsilon_{\mu\nu}}$ is not zero for optical frequencies. Notice that **D**, as given by (5.28), is not parallel to **E**, unless the electric field is parallel to one of the principal directions.

The index of refraction $\nu^{(i)}$ for linearly polarized light with the electric field vector parallel to a principal direction $\mathbf{e}^{(i)}$, is linked with the principal value $\varepsilon^{(i)}$ of the dielectric tensor by the Maxwell relation

$$\nu^{(i)} = \sqrt{\varepsilon^{(i)}}, \quad i = 1, 2, 3. \tag{5.29}$$

Birefringence occurs, whenever, at least two of the principal indices of refraction are different. The differences $\delta\nu_{12} = \nu^{(1)} - \nu^{(2)}$ or $\delta\nu_{13} = \nu^{(1)} - \nu^{(3)}$ quantify the size of the birefringence for light propagating in the 3-direction or in the 2-direction, respectively.

One way to detect birefringence is to measure the transmission of light through a medium, between crossed polarizer and analyzer. The effect is largest, when the incident light is linearly polarized with the electric field vector oriented under 45° between two principal directions, say between $\mathbf{e}^{(1)}$ and $\mathbf{e}^{(3)}$. The vector obviously has components to both these principal directions for which the speed of light is different, because the indices of refraction are different. After a propagation over the distance L through the birefringent medium, the two components have a phase shift $\delta\phi = \frac{L}{\lambda}\delta\nu_{13}$, where λ is the wave length of the light. As a consequence of this phase shift, the light is elliptically polarized and has a component perpendicular to the direction of the incident linear polarization. The intensity I of the light which passes through an analyzer oriented parallel to $\mathbf{e}^{(1)} - \mathbf{e}^{(3)}$, is proportional to

$$I \sim \sin^2(\delta\phi) = \sin^2\left(\frac{L}{\lambda}\delta\nu_{13}\right). \tag{5.30}$$

Clearly, this intensity of the light passing through the crossed analyzer can be used to measure the difference $\delta\nu_{13}$ between the indices of refraction which, in turn, is caused by the symmetric traceless part of the dielectric tensor. Its microscopic origin, described by an average of the molecular polarizability tensor and the alignment of optically anisotropic molecules, as well as a non-isotropic arrangement of atoms, will be discussed later.

5.3.5 Electric and Magnetic Torques

An electric field $\mathbf{E}$ exerts a torque on an electric dipole moment $\mathbf{p}^{\mathrm{el}}$. Similarly, a magnetic field $\mathbf{B}$ causes a torque on a magnetic dipole moment $\mathbf{m}$. These torques, denoted by $\mathbf{T}^{\mathrm{el}}$ and $\mathbf{T}^{\mathrm{mag}}$, are given by

$$T_\mu^{\mathrm{el}} = \varepsilon_{\mu\lambda\nu}\, p_\lambda\, E_\nu, \quad T_\mu^{\mathrm{mag}} = \varepsilon_{\mu\lambda\nu}\, m_\lambda\, B_\nu. \tag{5.31}$$

In general, the electric dipole moment is the sum of a permanent and an induced part, viz. $\mathbf{p}^{\mathrm{el}} = \mathbf{p}^{\mathrm{perm}} + \mathbf{p}^{\mathrm{ind}}$, cf. Sect. 5.3.3. The computation of the permanent moment for a given charge distribution is presented in Sect. 10.3. As discussed above, the induced dipole moment is proportional to the electric field, when the field strength is small enough, such that terms nonlinear in the applied electric field can be disregarded. In this linear regime, one has $p_\nu^{\text{ind}} = \varepsilon_0 \alpha_{\nu\kappa} E_\kappa$, cf. (5.25). Here $\alpha_{\nu\kappa}$ is the molecular polarizability tensor and ε_0 is the dielectric permeability of the vacuum. The α-tensor is symmetric, it can be decomposed into its isotropic and symmetric traceless parts: $\alpha_{\nu\kappa} = \frac{1}{3}\alpha_{\tau\tau}\delta_{\nu\kappa} + \overline{\alpha_{\nu\kappa}}$. It is only the anisotropic, i.e. irreducible part of the α-tensor, which contributes to the torque

$$T_\mu^{\text{el,ind}} = \varepsilon_0 \, \varepsilon_{\mu\lambda\nu} \, \overline{\alpha_{\lambda\kappa}} \, E_\kappa \, E_\nu. \tag{5.32}$$

This equation can also be written as

$$\mathbf{T}^{\text{el,ind}} = \varepsilon_0 \left(\overline{\alpha} \cdot \mathbf{E} \right) \times \mathbf{E}.$$

An expression analogous to (5.32) applies for the induced magnetic dipole moment. Again, only the anisotropic part of the magnetic polarizability tensor gives rise to the torque.

Similar relations follow from the angular momentum balance of the Maxwell equations, see Sect. 7.5. In particular, the torque density associated with the electromagnetic fields is $\mathbf{E} \times \mathbf{D} + \mathbf{H} \times \mathbf{B}$. Due to $\mathbf{D} = \varepsilon_0 \mathbf{E} + \mathbf{P}$ and $\mathbf{B} = \mu_0(\mathbf{H} + \mathbf{M})$, the torque density is also equal to $\mathbf{E} \times \mathbf{P} + \mu_0 \mathbf{H} \times \mathbf{M}$. Here μ_0 is the magnetic permeability, also called magnetic induction constant, of the vacuum. In SI-units, it is given by $\mu_0 = 4\pi 10^{-7}$ As/Vm, where A s/V m stands for "Ampere seconds/Volt meter". On the other hand, the torque density exerted by the fields on the matter is $\mathbf{P} \times \mathbf{E} + \mathbf{M} \times \mathbf{B}$, which involves the electric polarization $\mathbf{P}$ and the magnetization $\mathbf{M}$.

For the induced part, one has, in the linear regime,

$$P_\lambda^{\text{ind}} = \varepsilon_0 \, \chi_{\lambda\kappa}^{\text{el}} \, E_\kappa, \quad M_\lambda^{\text{ind}} = \mu_0^{-1} \chi_{\lambda\kappa}^{\text{mag}} \, B_\kappa, \tag{5.33}$$

where $\chi_{\lambda\kappa}^{\text{el}}$ and $\chi_{\lambda\kappa}^{\text{mag}}$ are the electric and magnetic susceptibility tensors. By analogy to (5.32), the torque density is determined by

$$\varepsilon_{\mu\lambda\nu} \left(\varepsilon_0 \, \overline{\chi_{\lambda\kappa}^{\text{el}}} \, E_\kappa \, E_\nu + \mu_0^{-1} \, \overline{\chi_{\lambda\kappa}^{\text{mag}}} \, B_\kappa \, B_\nu \right), \tag{5.34}$$

where just the anisotropic parts of the susceptibility tensors contribute to the torque.

5.4 Geometric Interpretation of Symmetric Tensors

5.4.1 Bilinear Form

Symmetric second rank tensors can be visualized through geometric interpretations. In one, the tensor is used as the *coefficient matrix of a bilinear form*, e.g.

$$x_\mu S_{\mu\nu} x_\nu = X^2, \tag{5.35}$$

with some positive number X. In many applications, all principal values are positive, $S^{(i)} > 0$. Then (5.35) represents an ellipsoid in "x-space" with the semi-axes $X/\sqrt{S^{(i)}}$. In the uniaxial case it is an ellipsoid of revolution. It reduces to a sphere with radius X, when all principal values are equal.

In many applications in physics, bilinear forms are encountered, where the principal values of the second rank tensor cannot be negative. An example is the kinetic energy T of a solid body rotating with angular velocity $\mathbf{w}$:

$$T = \frac{1}{2} w_\mu \Theta_{\mu\nu} w_\nu \geq 0. \tag{5.36}$$

This inequality has to hold true for any direction of the angular velocity, in particular for $\mathbf{w}$ parallel to any one of the principal axes. All principal values must not be negative. Thus the equation (5.36), with $T = \mathrm{const.} > 0$, describes an ellipsoid in w-space.

5.4.2 Linear Mapping

A second geometric interpretation of a symmetric tensor is based on the *linear mapping*

$$y_\mu = S_{\mu\nu} x_\nu \tag{5.37}$$

from $\mathbf{x}$-space into $\mathbf{y}$-space.

Firstly, notice that the vector $\mathbf{y}$ as related to $\mathbf{x}$ via (5.37), in general, is not parallel to $\mathbf{x}$, unless $\mathbf{x}$ is parallel to one of the principal directions. Thus the cross product $\mathbf{x} \times \mathbf{y}$ is not zero, provided that the symmetric traceless part of the tensor $\mathbf{S}$ is not isotropic. More specifically, one has

$$(\mathbf{x} \times \mathbf{y})_\mu = \varepsilon_{\mu\nu\lambda} x_\nu y_\lambda = \varepsilon_{\mu\nu\lambda} x_\nu S_{\lambda\kappa} x_\kappa = \varepsilon_{\mu\nu\lambda} x_\nu \overline{S_{\lambda\kappa}} x_\kappa. \tag{5.38}$$

In the uniaxial case, cf. (5.8), with the symmetry axis parallel to the unit vector $\mathbf{e}$, this relation reduces to

$$(\mathbf{x}\times\mathbf{y})_\mu = \varepsilon_{\mu\nu\lambda} x_\nu y_\lambda = (S_\parallel - S_\perp) \varepsilon_{\mu\nu\lambda} x_\nu e_\lambda e_\nu x_\kappa = (S_\parallel - S_\perp) \mathbf{x} \cdot \mathbf{e} (\mathbf{x}\times\mathbf{e})_\mu. \tag{5.39}$$

Clearly, this expression vanishes when the principal values $S_\parallel$ and $S_\perp$ are equal to each other. Furthermore, (5.39) gives zero both for $\mathbf{x}$ parallel and perpendicular to $\mathbf{e}$, and it assumes extremal values, when $\mathbf{x}$ encloses an angle of $45°$ with the symmetry axis.

Secondly, when the end point of the vector $\mathbf{x}$ scans a unit sphere, the end point of the vector $\mathbf{y}$ will be on an ellipsoid provided that all principal values are positive. In this case the semi-axes are equal to $S^{(i)}$. Again biaxial and uniaxial ellipsoids are generated by biaxial and uniaxial tensors, and the ellipsoid degenerates to a sphere

Fig. 5.1 Perspective view of uniaxial (*left*) and biaxial (*right*) ellipsoids generated by linear mappings with $S^{(1)} = S^{(2)} = 0.8$, $S^{(3)} = 1.4$ and $S^{(1)} = 1.4$, $S^{(2)} = 0.6$, $S^{(3)} = 1$, respectively

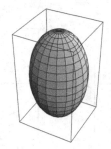
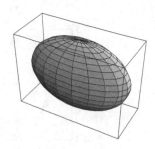

when the tensor is isotropic. In Fig. 5.1, uniaxial (left) and biaxial (right) ellipsoids are presented in a perspective view. The principal axes of the ellipsoids are parallel to the edges of the boxes, their lengths are proportional to the semiaxes.

In the uniaxial case where $S^{(1)} = S^{(2)} \neq S^{(3)}$ the ellipsoids generated by the linear mapping are cigar-like when $S^{(3)} > S^{(1)}$ holds true, and disk-like when one has $S^{(3)} < S^{(1)}$. They are referred to as *prolate* and *oblate* ellipsoids, respectively.

Now a geometric interpretation can be given to the decomposition of the symmetric tensor **S** according to (3.4) into an isotropic part and the symmetric traceless (anisotropic) part $\overline{\overline{\mathbf{S}}}$. In the mapping (5.37), the isotropic part of tensor **S** yields a sphere with its radius equal to the mean value $\frac{1}{3}(S^{(1)} + S^{(2)} + S^{(3)})$ of the principal values. The symmetric traceless part $\overline{\overline{\mathbf{S}}}$ characterizes the deviation of the ellipsoid generated by **S** from that sphere.

For the uniaxial case depicted on the left hand side in Fig. 5.1, the cross section of the ellipsoid in the 1–3-plane and of the pertaining sphere are shown on the left side of Fig. 5.2 as thick and dashed curves. The area between them is a measure for the anisotropic (symmetric traceless) part of the tensor **S**. The cross section of the ellipsoid in the 2–3-plane has the same appearance as in the 1–3-plane whereas it is a circle in the 1–2-plane.

The cross sections of the biaxial ellipsoid on the right hand side of Fig. 5.1 are ellipses in all three planes containing two of the coordinate axes. On the right hand side of Fig. 5.2, the cross sections of the ellipsoid (and the sphere pertaining to the isotropic part of **S**) are shown for the 1–3-plane (thick curve) and the 2–3-plane (thin curve) in the upper diagram. The lower diagram is for the 1–2-plane. Again, the area between the ellipses and the dashed circles is a measure for the deviation of the tensor **S** from being isotropic.

5.4.3 Volume and Surface of an Ellipsoid

The volume V and the surface area A of the **y**-ellipsoid generated by the linear mapping of a unit sphere in **x**-space according to (5.37), as discussed above for tensors with non-negative principal values, are given by

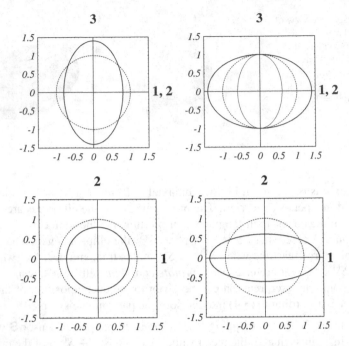

Fig. 5.2 Cross section of uniaxial (*left*) and biaxial (*right*) ellipsoids generated by linear mappings with $S^{(1)} = S^{(2)} = 0.8$, $S^{(3)} = 1.4$ and $S^{(1)} = 1.4$, $S^{(2)} = 0.6$, $S^{(3)} = 1$, respectively. The *upper* diagrams show the cross sections in the 1–3- and 2–3-planes, the *lower* ones are for the 1–2-plane. The *dashed circle* corresponds to the cross sections with the sphere generated by the isotropic part of the tensor

$$V = \frac{4}{3}\pi \det(\mathbf{S}), \quad A = \frac{2}{3}\pi \left(S_{\mu\mu} S_{\nu\nu} - S_{\mu\nu} S_{\nu\mu} \right). \tag{5.40}$$

The expression for A is equivalent to $A = \frac{4}{3}\pi(S^{(1)}S^{(2)} + S^{(2)}S^{(3)} + S^{(3)}S^{(1)})$. Insertion of the decomposition (3.4) of $\mathbf{S}$ into isotropic and anisotropic parts, i.e. of $S_{\mu\nu} = \bar{S}\delta_{\mu\nu} + \overline{S_{\mu\nu}}$ with $\bar{S} = \frac{1}{3}S_{\lambda\lambda}$ yields

$$\Delta A = -\frac{2}{3}\pi \; \overline{S_{\mu\nu}} \; \overline{S_{\mu\nu}}, \quad \Delta V = \frac{4}{9}\pi \; \overline{S_{\mu\nu}} \; \overline{S_{\nu\lambda}} \; \overline{S_{\lambda\mu}} + \bar{S} \, \Delta A \tag{5.41}$$

for the differences ΔA and ΔV between the area and the volume of the ellipsoid and the pertaining sphere with the radius $\bar{S}$ generated by the linear mapping with the isotropic part of $\mathbf{S}$.

5.5 Scalar Invariants of a Symmetric Tensor

5.5.1 Definitions

The trace $S_{\mu\mu}$, the square of the magnitude (norm) $S_{\mu\nu} S_{\mu\nu}$ and the determinant

$$\det(\mathbf{S}) = \frac{1}{3} S_{\mu\nu} S_{\nu\lambda} S_{\lambda\mu} + \frac{1}{6} \left(S_{\mu\mu} S_{\nu\nu} - 3 S_{\mu\nu} S_{\nu\mu} \right) S_{\lambda\lambda} \qquad (5.42)$$

of the symmetric tensor $\mathbf{S}$ are not affected by a rotation of the coordinate system and are therefore called *scalar invariants*. More specifically, they are called scalar invariants of first, second and third order, and denoted by I_1, I_2, I_3, respectively. In terms of the principal values, one has

$$S_{\mu\mu} = S^{(1)} + S^{(2)} + S^{(3)}, \quad S_{\mu\nu} S_{\mu\nu} = S^{(1)^2} + S^{(2)^2} + S^{(3)^2},$$

and

$$\det(\mathbf{S}) = S^{(1)} S^{(2)} S^{(3)}. \qquad (5.43)$$

To check the validity of (5.42), denote the principal values of the tensor by a, b, c for simplicity. Then $\det(\mathbf{S})$ reads

$$\frac{1}{3} \left(a^3 + b^3 + c^3 \right) + \frac{1}{6} \left((a + b + c)^2 - 3 \left(a^2 + b^2 + c^2 \right) \right) (a + b + c)$$

which is equal to abc, the result obtained directly from the determinant.

For symmetric traceless tensors, the symbols I_2 and I_3 are used in the following, for the norm and for the determinant, multiplied by 3:

$$I_2 = \overline{S_{\mu\nu}} \, \overline{S_{\nu\mu}}, \quad I_3 = 3 \det \left(\overline{S} \right) = \overline{S_{\mu\nu}} \, \overline{S_{\nu\lambda}} \, \overline{S_{\lambda\mu}}. \qquad (5.44)$$

A derivation of (5.42) and (5.44), based on the Hamilton-Cayley theorem, is presented in Sect. 5.6.

The third order invariant can be used to decide whether a tensor is uniaxial or biaxial without diagonalizing the tensor. A measure for biaxiality is introduced next.

5.5.2 Biaxiality of a Symmetric Traceless Tensor

As discussed above, a biaxial symmetric traceless tensor can be decomposed into a uniaxial part and a planar biaxial part, cf. (5.12), characterized by coefficients s and q. More specifically, the principal values are $-\frac{1}{3}s + q$, $-\frac{1}{3}s - q$, and $\frac{2}{3}s$, cf. (5.13). Thus one has, according to (5.44),

$$I_2 = \frac{2}{3}s^2 + 2q^2, \quad I_3 = 3 \det\left(\overset{\frown}{S}\right) = 2s\left(\frac{1}{9}s^2 - q^2\right). \tag{5.45}$$

The special planar biaxial case considered in (5.14) means $s = 0$, then one has $I_3 = 0$. On the other hand, I_3 is also zero for $q = \pm\frac{1}{3}s$. In this case the tensor is also planar biaxial, but the roles of the principal axes $1, 2, 3$ are interchanged. For a uniaxial tensor corresponding to $q = 0$, the ratio I_3^2/I_2^3 is equal to $1/6$. Thus a biaxiality parameter b can be defined by

$$b^2 = 1 - 6I_3^2/I_2^3. \tag{5.46}$$

Clearly, one has $b^2 = 1$ for the planar biaxial case corresponding to $s = 0$ and $b^2 = 0$ for a uniaxial tensor with $q = 0$.

For the dyadic tensor $2\,\overline{u_\mu v_\nu}$ constructed from the components of two unit vectors **u** and **v**, as considered in Sect. 5.2.5, one has $s = -c$ and $q = 1$, with $c := \mathbf{u} \cdot \mathbf{v} = \cos\vartheta$, where ϑ is the angle between the two vectors. Then b^2 is equal to

$$1 - 3c^2\left(1 - c^2/9\right)^2\left(c^2/3 + 1\right)^{-3}.$$

In Fig. 5.3, the biaxiality parameter b is plotted as function of $\sigma = \sin^2\vartheta = 1 - c^2$, as given by

$$b = \left[1 - (1 - \sigma)(8 + \sigma)^2(4 - \sigma)^{-3}\right]^{1/2}. \tag{5.47}$$

This curve does not deviate strongly from the dashed straight line $b = \sigma$. As expected, the biaxiality parameter is zero for $\sigma = 0$ corresponding to $\vartheta = 0$. It assumes its maximum value 1 for $\sigma = 1$ which pertains to $\vartheta = \pm\pi/2$.

Fig. 5.3 The biaxiality parameter b as function of $\sigma = \sin^2\vartheta$, where ϑ is the angle between the vectors **u** and **v**, which form the symmetric traceless dyadic. The dashed line corresponds to $b = \sigma$

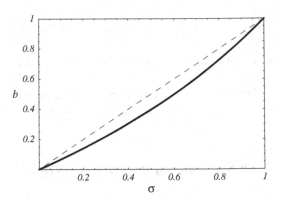

5.6 Hamilton-Cayley Theorem and Consequences

5.6.1 Hamilton-Cayley Theorem

Any symmetric second rank tensor $S_{\mu\nu}$, with principal values $S^{(1)}$, $S^{(2)}$, $S^{(3)}$ obeys the equation

$$\left(S_{\mu\nu} - S^{(1)}\delta_{\mu\nu}\right)\left(S_{\nu\lambda} - S^{(2)}\delta_{\nu\lambda}\right)\left(S_{\lambda\kappa} - S^{(3)}\delta_{\lambda\kappa}\right) = 0, \qquad (5.48)$$

which is essentially the *Hamilton-Cayley theorem*. In linear algebra, this theorem is applied to square matrices. To prove (5.48), notice firstly, that any vector $\mathbf{a}$, in 3D, can be written as linear combination of the three unit vectors $\mathbf{e}^{(1)}$, $\mathbf{e}^{(2)}$, $\mathbf{e}^{(3)}$ which are parallel to the principal directions: $a_\kappa = a^{(1)}e_\kappa^{(1)} + a^{(2)}e_\kappa^{(2)} + a^{(3)}e_\kappa^{(3)}$. Multiplication of the left hand side of (5.48) with a_κ yields zero. This can be checked, term by term, using the eigenvalue relation $S_{\mu\nu}e_\nu^{(i)} = e_\mu^{(i)}$, for $i = 1, 2, 3$. In particular, one has immediately $(S_{\lambda\kappa} - S^{(3)}\delta_{\lambda\kappa})e_\kappa^{(3)} = 0$. The multiplication $(S_{\nu\lambda} - S^{(2)}\delta_{\nu\lambda})(S_{\lambda\kappa} - S^{(3)}\delta_{\lambda\kappa})e_\kappa^{(2)} = (S_{\nu\lambda} - S^{(2)}\delta_{\nu\lambda})(S^{(2)} - S^{(3)})e_\lambda^{(2)} = 0$ again yields zero. Similarly, one finds $(S_{\mu\nu} - S^{(1)}\delta_{\mu\nu})(S_{\nu\lambda} - S^{(2)}\delta_{\nu\lambda})(S_{\lambda\kappa} - S^{(3)}\delta_{\lambda\kappa})e_\kappa^{(1)} = (S_{\mu\nu} - S^{(1)}\delta_{\mu\nu})(S^{(1)} - S^{(2)})(S^{(1)} - S^{(3)})e_\nu^{(1)} = 0$. Since this result is valid for all vectors, the tensorial relation (5.48) must hold true.

Explicit multiplication of the terms in (5.48) leads to

$$S_{\mu\nu}S_{\nu\lambda}S_{\lambda\kappa} - \left(S^{(1)} + S^{(2)} + S^{(3)}\right)S_{\mu\lambda}S_{\lambda\kappa}$$
$$+ \left(S^{(1)}S^{(2)} + S^{(2)}S^{(3)} + S^{(3)}S^{(1)}\right)S_{\mu\kappa} - S^{(1)}S^{(2)}S^{(3)}\delta_{\mu\kappa} = 0.$$

Since $S^{(1)}S^{(2)}S^{(3)} = \det(\mathbf{S})$, due to $S^{(1)} + S^{(2)} + S^{(3)} = S_{\alpha\alpha}$ and $S^{(1)}S^{(2)} + S^{(2)}S^{(3)} + S^{(3)}S^{(1)} = \frac{1}{2}((S^{(1)} + S^{(2)} + S^{(3)})^2 - ((S^{(1)})^2 + (S^{(2)})^2 + (S^{(3)})^2)) = \frac{1}{2}(S_{\alpha\alpha}S_{\beta\beta} - S_{\alpha\beta}S_{\beta\alpha})$, the Hamilton-Cayley theorem is equivalent to

$$\det(\mathbf{S})\,\delta_{\mu\kappa} = S_{\mu\nu}S_{\nu\lambda}S_{\lambda\kappa} - S_{\nu\nu}\,S_{\mu\lambda}S_{\lambda\kappa} + \frac{1}{2}(S_{\nu\nu}S_{\lambda\lambda} - S_{\nu\lambda}S_{\lambda\nu})\,S_{\mu\kappa}. \qquad (5.49)$$

In the case of a symmetric traceless tensor $\overline{\mathbf{S}}$, this relation reduces to

$$\det\left(\overline{\mathbf{S}}\right)\delta_{\mu\kappa} = \overline{S_{\mu\nu}}\,\overline{S_{\nu\lambda}}\,\overline{S_{\lambda\kappa}} - \frac{1}{2}\,\overline{S_{\nu\lambda}}\,\overline{S_{\lambda\nu}}\,\overline{S_{\mu\kappa}}. \qquad (5.50)$$

Consequences of the Hamilton-Cayley relations (5.49) and (5.50) are presented next. The contraction with $\mu = \kappa$ in the Hamilton-Cayley theorem (5.49) leads to the previously presented equation (5.42) for the calculation of the determinant in terms of products of the tensor. Likewise, the expression for the invariant I_3 of a symmetric traceless tensor, as given in (5.45), follows from (5.50) with $\mu = \kappa$. The symmetric traceless part of equation (5.50) leads to

$$\overline{S_{\mu\nu}}\ \overline{S_{\nu\lambda}}\ \overline{S_{\lambda\kappa}} = \frac{1}{2}\ \overline{S_{\nu\lambda}}\ \overline{S_{\lambda\nu}}\ \overline{S_{\mu\kappa}}, \tag{5.51}$$

which is a remarkable relation for the triple product of an irreducible second rank tensor. In matrix notation, this can be checked by writing the tensor in its principal axes coordinate system.

5.2 Exercise:
Verify the Relation (5.51) for the Triple Product of a Symmetric Traceless Tensor

Hint: use the matrix notation

$$\begin{pmatrix} a & 0 & 0 \\ 0 & b & 0 \\ 0 & 0 & c \end{pmatrix},$$

with $c = -(a + b)$, for the symmetric traceless tensor in its principal axes system. Compute the expressions on both sides of (5.51) and compare.

5.6.2 Quadruple Products of Tensors

Multiplication of (5.50) or (5.51) with $\overline{S_{\kappa\mu}}$ implies

$$\overline{S_{\kappa\mu}}\ \overline{S_{\mu\nu}}\ \overline{S_{\nu\lambda}}\ \overline{S_{\lambda\kappa}} = \frac{1}{2}\ \overline{S_{\nu\lambda}}\ \overline{S_{\lambda\nu}}\ \overline{S_{\kappa\mu}}\ \overline{S_{\mu\kappa}}. \tag{5.52}$$

This relation says: the trace of the four fold product of a symmetric traceless second rank tensor is equal to one half of the square of its norm squared. Similarly, multiplication of (5.49) with $S_{\kappa\mu}$ yields an expression for the fourth order product $S_{\kappa\mu}S_{\mu\nu}S_{\nu\lambda}S_{\lambda\kappa}$ in terms of the trace, of two fold and three fold products of the tensor.

Notice that the equations presented in Sects. 5.5 and 5.6 are specific for symmetric second rank tensors in three dimensional space. This is obvious in the formulation of the Hamilton-Cayley theorem (5.50), there are just three principal values in 3D. For second rank tensors in 2D or in 4D, analogous, but different relations apply, which give rise to different consequences.

5.7 Volume Conserving Affine Transformation

5.7.1 Mapping of a Sphere onto an Ellipsoid

Let $\mathbf{r}$ and $\mathbf{r}^A$ be the coordinates in the original and in an affine transformed space. The components are linked by

$$r^A_\mu r^A_\mu = r_\mu A_{\mu\nu} r_\nu, \quad r^A_\mu = A^{1/2}_{\mu\nu} r_\nu. \tag{5.53}$$

When the tensor $A_{\mu\nu}$ has positive eigenvalues, the relations (5.53) describe a mapping of a sphere $\mathbf{r}^A \cdot \mathbf{r}^A = $ const. onto an ellipsoid in $\mathbf{r}$-space. The affine transformation matrix, as considered in Sect. 2.3.2, is $A^{1/2}_{\mu\nu}$. The volume of the ellipsoid is equal to that of the sphere provided that the product of the eigenvalues $A_i, i = 1, 2, 3$ of $A_{\mu\nu}$ are equal to 1, viz.,

$$A_1 A_2 A_3 = 1. \tag{5.54}$$

For a uniaxial ellipsoid two of the eigenvalues are equal, e.g. $A_2 = A_3$.

5.7.2 Uniaxial Ellipsoid

Let $\mathbf{u}$ be a unit vector parallel to the symmetry axis of a uniaxial ellipsoid. In this case, one can make the ansatz $A_{\mu\nu} \sim \delta_{\mu\nu} + A \overline{u_\mu u_\nu}$ where the *non-sphericity parameter* A is bounded according to $-3/2 < A < 3$. The volume conserving condition (5.54) implies

$$A_{\mu\nu} = \left[\left(1 - \frac{1}{3} A\right)^2 \left(1 + \frac{2}{3} A\right) \right]^{-1/3} \left(\delta_{\mu\nu} + A \overline{u_\mu u_\nu}\right). \tag{5.55}$$

The equation (5.54) with $r^A_\mu r^A_\mu = r^2_A = $ const. describes an ellipsoid with the semi-axes $a = [(1 - \frac{1}{3}A)(1 + \frac{2}{3}A)^{-1}]^{1/3} r_A$ and $b = c = [(1 + \frac{2}{3}A)(1 - \frac{1}{3}A)^{-1}]^{1/6} r_A$. Thus the axes ratio is

$$Q = \frac{a}{b} = \left[\left(1 - \frac{1}{3} A\right) \left(1 + \frac{2}{3} A\right)^{-1} \right]^{1/2}, \tag{5.56}$$

and A is related to Q by

$$A = 3 \frac{1 - Q^2}{1 + 2 Q^2}. \tag{5.57}$$

The dependence of $A_{\mu\nu}$ and its inverse on Q are

$$A_{\mu\nu} = Q^{2/3} \left[\delta_{\mu\nu} + \left(Q^{-2} - 1 \right) u_\mu u_\nu \right], \quad A_{\mu\nu}^{-1} = Q^{-2/3} \left[\delta_{\mu\nu} + \left(Q^2 - 1 \right) u_\mu u_\nu \right].$$

$$(5.58)$$

Prolate and oblate ellipsoids correspond to $Q > 1$ and $Q < 1$, respectively. Applications of the volume conserving affine transformation to the interaction potential between perfectly oriented ellipsoidal particles and to the anisotropy of the viscosity, are presented in Sect. 16.4.2.

Chapter 6
Summary: Decomposition of Second Rank Tensors

Abstract This chapter provides a summary of formulae for the decomposition of a Cartesian second rank tensor into its isotropic, antisymmetric and symmetric traceless parts.

Any second rank tensor $A_{\mu\nu}$ can be decomposed into its isotropic part, associated with a scalar, its antisymmetric part, linked a vector, and its irreducible, symmetric traceless part:

$$A_{\mu\nu} = \frac{1}{3} A_{\lambda\lambda} \, \delta_{\mu\nu} + \frac{1}{2} \varepsilon_{\mu\nu\lambda} c_\lambda + \overline{A_{\mu\nu}} \, . \tag{6.1}$$

The dual vector **c** is linked with the antisymmetric part of the tensor by

$$c_\lambda = \varepsilon_{\lambda\sigma\tau} A_{\sigma\tau} = \varepsilon_{\lambda\sigma\tau} \frac{1}{2} (A_{\sigma\tau} - A_{\tau\sigma}). \tag{6.2}$$

The symmetric traceless second rank tensor, as defined previously, is

$$\overline{A_{\mu\nu}} = \frac{1}{2} (A_{\mu\nu} + A_{\nu\mu}) - \frac{1}{3} A_{\lambda\lambda} \, \delta_{\mu\nu}. \tag{6.3}$$

Similarly, for a dyadic tensor composed of the components of the two vectors **a** and **b**, the relations above give

$$a_\mu b_\nu = \frac{1}{3} (\mathbf{a} \cdot \mathbf{b}) \, \delta_{\mu\nu} + \frac{1}{2} \varepsilon_{\mu\nu\lambda} c_\lambda + \overline{a_\mu b_\nu} \, . \tag{6.4}$$

The isotropic part involves the scalar product $(\mathbf{a} \cdot \mathbf{b})$ of the two vectors. The antisymmetric part is linked with the cross product of the two vectors, here one has

$$c_\lambda = \varepsilon_{\lambda\sigma\tau} a_\sigma b_\tau = (\mathbf{a} \times \mathbf{b})_\lambda. \tag{6.5}$$

The symmetric traceless part of the dyadic tensor is

$$\overline{a_\mu b_\nu} = \frac{1}{2} (a_\mu b_\nu + a_\nu b_\mu) - \frac{1}{3} a_\lambda b_\lambda \, \delta_{\mu\nu}. \tag{6.6}$$

© Springer International Publishing Switzerland 2015   75
S. Hess, *Tensors for Physics*, Undergraduate Lecture Notes in Physics,
DOI 10.1007/978-3-319-12787-3_6

Chapter 7
Fields, Spatial Differential Operators

Abstract This chapter is devoted to the spatial differentiation of fields which are tensors of various ranks and to the properties of spatial differential operators. Firstly, scalar fields like potential functions are considered. The nabla operator is introduced and applications of gradient fields are discussed, e.g. force fields in Newton's equation of motion. Secondly, the differential change of vector fields is analyzed, the divergence and the curl or rotation of vector fields are defined. Special types of vector fields are studied: vorticity-free fields as derivatives of scalar potentials and divergence-free fields as derivatives of vector potentials. The Laplace operator, the Laplace and Poisson equations are introduced. The conventional classification of vector fields is listed. Thirdly, tensor fields are considered. A graphical representation of symmetric second rank tensors is given. Spatial derivatives of tensor fields are discussed. An application involves the pressure tensor in the local conservation law for linear momentum. Further applications are the Maxwell equations of electrodynamics in differential form. This chapter is concluded by rules for the nabla and Laplace operators, their decomposition into radial and angular parts, with applications to the orbital angular momentum and kinetic energy operators of Wave Mechanics.

A function $f = f(\mathbf{r})$ which determines a number at any space point $\mathbf{r}$ is called a *field*. In three dimensional space (3D), f is a function of the three components r_1, r_2, r_3 of the position vector. Such a function, in turn, can be a scalar, a component of a vector or of a tensor. Depending on the rank of the tensor, one talks of

1. *Scalar fields*, examples are:
   the potential energy $\Phi = \Phi(\mathbf{r})$ or the electrostatic potential $\phi = \phi(\mathbf{r})$.
2. *Vector fields*, like
   the force $\mathbf{F} = \mathbf{F}(\mathbf{r})$, the electric field $\mathbf{E} = \mathbf{E}(\mathbf{r})$, or
   the flow field $\mathbf{v} = \mathbf{v}(\mathbf{r})$ of hydrodynamics.
3. *Tensor fields*, an example for a second rank tensor field is
   the pressure tensor or the stress tensor.

© Springer International Publishing Switzerland 2015
S. Hess, *Tensors for Physics*, Undergraduate Lecture Notes in Physics,
DOI 10.1007/978-3-319-12787-3_7

7.1 Scalar Fields, Gradient

In physics, potential functions are important examples for scalar fields. The mathematical considerations presented in the following for potentials apply just as well for other scalar fields, such as the number density, the mass density and the charge density, or concentration and temperature fields.

7.1.1 Graphical Representation of Potentials

In the case of a two-dimensional space, the value of a potential field can be plotted into the third dimension, i.e. as the height above a plane. It can be visualized just like a panoramic map of a landscape, or in a 3D graphics plot. Alternatively, the information about the potential can be shown in equipotential lines, just like the lines of equal height in maps used for hiking.

The panoramic view of a potential function depending on a three-dimensional vector requires a four-dimensional space, which is beyond our visual experience. The surfaces of equal potential energy, also called *equal potential surfaces* can be visualized in 3D graphics. The surface where the value of the potential is equal to c, is determined by

$$\Phi(r_1, r_2, r_3) = c.$$

The solution of this equation for r_3 yields

$$r_3 = z(r_1, r_2, c),$$

where c is a curve parameter. In this way, equipotential surfaces of 3D fields can be represented geometrically with the same tools used for 2D potential functions.

Three simple special cases associated with planar, cylindrical and spherical geometry are discussed next.

1. *Planar Geometry.* Let a preferential direction be specified by the constant unit vector $\mathbf{e}$, and a special potential depending on $\mathbf{r}$ in the form $\Phi = \Phi(\mathbf{e} \cdot \mathbf{r}) = \Phi(x)$, where $x = \mathbf{e} \cdot \mathbf{r}$. In this case the equipotential surfaces are planes perpendicular to $\mathbf{e}$. The information in the variation of the potential is contained in the one-dimensional function $\Phi(x)$.
2. *Cylindrical Geometry.* Let the direction of a preferential axis be parallel to the constant unit vector $\mathbf{e}$. Now the case is considered, where the potential depends just on the components of $\mathbf{r}$ perpendicular to $\mathbf{e}$. Then one has $\Phi = \Phi(\mathbf{r}^\perp)$, with $\mathbf{r}^\perp = \mathbf{r} - \mathbf{e} \cdot \mathbf{r}\mathbf{e}$. Or, in other words, one has $\Phi = \Phi(x, y)$, where the components of $\mathbf{r}^\perp$ are denoted by x and y. When, even more special, the potential just depends on distance $\rho = \sqrt{\mathbf{r}^\perp \cdot \mathbf{r}^\perp} = \sqrt{x^2 + y^2}$ of a point from the axis,

the equipotential surfaces are coaxial cylinders. In this case, the values of the potential are determined by a function of a single variable, viz.: ρ.

3. *Spherical Geometry*. Of special importance are potential functions which depend on the position vector **r** just via its magnitude, i.e. via the distance $r = |\mathbf{r}| = \sqrt{\mathbf{r} \cdot \mathbf{r}}$ of the point **r** from the center $\mathbf{r} = 0$. The equipotential surfaces of such a *spherical potential* $\Phi = \Phi(r)$ are concentric spheres. In this special case, again, a function depending on one variable only, suffices to quantify the value of the potential.

7.1.2 Differential Change of a Potential, Nabla Operator

The change $d\Phi$ of the value of a potential function Φ, when one goes from the position **r** to an adjacent position $\mathbf{r}' = \mathbf{r} + d\mathbf{r}$ is given by the difference $\Phi(\mathbf{r} + d\mathbf{r}) - \Phi(\mathbf{r})$. Taking into account that one is dealing with a function depending on the three components of the position vector, one has explicitly

$$d\Phi = \Phi(r_1 + dr_1, r_2 + dr_2, r_3 + dr_3) - \Phi(r_1, r_2, r_3).$$

It is assumed that the potential function can be expanded in a power series with respect to the increment $d\mathbf{r}$. *Differential change* implies that the magnitude of $d\mathbf{r}$ is small enough, such that terms nonlinear in $d\mathbf{r}$ can be disregarded. Then

$$d\Phi = \frac{\partial \Phi}{\partial r_1} dr_1 + \frac{\partial \Phi}{\partial r_2} dr_2 + \frac{\partial \Phi}{\partial r_3} dr_3 = \frac{\partial \Phi}{\partial r_\mu} dr_\mu \tag{7.1}$$

is obtained. In 3D, the quantity $\frac{\partial \Phi}{\partial r_\mu}$ has three components and it is a vector, since the scalar product with the vector dr_μ yields the scalar $d\Phi$.

The partial differentiation with respect to the Cartesian components of the position vector is frequently denoted by the *nabla operator*

$$\nabla_\mu := \frac{\partial}{\partial r_\mu}. \tag{7.2}$$

Also the symbol ∂_μ is used for the spatial partial derivative. Here the nabla operator is preferred. Nabla applied on a scalar field is also referred to as the *gradient field*, and sometimes denoted by $\mathrm{grad}\,\Phi$.

7.1.3 Gradient Field, Force

A geometric interpretation of the gradient field $\nabla_\mu \Phi(\mathbf{r})$ is obtained as follows. Consider the case, where the differential change $d\mathbf{r}$ is tangential to an equipotential surface. Then one has $d\Phi = 0$, and consequently, in this case $\nabla_\mu \Phi dr_\mu = 0$. This means: the gradient field

$\nabla_\mu \Phi(\mathbf{r})$ *is perpendicular to the equipotential surface running through* $\mathbf{r}$.

It is recalled that a two-dimensional potential function can be visualized by analogy to a map showing the height of hills and dales. The gradient points in the direction of the steepest ascent. Hikers know, when the lines of equal height are close together, the ascent is very steep. The interrelation between the force and the potential is defined such that the force is determined by the negative gradient. In the landscape analogy, the force points into the direction of the steepest descent, just as water flows.

In mechanics, there are forces which can be derived from a potential and others, for which no potential function exists. Examples will be discussed later. When the force $\mathbf{F}$ acting on a single particle can indeed be derived from a potential, then it is determined by

$$F_\mu = F_\mu(\mathbf{r}) = -\nabla_\mu \Phi(\mathbf{r}). \tag{7.3}$$

Similarly, in electrostatics, the electric field $\mathbf{E} = \mathbf{E}(\mathbf{r})$ is the negative gradient of the electrostatic potential $\phi = \phi(\mathbf{r})$:

$$E_\mu = E_\mu(\mathbf{r}) = -\nabla_\mu \phi(\mathbf{r}). \tag{7.4}$$

The electric field is everywhere perpendicular to the surfaces of equal electrostatic potential. The field lines indicate these directions normal to the potential surfaces. The electric field, at a specific point, is parallel to the tangent vector of the electric field running through this point.

7.1.4 Newton's Equation of Motion, One and More Particles

In Newton's equation of motion the position $\mathbf{r} = \mathbf{r}(t)$ of a particle is a function of the time t. For a constant mass m and with the velocity $\mathbf{v}(t) = d\mathbf{r}/dt$, the equation of motion now reads

$$m\frac{dv_\mu}{dt} = m\frac{d^2 r_\mu}{dt^2} = F_\mu = -\nabla_\mu \Phi(\mathbf{r}). \tag{7.5}$$

So far, the dynamics of a single particle, subjected to an "external" force, was considered. The description of the dynamics of $N = 2, 3, \ldots$ particles with position vectors $\mathbf{r}^i$ and masses m_i where $i = 1, 2, \ldots, N$, is based on equations of motion for each one of these particles:

$$m_i\frac{dv_\mu^i}{dt} = m_i\frac{d^2 r_\mu^i}{dt^2} = F_\mu^i. \tag{7.6}$$

Clearly, $\mathbf{v}^i$ is the velocity of particle "i". When the force can be derived from a potential, this many-particle potential Φ is a function of the coordinates of all particles involved, viz.: $\Phi = \Phi(\mathbf{r}^1, \mathbf{r}^2, \ldots, \mathbf{r}^N)$. The force $\mathbf{F}^i$ acting on particle "i" is given by the negative partial derivative of the potential with respect to the vector $\mathbf{r}^i$:

$$F^i_\mu = -\frac{\partial}{\partial r^i_\mu} \Phi = -\nabla^i_\mu \Phi. \tag{7.7}$$

An important special case are two interacting particles, $N = 2$. When their dynamics can be treated as if they were isolated from the rest of the world, the relevant interaction potential depends on their positions only via the relative vector $\mathbf{r}^{12} = \mathbf{r}^1 - \mathbf{r}^2$. Then one has $\partial \Phi/\partial r^1_\mu = \partial \Phi/\partial r^{12}_\mu$ and $\partial \Phi/\partial r^2_\mu = -\partial \Phi/\partial r^{12}_\mu$. This implies $F^1_\mu = -F^2_\mu$, which corresponds to Newton's *actio equal reactio*. As a consequence, the motion of the center of mass of the two particles is "force free", and the total linear momentum $\mathbf{P} = m_1 \mathbf{v}^1 + m_2 \mathbf{v}^2$ is constant. The interesting motion is that one described by the relative vector $\mathbf{r}^{12}$ and the relative velocity $\mathbf{v}^{12} = \mathbf{v}^1 - \mathbf{v}^2$. The governing equation of this motion is

$$m_{12} \frac{dv^{12}_\mu}{dt} = m_{12} \frac{d^2 r^{12}_\mu}{dt^2} = F^1_\mu = -\frac{\partial \Phi(\mathbf{r}^{12})}{\partial r^{12}_\mu}, \tag{7.8}$$

with the reduced mass $m_{12} = m_1 m_2/(m_1 + m_2)$. Thus the two-particle dynamics is reduced to the force-free motion of its center of mass and an effective one-particle dynamics of the relative motion. With the replacements $m_{12} \to m, \mathbf{r}^{12} \to \mathbf{r}, \mathbf{F}^1 \to \mathbf{F}$, $\Phi(\mathbf{r}^{12}) \to \Phi(\mathbf{r})$ and $(\partial/\partial r^{12}_\mu) \to \nabla_\mu$ the equation of motion (7.8) is mathematically equivalent to (7.5). In this sense, some of the special potential and force functions to be mentioned below pertain to an effective one-particle problem rather than to a true one particle dynamics.

7.1.5 Special Force Fields

The application of the nabla operator to the vector $\mathbf{r}$ yields

$$\nabla_\mu r_\nu = \delta_{\mu\nu}. \tag{7.9}$$

The contraction with $\mu = \nu$ implies

$$\nabla_\mu r_\mu = 3, \tag{7.10}$$

in 3D. The fact that the scalar product of the nabla operator with the vector $\mathbf{r}$ yields a number, which is certainly a scalar, proves that the nabla operator is also a vector. The relation (7.9) is essential for the calculation of the force from a given potential

function. Some examples for the special forms of potential functions discussed in
Sect. 7.1, are presented next.

(i) Planar Geometry

Let the potential depend on $\mathbf{r}$ via $x = \mathbf{e} \cdot \mathbf{r} = e_\nu r_\nu$, where $\mathbf{e}$ is a constant unit vector.
Then the chain rule, with the help of (7.9), leads to,

$$\nabla_\mu \Phi(x) = \frac{d\Phi}{dx} \nabla_\mu e_\nu r_\nu = \frac{d\Phi}{dx} e_\mu. \tag{7.11}$$

Clearly, one has $F_\mu \sim e_\mu$, the direction of the force is constant. In this very special
case, where the potential is linear in x, also the force strength is constant. Such a
force field is referred to as a *homogeneous field*. The gravitational field above a flat
surface on earth is of this type, as long as the height over ground, corresponding to the
variable x, is very small compared with the diameter of the earth. Another example
of an approximately homogeneous field is the electric field between the charged flat
plates of an electric capacitor.

(ii) Cylindrical Geometry

Let Φ be a function of the distance $\rho = \sqrt{r_\nu^\perp r_\nu^\perp}$ of point $\mathbf{r}$ from an axis parallel the
the constant unit vector $\mathbf{e}$. Here $r_\nu^\perp = r_\nu - e_\nu e_\lambda r_\lambda$ is the projection of $\mathbf{r}$ onto a plane
perpendicular to $\mathbf{e}$. Use of the chain rule yields

$$\nabla_\mu \Phi(\rho) = \frac{d\Phi}{d\rho} \nabla_\mu \rho.$$

Furthermore, one obtains

$$\nabla_\mu \rho = \nabla_\mu \sqrt{r_\nu^\perp r_\nu^\perp} = \frac{1}{2} \rho^{-1} \nabla_\mu \left(r_\nu^\perp r_\nu^\perp \right)$$

$$= \rho^{-1} r_\nu^\perp \nabla_\mu r_\nu^\perp = \rho^{-1} r_\nu^\perp (\delta_{\mu\nu} - e_\mu e_\nu) = \rho^{-1} r_\mu^\perp.$$

Thus the gradient of the potential function is found to be

$$\nabla_\mu \Phi(\rho) = \frac{d\Phi}{d\rho} \rho^{-1} r_\mu^\perp = \frac{d\Phi}{d\rho} \widehat{r_\mu^\perp}, \tag{7.12}$$

where $\widehat{r_\mu^\perp}$ is the radial unit vector pointing outward from the cylinder axis.

Notice: both the planar and the cylindrical geometry as considered here have
cylindrical symmetry since there is a preferential direction parallel to the constant
vector $\mathbf{e}$. In both cases, the torque $T_\mu = \varepsilon_{\mu\nu\lambda} r_\nu F_\lambda = -\varepsilon_{\mu\nu\lambda} r_\nu \nabla_\lambda \Phi$ is proportional
to $\varepsilon_{\mu\nu\lambda} r_\nu e_\lambda$. Thus only the component of the orbital angular momentum $\mathbf{L}$ which is
parallel to $\mathbf{e}$ is conserved. The other two components of $\mathbf{L}$, which are perpendicular
to the preferential direction, do change in time.

(iii) Spherical Symmetry

Now the case is considered where the potential depends on **r** via the magnitude $r = \sqrt{r_\nu r_\nu}$, thus $\Phi = \Phi(r)$. Then one has

$$\nabla_\mu \Phi(r) = \frac{d\Phi}{dr} \nabla_\mu r.$$

Use of the definition of r leads to

$$\nabla_\mu r = \nabla_\mu \sqrt{r_\nu r_\nu} = \frac{1}{2} r^{-1} \nabla_\mu (r_\nu r_\nu) = r^{-1} r_\nu \nabla_\mu r_\nu = r^{-1} r_\nu \delta_{\mu\nu} = r^{-1} r_\mu.$$

The rule

$$\nabla_\mu r = r^{-1} r_\mu = \widehat{r_\mu} \tag{7.13}$$

is important for many applications. It can be remembered by observing that the direction of $\nabla_\mu r$ must be parallel to the μ-direction since r does not contain any directional information, and by dimensional considerations, the result of the application of the nabla operator on r must be a dimensionless unit vector.

Thus the gradient of a spherical potential function is

$$\nabla_\mu \Phi(r) = \frac{d\Phi}{dr} r^{-1} r_\mu = \frac{d\Phi}{dr} \widehat{r_\mu}. \tag{7.14}$$

The resulting force $F_\mu = -\nabla_\mu \Phi(r)$ is

$$F_\mu = -\Phi' \widehat{r_\mu}, \quad \Phi' = \frac{d\Phi}{dr}. \tag{7.15}$$

Such a force, which is parallel to **r**, is referred to as *central force*. The attractive gravitational force between two masses, like the sun and the earth, is of this type. The same applies to the electrostatic Coulomb force between two charges. Here the force is repulsive or attractive, when the signs of both charges are equal or opposite, respectively. The pertaining potential functions, both for gravitation and Coulomb, are proportional to r^{-1}. Notice that the interaction between two spherical particles is described by potential functions which have a dependence on r, in general. For example, the potential function of two electrically neutral atoms, e.g. Argon atoms, are of the type $(r_0/r)^{12} - (r_0/r)^6$, where r_0 is an effective diameter of the atom. The first term is responsible for the repulsion at short distances, the second one for the attraction at larger distances.

The torque **r** $\times$ **F** vanishes for a central force. This implies that the orbital angular momentum **L** is constant. It is recalled that **L** is both perpendicular to **r** and to the velocity. Thus the constant direction of **L** implies that the motion takes place in a plane. Such a motion is essentially two-dimensional although it takes place in 3D.

The interaction between two particles or extended objects can be described by a spherical potential function just depending on the inter-particle separation

$r = |\mathbf{r}_1 - \mathbf{r}_2|$ only, when these particles are effectively round and do not posses any internal directional properties, like electric dipole or quadrupole moments, which influence their interaction.

7.2 Vector Fields, Divergence and Curl or Rotation

Let $\mathbf{v}(\mathbf{r})$ be a vector field with Cartesian components $v_\mu(\mathbf{r})$. Here the symbol v can be associated with "vector", in general, or with the flow velocity field of hydrodynamics. The vector field might also be associated with the displacement induced by a deformation of an elastic solid. The mathematical considerations to be presented are invariant with respect to different interpretations in physics.

Vector fields can be visualized as a field of arrows. At each point $\mathbf{r}$, an arrow can be drawn, whose length and direction is determined by $\mathbf{v}(\mathbf{r})$. Firstly, some examples are considered.

7.2.1 Examples for Vector Fields

(i) Homogeneous Field

As previously mentioned, a vector field of the type $\mathbf{v} = $ const., where the vector everywhere has constant length and direction, is referred to as a homogeneous field. Let the direction be specified by the constant unit vector $\mathbf{e}$. Then one has, apart from a numerical factor,

$$v_\mu = e_\mu = \text{const.}$$

This field is the gradient of the simple potential function $\Phi = r_\nu e_\nu = x$ (Fig. 7.1).

Fig. 7.1 Homogeneous and linearly increasing vector fields

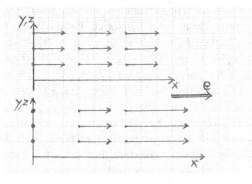

(ii) Linearly Increasing Field

Let $\mathbf{v}(\mathbf{r})$ be given by

$$v_\mu = e_\mu\, e_\nu r_\nu = x\, e_\mu. \tag{7.16}$$

In this case, the direction of the vector is still parallel to a constant unit vector, viz. to $\mathbf{e}$. This field can be derived from the potential function

$$\Phi = (1/2)\,(e_\nu r_\nu)^2 = (1/2)\, x^2, \tag{7.17}$$

which is that of an one-dimensional harmonic oscillator.

(iii) Radial and Cylindrical Fields

The vector field

$$v_\mu = r_\mu \tag{7.18}$$

has radial symmetry. It is the gradient of $\Phi = (1/2)r_\nu r_\nu = (1/2)r^2$, which has the functional form of the potential of an isotropic harmonic oscillator, in 3D.

The 2D version of a radial vector field is given by

$$v_\mu = r_\mu^\perp = r_\mu - e_\mu\, e_\nu r_\nu, \tag{7.19}$$

where the constant unit vector is perpendicular to the plane, in which the vector arrows lie. In this case the potential function

$$\Phi = (1/2)\, r_\nu^\perp r_\nu^\perp = (1/2)\,(x^2 + y^2) \tag{7.20}$$

is of the type of a 2D isotropic harmonic oscillator. Here the components of the position vector in the plane perpendicular to $\mathbf{e}$ have been denoted by x and y (Fig. 7.2).

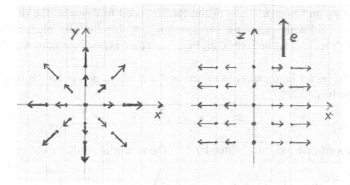

Fig. 7.2 Cylindrical vector field

Fig. 7.3 Uniaxial
squeeze-stretch field

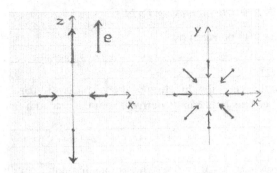

(iv) Uniaxial Squeeze-stretch Field

The vector field

$$v_\mu = 3\, e_\mu e_\nu r_\nu - r_\mu = 3z\, e_\mu - r_\mu,$$ (7.21)

with $z = e_\nu r_\nu$ can be used to describe a stretching in the direction parallel to the unit vector **e** and a squeezing in both directions perpendicular to **e**. This vector field is the gradient of the potential function

$$\Phi = \frac{3}{2}(e_\mu r_\mu)^2 - \frac{1}{2} r^2 = \frac{1}{2}\left(2z^2 - x^2 - y^2\right),$$ (7.22)

where the components of **r** perpendicular to **e** are denoted by x and y (Fig. 7.3).

(v) Planar Squeeze-stretch Field

Let **e** and **u** be two orthogonal unit vectors, $\mathbf{e} \cdot \mathbf{u} = 0$. The vector field

$$v_\mu = e_\mu u_\nu r_\nu + u_\mu e_\nu r_\nu = y\, e_\mu + x\, u_\mu,$$ (7.23)

with $x = e_\nu r_\nu$ and $y = u_\nu r_\nu$, is of the form needed to describe the deformation of an elastic solid with stretching and squeezing, within the x–y-plane, under 45° and 135° with respect to the x-axis, cf. Fig. 7.4. There is no deformation in the third direction.

Again, the vector field can be obtained as the gradient of a potential function, in this case one has

$$\Phi = e_\mu r_\mu u_\nu r_\nu = x\, y.$$ (7.24)

When the coordinate axis are rotated by 45°, the potential reads

$$\Phi = (1/2)\left((e_\nu r_\nu)^2 - (u_\nu r_\nu)^2\right) = (1/2)(x^2 - y^2).$$ (7.25)

Fig. 7.4 Planar
squeeze-stretch field

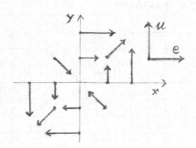

In this case, the gradient field is

$$v_\mu = e_\mu e_\nu r_\nu - u_\mu u_\nu r_\nu = x e_\mu - y u_\mu. \tag{7.26}$$

So far, all vector fields presented can be derived from a scalar potential function. However, there exist also vector fields for which this is not the case. A simple example is discussed next.

(vi) Solid-like Rotation or Vorticity Field

A circular flow with a constant angular velocity **w** is described by

$$v_\mu = \varepsilon_{\mu\nu\lambda} w_\nu r_\lambda. \tag{7.27}$$

This flow field is called *solid-like* since it corresponds to the motion of the points on a solid disc, rotating about an axis normal to the disc and running through the point **r** = 0. The axial vector **w** is parallel to the rotation axis. The field **v** has a non-zero vorticity $\nabla \times \mathbf{v}$. For this reason it is also referred to as *vorticity flow field* or just *vorticity field*. This kind of vector field cannot be represented as the gradient of a scalar potential function! (Fig 7.5).

(vii) Simple Shear Flow

Let **e** and **u** again be two orthogonal unit vectors, $\mathbf{e} \cdot \mathbf{u} = 0$. The simple vector field

$$v_\mu = e_\mu u_\nu r_\nu = y e_\mu, \tag{7.28}$$

with $x = e_\nu r_\nu$ and $y = u_\nu r_\nu$, is called a *simple shear* field. When the vector **v** is associated with the displacement of a part of a solid, the field describes a shear deformation. In fluids, such a flow field can be realized in a *plane Couette* geometry, where a fluid is confined between parallel flat plates, normal to **u**, and one plate moves parallel to **e**. Such a flow is also called *simple shear flow*. The vector field (7.28) is essentially a linear combination of the planar squeeze-stretch field (7.23) and the circular field (7.27). The simple vector field (7.28) cannot be obtained as the gradient of a scalar field (Fig. 7.6).

Fig. 7.5 Solid-like rotation field

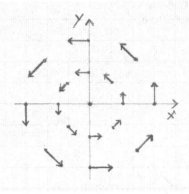

Fig. 7.6 Simple shear field

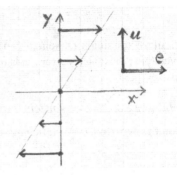

7.2.2 Differential Change of a Vector Fields

The difference $dv_\mu = v_\mu(\mathbf{r}+d\mathbf{r}) - v_\mu(\mathbf{r})$ of the vector field v_μ between the positions $\mathbf{r} + d\mathbf{r}$ and $\mathbf{r}$ can be expanded with respect to the small, differential change $d\mathbf{r}$. Up to linear terms, one has

$$dv_\mu = \frac{\partial v_\mu}{\partial r_\nu} dr_\nu = (\nabla_\nu v_\mu) dr_\nu. \qquad (7.29)$$

The quantity $\nabla_\nu v_\mu$ which is sometimes called *gradient* of $\mathbf{v}$, is a second rank tensor. It can be decomposed into its symmetric and antisymmetric parts or its isotropic, symmetric traceless and antisymmetric parts, just like any dyadic tensor, cf. (6.4). These decompositions are

$$\nabla_\nu v_\mu = \frac{1}{2}(\nabla_\nu v_\mu + \nabla_\nu v_\mu) + \frac{1}{2}(\nabla_\nu v_\mu - \nabla_\nu v_\mu), \qquad (7.30)$$

and

$$\nabla_\nu v_\mu = \frac{1}{3}(\nabla_\lambda v_\lambda)\delta_{\mu\nu} + \overline{\nabla_\nu v_\mu} + \frac{1}{2}\varepsilon_{\nu\mu\lambda}(\nabla \times \mathbf{v})_\lambda. \qquad (7.31)$$

The scalar $\nabla_\lambda v_\lambda = \nabla \cdot \mathbf{v} := \mathrm{div}\,\mathbf{v}$ is called the *divergence* of the vector $\mathbf{v}$. The cross product $(\nabla \times \mathbf{v})$ of the nabla operator and a vector $\mathbf{v}$ is also denoted by curl $\mathbf{v}$ or rot $\mathbf{v}$, pronounced as *curl* or *rotation*. In component notation it is given by

$$(\nabla \times \mathbf{v})_\lambda = \varepsilon_{\lambda\sigma\tau} \nabla_\sigma v_\tau. \tag{7.32}$$

When $\mathbf{v}$ stands for the flow velocity of a streaming fluid, the quantity $(1/2)(\nabla \times \mathbf{v})$ is referred to as the *vorticity* of the flow field. A sphere suspended in a plane Couette flow picks up an angular velocity equal to the vorticity, provided it is allowed to rotate freely.

The symmetric traceless part, also called deviatoric part of the gradient of the vector field, is given by

$$\overline{\nabla_\nu v_\mu} = \frac{1}{2}(\nabla_\nu v_\mu + \nabla_\nu v_\mu) - \frac{1}{3}(\nabla_\lambda v_\lambda)\delta_{\mu\nu}. \tag{7.33}$$

Vector fields with $\nabla \cdot \mathbf{v} \neq 0$, examples (ii) and (iii), have field lines with sources and sinks. Fields with $\nabla \cdot \mathbf{v} = 0$, all other cases above, are called *source free* vector fields. Vector fields with $\nabla \times \mathbf{v} = 0$, examples (i)–(v), are called *vortex-free*. Vectors of this kind can be derived from a scalar potential field. Frequently, vector fields are classified according to whether their divergence and their rotation are zero or not equal to zero, more details later.

The divergence, the rotation and $\overline{\nabla \mathbf{v}}$ can be calculated for the special vector fields treated above in Sect. 7.2.1. The cases (i)–(iii) and (vii) are considered next, the calculations for the vector fields (iv)–(vi) are transferred to the following exercise.

Divergence, Rotation and Symmetric Traceless Part of the Gradient Tensor for the vector fields (i)–(iii) and (vii) of Sect. 7.2.1:

(i) Homogeneous Field

$$v_\mu = e_\mu = \text{const.}$$

Obviously, one finds $\nabla_\nu v_\mu = 0$, consequently $\nabla \cdot \mathbf{v} = 0$, $\nabla \times \mathbf{v} = 0$, and $\overline{\nabla \mathbf{v}} = 0$.

(ii) Linearly Increasing Field

$$v_\mu = e_\mu e_\kappa r_\kappa = x\, e_\mu.$$

Due to $\nabla_\nu r_\kappa = \delta_{\nu\kappa}$ one obtains here, $\nabla_\nu v_\mu = e_\mu e_\nu$, and consequently

$$\nabla \cdot \mathbf{v} = 1, \quad \nabla \times \mathbf{v} = 0, \quad \overline{\nabla_\nu v_\mu} = \overline{e_\nu e_\mu}.$$

(iii) Radial and Cylindrical Fields

$$v_\mu = r_\mu.$$

For the three-dimensional radial field, one finds

$$\nabla \cdot \mathbf{v} = 3, \quad \nabla \times \mathbf{v} = 0, \quad \overline{\nabla_\nu v_\mu} = 0.$$

The 2D version of a radial vector field is

$$v_\mu = r_\mu^\perp = r_\mu - e_\mu\, e_\nu r_\nu,$$

with the unit vector $\mathbf{e}$, perpendicular to the plane, in which the radial vector lies. Here one obtains

$$\nabla \cdot \mathbf{v} = 2, \quad \nabla \times \mathbf{v} = 0, \quad \overline{\nabla_\nu v_\mu} = -\overline{e_\nu e_\mu}.$$

(vii) Simple Shear Flow

$$v_\mu = e_\mu\, u_\nu r_\nu,$$

where $\mathbf{e}$ and $\mathbf{u}$ are two orthogonal unit vectors, $\mathbf{e} \cdot \mathbf{u} = 0$. In this case, one finds $\nabla_\nu v_\mu = u_\nu e_\mu$, and

$$\nabla \cdot \mathbf{v} = 0, \quad (\nabla \times \mathbf{v})_\lambda = \varepsilon_{\lambda\nu\mu} u_\nu e_\mu, \quad \overline{\nabla_\nu v_\mu} = \overline{u_\nu e_\mu}.$$

The calculations here and in the following exercise show: the symmetric traceless part $\overline{\nabla \mathbf{v}}$ is zero for the flow fields (i), the 3D radial field of (iii) and for (vi). The tensor $\overline{\nabla \mathbf{v}}$ is uniaxial for the examples (ii), the 2D radial field of (iii), and for (iv), it is planar biaxial for (v) and (vii).

7.1 Exercise: Compute the Spatial Derivatives of Special Vector Fields
Compute the divergence $\nabla \cdot \mathbf{v}$, the curl or rotation $\nabla \times \mathbf{v}$ and the symmetric traceless part $\overline{\nabla_\nu v_\mu}$ of the gradient tensor $\nabla_\nu v_\mu$ for the vector fields (iv)–(vi) of Sect. 7.2.1.

7.3 Special Types of Vector Fields

7.3.1 Vorticity Free Vector Fields, Scalar Potential

Let $\mathbf{v}$ be a vector field, which can be represented as the gradient of a scalar potential: $\mathbf{v} = \nabla \Phi(\mathbf{r})$, then the rotation $\nabla \times \mathbf{v}$ of the vector is zero, thus

$$v_\lambda = \nabla_\lambda \Phi \;\Rightarrow\; (\nabla \times \mathbf{v})_\mu = \varepsilon_{\mu\nu\lambda} \nabla_\nu v_\lambda = \varepsilon_{\mu\nu\lambda} \nabla_\nu \nabla_\lambda \Phi = 0. \qquad (7.34)$$

Of course, it is assumed, that the scalar function Φ can be differentiated twice and that $\nabla_\nu \nabla_\lambda \Phi = \nabla_\lambda \nabla_\nu \Phi$.

The arguments just presented here, can be reverted. When $\nabla \times \mathbf{v} = 0$ holds true for a vector field $\mathbf{v}$, then it can be derived from a scalar potential Φ, such that $\mathbf{v} = \nabla \Phi$. The condition of a vanishing curl or rotation of the vector field is equivalent to the *integrability condition* $\nabla_\mu v_\nu = \nabla_\nu v_\mu$ or

$$\frac{\partial v_\nu}{\partial r_\mu} = \frac{\partial v_\mu}{\partial r_\nu}. \tag{7.35}$$

When the integrability condition holds true for a vector field, it does posses an "integral", viz.: a scalar potential function. Since nabla applied to a constant yields zero, the potential is only determined by its gradient up to an additive constant.

7.3.2 Poisson Equation, Laplace Operator

Let the divergence $\nabla \cdot \mathbf{v}$ of the vector field be equal to a given "density" function $\rho = \rho(\mathbf{r})$

$$\nabla_\nu v_\nu = \rho(\mathbf{r}). \tag{7.36}$$

The function $\rho(\mathbf{r})$ is the "source" for the vector field. When $\mathbf{v} = \nabla \Phi$ holds true, (7.36) implies, that the potential Φ obeys the *Poisson equation*

$$\nabla_\nu \nabla_\nu \Phi := \Delta \Phi = \rho(\mathbf{r}). \tag{7.37}$$

The symbol Δ stands for the *Laplace operator*. This second spatial derivative is defined by

$$\Delta := \nabla_\nu \nabla_\nu = \frac{\partial^2}{\partial r_\nu \partial r_\nu}. \tag{7.38}$$

By definition, the Laplace operator is a scalar, i.e. invariant under a rotation of the coordinate system.

7.3.3 Divergence Free Vector Fields, Vector Potential

A vector field $\mathbf{v}$ is called divergence-free or source-free when $\nabla \cdot \mathbf{v} = 0$, or equivalently,

$$\nabla_\mu v_\mu = 0$$

holds true. Such a field can be derived from a *vector potential* $\mathbf{A}$ according to

$$v_\mu = (\nabla \times \mathbf{A})_\mu = \varepsilon_{\mu\nu\lambda} \nabla_\nu A_\lambda. \tag{7.39}$$

Clearly, due to $\nabla_\mu v_\mu = \varepsilon_{\mu\nu\lambda}\nabla_\mu\nabla_\nu A_\lambda = 0$, the divergence of a vector field given by (7.39) vanishes, when the symmetry $\nabla_\mu\nabla_\nu = \nabla_\nu\nabla_\mu$ applies for the second spatial derivatives of the vector potential $\mathbf{A}$.

The vector potential $\mathbf{A}$ is not unique, in the sense that also the vector potential $\mathbf{A}' = \mathbf{A} + \nabla\varphi(\mathbf{r})$, with a scalar function $\varphi(\mathbf{r})$, used (7.39), yields the same vector field $\mathbf{v}$. The reason is that $\nabla \times \nabla\varphi(\mathbf{r}) = 0$. The freedom for the choice of a vector potential is by far greater than that of a scalar potential, which is determined except for an additive constant. In some applications, a source-free vector potential is required, i.e. the extra condition $\nabla \cdot \mathbf{A} = 0$ is imposed.

Some Examples for Vector Potentials

(i) Homogeneous Vector Field

A homogeneous vector field $\mathbf{v} = $ const. is obtained as the rotation of a vector potential $\mathbf{A}$, which should be linear in the position vector $\mathbf{r}$. Furthermore, $\mathbf{A}$ must contain the information on the constant $\mathbf{v}$. A plausible ansatz is $\mathbf{A} = c\mathbf{v} \times \mathbf{r}$, with a coefficient c, which has to be determined. To this purpose, one computes, with the help of the properties of the epsilon-tensor, cf. Sect. 4.1.2,

$$v_\mu = (\nabla \times \mathbf{A})_\mu = \varepsilon_{\mu\nu\lambda}\nabla_\nu c\,\varepsilon_{\lambda\sigma\tau}v_\sigma r_\tau = c\,\varepsilon_{\mu\nu\lambda}\varepsilon_{\lambda\sigma\tau}\delta_{\nu\tau}v_\sigma$$
$$= c\,\varepsilon_{\mu\nu\lambda}\varepsilon_{\lambda\sigma\nu}v_\sigma = c\,2\,\delta_{\mu\sigma}v_\sigma = 2\,c\,v_\mu$$

and consequently, $c = 1/2$. Thus the constant vector field $\mathbf{v}$ is represented as the rotation of the vector potential

$$A_\mu = \frac{1}{2}\varepsilon_{\mu\nu\lambda}\,v_\nu r_\lambda. \tag{7.40}$$

Notice that the constant vector field can be derived both from a scalar potential and from a vector potential. A scalar potential does not exist for the next example.

(ii) Solid-like Rotational Flow

The solid-like rotational flow is described by the vector field

$$v_\mu = \varepsilon_{\mu\nu\lambda}w_\nu r_\lambda,$$

with the constant angular velocity $\mathbf{w}$. The pertaining vector potential must be linear in $\mathbf{w}$ and of second order in $\mathbf{r}$. A guess is $A_\lambda = c_1 r^2 w_\lambda + c_2 r_\lambda r_\kappa w_\kappa$, with two coefficients c_1 and c_2, which have to be determined such that $\varepsilon_{\mu\nu\lambda}\nabla_\nu A_\lambda = v_\mu$. The direct computation gives

$$\varepsilon_{\mu\nu\lambda}\nabla_\nu A_\lambda = \varepsilon_{\mu\nu\lambda}(2\,c_1\,r_\nu + c_2\,\delta_{\nu\lambda}r_\kappa w_\kappa + c_2\,r_\lambda\delta_{\nu\kappa}w_\kappa) = (-2\,c_1 + c_2)\,\varepsilon_{\mu\nu\lambda}w_\nu r_\lambda.$$

Thus the yet undetermined coefficients must obey the relation $c_2 - 2c_1 = 1$. Clearly, there is no unique solution. With c_1 as open parameter, the vector potential can be written as

$$A_\lambda = c_1 \left(r^2 \, w_\lambda + 2 \, r_\lambda r_\kappa w_\kappa \right) + r_\lambda r_\kappa w_\kappa.$$

The first term, multiplied by c_1, is the gradient $\nabla_\lambda \varphi(\mathbf{r})$ of the scalar function $\varphi(\mathbf{r}) = r^2 w_\kappa r_\kappa$. Since $\nabla \times \nabla \varphi(\mathbf{r}) = 0$, the term proportional to c_1 does not contribute in the calculation of $\mathbf{v}$ via $\mathbf{v} = \nabla \times \mathbf{A}$. Thus one may chose $c_1 = 0$, or $c_1 = -1/2$. In the latter case, one has

$$A_\mu = -\frac{1}{2} r^2 \, w_\mu. \qquad (7.41)$$

When one requires that the divergence of the vector potential vanishes, i.e. that $\nabla_\mu A_\mu = 0$ holds true, the coefficients c_1 and c_2 are determined uniquely: $c_1 = -2/5$ and $c_2 = 1/5$. Then the vector potential reads

$$A_\mu = -\frac{2}{5} r^2 \, w_\mu + \frac{1}{5} r_\mu r_\kappa w_\kappa. \qquad (7.42)$$

Application in Electrodynamics

One of the Maxwell equations of electrodynamics is $\nabla_\mu B_\mu = 0$. Thus the magnetic flux density $\mathbf{B}$, in general, can be represented by a vector potential according to $B_\mu = \varepsilon_{\mu\nu\lambda} \nabla_\nu A_\lambda$. For the special case of a constant $\mathbf{B}$-field, $A_\lambda = (1/2)\varepsilon_{\lambda\sigma\tau} B_\sigma r_\tau$ is the pertaining vector potential.

7.3.4 Vorticity Free and Divergence Free Vector Fields, Laplace Fields

When a vector field $\mathbf{v}$ is vorticity free, $\nabla \times \mathbf{v} = 0$, there exists a potential function Φ. When furthermore, the vector field is divergence free, or "source free", $\nabla \cdot \mathbf{v} = 0$, the potential obeys the *Laplace equation*

$$\nabla \cdot \nabla \Phi \equiv \nabla_\nu \nabla_\nu \Phi = \Delta \Phi = 0. \qquad (7.43)$$

Scalar fields of this kind are called *Laplace fields*.

For the examples of vector fields with potentials, listed in Sect. 7.2.1, one finds $\Delta \Phi = 0$ for the cases (i), (iv) and (v), pertaining to the homogeneous field, the uniaxial and planar biaxial squeeze-stretch fields. The Laplace operator applied on Φ yields nonzero constant values for the cases (ii) and (iii).

Application in Electrostatics

For static electric fields $\mathbf{E}$, the curl vanishes: $\nabla \times \mathbf{E} = 0$. The $\mathbf{E}$-field is the negative gradient of electrostatic potential $\phi(\mathbf{r})$, which obeys the Laplace equation:

$$E_\mu = -\nabla_\mu \phi \Rightarrow \Delta \phi = 0. \tag{7.44}$$

Usually, in applications to specific problems, the electrostatic potential has to obey certain boundary conditions, e.g. ϕ must be constant on an electrically conducting (metal) surface.

7.3.5 Conventional Classification of Vector Fields

Based on the previous discussions, vector fields can be classified, as shown in the Table 7.1.

This conventional classification scheme does not include the information on the symmetric traceless part $\overline{\nabla \mathbf{v}}$ of the gradient of the vector field. There are, however, many applications in physics, where $\overline{\nabla \mathbf{v}}$ matters. Examples shall be presented later.

7.3.6 Second Spatial Derivatives of Spherically Symmetric Scalar Fields

From the examples presented in Sect. 7.2.2, on might assume, that $\overline{\nabla \mathbf{v}}$ is nonzero only when the pertaining potential function involves special directions, e.g. specified by constant unit vectors in the examples (ii), the 2D version of (iii), in (iv) and (v). However, also the general *spherical potential*, which depends on the position vector $\mathbf{r}$ only via its magnitude $r = \sqrt{r_\kappa r_\kappa}$, leads to non-vanishing values for $\overline{\nabla \mathbf{v}}$. Of course, as discussed before, the antisymmetric part of $\nabla \mathbf{v}$, associated with the vorticity, is zero when a scalar potential exists.

Table 7.1 The conventional classification of vector fields

	$\nabla \times \mathbf{v} = 0$	$\nabla \times \mathbf{v} \neq 0$
$\nabla \cdot \mathbf{v} = 0$	vorticity and source free Laplace field $\mathbf{v} = \nabla \Phi, \Delta \Phi = 0$	source free vorticity field with vector potential $\mathbf{A}$ $\mathbf{v} = \nabla \times \mathbf{A}$
$\nabla \cdot \mathbf{v} \neq 0$	vorticity free Poisson field with source density ρ $\mathbf{v} = \nabla \Phi, \Delta \Phi = \rho$	general vector field $\mathbf{v} = \nabla \Phi + \nabla \times \mathbf{A}$

Let v_μ be given by $v_\mu = \nabla_\mu \Phi$, with $\Phi = \Phi(r)$. Then

$$v_\mu = \frac{d\Phi}{dr} \nabla_\mu r = \frac{d\Phi}{dr} r^{-1} r_\mu = r^{-1} \Phi' r_\mu,$$

cf. (7.13). The prime indicates the derivative with respect to r. Nabla applied to the vector field yields

$$\nabla_\nu v_\mu = \nabla_\nu \nabla_\mu \Phi(r) = r_\mu \nabla_\nu (r^{-1} \Phi') + r^{-1} \Phi' \delta_{\mu\nu} = r^{-1} (r^{-1} \Phi')' r_\nu r_\mu + r^{-1} \Phi' \delta_{\mu\nu}.$$
(7.45)

The special case $\Phi = (1/2)r^2$, treated previously, yields $\nabla_\nu \nabla_\mu \Phi = \delta_{\mu\nu}$, and consequently, $\overline{\nabla_\nu v_\mu} = 0$. On the other hand, for $\Phi = r^{-1}$, one finds

$$\nabla_\nu \nabla_\mu r^{-1} = 3 r^{-5} r_\nu r_\mu - r^{-3} \delta_{\mu\nu} = 3 r^{-3} \overline{\widehat{r_\mu \widehat{r_\nu}}},$$
(7.46)

with the unit vector $\widehat{\mathbf{r}} = r^{-1} \mathbf{r}$. In this case, $\nabla \mathbf{v}$ is "automatically" symmetric traceless. In general, $\nabla_\nu v_\mu = \nabla_\nu \nabla_\mu \Phi(r)$ has an isotropic part, proportional to $\delta_{\mu\nu}$ and a symmetric traceless part, proportional to $\overline{\widehat{r_\mu \widehat{r_\nu}}}$. Relation (7.45) implies

$$\overline{\nabla_\nu v_\mu} = \overline{\nabla_\nu \nabla_\mu} \Phi(r) = r(r^{-1} \Phi')' \overline{\widehat{r_\mu \widehat{r_\nu}}}.$$
(7.47)

On the other hand, setting $\mu = \nu$ in (7.45), one finds

$$\nabla_\mu \nabla_\mu \Phi(r) = \Delta \Phi(r) = r(r^{-1} \Phi')' + 3 r^{-1} \Phi' = \Phi'' + 2 r^{-1} \Phi',$$
(7.48)

for the Laplace operator applied to a function, which depends on $\mathbf{r}$ via $r = |\mathbf{r}|$ only.

7.4 Tensor Fields

7.4.1 Graphical Representations of Symmetric Second Rank Tensor Fields

Examples for second rank tensor fields are the pressure or the stress tensor, as well as the *alignment tensor* of molecular fluids or liquid crystals, which describes the the orientation of molecules or of non-spherical particles. Sometimes, it is desirable to have a graphical representation of such a tensor field. In the case of a tensor with uniaxial symmetry, this can be accomplished by displaying the direction of the principal axis of the tensor, which is associated with the tensor's symmetry axis, i.e. with its largest or its smallest principal value. Such a representation then looks like that of a vector field, but here the directions are indicated by lines without arrowheads.

Fig. 7.7 Uniaxial tensor
fields for the director in a
nematic liquid crystal

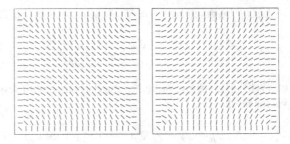

Fig. 7.8 Bricks indicating
the alignment tensor of a
nematic liquid crystal

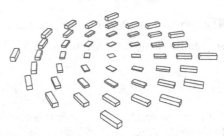

Examples are shown in Fig. 7.7, adapted from [23]. The 'defect' structure in the lower
left corner of the right figure is typical for a nematic substance, cf. Chap. 15.

In the general biaxial case, ellipsoids associated the tensor, as discussed in
Sect. 5.4, could be used to to visualize a second rank tensor field. However, "bricks"
with their sides proportional to the principal semi-axes can more easily convey the
information about the different principal values of the tensor at different space points.
As an example, the alignment tensor in the vicinity of a "defect" in a nematic liquid
crystal is shown in Fig. 7.8, adapted from [24, 83].

7.4.2 Spatial Derivatives of Tensor Fields

Let $T_{\mu\nu}(\mathbf{r})$ be a tensor field. Application of the nabla operator ∇_λ yields the third
rank tensor $\nabla_\lambda T_{\mu\nu}$. By analogy to (7.31), the tensor of rank three can be decomposed
into parts associated with a vector, with a second rank tensor, and with the pertaining
irreducible symmetric traceless third rank tensor. The first one of these parts involves
the *tensor divergence*

$$\nabla_\lambda T_{\lambda\nu},$$

which is a vector. Notice that this expression has to be distinguished from $\nabla_\lambda T_{\nu\lambda}$,
when one has $T_{\mu\nu} \neq T_{\nu\mu}$.

An application of the tensor divergence used for the pressure tensor, occurs in the
local conservation equation for the linear momentum.

7.4.3 Local Mass and Momentum Conservation, Pressure Tensor

Let $\rho = \rho(\mathbf{r})$ and $\mathbf{v} = \mathbf{v}(\mathbf{r})$ be the mass density and the local velocity field of a fluid. The *conservation of mass* implies the *continuity equation*

$$\frac{\partial \rho}{\partial t} + \nabla_\nu (\rho\, v_\nu) = 0. \tag{7.49}$$

With the help of the substantial time derivative

$$\frac{d}{dt} := \frac{\partial}{\partial t} + v_\nu \nabla_\nu, \tag{7.50}$$

the continuity equation is equivalent to

$$\frac{d\rho}{dt} + \rho\, \nabla_\nu v_\nu = 0. \tag{7.51}$$

The local conservation equation for the linear momentum density ρv_μ, in the absence of external forces, can be cast into the form

$$\rho\, \frac{dv_\mu}{dt} + \nabla_\nu p_{\nu\mu} = 0. \tag{7.52}$$

Here $p_{\nu\mu}$ is the pressure tensor. It characterizes the transport of momentum, which is not of convective type. The convective transport is described by the term $\rho v_\nu \nabla_\nu v_\mu$, which occurs in connection with the substantial derivative. The gradient $\nabla_\nu p_{\nu\mu} = k_\mu$ describes an internal force density.

In thermal equilibrium, the pressure tensor of a fluid reduces to the isotropic tensor $P\delta_{\mu\nu}$, where P is the hydrostatic pressure. In general, the pressure tensor can be decomposed into its isotropic, its symmetric traceless and its antisymmetric parts, cf. Chap. 6. Thus

$$p_{\nu\mu} = (P + \tilde{p})\, \delta_{\mu\nu} + \overline{p_{\nu\mu}} + \frac{1}{2}\varepsilon_{\nu\mu\lambda} p_\lambda. \tag{7.53}$$

In thermal equilibrium, the part $\tilde{p}$ of the scalar pressure is zero, just as $\overline{p_{\nu\mu}}$ and the axial vector $p_\lambda = \varepsilon_{\lambda\alpha\beta} p_{\alpha\beta}$ which is associated with the antisymmetric part of the pressure tensor.

The time change of the orbital angular momentum $\ell_\lambda = \varepsilon_{\lambda\kappa\mu} r_\kappa v_\mu$ can be inferred from the momentum conservation equation. More specifically, multiplication of (7.52) by $\varepsilon_{\lambda\kappa\mu} r_\kappa$ and use of $r_\kappa \nabla_\nu p_{\nu\mu} = \nabla_\nu (r_\kappa p_{\nu\mu}) - p_{\nu\mu} \nabla_\nu r_\kappa$ leads to

$$\rho\, \frac{d\ell_\lambda}{dt} + \varepsilon_{\lambda\kappa\mu} \nabla_\nu (r_\kappa p_{\nu\mu}) = \varepsilon_{\lambda\nu\mu} p_{\nu\mu}. \tag{7.54}$$

A fluid composed of particles with an internal rotational degree of freedom, in general, also possesses an internal angular momentum or spin density. The balance equation for this quantity contains the axial vector $p_\lambda = \varepsilon_{\lambda\nu\mu} p_{\nu\mu}$ in such a way, that it cancels in the sum of the equations of change for the orbital and the internal angular momenta. This is due to the conservation of the total angular momentum. As a consequence, the antisymmetric part of the pressure tensor is identical to zero for fluids composed of particles which have no rotational degree of freedom, like gaseous or liquid Argon. In the hydrodynamic description of flow processes in molecular fluids, and this includes water, the antisymmetric part of the pressure tensor relaxes to zero on a time scale fast compared with typical hydrodynamical time changes, such that the pressure tensor can be treated as being symmetric. Then constitutive laws are needed for $\tilde{p}$ and $\overline{p_{\nu\mu}}$ only. In hydrodynamics, the relations

$$\tilde{p} = -\eta_V \nabla_\lambda v_\lambda, \quad \overline{p_{\nu\mu}} = -2\eta \overline{\nabla_\nu v_\mu}, \tag{7.55}$$

are used. The non-negative coefficients η and η_V are the *shear viscosity* and the *volume viscosity*, respectively. A justification of these constitutive laws and generalizations thereof are treated in Sects. 16.3, 16.4, 17.3, and 17.4. Insertion of (7.55) into the momentum balance yields a closed equation for the flow velocity $\mathbf{v}$. For $\eta_V = 0$, this corresponds to the *Navier-Stokes equations*.

7.5 Maxwell Equations in Differential Form

7.5.1 Four-Field Formulation

The full Maxwell equations, in differential form, and in the conventional *four-field formulation*, are

$$\nabla_\mu D_\mu = \rho, \quad \varepsilon_{\mu\nu\lambda} \nabla_\nu H_\lambda = j_\mu + \frac{\partial}{\partial t} D_\mu, \tag{7.56}$$

and

$$\varepsilon_{\mu\nu\lambda} \nabla_\nu E_\lambda = -\frac{\partial}{\partial t} B_\mu, \quad \nabla_\mu B_\mu = 0. \tag{7.57}$$

The first pair of equations, referred to as the *inhomogeneous Maxwell equations*, involve the density ρ of electric charges and the electric current density $\mathbf{j}$. The second pair are the *homogeneous Maxwell equations*. Here $\mathbf{E}$ is the electric field, $\mathbf{D}$ is the electric displacement field. Frequently, both $\mathbf{H}$ and the *magnetic induction* $\mathbf{B}$ are called *magnetic field*. For charges and currents in vacuum, one has $\mathbf{D} = \varepsilon_0 \mathbf{E}$ and $\mathbf{B} = \mu_0 \mathbf{H}$, where ε_0 and μ_0 are the *dielectric permeability* and the *magnetic susceptibility* of the vacuum. When all charges and currents are represented by ρ and $\mathbf{j}$, the two fields $\mathbf{E}$ and $\mathbf{B}$ would suffice for electrodynamics, and one could use $\varepsilon_0 = 1, \mu_0 = 1,$

as in the Gaussian *cgs*-system of physical units. Such a description is inconvenient for electrodynamics applied to macroscopic matter. There, ρ and $\mathbf{j}$ stand for the density of *free charges* and *free currents*, whereas *bound charges*, *internal currents* and *magnetic moments* associated with the spin of particles are incorporated into the *electric polarization* $\mathbf{P}$ and the *magnetization* $\mathbf{M}$, respectively. These quantities occur in the relations

$$D_\mu = \varepsilon_0 \, E_\mu + P_\mu, \quad B_\mu = \mu_0 \, (H_\mu + M_\mu). \tag{7.58}$$

Vacuum corresponds to $\mathbf{P} = 0$ and $\mathbf{M} = 0$. In matter, constitutive relations are needed for $\mathbf{P}$ and $\mathbf{M}$ in order to obtain a closed set of equations. These constitutive relations are specific for the materials considered.

Remarks on Parity and Time Reversal The $\mathbf{E}$- and $\mathbf{D}$-fields are polar vectors, $\mathbf{B}$ and $\mathbf{H}$ are axial vectors. Since ρ is a true scalar, and ∇ as well as $\mathbf{j}$ are polar vectors, the Maxwell equations (7.56) and (7.57) conserve parity. The constitutive relations for $\mathbf{P}$ and $\mathbf{M}$, however, have to obey certain restrictions, when parity conservation should not be violated.

Furthermore, ρ and the fields $\mathbf{E}$, as well as $\mathbf{D}$, do not change sign under the time reversal operation, whereas $\mathbf{j}$ and the fields $\mathbf{B}$, as well as $\mathbf{H}$, do change sign. Thus the Maxwell equations (7.56) and (7.57) are invariant under time reversal. Time reversal invariance, however, can be broken by constitutive relations. An example is *Ohm's law*, in differential form, $\mathbf{j} = \sigma \mathbf{E}$. The non-negative coefficient σ is the electrical conductivity.

Parity (P) and time reversal (T) invariance hold true for charges, currents and fields in vacuum, where the relations $\mathbf{D} = \varepsilon_0 \mathbf{E}$ and $\mathbf{B} = \mu_0 \mathbf{H}$ apply. Parity conservation implies: when

$$\mathbf{E}(t, \mathbf{r}), \mathbf{D}(t, \mathbf{r}), \mathbf{B}(t, \mathbf{r}), \mathbf{H}(t, \mathbf{r})$$

are solutions of the Maxwell equations for given charge density and current density

$$\rho(t, \mathbf{r}), \mathbf{j}(t, \mathbf{r}),$$

then

$$-\mathbf{E}(t, -\mathbf{r}), -\mathbf{D}(t, -\mathbf{r}), \mathbf{B}(t, -\mathbf{r}), \mathbf{H}(t, -\mathbf{r})$$

are solutions for given

$$\rho(t, -\mathbf{r}), -\mathbf{j}(t, -\mathbf{r}).$$

Similarly, from T-invariance follows:

$$\mathbf{E}(-t, \mathbf{r}), \mathbf{D}(-t, \mathbf{r}), -\mathbf{B}(-t, \mathbf{r}), -\mathbf{H}(-t, \mathbf{r})$$

are solutions for

$$\rho(-t, \mathbf{r}), \quad -\mathbf{j}(-t, \mathbf{r}).$$

As a consequence of the combined PT-invariance

$$-\mathbf{E}(-t, -\mathbf{r}), -\mathbf{D}(-t, -\mathbf{r}), -\mathbf{B}(-t, -\mathbf{r}), -\mathbf{H}(-t, -\mathbf{r})$$

are solutions for

$$\rho(-t, -\mathbf{r}), \quad \mathbf{j}(-t, -\mathbf{r}).$$

7.5.2 Special Cases

Application of ∇_μ to the second equation of (7.56) and use of the first of these equations and of $\nabla_\mu \varepsilon_{\mu\nu\lambda} \nabla_\nu H_\lambda = 0$ yields the continuity equation for the time change of the charge density:

$$\frac{\partial}{\partial t} \rho + \nabla_\mu j_\mu = 0. \tag{7.59}$$

Without Maxwell's current density $\frac{\partial}{\partial t} D_\mu$ in the second inhomogeneous equation of (7.56), one would just have $\nabla_\mu j_\mu = 0$, which is true for stationary processes, but not in general. More important, the existence of electromagnetic waves hinges on the term $\frac{\partial}{\partial t} D_\mu$ in (7.56), see the next section.

Special cases, for a stationary situation, where the time derivatives vanish in the Maxwell equations, are the equations of *electrostatics, magnetostatics*, and the equations determining the magnetic field caused a steady current. For electrostatics, one has

$$\nabla_\mu D_\mu = \rho, \quad \varepsilon_{\mu\nu\lambda} \nabla_\nu E_\lambda = 0.$$

The first of these equations is referred to as the *Gauss law*.

The equations for magnetostatics, applicable to fields of permanent magnets, are mathematically equivalent to those of electrostatics with $\mathbf{D}$ and $\mathbf{E}$ replaced by $\mathbf{B}$ and $\mathbf{H}$, respectively, and $\rho = 0$, because there are no magnetic monopoles and consequently there is no magnetic charge density. In the equations ruling electrostatics and magnetostatics, there is no coupling between electric and magnetic fields, unless the constitutive relations for the electric polarization and the magnetization contain such terms.

The equations for the magnetic field caused by a stationary electric current are associated with the names *Oersted* and *Ampère*, who discovered and studied an effect which reveals a coupling between electric and magnetic phenomena. The relevant equations, the first one of which is referred to as *Oersted law*, are

$$\varepsilon_{\mu\nu\lambda}\,\nabla_\nu H_\lambda = j_\mu, \quad \nabla_\mu B_\mu = 0.$$

Applications of the stationary equations are given in Sects. 8.2.7 and 8.3.5.
The first of the homogeneous Maxwell equations (7.57), viz.

$$\varepsilon_{\mu\nu\lambda}\,\nabla_\nu E_\lambda = -\frac{\partial}{\partial t}B_\mu,$$

is referred to as *Faraday law*. It underlies the coupling between electric and magnetic fields discovered by *Faraday*: a time-dependent **B**-field induces an electric field **E**. For the application of this differential equation to the *Faraday induction* see Sect. 8.2.8.

7.5.3 Electromagnetic Waves in Vacuum

In vacuum, where $\mathbf{D} = \varepsilon_0\mathbf{E}$ and $\mathbf{B} = \mu_0\mathbf{H}$, and for $\rho = 0$, $j_\mu = 0$, application of $\varepsilon_{\alpha\beta\mu}\nabla_\beta$ on the first equation of (7.57), use of (4.10) for the double cross product, and of $\nabla_\nu E_\nu = 0$, yields $-\Delta E_\alpha = -\mu_0\frac{\partial}{\partial t}\varepsilon_{\alpha\beta\mu}H_\mu$. The second equation of (7.56) links the curl of the **H** field with the time derivative of $\varepsilon_0\mathbf{E}$. This then leads to the wave equation

$$\Box\mathbf{E} \equiv \Delta\mathbf{E} - \frac{1}{c^2}\frac{\partial^2}{\partial t^2}\mathbf{E} = 0, \tag{7.60}$$

with the speed of light, in vacuum, determined by

$$c^2 = (\varepsilon_0\,\mu_0)^{-1}. \tag{7.61}$$

The magnetic field **H** obeys the same type of wave equation. The symbol

$$\Box \equiv \Delta - \frac{1}{c^2}\frac{\partial^2}{\partial t^2} \tag{7.62}$$

is the *d'Alembert operator*. A solution of (7.60) is

$$E_\mu = E_\mu^{(0)}\,f(\xi), \quad \xi = \widehat{k_\nu}\,r_\nu - c\,t, \tag{7.63}$$

where $E_\mu^{(0)}$ is a constant vector characterizing the polarization of the field, $\widehat{k_\nu}$ is a unit vector parallel to the wave vector, pointing in the direction of propagation of the radiation, and f is any function which can be differentiated twice. Notice that $\widehat{k_\nu}E_\nu^{(0)} = 0$, i.e. the electromagnetic radiation, in vacuum, is a *transverse wave*.

The plane wave is a special case of (7.63). In complex notation, this solution of the wave equation reads

$$E_\mu = E_\mu^{(0)} \exp[i\, k_\nu\, r_\nu - i\,\omega\, t]. \qquad (7.64)$$

The wave vector k_ν and the circular frequency ω are linked by the *dispersion relation* $k_\nu k_\nu = \omega^2/c^2$ or

$$\omega = k\, c, \qquad (7.65)$$

where k is the magnitude of the wave vector.

7.2 Exercise: Test Solutions of the Wave Equation

Proof that both the ansatz (7.63) and the plane wave (7.64) obey the wave equation. Furthermore, show that the **E**-field is perpendicular to the wave vector, and that the **B**-field is perpendicular to both.

7.5.4 Scalar and Vector Potentials

The electric field **E** and the **B**-field can be expressed as derivatives of the electroscalar potential ϕ and a magnetic vector potential **A** according to

$$E_\mu = -\nabla_\mu \phi - \frac{\partial}{\partial t} A_\mu, \quad B_\mu = \varepsilon_{\mu\nu\lambda} \nabla_\nu A_\lambda. \qquad (7.66)$$

With the ansatz (7.66), the homogeneous Maxwell equations (7.57) are fulfilled automatically.

The electromagnetic potential functions, however, are not unique. More specifically, the same fields **E** and **B** follow from (7.66), when ϕ and **A** are replaced by

$$\phi' = \phi - \frac{\partial}{\partial t} f, \quad A_\lambda' = A_\lambda + \nabla_\lambda f,$$

where $f = f(t, \mathbf{r})$ is a scalar function. This allows, e.g. to require $\phi = 0$ or $\nabla_\nu A_\nu = 0$.

For charges and currents in vacuum, where $\mathbf{D} = \varepsilon_0 \mathbf{E}$ and $\mathbf{B} = \mu_0 \mathbf{H}$, insertion of (7.66) into the inhomogeneous Maxwell equations (7.56) and use of the scaling

$$\frac{\partial}{\partial t} \phi + \nabla_\lambda A_\lambda = 0, \qquad (7.67)$$

leads to

$$\Box A_\nu = \Delta A_\nu - \frac{1}{c^2} \frac{\partial^2}{\partial t^2} A_\nu = \mu_0\, j_\nu, \quad \Box \phi = \Delta \phi - \frac{1}{c^2} \frac{\partial^2}{\partial t^2} \phi = \frac{1}{\varepsilon_0} \rho. \qquad (7.68)$$

The current density and the charge density are the sources for the vector potential and the scalar potential. Electromagnetic radiation is caused by accelerated charges, then the time derivatives in (7.68) are essential. For a stationary situation, on the other hand, the time derivatives vanish and the second of these equations reduces to the Poisson equation, cf. Sect. 7.3.2. The first of the equations (7.68), linking the vector potential with the current density is mathematically equivalent to the Poisson equation, when a stationary situation is considered. It determines the magnetic field generated by a steady current, as formulated in the *Biot-Savart relation*.

7.5.5 Magnetic Field Tensors

In 3D, there exists a dual relation between an antisymmetric second rank tensor and a vector, cf. Sects. 3.3 and 4.1.3. This allows to replace the magnetic field vectors **B** and **H** by antisymmetric tensors. To see what this means, consider the homogeneous Maxwell equation $\varepsilon_{\lambda\sigma\tau}\nabla_\sigma E_\tau = -\frac{\partial}{\partial t}B_\lambda$. Multiplication of this equation by $\varepsilon_{\mu\nu\lambda}$ and use of $\varepsilon_{\mu\nu\lambda}\varepsilon_{\lambda\sigma\tau} = \delta_{\mu\sigma}\delta_{\nu\tau} - \delta_{\mu\tau}\delta_{\nu\sigma}$ yields

$$\nabla_\mu E_\nu - \nabla_\nu E_\mu = -\frac{\partial}{\partial t}\varepsilon_{\mu\nu\lambda}B_\lambda.$$

Both sides of this equation, which is equivalent to the first of the Maxwell equations (7.57), are antisymmetric tensors. The right hand side can be expressed in terms of the magnetic field tensor $B_{\mu\nu}$, which is related to the vector field B_λ, by

$$B_{\mu\nu} \equiv \varepsilon_{\mu\nu\lambda}B_\lambda. \tag{7.69}$$

In matrix notation, this relation is equivalent to

$$B_{\mu\nu} := \begin{pmatrix} 0 & B_3 & -B_2 \\ -B_3 & 0 & B_1 \\ B_2 & -B_1 & 0 \end{pmatrix}. \tag{7.70}$$

The reciprocal relation between the vector and the antisymmetric tensor is

$$B_\lambda = \frac{1}{2}\varepsilon_{\lambda\mu\nu}B_{\mu\nu}. \tag{7.71}$$

The field tensor is linked with the vector potential **A** via

$$B_{\mu\nu} = \nabla_\mu A_\nu - \nabla_\nu A_\mu. \tag{7.72}$$

The Lorentz Force **F**, viz.

$$F_\mu = e\,E_\mu + e\,\varepsilon_{\mu\lambda\nu}B_\lambda v_\nu,$$

acting on a particle with charge e, moving with velocity $\mathbf{v}$, in the presence of an electric field $\mathbf{E}$ and a magnetic field $\mathbf{B}$, cf. Sect. 3.4.6, is equivalent to

$$F_\mu = e\,E_\mu + e\,v_\nu\,B_{\nu\mu}. \tag{7.73}$$

The antisymmetric field tensor $H_{\mu\nu}$ pertaining to the magnetic field $\mathbf{H}$ is defined by analogy to (7.69). Using the magnetic field tensors, the Maxwell equations read

$$\nabla_\mu D_\mu = \rho, \quad -\nabla_\nu H_{\nu\mu} = j_\mu + \frac{\partial}{\partial t} D_\mu, \tag{7.74}$$

and

$$\nabla_\mu E_\nu - \nabla_\nu E_\mu = -\frac{\partial}{\partial t} B_{\mu\nu}, \quad \nabla_1 B_{23} + \nabla_2 B_{31} + \nabla_3 B_{12} = 0. \tag{7.75}$$

Notice that the last equation stems from

$$\varepsilon_{\lambda\mu\nu} \nabla_\lambda\,B_{\mu\nu} = 0,$$

and $B_{\nu\mu} = -B_{\mu\nu}$ was used.

Why should one bother to look at the alternative version of the Maxwell equations, rather than stick to the vector equations (7.56) and (7.57)? There are two answers to this question.

First, in the 4D formulation of electrodynamics, which reflects the Lorentz invariance of the Maxwell equations, the 3×3 field tensor (7.70) is enlarged to a the 4×4 field tensor, which also comprises the 3 components of the electric field. In 4D, an antisymmetric second rank tensor has 6 independent components, just like the two vectors $\mathbf{B}$ and $\mathbf{E}$ combined. The 3D tensorial notation of the Maxwell equations makes it easier to see their connection with the 4D version, discussed in Chap. 18. Notice, the non-euklidian metric of special relativity is used for that 4D-space.

A second, more mathematical reason is: the first three of the equations (7.74) and (7.75) can be adapted to any D-dimensional space with Euklidian metric, and $D \geq 2$. Thus it is possible to answer the question: do electromagnetic waves exist for 2D?

7.3 Exercise: Electromagnetic Waves in Flatland?

In flatland, one has just 2 dimensions. Cartesian components are denoted by Latin letters $i, j, \ldots; i = 1, 2; j = 1, 2$. The summation convention is used. In vacuum, and for zero charges and currents, the adapted Maxwell equations are

$$\nabla_i E_i = 0, \quad -\nabla_i H_{ij} = \varepsilon_0 \frac{\partial}{\partial t} E_j, \quad \nabla_i E_j - \nabla_j E_i = -\mu_0 \frac{\partial}{\partial t} H_{ij}.$$

Derive a wave equation for the electric field to proof that one can have electromagnetic waves in 2D. How about 1D?

7.6 Rules for Nabla and Laplace Operators

7.6.1 Nabla

The application of the nabla operator ∇_μ to a field, which is a tensor of rank ℓ, yields a tensor of rank $\ell + 1$. The appropriate decomposition shall be discussed latter. Here the obvious product and chain rules are listed, which have already been used above.

Let f and g be components of tensors, which depend on $\mathbf{r}$. Then one has

$$\nabla_\mu(f\,g) = g\,\nabla_\mu f + f\,\nabla_\mu g. \tag{7.76}$$

Now, let f be the component of a tensor, which is a function of the scalar g, which in turn, depends on $\mathbf{r}$. Then the chain rule applies:

$$\nabla_\mu(f(g)) = \frac{\partial f}{\partial g}\,\nabla_\mu g. \tag{7.77}$$

The position vector $\mathbf{r}$ is equal to the product of its magnitude r and of the unit vector $\hat{\mathbf{r}}$, viz.: $r_\mu = r\hat{r}_\mu$, with $\hat{r}_\mu = r^{-1}r_\mu$. In some applications, it may be convenient and useful to decompose the spatial derivative into differentiations with respect to r and with respect to $\hat{r}_\mu$. This is accomplished by observing

$$\nabla_\mu \equiv \frac{\partial}{\partial r_\mu} = \frac{\partial r}{\partial r_\mu}\frac{\partial}{\partial r} + \frac{\partial \hat{r}_\nu}{\partial r_\mu}\frac{\partial}{\partial \hat{r}_\nu}.$$

Due to

$$\frac{\partial r}{\partial r_\mu} = \hat{r}_\mu, \quad \frac{\partial \hat{r}_\nu}{\partial r_\mu} = r^{-1}(\delta_{\mu\nu} - \hat{r}_\mu\hat{r}_\nu),$$

the radial and the angular parts of the spatial derivative are

$$\nabla_\mu = \hat{r}_\mu\frac{\partial}{\partial r} + r^{-1}(\delta_{\mu\nu} - \hat{r}_\mu\hat{r}_\nu)\frac{\partial}{\partial \hat{r}_\nu}. \tag{7.78}$$

Multiplication of (7.78) by r_μ leads to

$$r_\mu\nabla_\mu = r\frac{\partial}{\partial r}. \tag{7.79}$$

With the help of the anti-hermitian operator

$$\mathscr{L}_\mu = \varepsilon_{\mu\nu\lambda}r_\nu\nabla_\lambda = \varepsilon_{\mu\nu\lambda}\hat{r}_\nu\frac{\partial}{\partial \hat{r}_\lambda}, \tag{7.80}$$

the equation (7.78) can be written as

$$\nabla_\mu = \widehat{r}_\mu \frac{\partial}{\partial r} - r^{-1} \varepsilon_{\mu\nu\lambda} \widehat{r}_\nu \mathscr{L}_\lambda. \tag{7.81}$$

Notice: the differential operator $\mathscr{L} = \mathbf{r} \times \nabla$ only acts on the angular part of a function depending on the vector $\mathbf{r}$. In particular, one has $\mathscr{L} f(r) = 0$ when f is only a function of $r = |\mathbf{r}|$.

The Cartesian components of the differential operator do not commute. More specifically, one finds the commutation relation

$$\mathscr{L}_\mu \mathscr{L}_\nu - \mathscr{L}_\nu \mathscr{L}_\mu = -\varepsilon_{\mu\nu\lambda} \mathscr{L}_\lambda, \tag{7.82}$$

or equivalently,

$$\varepsilon_{\lambda\mu\nu} \mathscr{L}_\mu \mathscr{L}_\nu = \mathscr{L}_\lambda. \tag{7.83}$$

The differential operator $\mathscr{L}$ is closely related to the quantum mechanical angular momentum operator, cf. Sect. 7.6.2.

7.4 Exercise: Radial and Angular Parts of the Nabla Operator, Compare Equation (7.81) with (7.78)

7.5 Exercise: Prove the Relations (7.82) and (7.83) for the Angular Nabla Operator

Hint: use (4.10) and observe that the names of summation indices can be changed conveniently, as long as none appears more than twice.

7.6.2 Application: Orbital Angular Momentum Operator

The quantum mechanical angular momentum operator $\mathbf{L}^{op}$, in spatial representation, is given by

$$L_\mu^{op} = \frac{\hbar}{i} \mathscr{L}_\mu = \frac{\hbar}{i} \varepsilon_{\mu\nu\lambda} r_\nu \nabla_\lambda. \tag{7.84}$$

Here $\hbar$ is the Planck constant h, divided by 2π, and i is the imaginary unit, with the property $i^2 = -1$. The expression (7.84) follows the definition $\mathbf{r} \times \mathbf{p}$ for the orbital angular momentum, cf. Sect. 3.4.1, when the linear momentum $\mathbf{p}$ is replaced by the operator

$$\mathbf{p}^{op} = \frac{\hbar}{i} \nabla, \tag{7.85}$$

in spatial representation.

Side Remark:
Where Does the Expression for the Linear Momentum Operator Come From?

A plane wave ψ with the wave vector $\mathbf{k}$ and the (circular) frequency ω is proportional to $\exp(i\mathbf{k}\cdot\mathbf{r} - i\omega t)$. Application of the nabla operator yields $\nabla\psi = i\mathbf{k}\psi$, which is just a mathematical identity. By analogy to the Einstein relation $E = \hbar\omega$ between the energy E and the frequency, de Broglie suggested that the linear momentum $\mathbf{p}$ of a particle should be associated with a wave vector according to $\mathbf{p} = \hbar\mathbf{k}$. Schrödinger took up this idea and invented *wave mechanics*. Later it became clear that the wave function ψ is a probability amplitude, its absolute square characterizes the probability to find a particle in a volume element at a specific position $\mathbf{r}$. For a free particle with linear momentum $\mathbf{p}$, the ψ-function is proportional to $\exp(i\mathbf{p}\cdot\mathbf{r}/\hbar)$, hence $\nabla\psi = \frac{i}{\hbar}\mathbf{p}\psi$. This corresponds to (7.85). The expression for the linear momentum operator derived for the special case of a plane wave holds true in general, in spatial representation.

Dimensionless Angular Momentum Operator

It is convenient to introduce angular momentum operators in units of $\hbar$ and to denote them by the same symbol $\mathbf{L}$ as the usual angular momentum, as long as no danger of confusion exists. Then one has

$$L_\mu = \frac{1}{i}\,\mathscr{L}_\mu = -i\,\varepsilon_{\mu\nu\lambda}\,r_\nu\,\nabla_\lambda. \tag{7.86}$$

Thanks to the imaginary unit i which is introduced in the definitions (7.84) and (7.86), the angular momentum operator is a *hermitian operator* with real eigenvalues.

The commutation relations (7.82) for the components of the differential operator $\mathscr{L}$ now lead to the angular momentum commutation relations

$$L_\mu L_\nu - L_\nu L_\mu = i\,\varepsilon_{\mu\nu\lambda}\,L_\lambda. \tag{7.87}$$

Similarly, relation (7.83) implies

$$\varepsilon_{\lambda\mu\nu}\,L_\mu L_\nu = iL_\lambda. \tag{7.88}$$

The commutation relations for the orbital angular momentum hold true in general, not only in the space representation which was used here to derive them.

Notice that (7.88) is equivalent to

$$\mathbf{L}\times\mathbf{L} = i\,\mathbf{L}. \tag{7.89}$$

This reflects the fact that the components of the quantum mechanical angular momentum do not commute, in contradistinction to the components of the classical angular momentum for which the corresponding cross product vanishes.

7.6.3 Radial and Angular Parts of the Laplace Operator

The Laplace operator can be decomposed into a radial and angular parts, by analogy to the decomposition (7.81) of the nabla operator. With the help of the angular differential operator $\mathscr{L}$, one has

$$\Delta = \Delta_{\mathrm{r}} + r^{-2} \mathscr{L}_\mu \mathscr{L}_\mu. \tag{7.90}$$

The radial part Δ_{r} is given by

$$\Delta_{\mathrm{r}} = r^{-2} \frac{\partial}{\partial r} \left(r^2 \frac{\partial}{\partial r} \right) = r^{-1} \frac{\partial^2}{\partial r^2} r = \frac{\partial^2}{\partial r^2} + 2 r^{-1} \frac{\partial}{\partial r}. \tag{7.91}$$

To prove the relation (7.90) with (7.91), one can compute $\mathscr{L}_\mu \mathscr{L}_\mu$, starting from the definition (7.80). One finds

$$\mathscr{L}_\mu \mathscr{L}_\mu = \varepsilon_{\mu\nu\lambda}\, \varepsilon_{\mu\alpha\beta}\, r_\nu\, \nabla_\lambda\, r_\alpha\, \nabla_\beta = \varepsilon_{\mu\nu\lambda}\, \varepsilon_{\mu\alpha\beta}\, (r_\nu\, r_\alpha\, \nabla_\lambda\, \nabla_\beta + r_\nu\, \delta_{\lambda\alpha}\, \nabla_\beta).$$

Now use of $\varepsilon_{\mu\nu\lambda}\varepsilon_{\mu\alpha\beta} = \delta_{\nu\alpha}\delta_{\lambda\beta} - \delta_{\nu\beta}\delta_{\lambda\alpha}$, cf. Sect. 4.1.2, leads to $\mathscr{L}_\mu \mathscr{L}_\mu = r^2 \nabla_\lambda \nabla_\lambda - r^2 \frac{\partial^2}{\partial r^2} + r \frac{\partial}{\partial r} - 3r \frac{\partial}{\partial r}$. Thus one obtains

$$\mathscr{L}_\mu \mathscr{L}_\mu = r^2 \left(\Delta - \frac{\partial^2}{\partial r^2} - 2 r^{-1} \frac{\partial}{\partial r} \right).$$

For $r > 0$, this relation is equivalent to (7.90).

Notice that $\Delta r^{-1} = 0$, for $r \neq 0$. This result applies just for 3D, the three-dimensional space we live in. In D dimensions one has $\Delta r^{(2-D)} = 0$, see the next exercise.

7.6 Exercise: Determine the Radial Part of the Laplace Operator in D Dimensions, Prove $\Delta r^{(2-D)} = 0$

Hint: Compute $\nabla_\mu \nabla_\mu f = \nabla_\mu (\nabla_\mu f)$, where the function $f = f(r)$ has no angular dependence, and use $\nabla_\mu r_\mu = D$.

Furthermore, make the ansatz $f = r^n$ and determine for which exponent n the equation $\Delta r^n = 0$ holds true.

7.6.4 Application: Kinetic Energy Operator in Wave Mechanics

The kinetic energy of a particle with mass m and with linear momentum $\mathbf{p}$ is $\mathbf{p} \cdot \mathbf{p}/(2m)$. In Schrödinger's wave mechanics, in spatial representation, the linear momentum corresponds to the operator $\mathbf{p}^{op} = \frac{\hbar}{i}\nabla$, cf. (7.85). Hence the Hamilton operator for the kinetic energy of a single particle is

$$H_{kin}^{op} = -\frac{\hbar^2}{2m}\nabla_\mu\nabla_\mu = -\frac{\hbar^2}{2m}\Delta. \tag{7.92}$$

In accord with the decomposition (7.90) of the Laplace operator into a radial part and an azimuthal or angular part involving the $\mathscr{L}$ operator, the kinetic energy operator is the sum a radial part and a part containing the dimensionless angular momentum operator $\mathbf{L}$, as defined in (7.86). Thus one has

$$H_{kin}^{op} = -\frac{\hbar^2}{2m}\Delta_r + \frac{\hbar^2}{2m}r^{-2}L_\mu L_\mu. \tag{7.93}$$

For the radial part Δ_r of the Laplace operator see (7.91).

Chapter 8
Integration of Fields

Abstract The integration of fields is treated in this chapter. Firstly, line integrals are considered and the computation of potential functions from vector fields is discussed. Secondly, surface integrals are introduced and the generalized Stokes law is derived. Applications are the magnetic field around an electric wire and the Faraday induction. Thirdly, volume integrals are treated and a generalized Gauss theorem is stated. The moment of inertia tensor is defined and computed for some examples. Applications of volume integrals in electrodynamics comprise the Gauss law and the Coulomb force, the formulation of balance equations for energy, linear and angular momentum and the definition of the Maxwell stress tensor. Further applications concern the continuity equation and the flow through a pipe, the momentum balance and the force on a solid body, the derivation of the Archimedes principle and the computation of the torque on a rotating solid body.

The differentiation of a field provides a local information about the changes of a function caused by small changes of the position considered. Integrals contain a more global information since the behavior of a function over a larger region of space is involved. These regions can be lines, surfaces or volumes, in 3D. All three types of integrals are needed for applications in physics. They are referred to as *line integrals, surface integrals, and volume integrals.*

8.1 Line Integrals

8.1.1 Definition, Parameter Representation

Let $f = f(\mathbf{r})$ be a well defined, smooth function within a region B of the 3D space. Furthermore, let C be a continuous, piecewise smooth curve within the region B with start point $\mathbf{r}_1$ and end point $\mathbf{r}_2$. The *line integral* of $f(\mathbf{r})$ along the curve C is defined by

$$\mathscr{I}_\mu = \int_C f(\mathbf{r}) \mathrm{d}r_\mu. \tag{8.1}$$

© Springer International Publishing Switzerland 2015   111
S. Hess, *Tensors for Physics*, Undergraduate Lecture Notes in Physics,
DOI 10.1007/978-3-319-12787-3_8

Here dr_μ is the Cartesian component of the differential change $d\mathbf{r}$ along the curve C. The line integral is also called *curve integral* or *path integral*.

When the curve is determined by a *parameter representation*

$$r_\mu = r_\mu(p), \quad p_1 < p < p_2,$$

the line integral (8.1) can be expressed as the ordinary Riemann integral

$$\mathscr{I}_\mu = \int_{p_1}^{p_2} f(\mathbf{r}(p)) \frac{dr_\mu}{dp} dp. \tag{8.2}$$

The parameter values p_1 and p_2 correspond to the start and end points of the curve C, i.e. $\mathbf{r}(p_i) = \mathbf{r}_i$, $i = 1, 2$.

Remark: in some applications, it may be convenient to use piecewise different parameter representations for the curve C. A simple example is a curve depicted in the Fig. 8.3.

The sign of a line integral changes, when the integration is performed backwards along the curve considered. This is obvious in the parameter representation since

$$\int_{p_1}^{p_2} \ldots dp = -\int_{p_2}^{p_1} \ldots dp.$$

Notice: the line integral is a vector, provided that f is a scalar. When the function f is the component of a tensor of rank ℓ, e.g. $f \equiv g_{\nu_1 \ldots \nu_\ell}$, the resulting line integral $\mathscr{I}_{\mu\nu_1\ldots\nu_\ell}$ is a component of a tensor of rank $\ell + 1$. Some examples are considered in Sect. 8.1.3.

Remark: in the literature, the term "line integral" is also used for an integral with the scalar integration element ds where s is the arc length of the curve. In that case, the integral is a tensor of the same rank as that of the integrand f. Such integrals are not considered here.

8.1.2 Closed Line Integrals

The symbol $\oint$ is used when the line integration is performed along a closed curve C:

$$\mathscr{I}_\mu = \oint_C f(\mathbf{r}) dr_\mu. \tag{8.3}$$

This type of line integral is also called *loop integral* or *contour integral*.

Next two curves C_1 and C_2 are considered, which have common start and end points, see Fig. 8.1. In general, one has $\int_{C_1} f(\mathbf{r}) dr_\mu \neq \int_{C_2} f(\mathbf{r}) dr_\mu$ and consequently

Fig. 8.1 Two curves C_1 and
C_2 starting and ending at the
same points

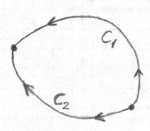

$$\int_{C_1} f(\mathbf{r}) dr_\mu - \int_{C_2} f(\mathbf{r}) dr_\mu = \oint_C f(\mathbf{r}) dr_\mu \neq 0.$$

Here $C = C_1 - C_2$ is a closed curve.

On the other hand, when $\int_{C_1} f(\mathbf{r}) dr_\mu = \int_{C_2} f(\mathbf{r}) dr_\mu$ holds true, in a special case, for arbitrary curves C_1 and C_2 within the region B, then one has

$$\oint_C f(\mathbf{r}) dr_\mu = 0.$$

This then applies to any closed curve within B, provided that B is a *simply connected region*, i.e. when there are no "holes" in B.

8.1.3 Line Integrals for Scalar and Vector Fields

(i) Scalar Fields

As mentioned before, the line integral (8.1) is a vector, when the field function f is a scalar.

A simple example is $f = 1$. Then one obtains

$$\mathscr{I}_\mu = \int_{p_1}^{p_2} r_\mu(p) dp = r_\mu(p_2) - r_\mu(p_1),$$

which is the vector pointing from $\mathbf{r}_1$ to $\mathbf{r}_2$, cf. Fig. 8.2.

The line integral, for $f = 1$, should not be confused with the arc length s, which is given by

$$s = \int_C |d\mathbf{r}| = \int_{p_1}^{p_2} \left(\frac{dr_\mu}{dp} \frac{dr_\mu}{dp}\right)^{1/2} dp.$$

The quantity s is a scalar.

Fig. 8.2 Line integral yields
a relative position vector

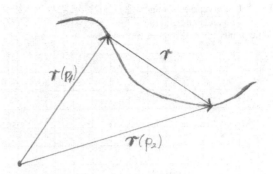

(ii) Vector Fields

Now let f be the component v_ν of a vector field $\mathbf{v} = \mathbf{v}(\mathbf{r})$. Then the line integral

$$\mathscr{I}_{\mu\nu} = \int_C v_\nu(\mathbf{r})\mathrm{d}r_\mu \qquad (8.4)$$

is a second rank tensor. In general, it can be decomposed into an isotropic part which is proportional to its trace times the unit tensor $\delta_{\mu\nu}$, an antisymmetric part, and a symmetric traceless part, cf. Chap. 6.

In some applications, the trace

$$\mathscr{I} \equiv \mathscr{I}_{\mu\mu} = \int_C v_\mu(\mathbf{r})\mathrm{d}r_\mu \qquad (8.5)$$

is needed. The scalar quantity $\mathscr{I}$ is referred to as *the curve integral of a vector field*.

8.1.4 Potential of a Vector Field

Now consider the special case where a vector field $\mathbf{v}$ is given by the gradient of a scalar potential field $\Phi = \Phi(\mathbf{r})$,

$$v_\mu = \nabla_\mu \Phi.$$

The scalar line integral (8.5) of such a vector field is computed according to

$$\mathscr{I} = \int_C v_\mu(\mathbf{r})\mathrm{d}r_\mu = \int_{p_1}^{p_2} \frac{\partial\Phi}{\partial r_\mu}\frac{\mathrm{d}r_\mu}{\mathrm{d}p}\mathrm{d}p = \int_{p_1}^{p_2} \frac{\mathrm{d}\Phi}{\mathrm{d}p}\mathrm{d}p = \Phi(\mathbf{r}(p_2)) - \Phi(\mathbf{r}(p_1)),$$
$$(8.6)$$

or

$$\mathscr{I} = \Phi(\mathbf{r}_2) - \Phi(\mathbf{r}_1). \tag{8.7}$$

For the case of a vector field obtained as a gradient of a potential, the line integral is given by the difference of the potential between the end and the start points of the curve, irrespective of its path in between these points. As consequence, the integral along a closed curve vanishes:

$$\oint_C v_\mu(\mathbf{r}) dr_\mu = 0. \tag{8.8}$$

For (8.8) to be valid, the region in which the curve C lies, has to be compact, it should not have any holes.

8.1.5 Computation of the Potential for a Vector Field

For a vector field $\mathbf{v}$ which obeys the integrability condition (7.35), or equivalently $\nabla \times \mathbf{v} = 0$, the pertaining potential function can be computed with the help of a line integral. A convenient integration path can be chosen, starting from an arbitrary point $\mathbf{r}_1$ and ending at the variable position $\mathbf{r}$. Then the path integral $\mathscr{I}$ is a function of $\mathbf{r}$. More specifically, one has

$$\mathscr{I} = \mathscr{I}(\mathbf{r}) = \int_{\mathbf{r}_1}^{\mathbf{r}} v_\mu dr_\mu = \Phi(\mathbf{r}) - \Phi(\mathbf{r}_1), \tag{8.9}$$

or equivalently

$$\Phi(\mathbf{r}) = \int_{\mathbf{r}_1}^{\mathbf{r}} v_\mu dr_\mu + \text{const.} \tag{8.10}$$

Obviously, the potential function Φ is only determined up to a constant const.

A Simple Example: Homogeneous Vector Field

As a simple special case the constant homogeneous vector field $\mathbf{v} = \text{const.}$ is considered. Then one has

$$\mathscr{I}(\mathbf{r}) = v_\mu \int_{\mathbf{r}_1}^{\mathbf{r}} dr_\mu = \mathbf{v} \cdot (\mathbf{r} - \mathbf{r}_1),$$

and consequently

$$\Phi(\mathbf{r}) = \mathbf{v} \cdot \mathbf{r} + \text{const.}$$

Remark: in physics, forces **F** associated with a potential are given by the negative gradient of the potential function. Thus the mechanical potential Φ is computed as

$$\Phi(\mathbf{r}) = -\int_{\mathbf{r}_1}^{\mathbf{r}} F_\mu dr_\mu + \text{const.} \qquad (8.11)$$

8.1 Exercise: Compute Path Integrals along a Closed Curve

Consider a special closed path composed of the curve C_1 with **r** given by

$$\{x, 0, 0\}, \quad -\rho \leq x \leq \rho,$$

and the curve C_2 with **r** determined by

$$\{x, y, 0\}, \quad x = \rho \cos\varphi, \quad y = \rho \sin\varphi, \quad 0 \leq \varphi \leq \pi.$$

The curve C_1 is a straight line, C_2 is a semi-circle with the constant radius ρ, cf. Fig. 8.3. The differential d**r** needed for the integration is equal to $dx\{1, 0, 0\}$ and $\rho d\varphi\{-\sin\varphi, \cos\varphi, 0\}$ for the curves C_1 and C_2, respectively. Compute the loop integral $\mathscr{I} = \oint_C v_\mu dr_\mu$ along the closed curve defined here for the following three vector fields:

(i)   *homogeneous field*, where $\mathbf{v} = \mathbf{e} = \text{const.}$, with **e** parallel to the x-axis;
(ii) *radial field*, where $\mathbf{v} = \mathbf{r}$;
(iii) *solid-like rotation field*, where $\mathbf{v} = \mathbf{w} \times \mathbf{r}$, with the constant axial vector **w** parallel to the z-axis.

Hint: guess whether $\mathscr{I} = 0$ or $\mathscr{I} \neq 0$ is expected for these vector fields, before you begin with the explicit calculation. Denote the line integrals along the curves C_1 and C_2 by $\mathscr{I}_1$ and $\mathscr{I}_2$. The desired integral $\mathscr{I}$ along the closed curve is the sum $\mathscr{I}_1 + \mathscr{I}_2$.

Fig. 8.3 Closed curve composed of a semi-circle and a straight line

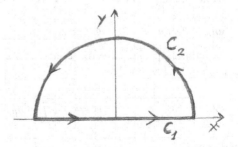

8.2 Surface Integrals, Stokes

8.2.1 Parameter Representation of Surfaces

Surfaces in 3D space can e.g. be described by a relation of the form $z = z(x, y)$, where the components of the position vector on the surface are denoted by x, y, z. Implicitly, the surface can also be determined by $\Phi = \Phi(x, y, z) = \text{const.}$, where Φ is a scalar function, by analogy to potentials, cf. Sect. 7.1.1. Sometimes, it is advantageous to use a parameter presentation of the form

$$r_\mu = r_\mu(p, q), \tag{8.12}$$

for the Cartesian components of the position vector $\mathbf{r}$ located on the surface, with the two parameters p and q.

For a constant q, e.g. $q = q_0$, the relation $r_\mu = r_\mu(p, q_0)$ describes a curve with the curve parameter p. Likewise, for $p = p_0$, the relation $r_\mu = r_\mu(p_0, q)$ is a curve with the curve parameter q. Different values $p = p_0, p_1, p_2, \ldots$ and $q = q_0, q_1, q_2, \ldots$ yield a (p, q)-mesh of curves on the surface, cf. Fig. 8.4.

Provided that the tangential vectors

$$t_\mu^p = \frac{\partial r_\mu}{\partial p}, \quad t_\mu^q = \frac{\partial r_\mu}{\partial q}, \tag{8.13}$$

in the p- and q-directions are not parallel to each other, the curves $\mathbf{r} = \mathbf{r}(p, q = \text{const.})$ and $\mathbf{r} = \mathbf{r}(p = \text{const.}, q)$ cover the surface. Subject to this condition, a vector normal to the surface is inferred from

$$\varepsilon_{\lambda\mu\nu} t_\mu^p t_\nu^q = \varepsilon_{\lambda\mu\nu} \frac{\partial r_\mu}{\partial p} \frac{\partial r_\nu}{\partial q} \neq 0. \tag{8.14}$$

Fig. 8.4 Schematic view of a surface generated by a mesh of curves

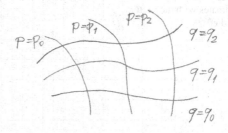

8.2.2 Examples for Parameter Representations of Surfaces

(i) Plane

Let **e** and **u** be two orthogonal unit vectors. Then

$$\mathbf{r} = p\mathbf{e} + q\mathbf{u}$$

represents the plane spanned by the vectors **e** and **u**. With the x- and y-axes of a coordinate system chosen parallel to **e** and **u**, the plane is represented by $x = p$, $y = q$, $z = 0$. The vector normal to the plane is parallel to the z-direction. The (p, q)-mesh covering the plane are orthogonal straight lines parallel to the x- and y-axes.

Alternatively, the *planar polar coordinates* ρ and φ can be used as parameters to represent the plane. Here the position vector within the x–y-plane is expressed as $\mathbf{r} = \{\rho \cos \varphi, \rho \sin \varphi, 0\}$, and consequently

$$\frac{\partial \mathbf{r}}{\partial \rho} = \{\cos \varphi, \sin \varphi, 0\} = \mathbf{e}^{\rho}, \quad \frac{\partial \mathbf{r}}{\partial \varphi} = \{-\rho \sin \varphi, \rho \cos \varphi, 0\} = \rho\, \mathbf{e}^{\varphi}. \quad (8.15)$$

Unit vectors in ρ- and φ-directions are denoted by $\mathbf{e}^{\rho}$ and $\mathbf{e}^{\varphi}$, see Fig. 8.5. These vectors are orthogonal. The plane is covered by a mesh of straight lines starting at the origin and concentric circles, corresponding to $\varphi = $ const. and $\rho = $ const.

(ii) Cylinder Mantle

For a circular cylinder with constant radius ρ, its mantle surface is described by

$$\mathbf{r} = \{\rho \cos \varphi, \rho \sin \varphi, z\}, \quad (8.16)$$

Fig. 8.5 Planar polar coordinates with the vectors $\mathbf{e}^{\rho}$ and $\mathbf{e}^{\varphi}$

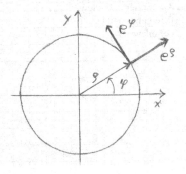

Fig. 8.6 Cylinder coordinates with the vectors $\mathbf{e}^\varphi$, $\mathbf{e}^z$ and $\mathbf{e}^\rho$

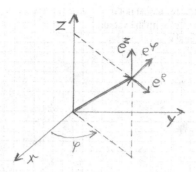

with the two parameters φ and z. The tangential vectors are

$$\frac{\partial \mathbf{r}}{\partial \varphi} = \{-\rho \sin \varphi, \rho \cos \varphi, 0\} = \rho\, \mathbf{e}^\varphi, \qquad \frac{\partial \mathbf{r}}{\partial z} = \{0, 0, 1\} = \mathbf{e}^z. \tag{8.17}$$

The unit vectors $\mathbf{e}^\varphi$ and $\mathbf{e}^z = \mathbf{e}^\rho$ are orthogonal, cf. Fig. 8.6. Their cross product $\mathbf{e}^\varphi \times \mathbf{e}^z$ is parallel to the outer normal $\mathbf{n}$ of the cylinder mantle.

(iii) Surface of a Sphere

The surface of a sphere with constant radius R parameterized with the ansatz

$$\mathbf{r} = \{R \cos \varphi \sin \theta, R \sin \varphi \sin \theta, R \cos \theta\}. \tag{8.18}$$

Here the parameters are the polar angles θ and φ. Now the tangential vectors are

$$\frac{\partial \mathbf{r}}{\partial \varphi} = \{-R \sin \varphi \sin \theta, R \cos \varphi \sin \theta, 0\} = R \sin \theta\, \mathbf{e}^\varphi, \tag{8.19}$$

and

$$\frac{\partial \mathbf{r}}{\partial \theta} = \{R \cos \varphi \cos \theta, R \sin \varphi \cos \theta, -R \sin \theta\} = R\, \mathbf{e}^\theta, \tag{8.20}$$

The cross product of these tangential vectors yields

$$\frac{\partial \mathbf{r}}{\partial \theta} \times \frac{\partial \mathbf{r}}{\partial \varphi} = R^2 \sin \theta\, \mathbf{e}^\theta \times \mathbf{e}^\varphi = R^2 \sin^2 \theta\, \widehat{r}, \tag{8.21}$$

with the unit vector $\widehat{r}$ pointing in radial direction. The mesh on the surface consists of circles around the polar axis with radius $R \sin \theta$, for $\theta = $ const., and grand semi-circles running from the North to the South pole, for $\varphi = $ const.. The unit vectors $\mathbf{e}^\theta$, $\mathbf{e}^\varphi$ and $\widehat{r}$ are mutually orthogonal, see Fig. 8.7.

Fig. 8.7 Spherical polar
coordinates with the vectors
$\mathbf{e}^\varphi$, $\mathbf{e}^\theta$ and $\mathbf{e}^r$

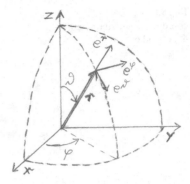

8.2.3 Surface Integrals as Integrals Over Two Parameters

Consider a finite surface A which has a rim and which is simply connected, i.e. A is without any holes. Furthermore, the surface should everywhere have a well oriented normal direction. A counter example is the *Moebius strip*.

The surface A is described by a parameter representation $\mathbf{r} = \mathbf{r}(p, q)$ where the parameters p and q vary between the values p_1, p_2 and q_1, q_2. Thus the rim of the surface in R^3 corresponds to the rim of a rectangle in the p–q-plane, cf. Fig. 8.8.

Now let $f = f(\mathbf{r})$ be a function of $\mathbf{r}$ which depends on the parameters q and q via $\mathbf{r} = \mathbf{r}(p, q)$. The surface integral of this function over the surface A is defined by

$$\mathscr{S}_\mu = \int_A f(\mathbf{r}) \, ds_\mu. \tag{8.22}$$

The *surface element* ds_μ is the cross product of the tangential vectors (8.13), see also (8.14). The axial vector ds_μ is perpendicular to the surface. More specifically, one has

$$ds_\mu = \varepsilon_{\mu\nu\lambda} \frac{\partial r_\nu}{\partial p} \frac{\partial r_\lambda}{\partial q} dp\,dq = \widehat{s}_\mu(p, q) d^2 s. \tag{8.23}$$

Here $\widehat{s}$ is a unit vector orthogonal to the surface, and $d^2 s$ is the magnitude of the surface element. It quantifies the change of $\mathbf{r}$ with the change of p and q. The exponent 2 in the symbol "$d^2 s$" used here indicates that the surface integration is "two dimensional". It is clearly distinguished from the arc length element ds occurring sometimes in "one dimensional" line integrals. Notice that the surface element ds, being defined in (8.23) as the cross product of two tangential vectors, is an axial vector. The same applies to the unit vector $\widehat{s}$.

Due to (8.23), the surface integral (8.22) can be computed as a double integral over p and q:

$$\mathscr{S}_\mu = \int_{p_1}^{p_2} dp \int_{q_1}^{q_2} dq \, f(\mathbf{r}) \varepsilon_{\mu\nu\lambda} \frac{\partial r_\nu}{\partial p} \frac{\partial r_\lambda}{\partial q}. \tag{8.24}$$

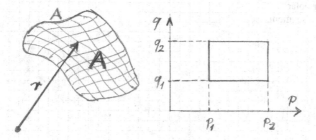

Fig. 8.8 Surface in real space and in the p–q parameter plane

Often the symbol $\oint$ is used to indicate that the integral is extended over a closed surface which is topologically equivalent to the surface of a sphere. Notice that the area in the parameter space has a well defined rim or border line even when the closed surface has none in the 3D space. This is obvious for the parameter representation of the surface of a sphere. There the polar angles θ and φ are within the intervals $[0, \pi]$ and $[0, 2\pi]$.

Surface integrals are discussed next for the examples of parameter presentations of surfaces shown in Sect. 8.2.2.

8.2.4 Examples for Surface Integrals

(i) Plane

Firstly, consider as area A over which the surface integral shall be evaluated a rectangle in the x–y-plane where the variables are within the intervals $[x_1, x_2]$ and $[y_1, y_2]$. The vector normal to the plane is parallel to the z-direction. The unit vector in this direction is denoted by $\mathbf{e}^z$. The surface element is $\mathrm{d}s_\mu = e_\mu^z \mathrm{d}x\mathrm{d}y$. Thus the surface integral of a function $f = f(\mathbf{r})$ with $\mathbf{r} = \{x, y, 0\}$, located within the plane, is evaluated according to

$$\mathscr{S}_\mu = e_\mu^z \int_A f(\mathbf{r}(x, y))\mathrm{d}x\mathrm{d}y = e_\mu^z \int_{x_1}^{x_2} \mathrm{d}x \int_{y_1}^{y_2} \mathrm{d}y\, f(\mathbf{r}(x, y)). \qquad (8.25)$$

The representation by the planar polar coordinates ρ, φ is appropriate for an area A whose border lines are parts of two circular arcs and two radial lines, see Fig. 8.9. Here the position vector within the plane is given by $\mathbf{r} = \{\rho \cos \varphi, \rho \sin \varphi, 0\}$.

The variables are within the intervals $[\rho_1, \rho_2]$ and $[\varphi_1, \varphi_2]$. Now the surface element is $\mathrm{d}s_\mu = e_\mu^z \rho \mathrm{d}\rho \mathrm{d}\varphi$ and the surface integral is to be evaluated according to

$$\mathscr{S}_\mu = e_\mu^z \int_A f(\mathbf{r}(\rho, \varphi))\rho \mathrm{d}\rho \mathrm{d}\varphi = e_\mu^z \int_{\rho_1}^{\rho_2} \rho \mathrm{d}\rho \int_{\varphi_1}^{\varphi_2} \mathrm{d}\varphi\, f(\mathbf{r}(\rho, \varphi)). \qquad (8.26)$$

Fig. 8.9 Planar polar
coordinates for a segment
of a planar ring

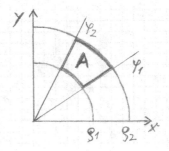

For the simple case $f = 1$ and an integration over the full disc with the constant radius R, the integration over φ yields 2π, that over ρ gives $(1/2)R^2$. Then (8.26) leads to $\mathscr{S}_\mu = e_\mu^z R^2 \pi$. As expected, in this case the surface integral gives the area $R^2\pi$ of the circular disc.

In both examples considered so far, the unit vector normal to the surface is constant. As a consequence, it could be put outside the integral, just as a factor. This is no longer the case when the integration is to be taken over curved surfaces.

(ii) Cylinder Mantle

For a circular cylinder with constant radius $\rho = R$, its mantle surface is described by $\mathbf{r} = \{R\cos\varphi, R\sin\varphi, z\}$. The two parameters are φ and z. Here the surface element is

$$\mathrm{d}s_\mu = R\, n_\mu(\varphi)\mathrm{d}\varphi\mathrm{d}z, \tag{8.27}$$

where vector normal to the cylinder mantle is given by $\mathbf{n} = \{\cos\varphi, \sin\varphi, 0\}$. The surface integral over a region A located on the cylinder mantle is

$$\mathscr{S}_\mu = R \int_A f(\mathbf{r}(\varphi, z))\, n_\mu(\varphi)\mathrm{d}\varphi\mathrm{d}z. \tag{8.28}$$

(iii) Surface of a Sphere

The surface of a sphere with constant radius R is described by

$$\mathbf{r} = \{R\cos\varphi\sin\theta, R\sin\varphi\sin\theta, R\cos\theta\},$$

with the polar angles θ and φ as parameters. The unit vector normal to the surface is the radial unit vector $\widehat{\mathbf{r}} = R^{-1}\mathbf{r}$. Here the surface element is

$$\mathrm{d}s_\mu = R^2 \widehat{r}_\mu(\theta, \varphi)\, \sin\theta\mathrm{d}\theta\mathrm{d}\varphi. \tag{8.29}$$

The surface integral over a region A on the surface of the sphere is

$$\mathscr{S}_\mu = R^2 \int_A f(\mathbf{r}(\theta, \varphi)) \widehat{r}_\mu(\theta, \varphi) \sin\theta d\theta d\varphi. \tag{8.30}$$

In the following, the abbreviation

$$d^2\widehat{r} = \sin\theta d\theta d\varphi \tag{8.31}$$

is used for the scalar part of the surface element pertaining to the surface of a sphere with radius $R = 1$. This sphere is referred to as *unit sphere*.

The symbol

$$\int d^2\widehat{r} \ldots, \tag{8.32}$$

without any indication of a specific area, is used for integrals over the whole unit sphere.

8.2.5 Flux of a Vector Field

The surface integral $\mathscr{S}_\mu = \int_A f(\mathbf{r})ds_\mu$, defined in (8.22) with (8.23), is a tensor of rank $\ell + 1$ when f stands for the components of a tensor of rank ℓ. In particular, for $f = v_\nu$ where $\mathbf{v} = \mathbf{v}(\mathbf{r})$ is a vector field, the corresponding integral over a surface A is the second rank tensor

$$\mathscr{S}_{\mu\nu} = \int_A v_\nu ds_\mu. \tag{8.33}$$

The isotropic part of this tensor, cf. Chap. 6, involves its trace $\mathscr{S} = \mathscr{S}_{\mu\mu}$. This scalar quantity, viz.

$$\mathscr{S} = \int_A v_\mu ds_\mu = \int_A \mathbf{v} \cdot d\mathbf{s} \tag{8.34}$$

is referred to as the *flux of the vector field* $\mathbf{v}$ *through the surface A.*

A simple example demonstrates the meaning of the term "flux". Let A be a plane surface with a fixed normal vector $\mathbf{n}$ and $\mathbf{v} = $ const. a homogeneous vector field. With $ds_\mu = n_\mu ds$ one obtains

$$\mathscr{S} = v_\mu n_\mu \int ds = v_\mu n_\mu \mathscr{A},$$

where $\mathscr{A}$ stands for the area of the surface. With $v_\mu = v\widehat{v}_\mu$, this result can be written as

Fig. 8.10 Flux through an area A and side view of an effective area A_{eff}

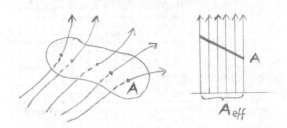

$$\mathscr{S} = v\,\mathscr{A}_{\text{eff}}.$$

Here v is the magnitude of the homogeneous vector field. The effective area

$$\mathscr{A}_{\text{eff}} = \widehat{v}_\mu\, n_\mu\, \mathscr{A}$$

is the actual area of the surface, reduced by the cosine of the angle between $\mathbf{v}$ and the normal to the plane, see Fig. 8.10.

8.2 Exercise: Surface Integrals over a Hemisphere

Consider a hemisphere with radius R, with the center at the origin. The unit vector pointing from its center to the North pole is denoted by $\mathbf{u}$.

Compute the surface integrals $\mathscr{S}_{\mu\nu} = \int v_\nu \mathrm{d}s_\mu$, over the hemisphere and the pertaining flux $\mathscr{S} = \mathscr{S}_{\mu\mu}$ for

(i)   the homogeneous vector field $v_\nu = v\widehat{v}_\nu = \text{const.}$ and
(ii) the radial field $v_\nu = r_\nu$.

Hint: use the symmetry arguments $\mathscr{S}_{\mu\nu} \sim u_\mu \widehat{v}_\nu$ (case i) and $\mathscr{S}_{\mu\nu} \sim u_\mu u_\nu$ (case ii), to simplify the calculations. Put the base of the hemisphere onto the x–y-plane, for the explicit integration.

8.2.6 Generalized Stokes Law

The *Stokes law* provides a relation between surface integrals of a certain type with a line integral along the closed rim of the surface. Thus the "dimension" of the integral is reduced from 2 to 1. This applies when the integrand of the surface integral is a spatial derivative of a function $f = f(\mathbf{r})$, which then occurs as integrand in the line integral. To be more specific, the *generalized Stokes law* reads:

$$\int_A \varepsilon_{\lambda\nu\mu}\, \nabla_\nu\, f(\mathbf{r})\mathrm{d}s_\lambda = \oint_C f(\mathbf{r})\mathrm{d}r_\mu. \tag{8.35}$$

It is understood that the closed curve $C = \partial A$ is the rim, or the contour line, of the surface A.

The standard Stokes law follows from (8.35) when the function f is identified with the Cartesian component v_μ of a vector field $\mathbf{v}$. Due to

$$\varepsilon_{\lambda\nu\mu} \nabla_\nu v_\mu = (\nabla \times \mathbf{v})_\lambda,$$

the *Stokes law* can be cast into the form

$$\int_A (\nabla \times \mathbf{v}) \cdot d\mathbf{s} = \oint_{\partial A} \mathbf{v} \cdot d\mathbf{r}. \qquad (8.36)$$

The line integral on the right hand side of (8.36) is referred to as the circulation of the vector field $\mathbf{v}$.

A remark on parity is in order. The nabla-operator ∇ and the line element $d\mathbf{r}$ occurring in (8.35) and (8.36) are polar vectors. Parity is conserved, i.e. both sides of the equation in the generalized Stokes law have the same parity behavior since the surface element $d\mathbf{s}$ is an axial vector.

Furthermore, when the integration in (8.35) and (8.36) is extended over a closed surface, there is no contour line and thus these integrals are equal to zero. Notice, however, that the generalized Stokes law applies to simply connected surfaces which have no holes. On the other hand, the circulation, i.e. the line integral, as it occurs on the right hand side of (8.36), can be non-zero, when the integration is around a hole in a surface, even when $\nabla \times \mathbf{v}$, occurring on the left hand side of (8.36), is zero.

A *proof* of (8.35), which includes the proof for the conventional Stokes law (8.36), is presented next. The surface is parameterized by $\mathbf{r} = \mathbf{r}(p, q)$. The area over which the integration is extended is assumed to be given by a rectangle in the p–q-parameter plane, see Fig. 8.11.

According to (8.23), the surface element can be written as

$$ds_\lambda = \varepsilon_{\lambda\alpha\beta} \frac{\partial r_\alpha}{\partial p} \frac{\partial r_\beta}{\partial q} dp dq.$$

Fig. 8.11 Area and rim curve in p–q-plane for the proof of the Stokes law

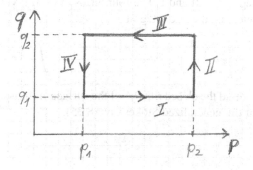

Insertion of this expression into $\mathscr{S}_\mu = \int_A \varepsilon_{\lambda\nu\mu} \nabla_\nu f(\mathbf{r}) ds_\lambda$, which is the left hand side of (8.35), leads to

$$\mathscr{S}_\mu = \int_A \varepsilon_{\lambda\nu\mu} \nabla_\nu f(\mathbf{r}) ds_\lambda = \int_{p_1}^{p_2} dp \int_{q_1}^{q_2} dq \left(\nabla_\nu f \frac{\partial r_\nu}{\partial p} \frac{\partial r_\mu}{\partial q} - \nabla_\nu f \frac{\partial r_\mu}{\partial p} \frac{\partial r_\nu}{\partial q} \right).$$

Here $\varepsilon_{\lambda\nu\mu}\varepsilon_{\lambda\alpha\beta} = \delta_{\nu\alpha}\delta_{\mu\beta} - \delta_{\nu\beta}\delta_{\mu\alpha}$, which corresponds to (4.10), has been used. Thanks to the chain rule, one has

$$\nabla_\nu f \frac{\partial r_\nu}{\partial p} = \frac{\partial f}{\partial p}, \quad \nabla_\nu f \frac{\partial r_\nu}{\partial q} = \frac{\partial f}{\partial q}.$$

Due to

$$\frac{\partial f}{\partial p} \frac{\partial r_\mu}{\partial q} = \frac{\partial}{\partial p}\left(f \frac{\partial r_\mu}{\partial q} \right) - f \frac{\partial^2 r_\mu}{\partial p \partial q}, \quad \frac{\partial f}{\partial q} \frac{\partial r_\mu}{\partial p} = \frac{\partial}{\partial q}\left(f \frac{\partial r_\mu}{\partial p} \right) - f \frac{\partial^2 r_\mu}{\partial q \partial p},$$

the integrand of the surface integral considered reduces to

$$\frac{\partial}{\partial p}\left(f \frac{\partial r_\mu}{\partial q} \right) - \frac{\partial}{\partial q}\left(f \frac{\partial r_\mu}{\partial p} \right).$$

The first term can immediately be integrated over p, the second one over q. This then yields

$$\mathscr{S}_\mu = \int_{q_1}^{q_2} dq \left(f(p_2, q) \frac{\partial r_\mu(p_2, q)}{\partial q} - f(p_1, q) \frac{\partial r_\mu(p_1, q)}{\partial q} \right)$$
$$- \int_{p_1}^{p_2} dp \left(f(p, q_2) \frac{\partial r_\mu(p, q_2)}{\partial p} - f(p, q_1) \frac{\partial r_\mu(p, q_1)}{\partial p} \right). \quad (8.37)$$

The two terms in the upper line of (8.37) are the line integrals $\mathscr{S}_\mu^{II} + \mathscr{S}_\mu^{IV}$ along the segments II and IV, those in the lower line are $\mathscr{S}_\mu^{III} + \mathscr{S}_\mu^{I}$ which pertain to the segments III and I. The four terms in (8.37), viz.: $\mathscr{S}_\mu = \mathscr{S}_\mu^{I} + \mathscr{S}_\mu^{II} + \mathscr{S}_\mu^{III} + \mathscr{S}_\mu^{IV}$ make up the line integral

$$\mathscr{S}_\mu \equiv \oint_{\partial A} f \, dr_\mu$$

around the closed rim ∂A of the surface A, thus $\mathscr{S}_\mu = \mathscr{S}_\mu$. This completes the proof of the generalized Stokes law (8.35).

8.3 Exercise: Verify the Stokes Law for a Vorticity Field

Compute the integrals on both sides of the Stokes law (8.36) for the vorticity vector field $\mathbf{v} = \mathbf{w} \times \mathbf{r}$, with a constant angular velocity $\mathbf{w}$. The surface integral should be evaluated for a circular disc with radius R. The disc is perpendicular to $\mathbf{w}$.

Hint: choose a coordinate system with its z-axis parallel to $\mathbf{w}$.

8.2.7 Application: Magnetic Field Around an Electric Wire

The Stokes law can be used to evaluate the strength of the magnetic field $\mathbf{H}$ outside a straight wire. The electric current density $\mathbf{j}$, inside the wire, is assumed to be steady, i.e. it does not change with time. In this stationary situation, one of the Maxwell equations reduces to

$$\nabla \times \mathbf{H} = \mathbf{j}. \tag{8.38}$$

This equation underlies the findings of Oersted and Ampere on the coupling between electricity and magnetism.

Next (8.38) is integrated over a circular surface, perpendicular to the wire. The radius R of the circle is larger than the diameter of the wire, cf. Fig. 8.12. The resulting "integral form" of (8.38) is

$$\int (\nabla \times \mathbf{H}) \cdot d\mathbf{s} = \int \mathbf{j} \cdot d\mathbf{s} \equiv I. \tag{8.39}$$

Here I is the electric current, notice that $\mathbf{j} = 0$, outside the wire. On the other hand, the Stokes law implies

$$\int (\nabla \times \mathbf{H}) \cdot d\mathbf{s} = \oint \mathbf{H} \cdot d\mathbf{r}. \tag{8.40}$$

Fig. 8.12 The magnetic field $\mathbf{H}$ around a long straight wire carrying the electric current I, due to the electric flux $\mathbf{j}$

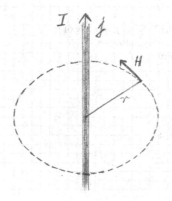

The line integral is over the circle with radius R. For symmetry reasons, the **H**-field is tangential, i.e. parallel to d**r**, and the magnitude H of the field is constant around the circle. Thus one has

$$\oint \mathbf{H} \cdot \mathrm{d}\mathbf{r} = HR \int_0^{2\pi} \mathrm{d}\varphi = 2\pi R H.$$

Consequently, the strength of the magnetic field, at the distance r from the center of the wire, with the previous R now called r, is

$$H = \frac{I}{2\pi r}. \tag{8.41}$$

Clearly, outside of the wire, the field strength H decreases with increasing distance r like $1/r$. The situation is different inside the wire. There the integral over a circular disc with radius r increases like its area, viz. like r^2, provided that the electric current density is homogeneous. The constant electric current I, in (8.41) is replaced by a term proportional to r^2. As a consequence, H increases linearly with R. Notice, however, that the charge density within a metal wire is not homogeneous, but rather confined to a surface layer.

Furthermore, the present considerations apply to long wires, end-effects are not taken into account.

8.2.8 Application: Faraday Induction

Consider an almost closed ring-like electrically conducting wire, placed into a magnetic **B** field, cf. Fig. 8.13.

Integration of the Faraday law

$$\varepsilon_{\mu\nu\lambda} \nabla_\nu E_\lambda = -\frac{\partial}{\partial t} B_\mu,$$

Fig. 8.13 Schematic view of the Faraday induction experiment

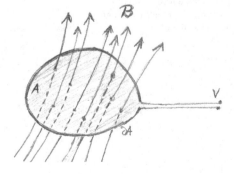

over the area A, bounded by the wire and application of the Stokes law (8.36) yields the equation governing the *Faraday induction*, viz.

$$\oint_{\partial A} \mathbf{E} \cdot \mathbf{dr} = \frac{d}{dt} \int_A \mathbf{B} \cdot \mathbf{ds} \equiv \frac{d}{dt} \Phi_A, \qquad (8.42)$$

where $\Phi_A = \int_A \mathbf{B} \cdot \mathbf{ds}$ is the flux of $\mathbf{B}$ through the area A, cf. Sect. 8.2.5. The left hand side of (8.42) quantifies the electric tension or the voltage $V = \oint_{\partial A} \mathbf{E} \cdot \mathbf{dr}$ generated by the change of the magnetic flux, underlying the *Faraday induction*. Notice, the time change of the magnetic flux Φ can be brought about by a time change of the field $\mathbf{B}$ or by a change of the area A, e.g. by a change of the surface normal of the area with respect to the direction of the magnetic field.

8.3 Volume Integrals, Gauss

8.3.1 Volume Integrals in R^3

The integral of a function $f = f(\mathbf{r})$ over a region V in R^3 is denoted by

$$\mathcal{V} = \int_V f(\mathbf{r}) d^3 r, \qquad (8.43)$$

where the scalar $d^3 r$ is the volume element. For $f = 1$, the volume integral yields the content or the size of the volume, which is also referred to as "volume" and denoted by V:

$$V = \int_V d^3 r, \qquad (8.44)$$

When the Cartesian components x, y, z are used for the integration, the volume integral is just the threefold integral

$$\mathcal{V} = \int_V f(\mathbf{r}) dx dy dz. \qquad (8.45)$$

Often it is advantageous to express the vector $\mathbf{r}$ in terms of three parameters p_1, p_2, p_3, viz.:

$$\mathbf{r} = \mathbf{r}(p_1, p_2, p_3),$$

and to use those as general coordinates for the integration. Then the prescription for the evaluation of the volume integral is

$$\mathcal{V} = \int_V f(\mathbf{r}(p_1, p_2, p_3)) |\mathcal{J}(p_1, p_2, p_3)| dp_1 dp_2 dp_3. \qquad (8.46)$$

Here $\mathscr{J}$ is the *Jacobi determinant*, also called *functional determinant*, which can be computed according to

$$\mathscr{J} = \varepsilon_{\mu\nu\lambda} \frac{\partial r_\mu}{\partial p_1} \frac{\partial r_\nu}{\partial p_2} \frac{\partial r_\lambda}{\partial p_3}. \tag{8.47}$$

Notice that $\mathscr{J} dp_1 dp_2 dp_3$ can also be written as

$$\mathscr{J} dp_1 dp_2 dp_3 = ds_\mu \frac{\partial r_\mu}{\partial p_1} dp_1, \tag{8.48}$$

where ds_μ is the surface element of a surface parameterized by p_2, p_3, with $p_1 = $ const., cf. Sect. 8.2.3.

Two examples for general coordinates are discussed next.

(i) Cylinder Coordinates

Consider a circular cylinder whose axis coincides with the z-axis. The presentation $\mathbf{r} = \{\rho \cos \varphi, \rho \sin \varphi, z\}$ is used, where ρ, φ, z are the parameters, see also (8.16). Now the volume integral is

$$\mathscr{V} = \int_V f(\mathbf{r}(\rho, \varphi, z)) \, \rho d\rho d\varphi dz. \tag{8.49}$$

When the region in space to be integrated over is a cylinder with radius R and length L, and furthermore the integrand is independent of the angle φ, the relation (8.49) reduces to

$$\mathscr{V} = 2\pi \int_0^R \rho d\rho \int_0^L dz \, f(\mathbf{r}(\rho, z)). \tag{8.50}$$

The factor 2π stems from the integration over φ. For $f = 1$, one obtains the volume $V = \pi R^2 L$ of the circular cylinder.

(ii) Spherical Coordinates

The standard parametrization with the spherical coordinates r, θ, φ corresponds to $\mathbf{r} = \{r \cos \varphi \sin \theta, r \sin \varphi \sin \theta, r \cos \theta\}$. The volume integral is given by

$$\mathscr{V} = \int_V f(\mathbf{r}(r, \theta, \varphi)) \, r^2 dr \, \sin\theta d\theta d\varphi = \int_V f(\mathbf{r}(r, \theta, \varphi)) \, r^2 dr d^2\hat{r}. \tag{8.51}$$

Here, as in (8.31), the symbol $d^2\hat{r} = \sin\theta d\theta d\varphi$ stands for the scalar surface element of the unit sphere. Sometimes it is advantageous to use $\zeta = \cos\theta$ as integration variable instead of θ. Then one has

$$d^2\hat{r} = -d\zeta d\varphi. \tag{8.52}$$

As a more specific example, a half-sphere with radius R, located above the x–y-plane, is chosen. When, furthermore, the function f does not depend on the angle φ, the volume integral reduces to

$$\mathcal{V} = 2\pi \int_0^R r^2 dr \int_0^1 d\zeta\, f(\mathbf{r}(r, \zeta)). \tag{8.53}$$

The minus sign occurring in (8.52) is taken care of by an exchange of the integration limits, $\theta = 0$ and $\pi/2$ correspond to $\zeta = 1$ and $\zeta = 0$.

For $f = 1$ the volume $V = (2/3)\pi R^3$ of the half-sphere is obtained.

8.3.2 Application: Mass Density, Center of Mass

The macroscopic description of matter, be it a gas, a liquid, or a solid, is based on the mass density $\rho = \rho(\mathbf{r})$. Its microscopic interpretation, for a substance composed of N particles with the mass m, is provided by

$$\rho(\mathbf{r})d^3r = m\,dN(\mathbf{r}), \tag{8.54}$$

where $dN(\mathbf{r})$ is the number of particles found within a small volume element d^3r, located at the position $\mathbf{r}$. Alternatively, and even more general, the mass density of a substance composed of particles with masses m_i, located at positions $\mathbf{r}^i$, with $i = 1, 2, \ldots, N$, is given by

$$\rho(\mathbf{r}) = \sum_{i=1}^{N} m_i\, \delta(\mathbf{r} - \mathbf{r}^i). \tag{8.55}$$

Here $\delta(\mathbf{r})$ is the three dimensional delta-distribution function $\delta(\mathbf{r})$, with the property

$$\int \delta(\mathbf{r} - \mathbf{s}) f(\mathbf{r})\, d^3r = f(\mathbf{s}), \tag{8.56}$$

which applies when the function f is single valued at the position $\mathbf{s}$. The integrals of both expressions for ρ, over a volume V, yield the mass M_V of the substance within this volume,

$$M_V = \int_V \rho(\mathbf{r})\, d^3r. \tag{8.57}$$

For (8.54) one finds $M_V = mN_V$, for (8.55) the result is

$$M_V = \sum_{i=1}^{N_V} m_i,$$

N_V is the number of particles located within the volume V.

In applications, where the atomistic structure of matter is not relevant, e.g. in hydrodynamics, the mass density $\rho = \rho(\mathbf{r})$ is treated as a continuous field function. For the dynamic phenomena discussed in Sects. 8.4.1 and 8.4.2, the density should also be a differentiable function. Differentiability plays no role for the global properties, which are obtained via volume integrals. The position of the center of mass and the moment of inertia tensor are of this type.

The position vector $\mathbf{R}$ of the *center of mass* of a substance characterized by the mass density ρ, and confined within the volume V, is determined by

$$M R_\mu = \int_V r_\mu \, \rho(\mathbf{r}) \mathrm{d}^3 r, \tag{8.58}$$

where $M = \int_V \rho(\mathbf{r}) \mathrm{d}^3 r$ is the total mass.

An example, instructive for the computation of volume integrals, is a homogeneous density, confined by a spherical cap. Its cross section is shown in Fig. 8.14.

The cap has uniaxial symmetry, characterized by the unit vector $\mathbf{u}$ which is parallel to the vector pointing from the geometric center to the North pole of the cap. In the figure, the geometric center is put at the origin of the coordinate system and the direction of the z-axis is chosen parallel to $\mathbf{u}$. The inner and outer radii are denoted by a_1 and a_2, respectively. As conventional, the angle θ is counted from the z-axis, its maximum is $\theta_{\max}$. With $\rho = \rho_0 = \text{const.}$ within the cap and $\rho = 0$ outside, the

Fig. 8.14 Cross section of a spherical cap

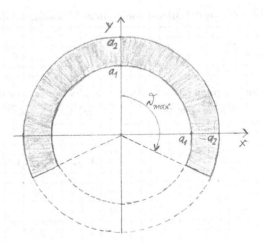

integrals to be evaluated are of the form

$$\int_V \ldots d^3 r = 2\pi \int_{a_1}^{a_2} r^2 dr \int_0^{\theta_{max}} \sin\theta d\theta \ldots = 2\pi \int_{a_1}^{a_2} r^2 dr \int_{\zeta_{min}}^1 d\zeta \ldots, \quad (8.59)$$

where the integrand ... is assumed to be independent of the polar angle φ. Furthermore, $\zeta = \cos\theta$ and $\zeta_{min} = \cos\theta_{max}$ are used.

First, the volume is computed with the help of (8.59), with the integrand 1. The result is

$$V = \frac{2\pi}{3} (a_2^3 - a_1^3)(1 - \zeta_{min}). \quad (8.60)$$

The volume of a sphere with radius a is recovered from this expression with $a_2 = a$, $a_1 = 0$ and $\zeta_{min} = \cos\pi = -1$. The mass of the cap is $M = \rho_0 V$.

Due to the uniaxial symmetry, the vector $\mathbf{R}$ is parallel (or anti-parallel) to the unit vector $\mathbf{u}$. The calculation of the center of mass is simplified with the help of this argument. The ansatz

$$R_\mu = c u_\mu, \quad (8.61)$$

is made. The coefficient c is inferred from the scalar multiplication of this equation with u_μ, thus $Mc = M u_\mu R_\mu$. The vector $\mathbf{u}$ is constant, so it can be put inside the integral (8.58) used for MR_μ. Then the integrand is $u_\mu r_\mu = r \cos\theta = r\zeta$. With the help of (8.59),

$$Vc = \frac{2\pi}{8} (a_2^4 - a_1^4)(1 - \zeta_{min}^2) \quad (8.62)$$

is obtained. Clearly, for $\theta_{max} = \pi$, corresponding to $\zeta_{min} = -1$, the coefficient c is zero. As expected, in this case, the center of mass coincides with the geometric center. For a half-sphere, with radius $a = a_2$, $a_1 = 0$, and $\zeta_{min} = 0$, on the other hand, the center of mass

$$\mathbf{R} = \frac{3}{8} a \mathbf{u} \quad (8.63)$$

is shifted "upwards".

The case $a_1 = a$, $a_2 = a + \delta a$, with $0 < \delta a \ll a$ corresponds to thin shell structure with thickness δa. Then the factors $(a_2^3 - a_1^3)$ and $(a_2^4 - a_1^4)$, occurring in (8.60) and (8.62), reduce to $3a^2\delta a$ and $4a^3\delta a$, respectively. As a consequence,

$$\mathbf{R} = \frac{1}{2} a \mathbf{u} \quad (8.64)$$

is found for the hollow hemisphere.

8.3.3 Application: Moment of Inertia Tensor

The moment of inertia tensor $\Theta_{\mu\nu}$, introduced in Sect. 4.3.2, can also be expressed as a volume integral over the mass density ρ of a solid body, viz.:

$$\Theta_{\mu\nu} = \int_V \rho(\mathbf{r})\,(r^2\,\delta_{\mu\nu} - r_\mu r_\nu)\mathrm{d}^3 r. \tag{8.65}$$

The moment of inertia tensor, evaluated either as a sum over discrete masses or a volume integral, depends on the choice of the origin. When one talks about *the* moment of inertia tensor of a mass distribution, it is understood that $\mathbf{r} = 0$, in (8.65), corresponds to its center of mass. In the general case, the effective moment of inertia tensor, entering the linear relation between the angular momentum and the angular velocity, is

$$\Theta_{\mu\nu}^{\mathrm{eff}} = M(R^2\,\delta_{\mu\nu} - R_\mu R_\nu) + \Theta_{\mu\nu}^{\mathrm{cm}}, \tag{8.66}$$

where M is the total mass, $\mathbf{R}$ is the position of the center of mass, and it is understood that $\Theta_{\mu\nu}^{\mathrm{cm}}$ is for a rotation around the center of mass. Relation (8.66) is referred to as *Steiner's law*. It can be derived from (8.65) with $\mathbf{r} = \mathbf{R} + \mathbf{r}'$. For the proof, use $\int_V \rho\mathbf{r}'\mathrm{d}^3 r' = 0$. When there is no danger of confusion $\Theta_{\mu\nu}^{\mathrm{cm}}$ will be denoted by $\Theta_{\mu\nu}$, in the following.

As also pointed out in Sect. 4.3.2, the *moment of inertia* for a rotation about a fixed axis is defined via the linear relation between the component of the angular momentum parallel to this axis and the magnitude w of the angular velocity. With the unit vector $\widehat{w}_\mu$, parallel to the axis of rotation, the moment of inertia is $\Theta = \widehat{w}_\mu \Theta_{\mu\nu} \widehat{w}_\nu$, and consequently

$$\Theta = \int_V \rho(\mathbf{r})\, r_\perp^2 \mathrm{d}^3 r, \tag{8.67}$$

where

$$r_\perp^2 = r^2 - \widehat{w}_\mu r_\mu\,\widehat{w}_\nu r_\nu = r^2 - (\widehat{\mathbf{w}}\cdot\mathbf{r})^2$$

is the square of the shortest distance of a mass element at $\mathbf{r}$, from the rotation axis. The origin of the position vector is a point on the rotation axis. With the z-axis chosen parallel to the rotation axis, $r_\perp^2$ is just $r^2 - z^2 = x^2 + y^2$.

By definition, the moment of inertia tensor is symmetric and positive definite. In general, it has three eigenvalues $\Theta^{(1)}$, $\Theta^{(2)}$ and $\Theta^{(3)}$, which are the moments of inertia for rotations about the three principal axes. An object with three different principal moments of inertia is referred to as asymmetric top. A symmetric top has two equal eigenvalues, e.g. $\Theta^1 = \Theta^{(2)} \neq \Theta^{(3)}$, for the spherical top, all three are equal. These different types of symmetry of the moment of inertia tensor result from

Fig. 8.15 Brick stone with sides a, b, c. The coordinate axes coincide with the principal axes of the moment of inertia tensor

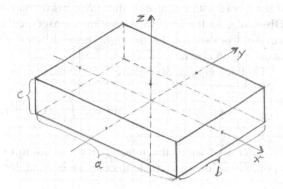

a biaxial, uniaxial and spherical symmetry, respectively, of the mass density. For a constant density inside a solid body, its shape determines the symmetry.

As a simple example, a brick stone with constant mass density ρ_0 and with edges a, b, c, is considered, cf. Fig. 8.15. The center of mass is in the middle, the principal axes go through it and they are perpendicular to the side planes of the brick. The volume element is $dx\,dy\,dz$, the integration goes over the intervals $[-a/2, a/2]$, $[-b/2, b/2]$, $[-c/2, c/2]$. The mass is $M = \rho_0 abc$. The moment of inertia $\Theta^{(3)}$ is associated with the rotation about the z-axis. Then the square of the distance from the axis is $r_\perp^2 = x^2 + y^2$. The integral (8.67) leads to

$$\Theta^{(3)} = \rho_0 \frac{1}{12}(a^3 b + a\,b^3)\,c = \frac{1}{12} M\,(a^2 + b^2),$$

which is the mass times the square of the length of the diagonal of the side perpendicular to the principal axis. The two other principal moments are

$$\Theta^{(1)} = \frac{1}{12} M\,(b^2 + c^2), \quad \Theta^{(2)} = \frac{1}{12} M\,(a^2 + c^2).$$

In general, a brick stone is an asymmetric top. For, e.g. $a = b$, it becomes a symmetric top. Then the moment of inertia tensor is, as in (5.20),

$$\Theta_{\mu\nu} = \Theta_\| \, u_\mu u_\nu + \Theta_\perp \, (\delta_{\mu\nu} - u_\mu u_\nu), \tag{8.68}$$

with $\Theta_\| = \Theta^{(3)}$, $\Theta_\perp = \Theta^{(1)} = \Theta^{(2)}$, and $\mathbf{u}$ is a unit vector parallel to the symmetry axis.

A cube corresponds to $a = b = c$ which implies three equal moments of inertia. So the cube is a spherical top with an isotropic moment of inertia tensor

$$\Theta_{\mu\nu} = \Theta\,\delta_{\mu\nu}, \quad \Theta = \frac{1}{6}Ma^2.$$

For a sphere with radius $a/2$, the tensor has the same form, just with $\Theta = \frac{1}{10}M a^2$. Obviously, on the level of second rank tensors, cubic symmetry cannot be distinguished from spherical symmetry. This is different for tensors of rank four, as they occur, e.g. in elasticity.

The trace $\Theta_{\mu\mu} = 3\bar{\Theta}$, where $\bar{\Theta}$ is the mean moment of inertia, is given by

$$3\bar{\Theta} = 2\int_V \rho(\mathbf{r})\,r^2 \mathrm{d}^3 r. \tag{8.69}$$

In the case of uniaxial symmetry, $3\bar{\Theta}$ is equal to $\Theta_\parallel + 2\Theta_\perp$, see (8.68). In some applications, it is preferable to compute $\Theta_\parallel$ and $3\bar{\Theta}$ and to infer $\Theta_\perp$ from $(3\bar{\Theta} - \Theta_\parallel)/2$.

8.4 Exercise: Moment of Inertia Tensor of a Half-Sphere
Compute the moment of inertia tensor of a half-sphere with radius a and constant mass density ρ_0. The orientation is specified by the unit vector $\mathbf{u}$, pointing from the center of the sphere to the center of mass of the half-sphere.

Hint: make use of the symmetry. First calculate the moments of inertia with respect to the center of the sphere, then use the Steiner law (8.66) to find the moments of inertia with respect to the center of mass of the half-sphere, see also Sect. 8.3.2.

8.3.4 Generalized Gauss Theorem

When the integrand of a volume integral is the spatial derivative of a function, the integral over the volume V can be transformed into a surface integral over the bounding surface $A = \partial V$. Thus the "three dimensional" integration is reduced to a "two dimensional" one. Here, a generalized version of the Gauss theorem is stated first and the standard Gauss theorem is obtained as a special case.

Let $f = f(\mathbf{r})$ be a differentiable function, V a volume with a well defined surface ∂V in R^3. The *generalized Gauss theorem* reads

$$\mathscr{V}_\mu \equiv \int_V \nabla_\mu f \mathrm{d}^3 r = \oint_{\partial V} f n_\mu \mathrm{d}^2 s, \tag{8.70}$$

where $\mathbf{n}$ is the outer normal of V, cf. Fig. 8.16 at its bounding surface ∂V and $\mathrm{d}^2 s$ is the scalar surface element.

When f occurring in (8.70) is the component of a tensor of rank ℓ the integrals on both sides of the equation are tensors of rank $\ell + 1$. In particular, for f being the component v_ν of a vector field $\mathbf{v}$, the quantities occurring on both sides of (8.70) are second rank tensors:

$$\mathscr{V}_{\mu\nu} \equiv \int_V \nabla_\mu v_\nu \mathrm{d}^3 r = \oint_{\partial V} v_\nu n_\mu \mathrm{d}^2 s. \tag{8.71}$$

Fig. 8.16 The potato of
Prof. Muschik as integration
volume, **n** is the outer
normal vector, perpendicular
to its peel

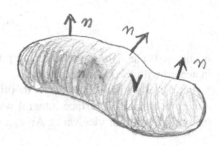

The isotropic part of the tensor equation involves the trace

$$\mathscr{V}_{\mu\mu} \equiv \int_V \nabla_\mu v_\mu d^3r = \oint_{\partial V} v_\mu n_\mu d^2s. \tag{8.72}$$

This equation can also be written as

$$\int_V \nabla \cdot \mathbf{v} d^3r = \oint_{\partial V} \mathbf{v} \cdot \mathbf{n} d^2s, \tag{8.73}$$

which is the standard *Gauss theorem*.

Again a remark on parity is in order. The nabla-operator ∇ and the outer normal **n** both are polar vectors. Thus parity is "conserved" in the generalized Gauss theorem (8.70) and its special cases, e.g. in (8.73). Notice that the surface element $\mathbf{n}d^2s$ occurring in connection with the Gauss theorem is a polar vector, whereas the surface element d**s** occurring in the Stokes law (8.35) is a pseudo vector.

A proof for the Gauss theorem is not given here, however, its validity shall be verified next with a simple example. A sphere with radius R is chosen as integration volume. The origin of the coordination system coincides with the center of the sphere. The radial vector field $\mathbf{v} = \mathbf{r}$ is inserted in (8.71). Due to $\nabla_\mu r_\nu = \delta_{\mu\nu}$, the left hand side of this equation yields the isotropic unit tensor times the volume of the sphere:

$$\mathscr{V}_{\mu\nu} \equiv \int_V \nabla_\mu r_\nu d^3r = \delta_{\mu\nu} \int_V d^3r = \delta_{\mu\nu} \frac{4\pi}{3} R^3.$$

Since $\mathbf{n} = \hat{\mathbf{r}}$ for the sphere, the surface integral standing on the right hand side of the Gauss theorem is equal to

$$\oint_{\partial V} r_\nu \hat{r}_\mu R^2 d^2\hat{r} = R^3 \int \hat{r}_\nu \hat{r}_\mu d^2\hat{r}.$$

The sphere does not possess any preferential direction. Thus the integral $\int \hat{r}_\nu \hat{r}_\mu d^2\hat{r}$ over the unit sphere must be of the form $c\delta_{\mu\nu}$, due to symmetry arguments. The proportionality coefficient c is obtained from the trace relation $\int \hat{r}_\mu \hat{r}_\mu d^2\hat{r} = 4\pi = 3c$, notice that $\delta_{\mu\mu} = 3$. Thus

$$\int \hat{r}_\mu \hat{r}_\nu d^2\hat{r} = \frac{4\pi}{3} \delta_{\mu\nu}$$

is obtained. The expressions just computed for the volume and surface integrals are in accord with the Gauss theorem.

On the other hand, the Gauss theorem can be applied for the determination of a more complicated surface integral when the evaluation of the pertaining volume integral is easier, or vice versa. An example: the relation

$$\int_V \nabla_\mu r_\nu d^3r = \delta_{\mu\nu} \int_V d^3r = \delta_{\mu\nu} V$$

holds true for a well bounded volume V with any shape, not just for a sphere, as considered above. The Gauss theorem (8.71) now implies the remarkable result

$$\oint_{\partial V} r_\nu n_\mu d^2s = \delta_{\mu\nu} V, \tag{8.74}$$

irrespective of the shape of the surface ∂V, as long as the outer normal $\mathbf{n}$ is well defined everywhere on the surface of the volume. The trace part of this relation, viz.:

$$\oint_{\partial V} \mathbf{r} \cdot \mathbf{n} d^2s = 3\,V, \tag{8.75}$$

shows that the volume V can also be computed with the help of a surface integral.

8.3.5 Application: Gauss Theorem in Electrodynamics, Coulomb Force

In electrodynamics, the symbol ρ is used for the charge density. Then the integral over the volume V

$$\int_V \rho d^3r = Q_V, \tag{8.76}$$

is equal to the electric charge contained in this volume. One of the Maxwell equations links the divergence of the electric displacement field $\mathbf{D}$ with the charge density, viz.:

$$\nabla_\mu D_\mu = \rho. \tag{8.77}$$

The relation (8.77) is referred to as the differential form of the *Gauss law*. The integral of (8.77) over a volume V and use of the Gauss theorem, with $d^2 s_\mu = n_\mu d^2 s$, yields:

$$\oint_{\partial V} D_\mu d^2 s_\mu = \int_V \rho d^3 r = Q_V. \tag{8.78}$$

This is the Gauss law of electrodynamics. It means that the flux of the D-field through the closed surface ∂V is equal to the charge contained within the volume V. The Coulomb law for the force between two charges, located in vacuum, follows from the Gauss law (8.78). This is seen as follows.

In general, one has $\mathbf{D} = \varepsilon_0 \mathbf{E} + \mathbf{P}$, where ε_0 is the dielectric permeability of the vacuum. Its numerical value depends on the choice of the basic physical units for length, time, mass and charge. In the system of physical units originally introduced by Gauss, where no independent basic unit for the charge occurs, ε_0 is equal to 1. The vector field $\mathbf{P}$ is the electric polarization. In vacuum, $\mathbf{P} = 0$ applies. Thus in vacuum, the electric field $\mathbf{E}$ is related to the charge density via

$$\oint_{\partial V} E_\mu d^2 s_\mu = \frac{1}{\varepsilon_0} \int_V \rho d^3 r = \frac{1}{\varepsilon_0} Q_V. \tag{8.79}$$

Now let ρ be a charge density with spherical symmetry, centered around $\mathbf{r} = 0$. Then the electric field is parallel (or anti-parallel) to r_μ, thus it can be written as $E_\mu = E \hat{r}_\mu$.

Now the volume integration is performed over a sphere with radius r, then one has $\hat{r}_\mu d^2 s_\mu = d^2 s$, and E is constant on the surface of the sphere. The surface integral of (8.79) yields E times the surface $4\pi r^2$ of the sphere. Assuming that the charge density is completely contained within this sphere and denoting the total charge by Q, one obtains $4\pi r^2 E = \frac{1}{\varepsilon_0} Q$, and

$$E_\mu = \frac{Q}{4\pi \varepsilon_0} \frac{1}{r^2} \hat{r}_\mu = \frac{Q}{4\pi \varepsilon_0} \frac{1}{r^3} r_\mu. \tag{8.80}$$

This is the electric field $E_\mu(\mathbf{r})$ at the position $\mathbf{r}$, located in vacuum, caused by the charge Q at $\mathbf{r} = 0$. A "test" charge q, placed at $\mathbf{r}$, experiences the force $\mathbf{F} = q\mathbf{E}(\mathbf{r})$. Thus the force between these charges is the *Coulomb force*

$$\mathbf{F} = \frac{q Q}{4\pi \varepsilon_0} \frac{1}{r^3} \mathbf{r}. \tag{8.81}$$

The strength of the Coulomb force decreases with increasing distance r between the charges like $1/r^2$, just like the gravitational force. Gravitation is always attractive. The Coulomb force is repulsive or attractive, depending on whether the charges have equal or opposite sign.

8.3.6 Integration by Parts

Let $f = f(\mathbf{r})$ and $g = g(\mathbf{r})$ be two field functions. Due to the product rule $\nabla_\mu(gf) = g\nabla_\mu f + f\nabla_\mu g$, and with the help of the generalized Gauss theorem, the volume integral $\int_V g\nabla_\mu f$ is equal to

$$\int_V g\,\nabla_\mu f\mathrm{d}^3 r = -\int_V f\,\nabla_\mu g\mathrm{d}^3 r + \int_{\partial V} fgn_\mu \mathrm{d}^2 s. \tag{8.82}$$

In many applications, the surface integral $\int_{\partial V} fgn_\mu \mathrm{d}^2 s$ is taken over a surface, where at least one of the two functions f and g vanishes. Then, the integration by parts is equivalent to

$$\int_V g\,\nabla_\mu f\mathrm{d}^3 r = -\int_V f\,\nabla_\mu g\mathrm{d}^3 r. \tag{8.83}$$

With $f = -\nabla_\mu g$ and $\Delta = \nabla_\mu\nabla_\mu$, the relation (8.83) implies

$$-\int_V g\,\Delta\,g\mathrm{d}^3 r = \int_V (\nabla_\mu g)(\nabla_\mu g)\mathrm{d}^3 r \geq 0. \tag{8.84}$$

Thus, subject to the condition that the contribution of the surface integral, occurring in connection with the integration by parts, is zero, the negative Laplace operator $-\Delta$ is a positive definite operator. This point is of importance for the kinetic energy operator in wave mechanics, cf. Sect. 7.6.4.

8.4 Further Applications of Volume Integrals

8.4.1 Continuity Equation, Flow Through a Pipe

The mass density and the local velocity field of a fluid are denoted by $\rho = \rho(\mathbf{r})$ and $\mathbf{v} = \mathbf{v}(\mathbf{r})$, as in Sect. 7.4.3. The vector field $\mathbf{j}(\mathbf{r}) = \rho\mathbf{v}$ is the *flux density*. The continuity equation, cf. (7.49),

$$\frac{\partial\rho}{\partial t} + \nabla_\nu j_\nu = 0, \tag{8.85}$$

expresses the *local conservation* of mass. The integral of ρ over a volume V yields the mass $M_V = \int_V \rho\mathrm{d}^3 r$ contained within V. Upon integration of the continuity equation over a volume V which does not change with time, the first term of the equation is the time change of the mass M_V. The second term can be expressed as a surface integral over ∂V, due to the Gauss theorem. Thus (8.85) leads to

$$\frac{d}{dt}M_V + \int_{\partial V} n_\nu j_\nu d^2 s = 0,$$ (8.86)

where **n** is the outer normal vector on the surface of the volume V. This equation says: the mass contained in the volume changes with time according to the flux of mass which goes in and out of the volume, through the surface. There is no creation or annihilation of mass.

As an example, consider a fluid confined by a pipe, with spatially varying cross section and open ends. The normal vectors of the open parts of the volume V are denoted by $\mathbf{n}_1$ and $\mathbf{n}_2$, the pertaining areas are A_1 and A_2. For this geometry $\mathbf{n}_1 = -\mathbf{n}_2$ applies. Since the side walls are assumed to be impenetrable for the fluid, the time change of the mass is

$$\frac{dM_V}{dt} = I_1 + I_2,$$

where $I_1 = -\int_{\partial A_1} \mathbf{n}_1 \cdot \mathbf{j} d^2 s$ and $I_2 = -\int_{\partial A_2} \mathbf{n}_2 \cdot \mathbf{j} d^2 s$ are the fluxes into and out of V, respectively. One has $I_1 > 0$ and $I_2 < 0$ when $\mathbf{n}_1$ and $\mathbf{n}_2$ are anti-parallel and parallel to $\mathbf{j}$.

With the help of the substantial time derivative (7.50), viz.

$$\frac{d}{dt} := \frac{\partial}{\partial t} + v_\nu \nabla_\nu,$$

the continuity equation can also be written as

$$\frac{d\rho}{dt} + \rho \nabla_\nu v_\mu = 0.$$

The quantity $\mathcal{V} = \rho^{-1}$ is the volume per mass, also called specific volume. The continuity equation is equivalent to

$$\frac{d}{dt} \ln \mathcal{V} = \nabla_\nu v_\nu.$$ (8.87)

This shows: the specific volume and hence the density does not change with time when the divergence of the velocity vanishes. A flow with $\nabla \cdot \mathbf{v} = 0$ is referred to as *incompressible flow*.

In electrodynamics, the symbols ρ and $\mathbf{j}$ are used for the charge density and the electric flux density. The integral of ρ over a volume is the charge Q contained in this volume. The continuity equation has the same form as (8.85), provided that no charges are created or annihilated. Then the continuity equation describes the local charge conservation.

8.4.2 Momentum Balance, Force on a Solid Body

The local conservation equation for the linear momentum density ρv_μ, in the absence of external forces, cf. (7.52), reads

$$\rho \frac{dv_\mu}{dt} + \nabla_\nu p_{\nu\mu} = 0. \tag{8.88}$$

The pressure tensor $p_{\nu\mu}$ characterizes the part of the momentum transport, which is not of convective type. The gradient $\nabla_\nu p_{\nu\mu} = k_\mu$ is an internal force density.

The total momentum associated with a fluid in a volume V is

$$P_\mu^V \equiv \int_V \rho\, v_\mu d^3 r. \tag{8.89}$$

Integration of the local conservation equation (8.88) over this volume and application of the Gauss theorem leads to

$$\frac{d}{dt} P_\mu^V = -\int_{\partial V} n_\nu^{\mathrm{fl}}\, p_{\nu\mu} d^2 s, \tag{8.90}$$

where $\mathbf{n}^{\mathrm{fl}}$ is the outer normal of the volume containing the fluid. The term on the right hand side of this equation is the force F^{fl} acting on the fluid. Due to *actio equal reactio*, the force F^{s} exerted by the fluid on a solid wall or on a solid body immersed in the fluid has the same magnitude, but with opposite sign: $\mathbf{F}^{\mathrm{s}} = -\mathbf{F}^{\mathrm{fl}}$. When the normal vector $\mathbf{n}^{\mathrm{fl}}$ is replaced by $\mathbf{n}^{\mathrm{s}} = -\mathbf{n}^{\mathrm{fl}}$, which points from the solid into the fluid, the expression for F^{s} has the same form as that one in (8.90),

$$F_\mu^{\mathrm{s}} = -\int_{\partial V} n_\nu\, p_{\nu\mu} d^2 s. \tag{8.91}$$

Consider now a plane wall with the surface area A and the normal vector $\mathbf{n} \equiv \mathbf{n}^{\mathrm{s}}$. Assuming that the pressure tensor of the fluid is constant at the wall, the force F^{w} acting on the plane wall is

$$F_\mu^{\mathrm{w}} = -n_\nu\, p_{\nu\mu}\, A. \tag{8.92}$$

In thermal equilibrium, the pressure tensor of a fluid is just the isotropic tensor $P\delta_{\mu\nu}$, with the hydrostatic pressure P, cf. (7.53). Then (8.92) reduces to $\mathbf{F} = -\mathbf{n}PA$. The minus sign means that the fluid pushes against the wall, provided that $P > 0$, as valid in thermal equilibrium. In equilibrium, there is only a normal force, i.e. a force perpendicular to the wall.

In general, the force (8.92) has both normal and tangential components

$$F_\mu^{\mathrm{norm}} = -n_\mu\, (n_\nu\, p_{\nu\lambda}\, n_\lambda)\, A, \quad F_\mu^{\mathrm{tang}} = -n_\nu\, (p_{\nu\mu} - \delta_{\mu\nu}\, n_\kappa\, p_{\kappa\lambda}\, n_\lambda)\, A. \tag{8.93}$$

Fig. 8.17 The tangential force density due to non-diagonal elements of the pressure tensor. The force exerted by the fluid on the solid is in the directions shown when $p_{yx} < 0$ and $p_{xy} < 0$

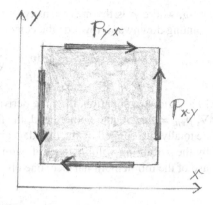

When the x–z-plane of a coordinate system is on the wall with the y-axis antiparallel to $\mathbf{n}$, the normal component of the force is $F_y = -p_{yy}A$. The tangential components are $F_x = -p_{yx}A$ and $F_z = -p_{yz}A$.

For a cube placed in the fluid, with its sides parallel to the coordinate axes, the tangential part of the vector $n_\nu p_{\nu\mu}$ occurring in (8.92), has directions indicated in Fig. 8.17, for the x–y-plane. Notice, when $p_{yx} \neq p_{xy}$, i.e. when the pressure tensor has a non-zero antisymmetric part, the cube experiences a torque, caused by the fluid.

By analogy to (8.91), the total force of the fluid, exerted on a stiff solid body is

$$F_\mu^s = -\oint_{\partial V} n_\nu \, p_{\nu\mu} \mathrm{d}^2 s. \tag{8.94}$$

Here ∂V is the closed surface of the solid body, $\mathbf{n}$ is its outer normal.

A remark on Fig. 8.17 is in order. The force exerted by the fluid on the solid cube, evaluated with (8.94), has tangential components in the directions indicated by the arrows, provided that p_{yx} and p_{xy} are negative. This happens, indeed, for a plane Couette flow with the geometry chosen as in Fig. 7.6.

8.4.3 The Archimedes Principle

The principle of Archimedes states: an impenetrable solid body immersed in a liquid experiences a lift force, against the direction of gravity. The magnitude of this buoyancy force is equal to the weight of the liquid in a volume, which is as large as that one occupied by the solid. Why is that so? Why does it apply to solid bodies of any shape, as long as the liquid does not penetrate into the solid?

Consider the local momentum conservation equation (8.88). In the presence of an external force, an extra force density has to be taken into account on the right hand side of the balance equation. In the case of the gravity on earth, this force density is $\rho \mathbf{g}$, where ρ is the mass density of the liquid, and $\mathbf{g}$ is the gravity acceleration, pointing downward, towards the center of the earth. Thus the force balance reads

$$\rho \frac{\mathrm{d}v_\mu}{\mathrm{d}t} + \nabla_\nu p_{\nu\mu} = \rho\, g_\mu. \tag{8.95}$$

In a stationary situation, the time derivative of the velocity vanishes. Then one has $\nabla_\nu p_{\nu\mu} = \rho g_\mu$. On the other hand, the force on the solid body, which is assumed to be totally surrounded by the liquid, is given by (8.91). Replacing the surface integral by the pertaining volume integral, with the help of the Gauss theorem, and making use of the momentum balance, one obtains for the buoyancy force

$$F_\mu^{\mathrm{buoy}} = - \int_{\mathrm{V}} \rho g_\mu \mathrm{d}^3 r, \tag{8.96}$$

where the integral is to be extended over the volume of the solid. When ρ and $\mathbf{g}$ are constant, the integral yields the volume V of the solid, irrespective of its shape, and $\rho V = M_{\mathrm{fl}}$ is the mass of the fluid, contained in such a volume. Thus one has

$$F_\mu^{\mathrm{buoy}} = - M_{\mathrm{fl}}\, g_\mu, \tag{8.97}$$

which is just the Archimedes principle. Due to the minus sign in (8.97), this force acts against gravity. The weight "felt" by the body with the mass M_{s}, immersed inside the liquid is $(M_{\mathrm{s}} - M_{\mathrm{fl}})\mathbf{g}$.

How about the proof for the Archimedes principle for a floating body, that is only partially immersed in the liquid, say in water? Imagine the floating body is cut at the water level and a mass equal to that of the part cut off is placed at the center of gravity of the part remaining under water. The force balance is still the same. Assuming that a very thin layer of water is above the body, the considerations given above now apply.

8.4.4 Torque on a Rotating Solid Body

The force F_ν, per surface element, exerted by a fluid on the surface of a solid body is proportional to $-n_\kappa p_{\kappa\nu}$, cf. (8.94). The torque T_μ exerted by a fluid on a stiff solid body is

$$T_\mu = -\varepsilon_{\mu\lambda\nu} \oint_{\partial V} r_\lambda n_\kappa\, p_{\kappa\nu} \mathrm{d}^2 s. \tag{8.98}$$

As before, ∂V indicates the closed surface of the solid body, $\mathbf{n}$ is its outer normal. It is understood, that $\mathbf{r} = 0$ coincides with the center of gravity of the body.

Examples for the computation of forces and torques are presented in Sects. 10.5.2 and 10.3

8.5 Further Applications in Electrodynamics

8.5.1 Energy and Energy Density in Electrostatics

The *Coulomb energy* U of charges q_i and q_j, located in vacuum, is

$$U = \frac{1}{4\pi \, \varepsilon_0} \sum_{i<j} \sum_j \frac{q_i \, q_j}{|\mathbf{r}^i - \mathbf{r}^j|} = \frac{1}{4\pi \, \varepsilon_0} \frac{1}{2} \sum_{i \neq j} \sum_j \frac{q_i \, q_j}{|\mathbf{r}^i - \mathbf{r}^j|}. \tag{8.99}$$

The last expression is equivalent to

$$U = \frac{1}{2} \sum_i q_i \, \phi(\mathbf{r}^i), \quad \phi(\mathbf{r}^i) = \frac{1}{4\pi \, \varepsilon_0} \sum_j \frac{q_j}{|\mathbf{r}^i - \mathbf{r}^j|}, \tag{8.100}$$

where ϕ is the electrostatic potential. With the continuous charge density

$$\rho(\mathbf{r}) = \sum_i q_i \, \delta(\mathbf{r} - \mathbf{r}^i),$$

the Coulomb energy reads

$$U = \frac{1}{2} \int \rho(\mathbf{r}) \, \phi(\mathbf{r}) \mathrm{d}^3 r. \tag{8.101}$$

Notice, the condition $i \neq j$ occurring in (8.99), is ignored in (8.100) and lost in (8.101).

Now use of the Gauss law $\rho = \nabla_\nu D_\nu$, cf. (7.56), of the relation $\phi \nabla_\nu D_\nu = \nabla_\nu (D_\nu \phi) - D_\nu \nabla_\nu \phi$ and $E_\nu = -\nabla_\nu \phi$, and the application of the Gauss theorem leads to

$$U = \frac{1}{2} \int \phi(\mathbf{r}) \nabla_\nu D_\nu \mathrm{d}^3 r = \frac{1}{2} \int E_\nu D_\nu \mathrm{d}^3 r + \oint n_\nu D_\nu \, \phi \mathrm{d}^2 s. \tag{8.102}$$

The last term vanishes, when the integral $\oint \ldots \mathrm{d}^2 s$ is performed over a far away surface where, at least $\phi = 0$, or $n_\nu D_\nu = 0$, holds true. Then the energy is given by

$$U = \frac{1}{2} \int u(\mathbf{r}) \mathrm{d}^3 r, \quad u = \frac{1}{2} E_\nu D_\nu, \tag{8.103}$$

where $u(\mathbf{r})$ is the energy density. So far charges in vacuum were considered, thus one has $D_\nu = \varepsilon_0 E_\nu$ and consequently $u = \frac{1}{2}\varepsilon_0 E_\nu E_\nu$. The relations still apply to a linear medium characterized by the dielectric tensor $\varepsilon_{\nu\mu}$ according to

$$D_\nu = \varepsilon_0 \, \varepsilon_{\nu\mu} \, E_\mu,$$

cf. Sects. 2.6.4 and 5.3.4. Then the energy density is equal to

$$u = \frac{1}{2} E_\nu D_\nu = \frac{1}{2} \varepsilon_0 \varepsilon_{\nu\mu} E_\mu E_\nu. \tag{8.104}$$

Clearly, the symmetric part of $\varepsilon_{\nu\mu}$ only contributes to the energy density.

The energy density for the more general case of a nonlinear relation between the **D** and **E** fields is treated in Sect. 8.5.3.

8.5.2 Force and Maxwell Stress in Electrostatics

The force F_μ acting on charges q_j, in vacuum, in the presence of a given electric field E_μ is

$$F_\mu = \sum_j q_j E_\mu(\mathbf{r}^j),$$

or, in terms of the charge density ρ, by

$$F_\mu = \int \rho(\mathbf{r}) E_\mu(\mathbf{r}) \mathrm{d}^3 r = \int k_\mu^{\mathrm{elstat}} \mathrm{d}^3 r. \tag{8.105}$$

With the help of the Gauss law $\rho = \nabla_\nu D_\nu$, cf. (7.56), the electrostatic force density k_μ^{elstat} can be rewritten as

$$k_\mu^{\mathrm{elstat}} = E_\mu \nabla_\nu D_\nu = \nabla_\nu (D_\nu E_\mu) - D_\nu \nabla_\nu E_\mu. \tag{8.106}$$

In electrostatics, one has $\nabla \times \mathbf{E} = 0$, and consequently $\nabla_\nu E_\mu = \nabla_\mu E_\nu$. Thus the last term of (8.106) is equal to $D_\nu \nabla_\mu E_\nu$. Provided that the interrelation between **D** and **E** is linear, this term can also be written as the total spatial derivative $(1/2)\nabla_\mu (E_\nu D_\nu)$. Thus in vacuum, and the same applies for any linear medium, the force density is given by

$$k_\mu^{\mathrm{elstat}} = \nabla_\nu (D_\nu E_\mu) - \frac{1}{2}\nabla_\mu (E_\lambda D_\lambda) = \nabla_\nu (D_\nu E_\mu) - \frac{1}{2}\nabla_\mu (E_\lambda D_\lambda \delta_{\mu\nu}) = \nabla_\nu T_{\nu\mu}^{\mathrm{elstat}}. \tag{8.107}$$

The electrostatic stress tensor for a linear medium is defined by

$$T_{\nu\mu}^{\mathrm{elstat}} \equiv D_\nu E_\mu - \frac{1}{2} E_\lambda D_\lambda \, \delta_{\mu\nu}. \tag{8.108}$$

This tensor is symmetric, i.e. $T_{\nu\mu}^{\mathrm{elstat}} = T_{\mu\nu}^{\mathrm{elstat}}$, in vacuum and for an isotropic linear medium, where $D_\nu = \varepsilon\varepsilon_0 E_\nu$ applies, with a scalar dielectric coefficient ε. In general however, and also in an anisotropic linear medium with $\overline{\varepsilon_{\nu\mu}} \neq 0$, the Maxwell stress tensor contains an antisymmetric part, which is associated with a torque. For a discussion of this point see Sect. 8.5.5.

8.5.3 Energy Balance for the Electromagnetic Field

Point of departure of the formulation of the energy balance for the electromagnetic field is the expression

$$j_\mu E_\mu$$

for the power, i.e. for the time change of the energy, delivered by the electric field $\mathbf{E}$ on the electric current density $\mathbf{j}$.

Side remark: A plausible argument for the power being given by $j_\mu E_\mu$.

In mechanics, the time change of the kinetic energy $\frac{d}{dt}\frac{1}{2}mv_\mu v_\mu = v_\mu \frac{d}{dt}mv_\mu$ of a particle moving with velocity $\mathbf{v}$ in a force field $\mathbf{F}$ is given by $v_\mu F_\mu$. For one type of carriers with the electric charge e, the current density is equal to $j_\mu = nev_\mu$, where n is the number density of the charges and $\mathbf{v}$ their average velocity. Then the relation stated above follows due to $F_\mu = eE_\mu$.

Now, multiplication of the inhomogeneous Maxwell equation involving the current density, cf. (7.56), by E_μ leads to

$$\varepsilon_{\mu\nu\lambda} E_\mu \nabla_\nu H_\lambda = j_\mu E_\mu + E_\mu \frac{\partial}{\partial t} D_\mu.$$

The term on the left hand side can be rewritten as

$$\varepsilon_{\mu\nu\lambda} E_\mu \nabla_\nu H_\lambda = \nabla_\nu (\varepsilon_{\mu\nu\lambda} E_\mu H_\lambda) - H_\lambda \varepsilon_{\mu\nu\lambda} \nabla_\nu E_\mu.$$

Due to the homogeneous Maxwell equation referred to as Faraday law, cf. (7.57), the last term of the equation above is equal to

$$-H_\lambda \varepsilon_{\mu\nu\lambda} \nabla_\nu E_\mu = H_\lambda \varepsilon_{\lambda\nu\mu} \nabla_\nu E_\mu = -H_\lambda \frac{\partial}{\partial t} B_\lambda.$$

Thus the energy balance equation reads

$$j_\mu E_\mu + E_\mu \frac{\partial}{\partial t} D_\mu + H_\mu \frac{\partial}{\partial t} B_\mu + \nabla_\mu S_\mu = 0, \tag{8.109}$$

with the energy flux density, also called *Poynting vector*,

$$S_\mu \equiv \varepsilon_{\mu\nu\lambda} E_\mu H_\lambda. \tag{8.110}$$

Notice, no specific form of an interrelation between **E** and **D** or between **B** and **H** has been used. So the energy balance (8.109) holds true in general.

Integration of the local balance equation over a volume V and use of the Gauss theorem yields

$$\int_V j_\mu E_\mu d^3r + \int_V \left(E_\mu \frac{\partial}{\partial t} D_\mu + H_\mu \frac{\partial}{\partial t} B_\mu \right) d^3r + \oint_{\partial V} n_\mu S_\mu d^2s = 0. \quad (8.111)$$

The first term is the power which the current extracts from the electromagnetic field and the second term stands for the power which the field takes from the current, both within the volume V. The last term describes the power given to the surroundings, e.g. by radiation.

The field energy can be defined for a nonlinear, hysteresis-free medium, where the **E** and **H** fields are uniquely determined by the **D** and **B** fields according to

$$\mathbf{E} = \mathbf{E}(\mathbf{r}, \mathbf{D}), \quad \mathbf{H} = \mathbf{H}(\mathbf{r}, \mathbf{B}). \quad (8.112)$$

The electric and magnetic energy densities are defined by

$$u^{el}(\mathbf{D}) = \int_0^{\mathbf{D}} E_\mu(\mathbf{D}') dD'_\mu, \quad u^{mag}(\mathbf{B}) = \int_0^{\mathbf{B}} H_\mu(\mathbf{B}') dB'_\mu. \quad (8.113)$$

The time derivatives of these energy densities are

$$\frac{\partial}{\partial t} u^{el} = E_\mu \frac{\partial}{\partial t} D_\mu, \quad \frac{\partial}{\partial t} u^{mag} = H_\mu \frac{\partial}{\partial t} B_\mu. \quad (8.114)$$

Thus, for the hysteresis-free medium, the local energy balance (8.109) is equivalent to

$$\frac{\partial}{\partial t}(u^{el} + u^{mag}) + \nabla_\mu S_\mu + j_\mu E_\mu = 0. \quad (8.115)$$

The definition (8.113) implies that **E** and **H** are derivatives of the electric and of the magnetic energy density with respect to **D** and **B**, respectively:

$$E_\lambda = \frac{\partial u^{el}}{\partial D_\lambda}, \quad u^{el} = u^{el}(\mathbf{D}), \quad H_\lambda = \frac{\partial u^{mag}}{\partial B_\lambda}, \quad u^{mag} = u^{mag}(\mathbf{B}). \quad (8.116)$$

For a linear medium characterized by the relative dielectric tensor $\varepsilon_{\lambda\kappa}$ and the magnetic permeability tensor $\mu_{\lambda\kappa}$ according to

$$D_\lambda = \varepsilon_0 \varepsilon_{\lambda\kappa} E_\kappa, \quad B_\lambda = \mu_0 \mu_{\lambda\kappa} H_\kappa, \quad (8.117)$$

one has

$$E_\lambda = \varepsilon_0^{-1}\, \varepsilon_{\lambda\kappa}^{-1}\, D_\kappa\,, \quad H_\lambda = \mu_0^{-1}\, \mu_{\lambda\kappa}^{-1}\, B_\kappa\,,$$

and consequently

$$u^{\text{el}} = \frac{1}{2}\, \varepsilon_0^{-1}\, D_\lambda\, \varepsilon_{\lambda\kappa}^{-1}\, D_\kappa = \frac{1}{2}\, D_\lambda\, E_\lambda$$

$$u^{\text{mag}} = \frac{1}{2}\, \mu_0^{-1}\, B_\lambda\, \mu_{\lambda\kappa}^{-1}\, B_\kappa = \frac{1}{2}\, B_\lambda H_\lambda\,. \tag{8.118}$$

For the special case of an isotropic linear medium, where $\varepsilon_{\lambda\kappa} = \varepsilon\delta_{\lambda\kappa}$ and $\mu_{\lambda\kappa} = \mu\delta_{\lambda\kappa}$, with the scalar coefficients ε and μ hold true, the equations for the electric and magnetic energy density reduce to

$$u^{\text{el}} = \frac{1}{2}\, (\varepsilon_0\, \varepsilon)^{-1}\, D^2\,, \quad u^{\text{mag}} = \frac{1}{2}\, (\mu_0\, \mu)^{-1}\, B^2\,. \tag{8.119}$$

8.5.4 Momentum Balance for the Electromagnetic Field, Maxwell Stress Tensor

The Lorentz force (3.47) describes the force, i.e. the time change of the linear momentum, experienced by a charge in the presence of $\mathbf{E}$ and $\mathbf{B}$ fields. When the "matter" is characterized by the charge density ρ and the current density j_ν, the force density exerted by the fields on the matter is

$$k_\mu = \rho E_\mu + \varepsilon_{\mu\nu\lambda}\, j_\nu\, B_\lambda\,.$$

With the help of the inhomogeneous Maxwell equations (7.56), this expression is equal to

$$\rho\, E_\mu + \varepsilon_{\mu\nu\lambda}\, j_\nu\, B_\lambda = E_\mu\, \nabla_\nu\, D_\nu + \varepsilon_{\mu\nu\lambda}\, \varepsilon_{\nu\kappa\tau}\, (\nabla_\kappa H_\tau)\, B_\lambda - \varepsilon_{\mu\nu\lambda}\, \left(\frac{\partial}{\partial t} D_\nu\right)\, B_\lambda\,.$$

The term involving the time derivative can be rewritten as

$$-\varepsilon_{\mu\nu\lambda}\, \left(\frac{\partial}{\partial t} D_\nu\right)\, B_\lambda = -\varepsilon_{\mu\nu\lambda}\, \frac{\partial}{\partial t}(D_\nu\, B_\lambda) + \varepsilon_{\mu\nu\lambda} D_\nu \frac{\partial}{\partial t} B_\lambda\,.$$

Due to the Faraday law, i.e. the homogeneous Maxwell equation (7.57) involving the time derivative of the $\mathbf{B}$ field, the last term is equal to

$$\varepsilon_{\mu\nu\lambda} D_\nu \frac{\partial}{\partial t} B_\lambda = -\varepsilon_{\mu\nu\lambda} D_\nu \varepsilon_{\lambda\kappa\tau} \nabla_\kappa E_\tau\,.$$

Putting terms together, one arrives at

$$\rho \, E_\mu + \varepsilon_{\mu\nu\lambda} \, j_\nu \, B_\lambda + \varepsilon_{\mu\nu\lambda} \, \frac{\partial}{\partial t} (D_\nu \, B_\lambda)$$
$$= E_\mu \, \nabla_\nu \, D_\nu - \varepsilon_{\mu\nu\lambda} D_\nu \varepsilon_{\lambda\kappa\tau} \nabla_\kappa E_\tau + \varepsilon_{\mu\nu\lambda} \, \varepsilon_{\nu\kappa\tau} (\nabla_\kappa H_\tau) \, B_\lambda.$$

The right hand side of this equation can be written as a total spatial derivative. First notice that, due to $\varepsilon_{\mu\nu\lambda}\varepsilon_{\lambda\kappa\tau} = \delta_{\mu\kappa}\delta_{\nu\tau} - \delta_{\mu\tau}\delta_{\nu\kappa}$,

$$-\varepsilon_{\mu\nu\lambda} D_\nu \varepsilon_{\lambda\kappa\tau} \nabla_\kappa E_\tau = -D_\nu \nabla_\mu E_\nu + D_\nu \nabla_\nu E_\mu.$$

With

$$-D_\nu \nabla_\mu E_\nu = -\nabla_\mu (D_\nu E_\nu) + E_\nu \nabla_\mu D_\nu$$

and

$$D_\nu \nabla_\nu E_\mu = \nabla_\nu (D_\nu E_\mu) - E_\mu \nabla_\nu D_\nu,$$

one obtains

$$-\varepsilon_{\mu\nu\lambda} D_\nu \varepsilon_{\lambda\kappa\tau} \nabla_\kappa E_\tau + E_\mu \nabla_\nu D_\nu = \nabla_\nu (D_\nu E_\mu - D_\kappa E_\kappa \, \delta_{\mu\nu}) + E_\nu \nabla_\mu D_\nu.$$

The last term on the right hand side is the gradient of the electric energy density u^{el}. By analogy to (8.114), one has

$$E_\nu \nabla_\mu D_\nu = \nabla_\mu \, u^{\text{el}},$$

provided that the medium is hysteresis-free. Then the terms involving the electric fields are equal to

$$-\varepsilon_{\mu\nu\lambda} D_\nu \varepsilon_{\lambda\kappa\tau} \nabla_\kappa E_\tau + E_\mu \nabla_\nu D_\nu = \nabla_\nu T^{\text{el}}_{\nu\mu},$$

where

$$T^{\text{el}}_{\nu\mu} = D_\nu E_\mu - (D_\kappa E_\kappa - u^{\text{el}}) \, \delta_{\mu\nu} \qquad (8.120)$$

is the electric part of the Maxwell stress tensor. By analogy, the term in the momentum balance involving the magnetic fields is equal to

$$\varepsilon_{\mu\nu\lambda} \, \varepsilon_{\nu\kappa\tau} (\nabla_\kappa H_\tau) \, B_\lambda = \nabla_\nu T^{\text{mag}}_{\nu\mu},$$

with the magnetic part

$$T^{\text{mag}}_{\nu\mu} = B_\nu H_\mu - (B_\kappa H_\kappa - u^{\text{mag}}) \, \delta_{\mu\nu} \qquad (8.121)$$

of the Maxwell stress tensor. The total stress tensor is

$$T_{\nu\mu} = T_{\nu\mu}^{\text{el}} + T_{\nu\mu}^{\text{mag}}. \tag{8.122}$$

Finally, the momentum balance is found to be

$$\nabla_\nu T_{\nu\mu} = \frac{\partial}{\partial t}(\varepsilon_{\mu\nu\lambda} D_\nu B_\lambda) + \rho E_\mu + \varepsilon_{\mu\nu\lambda} j_\nu B_\lambda. \tag{8.123}$$

The term on the left hand side is the divergence of the momentum flux density, the first term on the right side is the time change of the momentum density of the electromagnetic field. The other terms on the right hand side describe the time change of the momentum density of matter.

Integration of (8.123) over a volume V and application of the Gauss theorem yields

$$\oint_{\partial V} n_\nu T_{\nu\mu} d^2 s = \frac{d}{dt} \int_V \varepsilon_{\mu\nu\lambda} D_\nu B_\lambda d^3 r + \int_V (\rho E_\mu + \varepsilon_{\mu\nu\lambda} j_\nu B_\lambda) d^3 r. \tag{8.124}$$

The balance equations (8.123) and (8.124) show that

$$\mathbf{D} \times \mathbf{B} \tag{8.125}$$

is the density of the linear momentum of the electromagnetic field. For fields in vacuum, this quantity is proportional to the energy flux density **S**, cf. (8.110), viz.

$$\mathbf{D} \times \mathbf{B} = \frac{1}{c^2} \mathbf{E} \times \mathbf{H} = \frac{1}{c^2} \mathbf{S}, \tag{8.126}$$

where c is the speed of light in vacuum.

8.5.5 Angular Momentum in Electrodynamics

Multiplication of the momentum balance equation (8.123) by $\varepsilon_{\kappa\tau\mu} r_\tau$ and use of

$$\varepsilon_{\kappa\tau\mu} r_\tau \nabla_\nu T_{\nu\mu} = \nabla_\nu(\varepsilon_{\kappa\tau\mu} r_\tau T_{\nu\mu}) - \varepsilon_{\kappa\tau\mu} T_{\nu\mu} \nabla_\nu r_\tau,$$

together with $\nabla_\nu r_\tau = \delta_{\nu\tau}$, leads to the angular momentum balance equation

$$\nabla_\nu(\varepsilon_{\kappa\tau\mu} r_\tau T_{\nu\mu}) = \varepsilon_{\kappa\nu\mu} T_{\nu\mu} + \varepsilon_{\kappa\tau\mu} r_\tau \left(\frac{\partial}{\partial t}(\varepsilon_{\mu\nu\lambda} D_\nu B_\lambda) + \rho E_\mu + \varepsilon_{\mu\nu\lambda} j_\nu B_\lambda \right). \tag{8.127}$$

Notice that

$$\varepsilon_{\kappa\nu\mu}T_{\nu\mu} = \varepsilon_{\kappa\nu\mu}\,D_\nu\,E_\mu + \varepsilon_{\kappa\nu\mu}\,B_\nu\,H_\mu = (\mathbf{D}\times\mathbf{E})_\kappa + (\mathbf{B}\times\mathbf{H})_\kappa. \qquad (8.128)$$

Thus integration of the local balance equation over a volume V leads to

$$\frac{\mathrm{d}}{\mathrm{d}t}\int_V \mathbf{r}\times(\mathbf{D}\times\mathbf{B})\mathrm{d}^3r = -\int_V (\mathbf{D}\times\mathbf{E}+\mathbf{B}\times\mathbf{H})\mathrm{d}^3r - \int_V \mathbf{r}\times(\rho\mathbf{E}+\mathbf{j}\times\mathbf{B})\mathrm{d}^3r, \quad (8.129)$$

provided that the surface integral vanishes. The term on the left hand side is the time change of the angular momentum of the electromagnetic field, compensated by torques exerted by matter. So the torque of the field on the matter is equal to the right hand side, just with the opposite sign.

For electrically neutral matter with $\rho = 0$ and $\mathbf{j} = 0$, the torque acting on matter is

$$\int_V (\mathbf{D}\times\mathbf{E}+\mathbf{B}\times\mathbf{H})\mathrm{d}^3r.$$

Due to

$$\mathbf{D} = \varepsilon_0\,\mathbf{E}+\mathbf{P}, \quad \mathbf{B} = \mu_0\,(\mathbf{H}+\mathbf{M}),$$

and $\mathbf{D}\times\mathbf{E} = \mathbf{P}\times\mathbf{E}$, as well as $\mathbf{B}\times\mathbf{H} = \mu_0\mathbf{M}\times\mathbf{H} = \mathbf{M}\times\mathbf{B}$, the corresponding torque $\mathbf{T}$ acting on matter characterized by the electric polarization $\mathbf{P}$ and the magnetization $\mathbf{M}$ is the sum of the electric and magnetic torques $\mathbf{T}^{\mathrm{el}}$ and $\mathbf{T}^{\mathrm{mag}}$ determined by

$$\mathbf{T} = \mathbf{T}^{\mathrm{el}} + \mathbf{T}^{\mathrm{mag}}, \quad \mathbf{T}^{\mathrm{el}} = \int_V \mathbf{P}\times\mathbf{E}\,\mathrm{d}^3r, \quad \mathbf{T}^{\mathrm{mag}} = \int_V \mathbf{M}\times\mathbf{B}\,\mathrm{d}^3r. \qquad (8.130)$$

For a spatially constant electric field, the integration of $\mathbf{P}$ over the volume yields the electric dipole moment $\mathbf{p}^{\mathrm{el}}$, cf. Sect. 10.4.2. Similarly, for a constant magnetic field, the integration of $\mathbf{M}$ gives the magnetic moment $\mathbf{m}$. The relation (8.130) then reduces to the expressions (5.31) for the electric and magnetic torques.

The pertaining torque density $\mathbf{t}$ acting on the electric polarization $\mathbf{P}$ and the magnetization $\mathbf{M}$ is

$$\mathbf{t} = \mathbf{P}\times\mathbf{E}+\mathbf{M}\times\mathbf{B}. \qquad (8.131)$$

Part II
Advanced Topics

Part II
Advanced Topics

Chapter 9
Irreducible Tensors

Abstract At the begin of the more advanced part of the book, irreducible, i.e. symmetric traceless tensors of any rank are treated in this chapter. Products of irreducible tensors and contractions, a relation to Legendre polynomials as well as spherical components of tensors are pointed out. Cubic tensors and cubic harmonics associated with cubic symmetry are presented.

9.1 Definition and Examples

An arbitrary tensor $A_{\mu_1\mu_2...\mu_{\ell-1}\mu_\ell}$ of rank ℓ can be reduced to a tensor of rank $\ell - 1$ with the help of the epsilon-tensor:

$$\varepsilon_{\mu\mu_{\ell-1}\mu_\ell} A_{\mu_1\mu_2...\mu_{\ell-1}\mu_\ell}. \tag{9.1}$$

Similarly, the contraction

$$A_{\mu_1\mu_2...\mu_{\ell-2}\mu\mu} = \delta_{\mu_{\ell-1}\mu_\ell} A_{\mu_1\mu_2...\mu_{\ell-1}\mu_\ell} \tag{9.2}$$

yields a tensor of rank $\ell - 2$. A reduction of the rank of a tensor is not possible, when the tensor is symmetric with respect to the interchange of any pair of subscripts and when its partial trace vanishes, i.e. whenever two of its indices are equal and summed over. Such a tensor, which cannot be reduced to a lower-rank tensor is an *irreducible* tensor, sometimes it is also referred to as *symmetric traceless*. Just as in the case of second rank tensors, cf. (3.3), the symbol $\overline{\cdots}$ indicates the irreducible part of a tensor, of rank ℓ:

$$\overline{A_{\mu_1\mu_2...\mu_{\ell-1}\mu_\ell}}. \tag{9.3}$$

Scalars and vectors, tensors of rank $\ell = 0$ and $\ell = 1$, are irreducible tensors, per se. The irreducible part of a second rank tensor $A_{\mu\nu}$ is

$$\overline{A_{\mu\nu}} = \frac{1}{2}\left(A_{\mu\nu} + A_{\nu\mu}\right) - \frac{1}{3}A_{\lambda\lambda}\,\delta_{\mu\nu},$$

© Springer International Publishing Switzerland 2015
S. Hess, *Tensors for Physics*, Undergraduate Lecture Notes in Physics,
DOI 10.1007/978-3-319-12787-3_9

see Sects. 3.1.2 and Chap. 6. The number of independent components of an irreducible tensor is of rank ℓ is

$$2\ell + 1.$$

This can be verified easily for the cases $\ell = 0, 1, 2$, treated so far. In general, an irreducible Cartesian tensor $A_{\mu_1\mu_2...\mu_{\ell-1}\mu_\ell}$ of rank ℓ, is isomorphic to a corresponding spherical tensor with components $A_{\ell m}$. The integer m, with $-\ell \le m \le \ell$, has $2\ell + 1$ possible values. The connection between Cartesian and spherical components is discussed later, in Sect. 9.4. Some examples for irreducible tensors of rank 3 and 4 are presented next.

Let a_μ be a vector and $A_{\nu\lambda} = \overline{A_{\nu\lambda}}$ an irreducible second rank tensor. The irreducible third rank tensor $\overline{a_\mu A_{\nu\lambda}}$ is explicitly given by

$$\overline{a_\mu A_{\nu\lambda}} = \frac{1}{3}(a_\mu A_{\nu\lambda} + a_\nu A_{\mu\lambda} + a_\lambda A_{\mu\nu}) - \frac{2}{5}a_\kappa(\delta_{\mu\nu}A_{\lambda\kappa} + \delta_{\mu\lambda}A_{\nu\kappa} + \delta_{\nu\lambda}A_{\mu\kappa}). \quad (9.4)$$

The third rank irreducible tensor constructed from the components of the vector a_μ is

$$\overline{a_\mu a_\nu a_\lambda} = a_\mu a_\nu a_\lambda - \frac{1}{5}a^2(a_\mu \delta_{\nu\lambda} + a_\nu \delta_{\mu\lambda} + a_\lambda \delta_{\mu\nu}), \quad (9.5)$$

with $a^2 = a_\kappa a_\kappa = \mathbf{a} \cdot \mathbf{a}$. Similarly, the explicit expression for the irreducible tensor of rank four, constructed from the components of the vector, is

$$\overline{a_\mu a_\nu a_\lambda a_\kappa} = a_\mu a_\nu a_\lambda a_\kappa$$
$$- \frac{1}{7}a^2(a_\mu a_\nu \delta_{\lambda\kappa} + a_\mu a_\lambda \delta_{\nu\kappa} + a_\mu a_\kappa \delta_{\nu\lambda} + a_\nu a_\lambda \delta_{\mu\kappa} + a_\nu a_\kappa \delta_{\mu\lambda} + a_\lambda a_\kappa \delta_{\mu\nu})$$
$$+ \frac{1}{35}a^4(\delta_{\mu\nu}\delta_{\lambda\kappa} + \delta_{\mu\lambda}\delta_{\nu\kappa} + \delta_{\mu\kappa}\delta_{\nu\lambda}). \quad (9.6)$$

Notice: the symbol $\overline{\cdots}$ is defined such that the factor in front of the term $a_{\mu_1}a_{\mu_2}\cdots a_{\mu_\ell}$ in the irreducible tensor $\overline{a_{\mu_1}a_{\mu_2}\ldots a_{\mu_\ell}}$ is 1.

Let $A_{\mu\nu}$ be an irreducible tensor. The irreducible fourth rank tensor constructed from the product of the second rank tensor is

$$\overline{A_{\mu\nu}A_{\lambda\kappa}} = \frac{1}{3}(A_{\mu\nu}A_{\lambda\kappa} + A_{\mu\lambda}A_{\nu\kappa} + A_{\mu\kappa}A_{\lambda\nu})$$
$$- \frac{2}{21}(A_{\mu\tau}A_{\tau\nu}\delta_{\lambda\kappa} + A_{\mu\tau}A_{\tau\lambda}\delta_{\nu\kappa} + A_{\mu\tau}A_{\tau\kappa}\delta_{\nu\lambda} + A_{\nu\tau}A_{\tau\lambda}\delta_{\mu\kappa} + A_{\nu\tau}A_{\tau\kappa}\delta_{\mu\lambda} + A_{\lambda\tau}A_{\tau\kappa}\delta_{\mu\nu})$$
$$+ \frac{2}{105}A_{\tau\sigma}A_{\tau\sigma}(\delta_{\mu\nu}\delta_{\lambda\kappa} + \delta_{\mu\lambda}\delta_{\nu\kappa} + \delta_{\mu\kappa}\delta_{\nu\lambda}). \quad (9.7)$$

In many cases, the explicit form of the symmetric traceless higher rank tensors are not needed. It is one of the advantages of tensor calculus that general properties of tensors often suffice in applications.

A remark on notation is appropriate. As mentioned before in Sect. 3.1.2, alternative ways to write and display irreducible tensors are found in the literature. In additions to the double arrow, the symbols $[\ldots]_0$ or $[[\ldots]]_0$, as well as $\{\ldots\}_0$ were used to indicate the symmetric traceless part of a tensor.

9.2 Products of Irreducible Tensors

Let $A_{\mu\nu}$ and $B_{\lambda\kappa}$ be irreducible tensors. Multiplication of the irreducible fourth rank tensor constructed from the product of the second rank tensor $A_{\mu\nu}$, cf. (9.7), by $B_{\lambda\kappa}$ yields

$$\overline{A_{\mu\nu}A_{\lambda\kappa}}\,B_{\lambda\kappa} = \frac{1}{3}\overline{A_{\mu\nu}(A_{\lambda\kappa}B_{\lambda\kappa})} + \frac{4}{105}\overline{B_{\mu\nu}(A_{\lambda\kappa}A_{\lambda\kappa})}$$
$$+ \frac{2}{3}\overline{A_{\mu\lambda}B_{\lambda\kappa}A_{\kappa\nu}} - \frac{8}{21}\overline{B_{\mu\lambda}A_{\lambda\kappa}A_{\kappa\nu}}. \tag{9.8}$$

For the special case $B = A$, (9.8), reduces to

$$\overline{A_{\mu\nu}A_{\lambda\kappa}}\,A_{\lambda\kappa} = \frac{13}{35}\overline{A_{\mu\nu}(A_{\lambda\kappa}A_{\lambda\kappa})} + \frac{2}{7}\overline{A_{\mu\lambda}A_{\lambda\kappa}A_{\kappa\nu}} = \frac{18}{35}\overline{A_{\mu\nu}(A_{\lambda\kappa}A_{\lambda\kappa})}. \tag{9.9}$$

The last equality follows from $\overline{A_{\mu\lambda}A_{\lambda\kappa}A_{\kappa\nu}} = \frac{1}{2}\overline{A_{\mu\nu}A_{\lambda\kappa}A_{\lambda\kappa}}$, cf. (5.51).

9.1 Exercise: Verify the Required Properties of the Third and Fourth Rank Irreducible Tensors (9.5) and (9.6)
First, verify by inspection, that the tensors explicitly given by (9.5) and (9.6) are symmetric against the interchange of indices, within any pair of components. Then put $\lambda = \nu$, in (9.5) to show $\overline{a_\mu a_\nu a_\nu} = 0$. Likewise, use $\kappa = \lambda$ in (9.6) to verify $\overline{a_\mu a_\nu a_\lambda a_\lambda} = 0$.

9.3 Contractions, Legendre Polynomials

The multiplication of two tensors of rank ℓ constructed from the components of two vectors **a** and **b** and their subsequent total contraction yields a scalar which is proportional to the Legendre polynomial P_ℓ, depending on the cosine of the angle between these two vectors. More specifically:

$$\mathscr{P}_\ell(\mathbf{a}, \mathbf{b}) \equiv \overline{a_{\mu_1}a_{\mu_2}\ldots a_{\mu_\ell}}\,\overline{b_{\mu_1}b_{\mu_2}\ldots b_{\mu_\ell}} = a^\ell\,b^\ell\,N_\ell\,P_\ell(\hat{\mathbf{a}}\cdot\hat{\mathbf{b}}), \tag{9.10}$$

with

$$N_\ell = \frac{\ell!}{(2\ell - 1)!!} = \frac{1 \cdot 2 \cdot 3 \cdots (\ell - 1) \cdot \ell}{1 \cdot 3 \cdot 5 \cdots (2\ell - 3) \cdot (2\ell - 1)}. \tag{9.11}$$

For a proof of this relation see Sect. 10.3.5. The special case $\mathbf{a} = \mathbf{b}$ leads to $\mathscr{P}_\ell(\mathbf{a}, \mathbf{a}) = a^{2\ell} N_\ell$, since $P_\ell(1) = 1$.

Examples for $\ell = 1, 2, 3, 4$ are listed explicitly:

$$\mathscr{P}_1(\mathbf{a}, \mathbf{b}) = a\, b\, P_1(\widehat{\mathbf{a}} \cdot \widehat{\mathbf{b}}), \quad P_1(x) = x, \tag{9.12}$$

$$\mathscr{P}_2(\mathbf{a}, \mathbf{b}) = \frac{2}{3} a^2\, b^2\, P_2(\widehat{\mathbf{a}} \cdot \widehat{\mathbf{b}}), \quad P_2(x) = \frac{3}{2}\left(x^2 - \frac{1}{3}\right), \tag{9.13}$$

$$\mathscr{P}_3(\mathbf{a}, \mathbf{b}) = \frac{2}{5} a^3\, b^3\, P_3(\widehat{\mathbf{a}} \cdot \widehat{\mathbf{b}}), \quad P_3(x) = \frac{5}{2}\left(x^3 - \frac{3}{5}x\right), \tag{9.14}$$

$$\mathscr{P}_4(\mathbf{a}, \mathbf{b}) = \frac{8}{35} a^4\, b^4\, P_4(\widehat{\mathbf{a}} \cdot \widehat{\mathbf{b}}), \quad P_4(x) = \frac{35}{8}\left(x^4 - \frac{6}{7}x^2 + \frac{3}{35}\right). \tag{9.15}$$

Some general properties of Legendre polynomials, in particular the prescription for their evaluation via a generating function, are presented in Sect. 10.3.5.

9.4 Cartesian and Spherical Tensors

9.4.1 Spherical Components of a Vector

Let $\mathbf{e}^{(x)}, \mathbf{e}^{(y)}, \mathbf{e}^{(z)}$ be unit vectors parallel to the coordinate axes. A vector $\mathbf{a}$ is given by the linear combination $\mathbf{a} = a_x \mathbf{e}^{(x)} + a_y \mathbf{e}^{(x)} + a_z \mathbf{e}^{(x)}$. The a_x, a_y, a_z are the standard Cartesian components. The *spherical unit vectors*

$$\mathbf{e}^{(0)} = \mathbf{e}^{(z)}, \quad \mathbf{e}^{(\pm 1)} = \mp \frac{1}{\sqrt{2}}\left(\mathbf{e}^{(x)} \mp i\, \mathbf{e}^{(y)}\right), \tag{9.16}$$

which have the properties

$$\left(\mathbf{e}^{(m)}\right)^* = (-1)^m \mathbf{e}^{(-m)}, \quad \left(\mathbf{e}^{(m)}\right)^* \cdot \mathbf{e}^{(m')} = \delta_{mm'}, \quad m, m' = -1, 0, 1, \tag{9.17}$$

can as well be used as basis vectors. Then the vector $\mathbf{a}$ is represented by

$$\mathbf{a} = a^{(1)}\, \mathbf{e}^{(1)} + a^{(0)}\, \mathbf{e}^{(0)} + a^{(-1)}\, \mathbf{e}^{(-1)} = \sum_{m=-1}^{1} a^{(m)}\, \mathbf{e}^{(m)}, \tag{9.18}$$

with the *spherical components*

$$a^{(m)} = \mathbf{a} \cdot (\mathbf{e}^{(m)})^* = (-1)^m \mathbf{a} \cdot \mathbf{e}^{(-m)}. \tag{9.19}$$

Explicitly, the relation between the spherical and Cartesian components is

$$a^{(0)} = a_z, \quad a^{(\pm 1)} = \mp \frac{1}{\sqrt{2}} (a_x \pm i\, a_y). \tag{9.20}$$

The unit vector $\hat{\mathbf{r}} = r^{-1}\mathbf{r}$ is represented in terms of the spherical polar angles ϑ and φ according to $\{\cos\varphi \sin\vartheta, \sin\varphi \sin\vartheta, \cos\vartheta\}$. Then the spherical components of the position vector $\mathbf{r}$ are

$$r^{(0)} = r\cos\vartheta, \quad r^{(\pm 1)} = \mp \frac{1}{\sqrt{2}} r \sin\vartheta \exp(\pm i\,\varphi). \tag{9.21}$$

Apart from the numerical factor $\sqrt{4\pi/3}$, the spherical components of $\hat{\mathbf{r}} = r^{-1}\mathbf{r}$ are equal to the first order spherical harmonics $Y_1^m(\hat{\mathbf{r}})$. Similarly, for any vector $\mathbf{a} = a\,\hat{\mathbf{a}}$, one has

$$a^{(m)} = a \sqrt{\frac{4\pi}{3}} Y_1^m(\hat{\mathbf{a}}). \tag{9.22}$$

The generalization of the interrelation between Cartesian and spherical components to tensors of rank $\ell > 1$ is discussed next.

9.4.2 Spherical Components of Tensors

Let $S_{\mu_1\mu_2\cdots\mu_\ell}$ be a symmetric traceless tensor of rank ℓ, given by its Cartesian components. By analogy to (9.19) the pertaining spherical components $S_\ell^{(m)}$ are obtained by the scalar multiplication with ℓ-fold product of the Cartesian components of the vectors $\mathbf{e}^{(m)*}$, viz.

$$S_\ell^{(m)} = \sum_{m_1=-1}^{1} \cdots \sum_{m_\ell=-1}^{1} (-1)^m$$
$$\times S_{\mu_1\mu_2\cdots\mu_\ell} e_{\mu_1}^{(-m_1)} e_{\mu_2}^{(-m_2)} \cdots e_{\mu_\ell}^{(-m_\ell)} \delta(m, m_1 + m_2 + \ldots + m_\ell). \tag{9.23}$$

Here, the notation $\delta(m, m_1 + m_2 + \ldots + m_\ell)$ is used for the Kronecker delta symbol, i.e. $\delta(\cdots) = 1$, for $m_1 + m_2 + \ldots + m_\ell = m$, and $\delta(\cdots) = 0$, otherwise. By definition, the possible values for m are the integer numbers $m = -\ell, -\ell + 1, \ldots, 0, \ldots, \ell - 1, \ell$. Clearly, the irreducible tensor of rank ℓ has $2\ell + 1$ spherical components.

The isomorphic Cartesian tensor has the same number of independent components, though it is not obvious from its notation.

The expression (9.23) can also be applied to the irreducible tensor $\overline{a_{\mu_1} a_{\mu_2} \cdots a_{\mu_\ell}}$ constructed from the components of the vector **a**. The resulting spherical components $a_\ell^{(m)}$ are related to the ℓ-th order spherical harmonic $Y_\ell^{(m)}(\widehat{\mathbf{a}})$ by

$$a_\ell^{(m)} = a^\ell \sqrt{\frac{4\pi \, \ell!}{(2\ell + 1)!!}} \, Y_\ell^{(m)}(\widehat{\mathbf{a}}). \tag{9.24}$$

The scalar product, i.e. the total contraction of the two tensors in (9.10) can also be expressed in terms of the pertaining spherical components, where the sum over m corresponds to a scalar product, viz.

$$\overline{a_{\mu_1} a_{\mu_2} \ldots a_{\mu_\ell}} \, \overline{b_{\mu_1} b_{\mu_2} \ldots b_{\mu_\ell}} = \sum_{m=-\ell}^{\ell} a_\ell^{(m)} \left(b_\ell^{(m)} \right)^* \tag{9.25}$$

$$= a^\ell b^\ell \frac{4\pi \, \ell!}{(2\ell + 1)!!} \sum_{m=-\ell}^{\ell} Y_\ell^{(m)}(\widehat{\mathbf{a}}) \left(Y_\ell^{(m)} \right)^* (\widehat{\mathbf{b}}).$$

Comparison of this equation with the right hand side of (9.10) yields the relation

$$P_\ell(\widehat{\mathbf{a}} \cdot \widehat{\mathbf{b}}) = \frac{4\pi}{(2\ell + 1)} \sum_{m=-\ell}^{\ell} Y_\ell^{(m)}(\widehat{\mathbf{a}}) \left(Y_\ell^{(m)} \right)^* (\widehat{\mathbf{b}}), \tag{9.26}$$

which expresses the Legendre polynomial with a scalar product of spherical harmonics.

For ease of reference, the first few spherical harmonics, for $\ell = 0, 1, 2$, are listed here, where the Cartesian components of **r** are denoted by $\{x, y, z\}$, and the coefficients $c_{(\ell)} = \sqrt{(2\ell + 1)!!/4\pi}$ are used:

$$Y_0^{(0)} = c_{(0)} = 1/\sqrt{4\pi},$$

$$Y_1^{(0)} = c_{(1)} \, r^{-1} z = c_{(1)} \cos \vartheta,$$

$$Y_1^{(\pm 1)} = \mp c_{(1)} \frac{1}{\sqrt{2}} r^{-1}(x \pm i \, y) = \mp c_{(1)} \frac{1}{\sqrt{2}} \sin \vartheta \, \exp(\pm i \, \varphi), \tag{9.27}$$

$$Y_2^{(0)} = c_{(2)} \frac{\sqrt{3}}{2} r^{-2} \left(z^2 - \frac{1}{3} r^2 \right) = c_{(2)} \frac{\sqrt{3}}{2} \left(\cos^2 \vartheta - \frac{1}{3} \right),$$

$$Y_2^{(\pm 1)} = \mp c_{(2)} \frac{1}{\sqrt{2}} r^{-2} z \, (x \pm i \, y) = \mp c_{(2)} \frac{1}{\sqrt{2}} \sin \vartheta \cos \vartheta \, \exp(\pm i \, \varphi),$$

$$Y_2^{(\pm 2)} = c_{(2)} \frac{1}{2\sqrt{2}} r^{-2} (x \pm i\, y)^2 = c_{(2)} \frac{1}{2\sqrt{2}} \sin^2 \vartheta \, \exp(\pm 2\, i\, \varphi). \qquad (9.28)$$

For all ℓ one has

$$Y_\ell^{(m)} \sim \exp(m\, i\, \varphi). \qquad (9.29)$$

Furthermore, the $Y_\ell^{(0)}$, which do not depend on φ, are proportional to the Legendre polynomials $P_\ell(\cos \vartheta)$, viz. $Y_\ell^{(0)} = \sqrt{(2\ell + 1)/4\pi}\, P_\ell$. The spherical harmonics $Y_\ell^{(0)}$, as well as $Y_\ell^{(m)} + Y_\ell^{(m)*}$ and $i(Y_\ell^{(m)} - Y_\ell^{(m)*})$ are real functions.

Notice: the name *spherical harmonics* does not indicate the symmetry of these functions, but rather their dependence on the polar angles of spherical coordinates. As far as symmetry is concerned, a preferential axis, usually chosen as the z-axis, is linked with the spherical harmonics. In Quantum Mechanics, this reference axis is also referred to as *quantization axis*.

9.5 Cubic Harmonics

9.5.1 Cubic Tensors

Irreducible Cartesian tensors, which reflect the symmetry of cubic crystals are tensors of ranks $\ell = 4, 6, \ldots$. Let $\mathbf{e}^{(i)}$, with $i = 1, 2, 3$ be unit vectors parallel to the axes of a cubic crystal. The first of these tensors with full cubic symmetry, as used in [25], are

$$H_{\mu\nu\lambda\kappa}^{(4)} \equiv \sum_{i=1}^{3} \overline{e_\mu^{(i)} e_\nu^{(i)} e_\lambda^{(i)} e_\kappa^{(i)}} = \sum_{i=1}^{3} e_\mu^{(i)} e_\nu^{(i)} e_\lambda^{(i)} e_\kappa^{(i)} - \frac{1}{5} (\delta_{\mu\nu}\delta_{\lambda\kappa} + \delta_{\mu\lambda}\delta_{\nu\kappa} + \delta_{\mu\kappa}\delta_{\nu\lambda}),$$

$$(9.30)$$

$$H_{\mu\nu\lambda\kappa\sigma\tau}^{(6)} \equiv \overline{e_\mu^{(1)} e_\nu^{(1)} e_\lambda^{(2)} e_\kappa^{(2)} e_\sigma^{(3)} e_\tau^{(3)}}, \qquad H_{\mu_1 \cdots \mu_8}^{(8)} \equiv \sum_{i=1}^{3} \overline{e_{\mu_1}^{(i)} \cdots e_{\mu_8}^{(i)}}. \qquad (9.31)$$

These tensors are invariant against the exchange of the cubic axes.

9.5.2 Cubic Harmonics with Full Cubic Symmetry

Multiplication of $H_{\mu\nu\lambda\kappa}^{(4)}$ with the irreducible fourth rank tensor $\overline{\hat{r}_\mu \hat{r}_\nu \hat{r}_\lambda \hat{r}_\kappa}$, constructed from the components of the unit vector $\hat{\mathbf{r}} = r^{-1} \mathbf{r}$, yields

$$H_4 \equiv H^{(4)}_{\mu\nu\lambda\kappa} \, \overline{\widehat{r}_\mu \widehat{r}_\nu \widehat{r}_\lambda \widehat{r}_\kappa} = x^4 + y^4 + z^4 - \frac{3}{5}. \tag{9.32}$$

Here x, y, z stand for the components of the unit vector $\widehat{\mathbf{r}}$ with respect to the cubic axes $\mathbf{e}^{(i)}$, e.g. $x = \widehat{r}_\mu e^{(1)}_\mu$. The function H_4 is proportional to the fourth order cubic harmonic K_4 with the full cubic symmetry. Cubic harmonics were introduced in [26], and also used in [27, 28]. These real functions can be expressed in terms of linear combinations of spherical harmonics, e.g.

$$K_4 \equiv \frac{5}{4}\sqrt{21}\, H_4 = \sqrt{4\pi}\left[\sqrt{\frac{7}{12}} Y_4^{(0)} + \sqrt{\frac{5}{6}}\frac{1}{2}\left(Y_4^{(4)} + Y_4^{(-4)}\right)\right]. \tag{9.33}$$

The other 8 cubic harmonics of order 4 can be found in [26–28]. Similarly, of the 13 and 17 cubic harmonics of order 6 and 8, only those are presented here, which are obtained by analogy to (9.32), from the tensors $H^{(6)}_{...}$ and $H^{(8)}_{...}$:

$$K_6 \equiv \frac{231}{8}\sqrt{26}\, H_6, \quad H_6 = x^2 y^2 z^2 - \frac{1}{105} + \frac{1}{22} H_4, \tag{9.34}$$

$$K_8 \equiv \frac{65}{16}\sqrt{561}\, H_8, \quad H_8 = x^8 + y^8 + z^8 - \frac{1}{3} - \frac{28}{5} H_6 - \frac{210}{143} H_4. \tag{9.35}$$

The numerical factors occurring in the relations between the K_ℓ and H_ℓ, for $\ell = 4, 6, 8$, are chosen such that the orientational averages of K_ℓ^2 are equal to 1. The function K_6 is proportional to a linear combination of $Y_6^{(0)}$ and $Y_6^{(4)} + Y_6^{(-4)}$. For K_8, it is $Y_8^{(0)}$, $Y_8^{(4)} + Y_8^{(-4)}$, and $Y_8^{(8)} + Y_8^{(-8)}$, cf. [28].

The values of the functions H_4, H_6, and H_8, taken at the positions of the first and second coordination shell of *simple cubic* (**sc**), *body centered cubic* (**bcc**), and *face centered cubic* (**fcc**) crystals, are characteristic for these different types of cubic crystals. The coordinates $\{x, y, z\}$ of a representative nearest neighbor are $\{1, 0, 0\}$ for the sc, $\{1/\sqrt{3}, 1/\sqrt{3}, 1/\sqrt{3}\}$ for the bcc, and $\{1/\sqrt{2}, 1/\sqrt{2}, 0\}$ for the fcc crystal. Then H_4, H_6, H_8 are equal to 2/5, 2/231, 2/65, respectively, for sc, the simple cubic crystal. The corresponding values for bcc, the body centered cubic crystal, are $-4/15$, 32/2079, 16/1755. For fcc, the face centered cubic structure, these values are $-1/10$, $-13/924$, 9/520.

Chapter 10
Multipole Potentials

Abstract In this chapter descending and ascending multipole potentials are introduced, their properties are discussed and the dipole, quadrupole and octupole potentials are considered in more detail. An application is the multipole expansion of electrostatics, the multipole moments, like electric dipole, quadrupole, octupole moments are defined. Further applications in electrodynamics are the calculation of the induced dipole moment of a metal sphere, the electric polarization expressed as dipole density, the determination of the energy of multipole moments in an external field, as well as the multipole-multipole interaction. An application of the multipole expansion for the pressure and velocity in hydrodynamics yields the Stokes force acting on sphere.

Multipole potentials are tensorial solutions of the Laplace equation $\Delta\phi_{...} = 0$. Depending on their behavior for $r \to \infty$ and at $r = 0$, *descending* and *ascending* multipole potentials are distinguished. More specifically,

descending multipole potentials approach 0 for $r \to \infty$, and diverge for $r \to 0$, *ascending multipole potentials* are 0 at $r = 0$ and diverge for $r \to \infty$.

10.1 Descending Multipoles

10.1.1 Definition of the Multipole Potential Functions

As noticed before, cf. Sect. 7.6.3, the spherical symmetric solution of the Laplace equation, which vanishes for $r \to \infty$, is

$$X_0 = r^{-1}. \tag{10.1}$$

It is understood that $r > 0$. The spatial derivative $\nabla_\mu r^{-1} \equiv \frac{\partial}{\partial r_\mu} r^{-1}$ is also a solution of the Laplace equation. This follows from

$$\Delta\left(\frac{\partial}{\partial r_\mu} r^{-1}\right) = \frac{\partial^2}{\partial r_\lambda \partial r_\lambda}\left(\frac{\partial}{\partial r_\mu} r^{-1}\right) = \frac{\partial}{\partial r_\mu}\Delta r^{-1} = 0.$$

© Springer International Publishing Switzerland 2015
S. Hess, *Tensors for Physics*, Undergraduate Lecture Notes in Physics,
DOI 10.1007/978-3-319-12787-3_10

The same applies for a ℓ-fold spatial differentiation of r^{-1}. In this spirit, Cartesian tensors of rank ℓ are defined by

$$X_{\mu_1\mu_2\cdots\mu_\ell} \equiv (-1)^\ell \frac{\partial^\ell}{\partial r_{\mu_1} \partial r_{\mu_2} \cdots \partial r_{\mu_\ell}} r^{-1} = (-1)^\ell \nabla_{\mu_1} \nabla_{\mu_2} \cdots \nabla_{\mu_\ell} r^{-1}. \quad (10.2)$$

The tensorial functions $X_{...}$ approach 0 for $r \to \infty$. By definition, these tensors are symmetric. Whenever two subscripts are equal and summed over, these two spatial derivatives are equivalent to Δ and consequently 0 is obtained. Thus the spatial differentiation (10.2) yields irreducible tensors of rank ℓ. These are the *descending multipole potentials*. Due to the definition (10.2), the ℓ-th multipole potential is related to the $\ell - 1$ function by

$$X_{\mu_1\mu_2\cdots\mu_{\ell-1}\mu_\ell} = -\frac{\partial}{\partial r_{\mu_\ell}} X_{\mu_1\mu_2\cdots\mu_{\ell-1}} = -\nabla_{\mu_\ell} X_{\mu_1\mu_2\cdots\mu_{\ell-1}}. \quad (10.3)$$

10.1.2 Dipole, Quadrupole and Octupole Potentials

Examples for multipole potential tensors of rank $\ell = 1, 2, 3$ are the

dipole potential

$$X_\mu = r^{-3} r_\mu = r^{-2} \widehat{r}_\mu, \quad (10.4)$$

the *quadrupole potential*

$$X_{\mu\nu} = 3 r^{-5} \left(r_\mu r_\nu - \frac{1}{3} r^2 \delta_{\mu\nu} \right) = 3 r^{-5} \overline{r_\mu r_\nu} = 3 r^{-3} \overline{\widehat{r}_\mu \widehat{r}_\nu}, \quad (10.5)$$

and the *octupole potential*

$$X_{\mu\nu\lambda} = 15 r^{-7} \overline{r_\mu r_\nu r_\lambda} = 15 r^{-4} \overline{\widehat{r}_\mu \widehat{r}_\nu \widehat{r}_\lambda}. \quad (10.6)$$

The reason for the names *dipole*, *quadrupole* and *octupole* potential is seen in Sect. 10.3.

10.1.3 Source Term for the Quadrupole Potential

The formulas presented here and in the following for the multipole potential tensors are valid for $r > 0$. When a source term is included at $r = 0$, the second rank multipole tensor obeys the relation

$$\nabla_\mu \nabla_\nu r^{-1} = X_{\mu\nu}(\mathbf{r}) - \frac{4\pi}{3}\delta_{\mu\nu}\delta(\mathbf{r}),\tag{10.7}$$

where $X_{\mu\nu}(\mathbf{r})$ is given by (10.5).

To verify (10.7), integrate it over a sphere with a finite radius R, centered at $\mathbf{r} = 0$. More specifically, use the Gauss theorem for $\int \nabla_\mu(\nabla_\nu r^{-1})d^3r$ to obtain

$$\int \nabla_\mu \nabla_\nu r^{-1} d^3r = R^2 \int \hat{r}_\mu \left(\nabla_\nu r^{-1}\right)_{r=R} d^2\hat{r} = -\int \hat{r}_\mu \hat{r}_\nu d^2\hat{r} = -(4\pi/3)\,\delta_{\mu\nu}.$$

Notice that $d^3r = r^2 dr d^2\hat{r}$, that the integral of the symmetric traceless tensor $X_{\mu\nu}$ over the sphere vanishes since $\int \overline{\hat{r}_\mu \hat{r}_\nu}\, d^2\hat{r} = 0$, and that the δ-function has the property $\int \delta(\mathbf{r})d^3r = 1$.

The trace of (10.7) yields

$$\Delta r^{-1} = -4\pi\,\delta(\mathbf{r}).\tag{10.8}$$

The Poisson equation of electrostatics or the electric potential $\phi \sim r^{-1}$ in vacuum, caused by a point charge q located at $r = 0$, viz. $\varepsilon_0 \Delta\phi = -4\pi q\delta(\mathbf{r})$, is mathematically equivalent to (10.8). The Poisson equation, in turn, is a consequence of the Gauss law of electrodynamics, cf. Sect. 7.5.

10.1.4 General Properties of Multipole Potentials

In general, the ℓ-th multipole potential can be written as

$$X_{\mu_1\mu_2\cdots\mu_\ell} = (2\ell - 1)!!\, r^{-(2\ell+1)}\, \overline{r_{\mu_1}r_{\mu_2}\cdots r_{\mu_\ell}},\tag{10.9}$$

or equivalently

$$X_{\mu_1\mu_2\cdots\mu_\ell}(\mathbf{r}) = r^{-(\ell+1)}\, Y_{\mu_1\mu_2\cdots\mu_\ell}(\hat{\mathbf{r}}),\tag{10.10}$$

with the tensors

$$Y_{\mu_1\mu_2\cdots\mu_\ell}(\hat{\mathbf{r}}) = (2\ell - 1)!!\, \overline{\hat{r}_{\mu_1}\hat{r}_{\mu_2}\cdots\hat{r}_{\mu_\ell}}.\tag{10.11}$$

The irreducible tensors $Y_{...}$ depend on the direction of $\mathbf{r}$ only. With the unit vector $\hat{\mathbf{r}}$ expressed in terms of the polar angles, the components of the ℓ-th rank Cartesian tensor are isomorphic to the spherical harmonics Y_ℓ^m.

Clearly, cf. (10.10), the ℓ-th rank multipole potential is proportional to

$$r^{-(\ell+1)}.$$

Multiplication of $X_{\mu_1\mu_2\cdots\mu_\ell}$ by the components of $\mathbf{r}$ yields a tensor of rank $\ell + 1$. The contraction with the position vector, on the other hand, gives a tensor of rank $\ell - 1$, which is proportional to a multipole potential, viz.

$$r_{\mu_\ell} X_{\mu_1\mu_2\cdots\mu_{\ell-1}\mu_\ell} = \ell\, X_{\mu_1\mu_2\cdots\mu_{\ell-1}}. \tag{10.12}$$

The proof is as follows. On account of (10.3) and $r_{\mu_\ell}\frac{\partial}{\partial r_{\mu_\ell}} = r\frac{\partial}{\partial r}$, the left hand side of (10.12) is equal to $-r\frac{\partial}{\partial r}X_{\mu_1\mu_2\cdots\mu_{\ell-1}}$. The multipole potential of rank $\ell - 1$ is proportional to $r^{-\ell}$, cf. (10.10). Thus the differentiation with respect to the magnitude r yields the result (10.12).

Let $g = g(r)$ be a function which depends on $\mathbf{r}$ only via the magnitude $r = |\mathbf{r}|$. Then the Laplace operator applied on the product of $g(r)$ and the multipole potential of rank ℓ is equal to

$$\Delta\left(g(r)X_{\mu_1\mu_2\cdots\mu_\ell}\right) = \left(g'' - 2\ell r^{-1} g'\right) X_{\mu_1\mu_2\cdots\mu_\ell}(\mathbf{r}) = r^{2\ell}\left(r^{-2\ell}g'\right)' X_{\mu_1\mu_2\cdots\mu_\ell}(\mathbf{r})$$
$$\ell \geq 1, \tag{10.13}$$

where the prime $'$ indicates the derivative with respect to r. When the case $\ell = 0$ is associated with the monopole function r^{-1}, then (10.13) also applies for $\ell = 0$. The proof of (10.13) is transferred to the following exercise.

10.1 Exercise: Prove the Product Rule (10.13) for the Laplace Operator
Hint: use $\Delta(fg) = f\Delta g + 2(\nabla_\kappa f)(\nabla_\kappa g) + g\Delta f$, for any two functions f and g.

10.2 Exercise: Multipole Potentials in D Dimensional Space
In D dimensions, $r^{(2-D)}$ is the radially symmetric solution of the Laplace equation, cf. Exercise 7.6, for $D \geq 3$. By analogy with (10.2), D dimensional multipole potential tensors are defined by

$$X^{D}_{\mu_1\mu_2\cdots\mu_\ell} \equiv (-1)^\ell \frac{\partial^\ell}{\partial r_{\mu_1}\partial r_{\mu_2}\cdots\partial r_{\mu_\ell}} r^{(2-D)} = (-1)^\ell \nabla_{\mu_1}\nabla_{\mu_2}\cdots\nabla_{\mu_\ell} r^{(2-D)}, \tag{10.14}$$

where now ∇ is the in D dimensional nabla operator. For $D = 2$, $r^{(2-D)}$ is replaced by $-\ln r$. Compute the first and second multipole potentials, for $D \geq 3$ and for $D = 2$.

10.2 Ascending Multipoles

The factor $(g'' - 2\ell r^{-1}g') = r^{2\ell}(r^{-2\ell}g')'$ in (10.13) is equal to zero not only for $g = 1$, but also for $g = r^{(2\ell+1)}$. This implies: the tensors

$$\tilde{X}_{\mu_1\mu_2\cdots\mu_\ell} \equiv r^{(2\ell+1)} X_{\mu_1\mu_2\cdots\mu_\ell} = r^\ell\, Y_{\mu_1\mu_2\cdots\mu_\ell} \tag{10.15}$$

are also solutions of the Laplace equation. These are the *ascending multipole potentials*. They are proportional to r^ℓ and thus 0 for $r = 0$. Apart from the factor $(2\ell - 1)!!$, the ascending multipoles are just the irreducible tensors constructed from the components of **r**, viz.

$$\tilde{X}_{\mu_1\mu_2\cdots\mu_\ell} = (2\ell - 1)!! \; \overline{r_{\mu_1} r_{\mu_2} \cdots r_{\mu_\ell}} . \tag{10.16}$$

Examples for $\ell = 0, 1, 2, 3$ are $\tilde{X} = 1$, $\tilde{X}_\mu = r_\mu$ and

$$\tilde{X}_{\mu\nu} = 3 \, \overline{r_\mu r_\nu} = 3 r_\mu r_\nu - r^2 \delta_{\mu\nu}, \tag{10.17}$$

$$\tilde{X}_{\mu\nu\lambda} = 15 \, \overline{r_\mu r_\nu r_\lambda} = 3 \left(5 r_\mu r_\nu r_\lambda - r^2 \left(r_\mu \delta_{\nu\lambda} + r_\nu \delta_{\mu\lambda} + r_\lambda \delta_{\mu\nu} \right) \right) . \tag{10.18}$$

By analogy to (10.12), the multiplication of the symmetric traceless tensor of rank ℓ, constructed from the components of the vector **r**, by a component of this vector and subsequent contraction, yields a corresponding tensor of rank $\ell - 1$. In particular, (10.12) implies

$$r_{\mu_\ell} \, \overline{r_{\mu_1} r_{\mu_2} \cdots r_{\mu_{\ell-1}} r_{\mu_\ell}} = \frac{\ell}{2\ell - 1} r^2 \, \overline{r_{\mu_1} r_{\mu_2} \cdots r_{\mu_{\ell-1}}} . \tag{10.19}$$

Similarly, from $\nabla_{\mu_\ell} X_{\mu_1 \mu_2 \cdots \mu_{\ell-1} \mu_\ell} = 0$ and (10.9) follows

$$\nabla_{\mu_\ell} \, \overline{r_{\mu_1} r_{\mu_2} \cdots r_{\mu_{\ell-1}} r_{\mu_\ell}} = \ell \frac{2\ell + 1}{2\ell - 1} \, \overline{r_{\mu_1} r_{\mu_2} \cdots r_{\mu_{\ell-1}}} . \tag{10.20}$$

For $\ell = 1$, the relations $r_\mu r_\mu = r^2$ and $\nabla_\mu r_\mu = 3$ are recovered. The nontrivial case $\ell = 2$ can be used to verify the factors occurring on the right hand side of (10.19) and (10.20).

10.3 Multipole Expansion and Multipole Moments in Electrostatics

10.3.1 Coulomb Force and Electrostatic Potential

The Coulomb force exerted on a charge q at the position **r**, by a charge Q, located at the position $\mathbf{r}' = 0$, is

$$\mathbf{F} = \frac{q \, Q}{4\pi\varepsilon_0} \frac{1}{r^3} \mathbf{r},$$

cf. (8.81). This force can be written as the negative gradient of $q\phi$, viz. $F_\mu = -\nabla_\mu q\phi$, where

$$\phi = \frac{Q}{4\pi\varepsilon_0}\frac{1}{r},$$

is the electrostatic potential. For N charges q_i, located at positions $\mathbf{r}^i$ the corresponding expression is

$$\phi = \frac{1}{4\pi\varepsilon_0}\sum_{i=1}^{N}\frac{q_i}{|\mathbf{r}-\mathbf{r}^i|}. \tag{10.21}$$

The generalization to a continuous charge density $\rho(\mathbf{r}')$ is the electrostatic potential $\phi(\mathbf{r})$ given by the integral

$$\phi(\mathbf{r}) = \frac{1}{4\pi\varepsilon_0}\int\frac{\rho(\mathbf{r}')}{|\mathbf{r}-\mathbf{r}'|}\,d^3r'. \tag{10.22}$$

By analogy to the mass density, cf. Sect. 8.3.2, the charge density can also be written as $\rho(\mathbf{r}') = \sum_i q_i\delta(\mathbf{r}'-\mathbf{r}^i)$. Insertion of this expression for the charge density into (10.22) yields (10.21).

10.3.2 Expansion of the Electrostatic Potential

In the following, it is understood that the charge distribution is centered around $\mathbf{r}' = 0$ and that $\mathbf{r}$ is a point further away from this center than any of the charges generating the electrostatic potential, cf. Fig. 10.1.

So it makes sense to expand $|\mathbf{r}-\mathbf{r}'|^{-1}$, occurring in (10.22), around $\mathbf{r}' = 0$. Due to

$$\frac{\partial}{\partial r'_\mu}|\mathbf{r}-\mathbf{r}'|^{-1} = -\frac{\partial}{\partial r_\mu}|\mathbf{r}-\mathbf{r}'|^{-1},$$

the Taylor series expansion of $|\mathbf{r}-\mathbf{r}'|^{-1}$ reads

Fig. 10.1 Charge cloud centered around $\mathbf{r}' = 0$. The point $\mathbf{r}$ is outside of the cloud

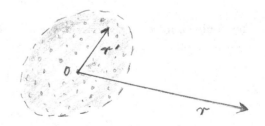

$$|\mathbf{r} - \mathbf{r}'|^{-1} = \sum_{\ell=0}^{\infty} \frac{1}{\ell!} \frac{(-1)^{\ell} \, \partial^{\ell} \, r^{-1}}{\partial r_{\mu_1} \partial r_{\mu_2} \cdots \partial r_{\mu_\ell}} \, r'_{\mu_1} r'_{\mu_2} \cdots r'_{\mu_\ell}. \tag{10.23}$$

The spatial derivatives of r^{-1} are the descending multipole potential tensors. Since the tensors $r'_{\mu_1} r'_{\mu_2} \cdots r'_{\mu_\ell}$, in (10.23), are contracted with irreducible tensors $X_{...}$, the irreducible part $\overline{r'_{\mu_1} r'_{\mu_2} \cdots r'_{\mu_\ell}}$ only contributes in the product. Thus (10.23) is equivalent to

$$|\mathbf{r} - \mathbf{r}'|^{-1} = \sum_{\ell=0}^{\infty} \frac{1}{\ell!} X_{\mu_1 \mu_2 \cdots \mu_\ell}(\mathbf{r}) \, \overline{r'_{\mu_1} r'_{\mu_2} \cdots r'_{\mu_\ell}} \,. \tag{10.24}$$

Insertion of this expansion into (10.22) leads to

$$\phi = \frac{1}{4 \pi \varepsilon_0} \sum_{\ell=0}^{\infty} \frac{1}{\ell! \, (2\ell - 1)!!} X_{\mu_1 \mu_2 \cdots \mu_\ell}(\mathbf{r}) \, Q_{\mu_1 \mu_2 \cdots \mu_\ell}. \tag{10.25}$$

Here

$$Q_{\mu_1 \mu_2 \cdots \mu_\ell} = \int \rho(\mathbf{r}')(2\ell - 1)!! \, \overline{r'_{\mu_1} r'_{\mu_2} \cdots r'_{\mu_\ell}} \, d^3 r' = \int \rho(\mathbf{r}) \tilde{X}_{\mu_1 \mu_2 \cdots \mu_\ell} d^3 r,$$
$$\tag{10.26}$$

is the 2^{ℓ}-pole moment of the charge distribution. The quantity $\tilde{X}_{\mu_1 \mu_2 \cdots \mu_\ell}$ is the ascending multipole defined in (10.16).

Due to (10.9), the expansion (10.25) is equivalent to

$$\phi = \frac{1}{4 \pi \varepsilon_0} \sum_{\ell=0}^{\infty} \frac{1}{\ell!} r^{-(2\ell+1)} \, \overline{r_{\mu_1} r_{\mu_2} \cdots r_{\mu_\ell}} \, Q_{\mu_1 \mu_2 \cdots \mu_\ell}. \tag{10.27}$$

With the integration variable denoted by $\mathbf{r}$ instead of $\mathbf{r}'$, the first four of these multipole moments are the

total charge or *monopole moment*

$$Q = \int \rho(\mathbf{r}) d^3 r,$$

the *dipole moment*

$$Q_\mu \equiv p_\mu^{\text{el}} = \int \rho(\mathbf{r}) \, r_\mu d^3 r, \tag{10.28}$$

the *quadrupole moment*

$$Q_{\mu\nu} = \int \rho(\mathbf{r}) \, 3 \, \overline{r_\mu r_\nu} \, d^3 r, \tag{10.29}$$

and the *octupole moment*

$$Q_{\mu\nu\lambda} = \int \rho(\mathbf{r})\, 15\, \overline{r_\mu r_\nu r_\lambda}\, \mathrm{d}^3 r. \tag{10.30}$$

Up to $\ell = 3$, the expansion (10.27) of the electrostatic potential reads

$$\phi = \frac{1}{4\pi\varepsilon_0}\left(r^{-1}Q + r^{-3}r_\mu Q_\mu + \frac{1}{2}r^{-5}\overline{r_\mu r_\nu}\, Q_{\mu\nu} + \frac{1}{6}r^{-7}\overline{r_\mu r_\nu r_\lambda}\, Q_{\mu\nu\lambda} + \cdots\right). \tag{10.31}$$

The next higher moment, pertaining to $\ell = 4$, is the *hexadecapole moment*. The parts of the potential associated with the monopole, dipole, quadrupole, octupole and hexadecapole moments are proportional to $r^{-1}, r^{-2}, r^{-3}, r^{-4}$ and r^{-5}, respectively. Thus at large distances from the charges, the contribution of the lowest order multipole moment will be most important for the electrostatic potential.

Just the lowest non-vanishing multipole moment is independent of the choice of the origin chosen in the integral (10.26) for the evaluation of the multipole moments. Notice, all multipole moments with $\ell \geq 1$ vanish, when the charge density has spherical symmetry. Furthermore, when the charge density is an even function of $\mathbf{r}$, i.e. when $\rho(-\mathbf{r}) = \rho(\mathbf{r})$, all odd multipoles with $\ell = 1, 3, \ldots$ are zero. Similarly, for an odd charge density where $\rho(-\mathbf{r}) = -\rho(\mathbf{r})$, all even multipoles with $\ell = 0, 2, 4, \ldots$ vanish.

10.3.3 Electric Field of Multipole Moments

The electric field E_μ, caused by the electrostatic potential ϕ, is determined by $E_\mu = -\nabla_\mu \phi$. Due to (10.3), the expansion (10.27) yields

$$E_\mu = \frac{1}{4\pi\varepsilon_0} \sum_{\ell=0}^{\infty} \frac{1}{\ell!\,(2\ell-1)!!}\, X_{\mu\mu_1\mu_2\cdots\mu_\ell}(\mathbf{r})\, Q_{\mu_1\mu_2\cdots\mu_\ell}. \tag{10.32}$$

Notice, that in this sum, the multipole potential tensor of rank $\ell + 1$ is multiplied with the ℓ-th multipole moment.

The first few terms in the expansion (10.32) for the electric field are

$$E_\mu = \frac{1}{4\pi\varepsilon_0}\left(r^{-3}r_\mu Q + 3r^{-5}\overline{r_\mu r_\nu}\, Q_\nu + \frac{5}{2}r^{-7}\overline{r_\mu r_\nu r_\lambda}\, Q_{\nu\lambda} + \cdots\right). \tag{10.33}$$

The parts of the electric field associated with the monopole, dipole and quadrupole moments are determined by the dipole, quadrupole and octupole potential tensors, and are proportional to r^{-2}, r^{-3} and r^{-4}, respectively. At large distances from the field-generating charges, the contribution of the lowest order multipole moment will be most important for the electric field.

10.3.4 Multipole Moments for Discrete Charge Distributions

The multipole expansion can also be applied to the electrostatic potential (10.21) of N charges q_i, located at positions $\mathbf{r}^i$. The formula for the evaluation of the multipole moment tensors, corresponding to (10.26), is

$$Q_{\mu_1\mu_2\cdots\mu_\ell}(\mathbf{r}) = \sum_{i=1}^N q_i \, (2\ell - 1)!! \, \overline{r^i_{\mu_1} r^i_{\mu_2} \cdots r^i_{\mu_\ell}} \, . \qquad (10.34)$$

Obviously, a single charge can just have a monopole. Two charges with opposite sign possess a dipole moment. For example, consider two charges separated by the distance a, the unit vector parallel the vector joining them is denoted by $\mathbf{u}$, viz. $q_1 = q$, $q_2 = -q$ and $\mathbf{r}^1 = (a/2)\mathbf{u}$, $\mathbf{r}^2 = -(a/2)\mathbf{u}$. Clearly one has $Q = 0$ and all other even multipole moments vanish due to the *dipolar symmetry*. The dipole moment is equal to

$$Q_\mu = p^{\text{el}}_\mu = q \, a \, u_\mu = p^{\text{el}} u_\mu,$$

where $p^{\text{el}} = qa$ is the magnitude of the electric dipole moment. All higher order odd multipole moments, with $\ell = 3, 5, \ldots$, are also non-zero. In particular, the octupole moment is

$$Q_{\mu\nu\lambda} = q \, a^3 \frac{15}{4} \overline{u_\mu u_\nu u_\lambda} = a^2 \, p^{\text{el}} \frac{15}{4} \overline{u_\mu u_\nu u_\lambda} \, .$$

At the distance r from the center of the charge distribution, the contribution from the octupole moment in the electrostatic potential has an extra factor $(a/r)^2$, compared with that from the dipole moment. In most applications this factor is exceedingly small. Hence, in this case, the octupole and higher moments can be disregarded.

Four charges, two positive and two negative ones, can be arranged such that they constitute a quadrupole. For example $q_1 = q$, $q_2 = q$, $q_3 = -q$, $q_4 = -q$ and $\mathbf{r}^1 = \mathbf{r}^2 = 0$, $\mathbf{r}^3 = (a/2)\mathbf{u}$, $\mathbf{r}^4 = -(a/2)\mathbf{u}$. In this case, one has $Q = 0$ and the dipole moment as well as all other odd multipole moments are zero. The quadrupole moment is

$$Q_{\mu\nu} = \frac{-3}{2} q \, a^2 \, \overline{u_\mu u_\nu} \, .$$

The hexadecapole moment, corresponding to $\ell = 4$, and the higher even moments are also non-zero. However, due to extra factors $(a/r)^2$, $(a/r)^4 \ldots$, they can be disregarded in the electrostatic potential, when $a/r \ll 1$. On the other hand, there exist charge distributions where the hexadecapole is the lowest non-vanishing multipole moment. Examples are charges on the corners of a cube or of a regular octahedron, compensated by an appropriate opposite charge in the center.

10.3.5 Connection with Legendre Polynomials

The reciprocal distance $|\mathbf{r} - \mathbf{r}'|^{-1}$ occurring in (10.24), is equal to

$$|\mathbf{r} - \mathbf{r}'|^{-1} = r^{-1} \left[1 - 2 \left(r'/r \right) \cos\theta + (r'/r)^2 \right]^{-1/2}, \qquad (10.35)$$

with $\cos\theta = \hat{\mathbf{r}} \cdot \hat{\mathbf{r}}'$. Clearly, the angle between $\mathbf{r}$ and $\mathbf{r}'$ is denoted by θ. Assuming $r > r'$, the expression $[\ldots]^{-1/2}$ in (10.35) is the generating function of the Legendre polynomials $P_\ell = P_\ell(\cos\theta)$. The resulting expansion of $|\mathbf{r} - \mathbf{r}'|^{-1}$ with respect to Legendre polynomials reads

$$|\mathbf{r} - \mathbf{r}'|^{-1} = r^{-1} \sum_{\ell=0}^{\infty} \left(\frac{r'}{r} \right)^\ell P_\ell(\cos\theta), \quad r > r'. \qquad (10.36)$$

Due to $X_{\mu_1\mu_2\cdots\mu_\ell} = (2\ell - 1)!! \, r^{-(2\ell+1)} \, \overline{r_{\mu_1} r_{\mu_2} \cdots r_{\mu_\ell}}$, see also (10.10) and (10.11), comparison with (10.24) yields

$$P_\ell(\cos\theta) = \frac{1}{\ell!} Y_{\mu_1\mu_2\cdots\mu_\ell} \overline{\hat{r}'_{\mu_1} \hat{r}'_{\mu_2} \cdots \hat{r}'_{\mu_\ell}} = \frac{(2\ell - 1)!!}{\ell!} \overline{\hat{r}_{\mu_1} \hat{r}_{\mu_2} \cdots \hat{r}_{\mu_\ell}} \, \overline{\hat{r}'_{\mu_1} \hat{r}'_{\mu_2} \cdots \hat{r}'_{\mu_\ell}}, \qquad (10.37)$$

which is equivalent to (9.10) with (9.11).

10.4 Further Applications in Electrodynamics

10.4.1 Induced Dipole Moment of a Metal Sphere

Consider a piece of metal placed into an electric field. The conduction electrons inside the metal feel the force caused by the electric field. When the piece of metal is electrically isolated, the 'free' charges are displaced just within the metal, such that an electric dipole is induced. This dipole modifies the surrounding electric field. For the special case of a metallic sphere with radius R, the induced electric dipole moment $\mathbf{p}^{\text{ind}}$ is computed as follows.

The center of the sphere is put at $\mathbf{r} = 0$. The electric field, far from the sphere, i.e. for $r \to \infty$ is the imposed field $\mathbf{E}^\infty$. The electric potential ϕ obeys the Laplace equation $\Delta \phi = 0$. The boundary condition is $\phi = $ const., on the surface of a conductor. Here $\phi = 0$ is chosen for $r = R$. The electrostatic potential is a scalar, and it should be linear in the imposed vectorial field $\mathbf{E}^\infty$. For $r > R$, the ansatz

$$\phi(\mathbf{r}) = a\, r_\mu\, E_\mu^\infty + b\, r^{-3} r_\mu\, E_\mu^\infty$$

is made. It obeys the required symmetry and it is a solution of the Laplace equation because it involves the ascending and descending vectorial multipole potential functions. The coefficients a and b are determined by the boundary conditions. First, notice that the electric field is given by

$$E_\mu = -\nabla_\mu\, \phi(\mathbf{r}) = -a\, E_\mu^\infty + 3\, b\, r^{-5} \overline{r_\mu r_\nu}\, E_\nu^\infty.$$

Now, the condition $\mathbf{E} \to \mathbf{E}^\infty$, for $r \to \infty$, implies $a = -1$. Then, $\phi = 0$ for $r = R$, leads to $b = R^3$. Thus the solution for the present potential problem is

$$\phi(\mathbf{r}) = -r_\mu\, E_\mu^\infty + \frac{R^3}{r^3}\, r_\mu\, E_\mu^\infty = -r_\mu\, E_\mu^\infty + \frac{1}{4\pi\varepsilon_0}\, r^{-3}\, r_\mu\, p_\mu^{\text{ind}}. \tag{10.38}$$

The second term of the computed potential has the form typical for a dipole potential. The induced dipole moment is introduced by

$$p_\mu^{\text{ind}} = 4\pi\, R^3\, \varepsilon_0\, E_\mu^\infty = \alpha\, \varepsilon_0 E_\mu^\infty. \tag{10.39}$$

The coefficient

$$\alpha = 4\pi\, R^3 = 3\, V_{\text{sph}}, \tag{10.40}$$

is the polarizability, cf. Sect. 5.3.3. It is proportional to the volume $V_{\text{sph}} = (4\pi/3)R^3$ of the sphere.

10.4.2 Electric Polarization as Dipole Density

The total charge density of a material is the sum of the density ρ of the free charges and the density ρ^i of the bound internal charges, i.e. of the atomic nuclei and the electrons bound in atoms and molecules. When all charges are counted, the Gauss law is

$$\varepsilon_0 \nabla_\mu E_\mu = \rho + \rho^i,$$

cf. Sect. 7.5. Thus the electric polarization $\mathbf{P} = \mathbf{D} - \varepsilon_0\mathbf{E}$, cf. (7.58) of dielectric material is determined by

$$\rho^i = -\nabla_\nu P_\nu. \tag{10.41}$$

The volume integral over ρ^i vanishes. Multiplication of ρ^i by r_μ and subsequent integration over a volume V which totally encloses the internal charges, yields the macroscopic electric dipole moment p_μ^{el} of this charge distribution, see (10.28). From (10.41) follows

$$p_\mu^{el} = \int_V \rho(\mathbf{r})^i\, r_\mu d^3r = -\int_V r_\mu \nabla_\nu P_\nu d^3r. \tag{10.42}$$

Due to

$$r_\mu \nabla_\nu P_\nu = \nabla_\nu (r_\mu P_\nu) - P_\nu\nabla_\nu r_\mu = \nabla_\nu (r_\mu P_\nu) - P_\mu,$$

and the application of the Gauss theorem, one finds

$$p_\mu^{el} = -\int_{\partial V} n_\nu P_\nu r_\mu d^2s + \int_V P_\mu d^3r.$$

The surface integral, taken over a surface outside the internal charge distribution, yields zero. Thus the macroscopic electric dipole moment is the volume integral over the electric polarization:

$$p_\mu^{el} = \int P_\mu d^3r. \tag{10.43}$$

This means, the electric polarization $\mathbf{P}$ is the density of electric dipole moments. These dipole moments can be permanent dipoles of molecules or dipoles induced by an electric field.

10.4.3 Energy of Multipole Moments in an External Field

The electrostatic energy of a cloud of particles with charges q_j located at the positions $\mathbf{r} + \mathbf{r}^j$, in the presence of an electrostatic potential ϕ is $W = \sum_j q_j\phi(\mathbf{r} + \mathbf{r}^j)$. It is assumed that the position $\mathbf{r}$ is in the center of the charge cloud and that ϕ is generated by other charges, which are far away. Alternatively, when the charges are characterized by a charge density ρ, the energy is

$$W = \int \rho(\mathbf{r}')\phi(\mathbf{r} + \mathbf{r}'). \tag{10.44}$$

An expansion of the potential in powers of $\mathbf{r}'$ yields

$$\phi(\mathbf{r} + \mathbf{r}') = \phi(\mathbf{r}) + r'_\mu \nabla_\mu \phi(\mathbf{r}) + \frac{1}{2} r'_\mu r'_\nu \nabla_\nu \nabla_\mu \phi(\mathbf{r}) + \dots ,$$

or, equivalently

$$\phi(\mathbf{r} + \mathbf{r}') = \phi(\mathbf{r}) - r'_\mu E_\mu(\mathbf{r}) - \frac{1}{2} r'_\mu r'_\nu \nabla_\nu E_\mu(\mathbf{r}) - \dots .$$

Notice that $\delta_{\mu\nu} \nabla_\nu \nabla_\mu \phi(\mathbf{r}) = -\delta_{\mu\nu} \nabla_\nu E_\mu = 0$, for electrostatic fields, $r'_\mu r'_\nu$ in these equations can be replaced by $r'_\mu r'_\nu - (1/3) r'_\kappa r'_\kappa \delta_{\mu\nu}$. Thus, the expansion of the energy (10.44) reads

$$W = \phi(\mathbf{r}) \int \rho(\mathbf{r}') \mathrm{d}^3 r' - E_\mu \int r'_\mu \rho(\mathbf{r}') \mathrm{d}^3 r'$$

$$- \frac{1}{2} \nabla_\nu E_\mu \int \left(r'_\mu r'_\nu - \frac{1}{3} r'_\kappa r'_\kappa \delta_{\mu\nu} \right) \rho(\mathbf{r}') \mathrm{d}^3 r' - \dots$$

The integrals over the charge density can be expressed in terms of the multipole moments. This leads to

$$W = Q \phi(\mathbf{r}) - p^{\mathrm{el}}_\mu E_\mu - \frac{1}{6} Q_{\mu\nu} \nabla_\nu E_\mu - \dots , \qquad (10.45)$$

where Q is recalled as the total charge, p^{el}_μ is the electric dipole moment, and $Q_{\mu\nu}$ is the quadrupole moment tensor.

10.4.4 Force and Torque on Multipole Moments in an External Field

The force on a cloud of particles with charges q_j located at the positions $\mathbf{r} + \mathbf{r}^j$, in the presence of an external electric field $\mathbf{E}$ is $F_\mu = -\sum_j q_j E_\mu(\mathbf{r} + \mathbf{r}^j)$, or, in terms of the charge density

$$F_\mu = -\int \rho(\mathbf{r}') E_\mu(\mathbf{r} + \mathbf{r}') \mathrm{d}^3 r'. \qquad (10.46)$$

A power series expansion with respect to $\mathbf{r}'$ yields, by analogy to (10.45),

$$F_\mu = Q E_\mu(\mathbf{r}) + p^{\mathrm{el}}_\nu \nabla_\nu E_\mu + \frac{1}{6} Q_{\nu\kappa} \nabla_\kappa \nabla_\nu E_\mu - \dots \qquad (10.47)$$

As expected, this force is also obtained as a spatial derivative of the energy W, as given by (10.45), viz.

$$F_\mu = -\nabla_\mu W(\mathbf{r}). \tag{10.48}$$

Notice, there is no force acting on an electric dipole moment subjected to a homogeneous electric field with $\nabla_\mu E_\nu = 0$.

A torque, however is exerted on an electric dipole in a spatially constant electric field. For discrete charges, the torque is $T_\mu = \varepsilon_{\mu\nu\lambda} \sum_j r_\nu^j E_\lambda(\mathbf{r}+\mathbf{r}^j)$. The corresponding expression for a continuous charge density is

$$T_\mu = \varepsilon_{\mu\nu\lambda} \int r_\nu' E_\lambda(\mathbf{r} + \mathbf{r}') \mathrm{d}^3 r'. \tag{10.49}$$

The torque can be expanded by analogy to (10.47). The leading term is

$$T_\mu = \varepsilon_{\mu\nu\lambda} \, p_\nu^{\mathrm{el}} \, E_\lambda(\mathbf{r}). \tag{10.50}$$

10.4.5 Multipole–Multipole Interaction

The multipole–multipole interaction of a cloud 1 of charges, located around the position $\mathbf{r}$, in a potential and electric field generated by a group of charges 2, centered around the position $\mathbf{r} = 0$, can be inferred from (10.31) with (10.33) and (10.45). The corresponding total charges are by Q_1, Q_2. The electric dipole moments and electric quadrupole moment tensors are denoted by $p_\mu^{(1)}$, $p_\mu^{(2)}$, and $Q_{\mu\nu}^{(1)}$, $Q_{\mu\nu}^{(2)}$, respectively.

The *pole–pole*, the *pole–dipole* and the *pole–quadrupole* interaction energies are

$$W^{\mathrm{pole-pole}} = \frac{1}{4\pi\varepsilon_0} Q_1 r^{-1} Q_2,$$

$$W^{\mathrm{pole-dip}} = \frac{1}{4\pi\varepsilon_0} Q_1 X_\mu p_\mu^{(2)} = \frac{1}{4\pi\varepsilon_0} Q_1 r^{-3} r_\mu p_\mu^{(2)}, \tag{10.51}$$

$$W^{\mathrm{pole-quad}} = \frac{1}{4\pi\varepsilon_0} \frac{1}{6} Q_1 X_{\mu\nu} Q_{\mu\nu}^{(2)} = \frac{1}{4\pi\varepsilon_0} \frac{1}{2} Q_1 r^{-5} \overline{r_\mu r_\nu} \, Q_{\mu\nu}^{(2)}. \tag{10.52}$$

Just as the pole–quadrupole interaction, the dipole–dipole interaction is governed by the second multipole potential tensor, which decreases, with increasing distance r, like r^{-3},

$$W^{\mathrm{dip-dip}} = -\frac{1}{4\pi\varepsilon_0} \frac{1}{2} p_\mu^{(1)} X_{\mu\nu} p_\nu^{(2)} = -\frac{1}{4\pi\varepsilon_0} \frac{3}{2} r^{-5} p_\mu^{(1)} \overline{r_\mu r_\nu} \, p_\nu^{(2)}. \tag{10.53}$$

The interaction between two magnetic dipoles has the same functional form, the factor $\frac{1}{4\pi\varepsilon_0}$, however is replaced by $\frac{\mu_0}{4\pi}$. The dipole–quadrupole and the quadrupole–quadrupole interactions involve the third and the fourth multipole potentials, which are proportional to r^{-4} and r^{-5}, respectively,

$$W^{\text{dip-quad}} = -\frac{1}{4\pi\varepsilon_0}\frac{1}{6}\, p_\mu^{(1)}\, X_{\mu\nu\lambda}\, Q_{\nu\lambda}^{(2)} = -\frac{1}{4\pi\varepsilon_0}\frac{5}{2}\, r^{-7}\, p_\mu^{(1)}\, \overline{r_\mu r_\nu r_\lambda}\, Q_{\nu\lambda}^{(2)}.$$
(10.54)

$$W^{\text{quad-quad}} = \frac{1}{4\pi\varepsilon_0}\frac{1}{36}\, Q_{\mu\nu}^{(1)}\, X_{\mu\nu\lambda\kappa}\, Q_{\lambda\kappa}^{(2)} = \frac{1}{4\pi\varepsilon_0}\frac{35}{12}\, r^{-9}\, Q_{\mu\nu}^{(1)}\, \overline{r_\mu r_\nu r_\lambda r_\kappa}\, Q_{\lambda\kappa}^{(2)}.$$
(10.55)

These multipole-multipole interactions are of importance for the anisotropy of the long range interaction between two molecules, labelled by 1 and 2. The multipole moments depend on the orientation of the molecules. For electrically neutral linear molecules, without permanent dipole moments, the quadrupole-quadrupole interaction is the lowest order contribution. The induced dipole-dipole interaction, referred to as *van der Waals interaction*, as well as the repulsive interaction, associated with the shape of the molecules, in general, possess anisotropic parts. The van der Waals interaction is proportional to $\alpha_{\mu\nu}^{(1)} X_{\nu\lambda}\alpha_{\lambda\kappa}^{(2)} X_{\kappa\mu}$, where $\alpha_{\mu\nu}^{(1)}$ and $\alpha_{\lambda\kappa}^{(2)}$ are the molecular polarizability tensors of particles 1 and 2.

10.5 Applications in Hydrodynamics

10.5.1 Stationary and Creeping Flow Equations

In a stationary situation, the local momentum conservation equation (7.52), for an isotropic fluid with the spatially constant shear viscosity η, cf. (7.55), is

$$\rho\, v_\nu\, \nabla_\nu v_\mu + \nabla_\mu\, p = \eta\, \Delta\, v_\mu.$$
(10.56)

Here ρ is the constant mass density, p is the hydrostatic pressure which may depend on the position $\mathbf{r}$. The right hand side of (10.56) is obtained from (7.52) with the Newton ansatz (7.55) for the friction pressure tensor

$$\overline{p_{\nu\mu}} = -2\eta\, \overline{\nabla_\nu v_\mu} = -\eta\, (\nabla_\nu v_\mu + \nabla_\nu v_\mu), \quad \nabla_\lambda v_\lambda = 0.$$

Due the incompressibility condition $\nabla_\mu v_\mu = 0$, application of ∇_μ to (10.56) implies

$$\rho\, \nabla_\mu v_\nu\, \nabla_\nu v_\mu + \Delta\, p = 0.$$
(10.57)

The solutions of these equations, for specific hydrodynamic applications, have to obey the relevant *boundary conditions*.

In the *creeping flow approximation*, applicable for slow velocities, the nonlinear terms involving $v_\nu \nabla_\nu v_\mu$ are disregarded in (10.56) and (10.57). Applications thereof are the calculation of the Stokes force acting on a solid body moving in a viscous fluid or the torque acting on a rotating body. The special case of a sphere is considered next.

10.5.2 Stokes Force on a Sphere

The friction force **F** acting on sphere with radius R, which undergoes a slow translational motion, with velocity **V** in a viscous fluid, is

$$F_\mu = -6\pi \eta R V_\mu. \tag{10.58}$$

The expression for the force, first derived by Stokes and referred to as *Stokes force*, is remarkable since is not proportional to the cross section πR^2 of the sphere, but rather linear in R. This is an effect typical for hydrodynamics. The minus sign in (10.58) means that the force is damping the motion.

The derivation of the Stokes force starts from the creeping motion version of (10.56). For convenience, the dimensionless position vector $\mathbf{r}^* = R^{-1}\mathbf{r}$ is used. When no danger of confusion exists, $\mathbf{r}^*$ is written as $\mathbf{r}$ and it is understood that ∇ stands for the derivative with respect to the dimensionless position vector. The differential equations then read

$$\nabla_\mu p = \eta R^{-1} \Delta v_\mu, \quad \Delta p = 0. \tag{10.59}$$

The equation is linear in the velocity. The situation of a sphere moving with the velocity $-\mathbf{V}$ and the fluid at rest, far away from the sphere, is equivalent to the sphere at rest, with its center corresponding to $\mathbf{r} = 0$, and the fluid flowing with the constant velocity **V**, far away, i.e. for $r \to \infty$. Furthermore $p \to$ const. for $r \to \infty$ should hold true. In the following p is associated with the deviation of the pressure from its constant value at $r \to \infty$. In addition to these boundary conditions, the behavior of the velocity at the surface of the sphere has to be prescribed for the solution of (10.59). The Stokes force (10.58) pertains to the *stick* or *no slip* boundary condition $v_\mu = 0$ at the surface of the sphere, corresponding to $r = 1$. Of course, for the sphere to stay fixed in the streaming fluid, it has to be held by a force which has to balance the Stokes force.

Solutions of (10.59) must be linear in **V**. Symmetry and parity considerations suggest the ansatz

$$p = A(r) X_\nu V_\nu, \quad v_\mu = a(r)V_\mu + b(r)X_{\mu\nu} V_\nu, \tag{10.60}$$

where the $X_{...}$ are multipole potential functions and the coefficients A, a, b have yet to be determined. The equation $\Delta p = 0$ implies $A =$ const. The spatial derivatives of the functions p and $\mathbf{v}$ are

$$\nabla_\mu p = -A \, X_{\mu\nu} \, V_\nu,$$

and

$$\nabla_\mu v_\nu = a' \widehat{r_\mu} V_\nu + b' \widehat{r_\mu} X_{\nu\lambda} V_\lambda - b X_{\mu\nu\lambda}.$$

The divergence of this expression for the velocity field is

$$\nabla_\mu v_\mu = a' \widehat{r_\mu} V_\mu + b' \widehat{r_\mu} X_{\mu\lambda} V_\lambda = (a' + 2r^{-3} b') \widehat{r_\mu} V_\mu = 0.$$

Thus $a' + 2r^{-3} b' = 0$ has to hold true. Due to

$$\Delta v_\mu = \Delta a \, V_\mu + (b'' - 4r^{-1} b') \, X_{\mu\nu} \, V_\nu,$$

cf. (10.13), the creeping flow equation (10.59) implies

$$a'' + 2r^{-1} a' = 0, \quad -A = \eta R^{-1} (b'' - 4r^{-1} b').$$

The boundary conditions impose

$$r \to \infty : \; r^{-2} A \to 0, \quad a(\infty) = 1, \quad r^{-3} b \to 0, \tag{10.61}$$

and

$$a(1) = 0, \quad b(1) = 0. \tag{10.62}$$

The solution of the differential equations (10.59), which obey the boundary conditions, are determined by

$$A = \frac{3}{2} \eta R^{-1}, \quad a = 1 - r^{-1}, \quad b = \frac{1}{4}(1 - r^2). \tag{10.63}$$

To verify that (10.60), with (10.63) are solutions of the creeping flow equation, make use of (10.13). The presence of the fixed sphere in the streaming fluid has far reaching consequences. There are parts of the distortion of the velocity field which decrease with increasing distance from the sphere, both upstream and downstream, only like r^{-1}. This is an effect typical for hydrodynamics.

The part of the velocity difference $V_\mu - v_\mu$ proportional to r^{-1} is $r^{-1}(\delta_{\mu\nu} + \frac{1}{4} r^3 X_{\mu\nu}) V_\nu$. The tensor $r^{-1}(\delta_{\mu\nu} + \frac{1}{4} r^3 X_{\mu\nu}) = \frac{3}{4} r^{-1}(\delta_{\mu\nu} + \widehat{r_\mu} \widehat{r_\nu})$ is proportional to the *Oseen tensor*

$$(8\pi \eta r)^{-1} (\delta_{\mu\nu} + \widehat{r_\mu} \widehat{r_\nu}),$$

which determines the hydrodynamic interaction between colloidal particles [29–33].

According to (8.91), the force acting on a solid body is given by the surface integral $-\oint_{\partial V} n_\nu p_{\nu\mu} d^2 s$. For the case of the sphere, the outer normal $\mathbf{n}$ is equal to $\hat{\mathbf{r}}$, and $d^2 s = R^2 d^2 \hat{r}$. As used before, cf. (7.55), the pressure tensor $p_{\nu\mu}$ is the sum of the isotropic part $p\delta_{\mu\nu}$ and the symmetric traceless part $\overline{p_{\nu\mu}} = -\eta R^{-1}(\nabla_\nu v_\mu + \nabla_\mu v_\nu)$. The part of the force associated with the distorted pressure field is

$$F_\mu^{\text{Stp}} \equiv \oint A\, X_\nu\, V_\nu\, \widehat{r_\mu}\, R^2 d^2\hat{r} = \frac{4\pi}{3} R^2 A\, V_\mu = 2\pi R\, \eta\, V_\mu. \tag{10.64}$$

Here $\oint \widehat{r_\mu}\widehat{r_\nu} d^2\hat{r} = \frac{4\pi}{3}\delta_{\mu\nu}$ was used.

The contribution of the force associated with $\mathbf{v}$ requires the computation of

$$\widehat{r_\nu}\nabla_\nu\, v_\mu + \widehat{r_\nu}\,\nabla_\mu v_\nu.$$

The first of these two terms is

$$\widehat{r_\nu}\,\nabla_\nu v_\mu = \frac{\partial}{\partial r}\, v_\mu = a'\, V_\mu + (b' - 3r^{-1}b)X_{\mu\nu}V_\nu.$$

On the other hand, one has

$$\nabla_\mu v_\nu = a'\widehat{r_\mu}V_\nu + b'\widehat{r_\mu}X_{\nu\lambda}V_\lambda - bX_{\mu\nu\lambda},$$

and

$$\widehat{r_\nu}\,\nabla_\mu v_\nu = a'\widehat{r_\mu}\widehat{r_\nu}\,V_\nu + b'\widehat{r_\mu}2\,r^{-3}\,\widehat{r_\lambda}V_\lambda - b\widehat{r_\mu}X_{\mu\nu\lambda}.$$

Due to the compressibility condition $\nabla_\mu v_\mu = 0$, one has and $a' + r^{-3}b' = 0$, and

$$\widehat{r_\nu}\,\nabla_\mu v_\nu = -b\widehat{r_\mu}X_{\mu\nu\lambda}.$$

The term involving $\widehat{r_\mu}X_{\mu\nu\lambda} \sim X_{\nu\lambda}$, however, vanishes in the integration over the surface of the sphere.

Thus the additional force is

$$F_\mu^{\text{Stv}} \equiv \eta\, R^{-1} \oint \widehat{r_\nu}\nabla_\nu\, v_\mu\, R^2 d^2\hat{r} = 4\pi\, \eta\, R\, V_\mu. \tag{10.65}$$

The term involving $X_{\mu\nu}$ gives no contribution to the integral over the surface of the sphere.

Notice, in the calculation, the sphere was at rest and the fluid, far away from the sphere, moved with the velocity $\mathbf{V}$. As mentioned before, this corresponds to a situation, where the sphere moves with the velocity $-\mathbf{V}$ and the fluid is at rest, far away from the obstacle. This explains the opposite sign in the expression for the friction force.

With the help of the multipoles in D dimensions, expressions analogous to the Stokes force and the torque acting on a rotating body can be computed for the D-dimensional case [38]. This is of particular interest for the case $D = 2$, cf. [39].

10.3 Exercise: Compute the Torque on a Rotating Sphere

A sphere rotating with the angular velocity Ω experiences a friction torque

$$T_\mu = -8\pi\,\eta\,R^3\,\Omega_\mu.$$

Derive this result from the creeping flow equation by considerations similar to those used for the Stokes force.

Chapter 11
Isotropic Tensors

Abstract This chapter deals with isotropic Cartesian tensors. Firstly, the isotropic Delta-tensors of rank 2ℓ are introduced which, when applied on a tensor of rank ℓ, project onto the symmetric traceless part of that tensor. The Δ-tensors can be expressed in terms of ℓ fold products of the second rank isotropic delta-tensor. Secondly, a generalized cross product between a vector and symmetric traceless tensor of rank ℓ is defined via the $\square$-tensors. These isotropic tensors of rank $2\ell + 1$ can be expressed in terms of the product of epsilon-tensors and $\ell - 1$-fold products of delta-tensors. They describe the action of the orbital angular momentum operator on tensors. Furthermore, isotropic tensors are defined in connection with the coupling of vectors and second rank tensors with tensors of rank ℓ. Scalar product of three irreducible tensors and their relevance for the interaction potential of uniaxial particles are discussed.

11.1 General Remarks on Isotropic Tensors

A tensor which does not change its form when the Cartesian coordinate system is replaced by a rotated one, is referred to as *isotropic tensor*. A scalar is isotropic, per definition. Except for the zero-vector, no isotropic tensor of rank 1 exists. Isotropic tensors of rank 2 and 3 are the delta- and epsilon-tensors $\delta_{\mu\nu}$ and $\varepsilon_{\mu\nu\lambda}$, discussed before.

Isotropic tensors of rank $\ell \geq 4$ can be constructed as products of δ-tensors and the ε-tensor. Notice, the product of two ε-tensors can be expressed by δ-tensors, cf. Sect. 4.1.2. Thus isotropic tensors of odd rank $\ell = 5, 7, \ldots$ can be expressed by one ε-tensor and products of δ-tensors. Isotropic tensors of even rank $\ell = 4, 6, \ldots$ are expressed by products of $\ell/2$ second rank δ-tensors. The projection tensors discussed in Sect. 3.1.5 are isotropic tensors of rank 4.

© Springer International Publishing Switzerland 2015
S. Hess, *Tensors for Physics*, Undergraduate Lecture Notes in Physics,
DOI 10.1007/978-3-319-12787-3_11

11.2 Δ-Tensors

11.2.1 Definition and Examples

Of special interest among the tensors of rank 2ℓ are those, which project onto the symmetric traceless, i.e. irreducible part $\overset{\frown}{\mathbf{A}}$ of a tensor $\mathbf{A}$ of rank ℓ. By analogy to the δ-tensor, these tensors are denoted as $\Delta^{(\ell)}$-tensors. They are defined by

$$\Delta^{(\ell)}_{\mu_1\mu_2\cdots\mu_\ell,\mu_1'\mu_2'\cdots\mu_\ell'}\, A_{\mu_1'\mu_2'\cdots\mu_\ell'} = \overset{\frown}{A_{\mu_1\mu_2\cdots\mu_\ell}}\,. \tag{11.1}$$

Examples for this type of isotropic tensors are

$$\Delta^{(0)} = 1, \quad \Delta^{(1)} = \delta_{\mu\nu},$$

and

$$\Delta^{(2)}_{\mu\nu,\mu'\nu'} \equiv \Delta_{\mu\nu,\mu'\nu'} = \frac{1}{2}\left(\delta_{\mu\mu'}\delta_{\nu\nu'} + \delta_{\mu\nu'}\delta_{\nu\mu'}\right) - \frac{1}{3}\delta_{\mu\nu}\delta_{\mu'\nu'}. \tag{11.2}$$

11.2.2 General Properties of Δ-Tensors

The $\Delta^{(\ell)}$-tensors are symmetric against the exchange of any fore pair or hind pair of subscript, as well as against the exchange of all fore against all hind indices, viz.

$$\Delta^{(\ell)}_{\mu_1\mu_2\cdots\mu_\ell,\mu_1'\mu_2'\cdots\mu_\ell'} = \Delta^{(\ell)}_{\mu_1'\mu_2'\cdots\mu_\ell',\mu_1\mu_2\cdots\mu_\ell}. \tag{11.3}$$

The $\Delta^{(\ell)}$-tensors obey the product rule

$$\Delta^{(\ell)}_{\mu_1\mu_2\cdots\mu_\ell,\nu_1\nu_2\cdots\nu_\ell}\Delta^{(\ell)}_{\nu_1\nu_2\cdots\nu_\ell,\mu_1'\mu_2'\cdots\mu_\ell'} = \Delta^{(\ell)}_{\mu_1\mu_2\cdots\mu_\ell,\mu_1'\mu_2'\cdots\mu_\ell'}, \tag{11.4}$$

which is the relation typical for a projection tensor. On the other hand, the following contraction of a $\Delta^{(\ell)}$-tensor with a $\Delta^{(\ell-1)}$-tensor yields zero, more specifically:

$$\Delta^{(\ell)}_{\mu_1\mu_2\cdots\mu_\ell,\nu_1\nu_2\cdots\nu_\ell}\Delta^{(\ell-1)}_{\nu_1\nu_2\cdots\nu_{\ell-1},\nu_\ell\mu_1'\cdots\mu_{\ell-2}'} = 0. \tag{11.5}$$

The $\Delta^{(\ell)}$-tensors are traceless in the sense that they vanish, whenever any pair of hind indices or any pair of fore indices are equal and summed over. On the other hand, when one fore subscript is equal to one hind subscript, and the standard summation convention is used, a Delta-tensor with $\ell - 1$ is obtained:

$$\Delta^{(\ell)}_{\mu_1\mu_2\cdots\mu_{\ell-1}\lambda,\mu'_1\mu'_2\cdots\mu'_{\ell-1}\lambda} = \frac{2\ell+1}{2\ell-1}\Delta^{(\ell-1)}_{\mu_1\mu_2\cdots\mu_{\ell-1},\mu'_1\mu'_2\cdots\mu'_{\ell-1}}. \tag{11.6}$$

The total contraction of the fore with the hind subscripts yields

$$\Delta^{(\ell)}_{\mu_1\mu_2\cdots\mu_\ell,\mu_1\mu_2\cdots\mu_\ell} = 2\ell+1, \tag{11.7}$$

which is the number of the independent components of an irreducible tensor of rank ℓ.

In the following product of two $\Delta^{(\ell)}$-tensors, fore and hind subscripts are mixed:

$$\Delta^{(\ell)}_{\mu_1\mu_2\cdots\mu_{\ell-1}\nu,\mu_\ell\nu_1\nu_2\cdots\nu_{\ell-1}}\Delta^{(\ell)}_{\nu\nu_1\nu_2\cdots\nu_{\ell-1},\mu'_1\mu'_2\cdots\mu'_\ell} = \frac{1}{\ell(2\ell-1)}\Delta^{(\ell)}_{\mu_1\mu_2\cdots\mu_\ell,\mu'_1\mu'_2\cdots\mu'_\ell}. \tag{11.8}$$

The special case $\ell = 2$ of this equation corresponds to

$$\Delta_{\mu\lambda,\nu\kappa}\,\Delta_{\lambda\kappa,\mu'\nu'} = \frac{1}{6}\Delta_{\mu\nu,\mu'\nu'}. \tag{11.9}$$

The total contraction $\{\mu_1\mu_2\cdots\mu_\ell\} = \{\mu'_1\mu'_2\cdots\mu'_\ell\}$ in (11.8) yields

$$\Delta^{(\ell)}_{\mu_1\mu_2\cdots\mu_{\ell-1}\nu,\mu_\ell\nu_1\nu_2\cdots\nu_{\ell-1}}\Delta^{(\ell)}_{\nu\nu_1\nu_2\cdots\nu_{\ell-1},\mu_1\mu_2\cdots\mu_\ell} = \frac{2\ell+1}{\ell(2\ell-1)}. \tag{11.10}$$

The contraction $\mu'_\ell = \mu_\ell$, renamed as μ, in (11.8) and use of (11.6), yields

$$\Delta^{(\ell)}_{\mu_1\mu_2\cdots\mu_{\ell-1}\nu,\mu\nu_1\nu_2\cdots\nu_{\ell-1}}\Delta^{(\ell)}_{\nu\nu_1\nu_2\cdots\nu_{\ell-1},\mu\mu'_1\mu'_2\cdots\mu'_{\ell-1}} = \frac{2\ell+1}{\ell(2\ell-1)^2}\Delta^{(\ell-1)}_{\mu_1\mu_2\cdots\mu_{\ell-1},\mu'_1\mu'_2\cdots\mu'_{\ell-1}}. \tag{11.11}$$

The special case $\ell = 2$ of (11.11) corresponds to

$$\Delta_{\mu\lambda,\nu\kappa}\,\Delta_{\lambda\kappa,\nu\mu'} = \frac{5}{18}\delta_{\mu\mu'}. \tag{11.12}$$

The next contraction $\mu' = \mu$ gives

$$\Delta_{\mu\lambda,\nu\kappa}\,\Delta_{\lambda\kappa,\nu\mu} = \frac{5}{6}. \tag{11.13}$$

The relations (11.12) and (11.13) can also be inferred from (11.9). For comparison, it is recalled that $\Delta_{\mu\lambda,\nu\kappa}\Delta_{\nu\kappa,\lambda\mu} = \Delta_{\mu\lambda,\lambda\mu} = 5$, cf. (11.7). The order of subscripts matters!

11.1 Exercise: Contraction Rules for Delta-Tensors
Verify (11.6) for $\ell = 2$.

11.2.3 Δ-Tensors as Derivatives of Multipole Potentials

The ℓ-fold spatial derivative of the ℓ-th rank irreducible tensor constructed from the components of $\mathbf{r}$ is proportional to the isotropic $\Delta^{(\ell)}$-tensor. More specifically, one has

$$\nabla_{\mu_1}\nabla_{\mu_2}\cdots\nabla_{\mu_\ell}\overline{r_{\nu_1}r_{\nu_2}\cdots r_{\nu_\ell}} = \ell!\,\Delta^{(\ell)}_{\mu_1\mu_2\cdots\mu_\ell,\nu_1\nu_2\cdots\nu_\ell}. \tag{11.14}$$

Notice, the tensor $\overline{\cdots}$ with the ν-indices imposes its property of being traceless onto the μ-indices. As an exercise, verify (11.14), for $\ell = 1$ and $\ell = 2$. Due to

$$\overline{r_{\mu_1}r_{\mu_2}\cdots r_{\mu_\ell}} = \frac{1}{(2\ell-1)!!}\,r^{(2\ell+1)}X_{\mu_1\mu_2\cdots\mu_\ell},$$

see (10.9), and with the multipole potential given by (10.2), (11.14) leads to

$$\ell!(2\ell-1)!!(-1)^\ell\Delta^{(\ell)}_{\mu_1\mu_2\cdots\mu_\ell,\nu_1\nu_2\cdots\nu_\ell} = \nabla_{\mu_1}\nabla_{\mu_2}\cdots\nabla_{\mu_\ell}\left(r^{(2\ell+1)}\nabla_{\nu_1}\nabla_{\nu_2}\cdots\nabla_{\nu_\ell}r^{-1}\right). \tag{11.15}$$

It is remarkable that the differentiation, with unrestricted subscripts, on the right hand side of (11.15), yields an expression, on the left hand side, which is traceless with respect to any pair of either μ- or ν-subscripts.

11.2 Exercise: Determine $\Delta^{(3)}_{\mu\nu\lambda,\,\mu'\nu'\lambda'}$

Hint: compute $\Delta^{(3)}_{\mu\nu\lambda,\mu'\nu'\lambda'}$, in terms of triple products of δ-tensors, from (11.15) for $\ell = 3$.

11.3 Generalized Cross Product, □-Tensors

11.3.1 Cross Product via the □-Tensor

The isotropic tensor of rank $2\ell + 1$, defined by

$$\Box^{(\ell)}_{\mu_1\mu_2\cdots\mu_\ell,\lambda,\mu_1'\mu_2'\cdots\mu_\ell'} \equiv \Delta^{(\ell)}_{\mu_1\mu_2\cdots\mu_\ell,\nu_1\nu_2\cdots\nu_{\ell-1}\nu_\ell}\,\varepsilon_{\nu_\ell\lambda\nu_\ell'}\,\Delta^{(\ell)}_{\nu_\ell'\nu_1\nu_2\cdots\nu_{\ell-1},\mu_1'\mu_2'\cdots\mu_\ell'}, \tag{11.16}$$

allows to link a vector with the symmetric traceless part of a tensor of rank ℓ such that the result is a symmetric traceless, i.e. irreducible tensor of rank ℓ. Here and in the following, it is understood that $\ell \geq 1$.

Let $\mathbf{w}$ be a vector and A a tensor of rank ℓ. The generalized cross product is given by

$$(\mathbf{w} \times \mathsf{A})_{\mu_1\mu_2\cdots\mu_\ell} = \Box^{(\ell)}_{\mu_1\mu_2\cdots\mu_\ell,\lambda,\mu_1'\mu_2'\cdots\mu_\ell'} \, w_\lambda \, A_{\mu_1'\mu_2'\cdots\mu_\ell'}. \tag{11.17}$$

For a symmetric traceless tensor S, the generalized cross product can also be written as

$$(\mathbf{w} \times \mathsf{S})_{\mu_1\mu_2\cdots\mu_\ell} = \Box^{(\ell)}_{\mu_1\mu_2\cdots\mu_\ell,\lambda,\mu_1'\mu_2'\cdots\mu_\ell'} \, w_\lambda \, S_{\mu_1'\mu_2'\cdots\mu_\ell'} = \overline{\varepsilon_{\mu_1\lambda\nu} \, w_\lambda \, S_{\nu\mu_2\cdots\mu_\ell}}. \tag{11.18}$$

For $\ell = 1$, one has

$$\Box^{(1)}_{\mu,\lambda,\nu} = \varepsilon_{\mu\lambda\nu}.$$

Thus, in this case, (11.17) and (11.18) correspond to the standard vector product, cf. Sects. 3.3 and 4. For $\ell = 2$, the □-tensor is explicitly expressed by a linear combination of products of epsilon- and delta-tensors, viz.

$$\Box^{(2)}_{\mu\nu,\lambda,\mu'\nu'} \equiv \Box_{\mu\nu,\lambda,\mu'\nu'} = \frac{1}{4}(\varepsilon_{\mu\lambda\mu'}\delta_{\nu\nu'} + \varepsilon_{\mu\lambda\nu'}\delta_{\nu\mu'} + \varepsilon_{\nu\lambda\mu'}\delta_{\mu\nu'} + \varepsilon_{\nu\lambda\nu'}\delta_{\mu\mu'}). \tag{11.19}$$

The properties $\Box_{\mu\mu,\lambda,\mu'\nu'} = 0$ and $\Box_{\mu\nu,\lambda,\mu'\mu'} = 0$ follow from the explicit expression given above, due to the antisymmetry of the epsilon-tensor, e.g. $\varepsilon_{\nu'\lambda\mu'} = -\varepsilon_{\mu'\lambda\nu'}$. The contraction $\mu' = \mu$, in (11.19), yields

$$\Box_{\mu\nu,\lambda,\mu\nu'} = \frac{5}{4}\varepsilon_{\nu\lambda\nu'}. \tag{11.20}$$

The cross product of a vector $\mathbf{w}$ with a symmetric traceless second rank tensor S is, in accord with (11.18), given by

$$(\mathbf{w} \times \mathsf{S})_{\mu\nu} = \Box_{\mu\nu,\lambda,\mu'\nu'} w_\lambda S_{\mu'\nu'} = \overline{\varepsilon_{\mu\lambda\kappa} w_\lambda S_{\kappa\nu}} = \frac{1}{2}(\varepsilon_{\mu\lambda\kappa} w_\lambda S_{\kappa\nu} + \varepsilon_{\nu\lambda\kappa} w_\lambda S_{\kappa\mu}). \tag{11.21}$$

As an alternative to (11.16), the $\Box^{(\ell)}$-tensor can also be defined in terms of the epsilon-tensor and one Δ-tensor of rank $\ell + 1$, more specifically:

$$\Box^{(\ell)}_{\mu_1\mu_2\cdots\mu_\ell,\lambda,\mu_1'\mu_2'\cdots\mu_\ell'} = \frac{\ell+1}{\ell} \frac{2\ell+1}{2\ell+3} \Delta^{(\ell+1)}_{\mu_1\mu_2\cdots\mu_\ell\mu,\mu'\mu_1'\mu_2'\cdots\mu_\ell'} \varepsilon_{\mu\mu'\lambda}. \tag{11.22}$$

11.3.2 Properties of □-Tensors

The □-tensors are antisymmetric against the exchange of the fore and hind subscripts, viz.

$$\Box^{(\ell)}_{\mu_1\mu_2\cdots\mu_\ell,\lambda,\mu'_1\mu'_2\cdots\mu'_\ell} = -\Box^{(\ell)}_{\mu'_1\mu'_2\cdots\mu'_\ell,\lambda,\mu_1\mu_2\cdots\mu_\ell}, \tag{11.23}$$

furthermore, they vanish whenever the middle index is equal to one of the fore or hind indices, e.g.

$$\Box^{(\ell)}_{\mu_1\mu_2\cdots\mu_\ell,\mu_\ell,\mu'_1\mu'_2\cdots\mu'_\ell} = 0. \tag{11.24}$$

For $\ell = 2$, in particular, one has

$$\Box_{\mu\nu,\mu,\sigma\tau} = 0.$$

Due to the product properties of epsilon-tensor, cf. Sect. 4, the multiplication of a □-tensor with an epsilon-tensor or the product of two □-tensors, yields Δ-tensors which, in turn, are products of δ-tensors. For example, one has

$$\varepsilon_{\mu_\ell\lambda\mu'_\ell}\,\Box^{(\ell)}_{\mu_1\mu_2\cdots\mu_\ell,\lambda,\mu'_1\mu'_2\cdots\mu'_\ell} = \frac{\ell+1}{\ell}\frac{2\ell+1}{2\ell-1}\Delta^{(\ell-1)}_{\mu_1\mu_2\cdots\mu_{\ell-1},\mu'_1\mu'_2\cdots\mu'_{\ell-1}}, \tag{11.25}$$

and

$$\Box^{(\ell)}_{\mu_1\mu_2\cdots\mu_\ell,\lambda,\nu_1\nu_2\cdots\nu_\ell}\,\Box^{(\ell)}_{\nu_1\nu_2\cdots\nu_\ell,\lambda,\mu'_1\mu'_2\cdots\mu'_\ell} = -\frac{\ell+1}{\ell}\Delta^{(\ell)}_{\mu_1\mu_2\cdots\mu_\ell,\mu'_1\mu'_2\cdots\mu'_\ell}. \tag{11.26}$$

The special case $\ell = 1$ of this relation corresponds to $\varepsilon_{\mu\lambda\nu}\varepsilon_{\nu\lambda\mu'} = -2\delta_{\mu\mu'}$, cf. Sect. 4.1.2.

Some additional relations are listed for the case $\ell = 2$, which follow from the explicit form of the □-tensor (11.18):

$$\Box_{\lambda\kappa,\nu,\sigma\tau}\Box_{\sigma\tau,\lambda,\mu\kappa} = \frac{5}{4}\delta_{\mu\nu}, \tag{11.27}$$

and

$$\Box_{\mu\nu,\lambda,\sigma\tau}\,\Box_{\mu'\nu,\lambda',\sigma\tau} = \frac{1}{8}\,(9\,\delta_{\mu\mu'}\delta_{\lambda\lambda'} - 6\,\delta_{\mu\lambda'}\delta_{\mu'\lambda} - \delta_{\mu\lambda}\delta_{\mu'\lambda'}). \tag{11.28}$$

The relation (11.27) is recovered from (11.28) with the contraction $\mu = \lambda'$, the use of (11.23) and the appropriate renaming of indices.

11.3.3 Action of the Differential Operator $\mathscr{L}$ on Irreducible Tensors

Application of the differential operator $\mathscr{L}_\lambda = \varepsilon_{\lambda\nu\kappa} r_\nu \nabla_\kappa$, cf. (7.80), on the components of the position vector $\mathbf{r}$ yields

$$\mathscr{L}_\lambda r_\mu = \varepsilon_{\mu\lambda\nu} r_\nu.$$

Likewise, the action of $\mathscr{L}$ on the symmetric traceless tensors of rank ℓ, constructed from the components of $\mathbf{r}$, can be expressed with the help of the □-tensors:

$$\mathscr{L}_\lambda \overline{r_{\mu_1} r_{\mu_2} \cdots r_{\mu_\ell}} = \ell \, \square^{(\ell)}_{\mu_1\mu_2\cdots\mu_\ell,\lambda,\mu_1'\mu_2'\cdots\mu_\ell'} \, \overline{r_{\mu_1'} r_{\mu_2'} \cdots r_{\mu_\ell'}}. \tag{11.29}$$

Application of the second derivative $\mathscr{L}_\lambda \mathscr{L}_\lambda$ on the ℓ-th rank tensor yields

$$\mathscr{L}_\lambda \mathscr{L}_\lambda \overline{r_{\mu_1} r_{\mu_2} \cdots r_{\mu_\ell}} = \ell^2 \, \square^{(\ell)}_{\mu_1\mu_2\cdots\mu_\ell,\lambda,\nu_1\nu_2\cdots\nu_\ell} \, \square^{(\ell)}_{\nu_1\nu_2\cdots\nu_\ell,\lambda,\mu_1'\mu_2'\cdots\mu_\ell'} \, \overline{r_{\mu_1'} r_{\mu_2'} \cdots r_{\mu_\ell'}}. \tag{11.30}$$

Use of (11.26) leads to

$$\mathscr{L}_\lambda \mathscr{L}_\lambda \overline{r_{\mu_1} r_{\mu_2} \cdots r_{\mu_\ell}} = -\ell(\ell+1) \overline{r_{\mu_1} r_{\mu_2} \cdots r_{\mu_\ell}}. \tag{11.31}$$

Thus the tensors $\overline{r_{\mu_1} r_{\mu_2} \cdots r_{\mu_\ell}}$ are eigenfunctions of the operator $\mathscr{L}^2$ with the eigenvalues $-\ell(\ell+1)$. The same applies to the corresponding tensors constructed from the components of the unit vector $\hat{\mathbf{r}}$, as well as to the multipole tensors $X_{\mu_1\mu_2\cdots\mu_\ell}$, cf. Sect. 10.1, since the differential operator $\mathscr{L}$ acts just on the angular part of $\mathbf{r}$, but not on its magnitude.

The result (11.31) can be derived via an alternative route. The irreducible tensors $\overline{r_{\mu_1} r_{\mu_2} \cdots r_{\mu_\ell}}$ are solutions of the Laplace equation, cf. Sect. 10.2. On the other hand, the Laplace operator Δ can be split into its radial part Δ_r and the angular part involving $\mathscr{L} \cdot \mathscr{L}$, cf. (7.90) with (7.91). Thus one has

$$(\Delta_r + r^{-2} \mathscr{L}_\lambda \mathscr{L}_\lambda) \overline{r_{\mu_1} r_{\mu_2} \cdots r_{\mu_\ell}} = 0.$$

Due to

$$\Delta_r = r^{-2} \frac{\partial}{\partial r} \left(r^2 \frac{\partial}{\partial r} \right),$$

and $\overline{r_{\mu_1} r_{\mu_2} \cdots r_{\mu_\ell}} \sim r^\ell$, the application of the radial part of the Laplace operator on the tensor yields

$$\Delta_r \overline{r_{\mu_1} r_{\mu_2} \cdots r_{\mu_\ell}} = \ell(\ell+1) r^{-2} \overline{r_{\mu_1} r_{\mu_2} \cdots r_{\mu_\ell}},$$

and consequently the result (11.31) is recovered. This, incidentally, proves the validity of the ℓ-dependent factors in (11.26) and (11.29).

11.3.4 Consequences for the Orbital Angular Momentum Operator

In spatial representation, the orbital angular momentum operator $\mathbf{L}$, in units of $\hbar$, is related to the differential operator $\mathscr{L}$ by

$$L_\mu = \frac{1}{i} \mathscr{L}_\mu,$$

cf. (7.86). The action of the operator $\mathbf{L}$ on irreducible tensors $\overline{r_{\mu_1} r_{\mu_2} \cdots r_{\mu_\ell}}$ is immediately inferred from (11.29). Thanks to the imaginary unit i occurring here, application of $\mathbf{L} \cdot \mathbf{L} = L_\lambda L_\lambda$ on $\overline{r_{\mu_1} r_{\mu_2} \cdots r_{\mu_\ell}}$ yields

$$L_\lambda L_\lambda \overline{r_{\mu_1} r_{\mu_2} \cdots r_{\mu_\ell}} = \ell(\ell+1) \overline{r_{\mu_1} r_{\mu_2} \cdots r_{\mu_\ell}}. \tag{11.32}$$

Thus the irreducible tensors of rank ℓ, constructed from the components of $\mathbf{r}$ or of its unit vector $\hat{\mathbf{r}}$, as well as the multipole potential tensors $X_{\mu_1 \mu_2 \cdots \mu_\ell}$ are eigenfunctions of the square of the angular momentum with the eigenvalues $\ell(\ell+1)$. Since ℓ is the rank of a tensor, the possible values for ℓ are non-negative, integer numbers. This underlies the quantization of the magnitude of the orbital angular momentum.

Spherical components of irreducible tensors, cf. Sect. 9.4.2, which involve the components of the position vector $\mathbf{r}$, like $\overline{r_{\mu_1} r_{\mu_2} \cdots r_{\mu_\ell}}$, the multipole potentials $X_{...}$, or the spherical harmonics $Y_\ell^{(m)}$, are eigenfunctions of the z-component L_z of the orbital angular momentum operator $\mathbf{L}$. For $\ell = 1$, with the components of $\mathbf{r}$ denoted by $\{x, y, z\}$, this is inferred from $\mathscr{L}_z(x \pm iy) = \pm ix - y$, which implies $L_z(x \pm iy) = \pm(x \pm iy)$, and $L_z z = 0 = 0 \cdot z$. Obviously, the eigenvalues occurring here are $m = \pm 1$ and $m = 0$. For general ℓ, the eigenvalue equation is

$$L_z Y_\ell^{(m)} = m Y_\ell^{(m)}. \tag{11.33}$$

Due to $L_z = (1/i)\partial/\partial\varphi$, this relation follows from $Y_\ell^{(m)} \sim \exp(mi\varphi)$, cf. (9.29). Since the operator $\mathbf{L}$ does not affect the magnitude r of $\mathbf{r}$, but only the angles, the spherical components of $\overline{r_{\mu_1} r_{\mu_2} \cdots r_{\mu_\ell}}$ and of the multipole potentials $X_{...}$ introduced in Sect. 10.1.1, also obey the eigenvalue equation (11.33).

11.4 Isotropic Coupling Tensors

11.4.1 Definition of $\Delta^{(\ell,2,\ell)}$-Tensors

Isotropic tensors $\Delta^{(\ell_3,\ell_2,\ell_1)}_{\cdots,\cdots,\cdots}$ can be constructed from δ-tensors and zero or one ε-tensor, such that they generate an irreducible tensor of rank ℓ_3 from the product of irreducible tensors of ranks ℓ_1 and ℓ_2. These tensors are also referred to as *Clebsch-Gordan tensors*, [16, 17].

Special cases of coupling tensors, discussed so far, are the $\Delta^{(\ell)}_{\cdots,\cdots}$-tensors of Sect. 11.2, and the $\square^{(\ell)}_{\cdots,\cdots}$-tensors of Sect. 11.3, pertaining to $\ell_3 = \ell_1 \equiv \ell, \ell_2 = 0$ and $\ell_3 = \ell_1 \equiv \ell, \ell_2 = 1$, respectively. The next and only other case to be considered here corresponds to $\ell_3 = \ell_1 \equiv \ell, \ell_2 = 2$. These tensors are defined by

$$\Delta^{(\ell,2,\ell)}_{\mu_1\mu_2\cdots\mu_\ell,\lambda\kappa,\mu'_1\mu'_2\cdots\mu'_\ell} = \Delta^{(\ell)}_{\mu_1\mu_2\cdots\mu_\ell,\nu_1\nu_2\cdots\nu_{\ell-1}\sigma}\,\Delta^{(2)}_{\sigma\tau,\lambda\kappa}\,\Delta^{(\ell)}_{\tau\nu_1\nu_2\cdots\nu_{\ell-1},\mu'_1\mu'_2\cdots\mu'_\ell}.$$
(11.34)

Let **S** and **A** be irreducible tensors of rank 2 and ℓ. The tensor $\Delta^{(\ell,2,\ell)}$ accomplishes their multiplicative coupling of these tensors to a tensor of rank ℓ. This can be expressed as

$$(\overset{\frown}{\mathbf{S}\cdot\mathbf{A}})_{\mu_1\mu_2\cdots\mu_\ell} \equiv \overset{\frown}{S_{\mu_1\lambda}A_{\lambda\mu_2\cdots\mu_\ell}} = \Delta^{(\ell,2,\ell)}_{\mu_1\mu_2\cdots\mu_\ell,\lambda\kappa,\mu'_1\mu'_2\cdots\mu'_\ell}\,S_{\lambda\kappa}\,A_{\mu'_1\mu'_2\cdots\mu'_\ell}.$$
(11.35)

For $\ell = 1$, the expression (11.34) reduces to

$$\Delta^{(1,2,1)}_{\mu,\lambda\kappa,\nu} = \Delta^{(2)}_{\mu\nu,\lambda\kappa} \equiv \Delta_{\mu\nu,\lambda\kappa}.$$

Of particular interest is the case $\ell = 2$. Here one has

$$\Delta^{(2,2,2)}_{\mu\nu,\lambda\kappa,\alpha\beta} \equiv \Delta_{\mu\nu,\lambda\kappa,\alpha\beta} = \Delta_{\mu\nu,\mu'\nu'}\,\Delta_{\mu'\lambda',\lambda\kappa}\,\Delta_{\nu'\lambda',\alpha\beta}.$$
(11.36)

This tensor is symmetric against the interchange of any pair of subscripts, e.g.

$$\Delta_{\mu\nu,\lambda\kappa,\alpha\beta} = \Delta_{\mu\nu,\alpha\beta,\lambda\kappa} = \Delta_{\lambda\kappa,\mu\nu,\alpha\beta}.$$
(11.37)

Further properties are:

$$\Delta_{\mu\nu,\lambda\kappa,\kappa\sigma} = \frac{7}{12}\,\Delta_{\mu\nu,\lambda\sigma},\quad \Delta_{\mu\nu,\lambda\kappa,\lambda\kappa} = 0,\quad \Delta_{\mu\nu,\nu\kappa,\kappa\mu} = \frac{35}{12},$$
(11.38)

$$\Delta_{\mu\nu,\lambda\kappa,\sigma\tau}\,\Delta_{\mu'\nu',\lambda\kappa,\sigma'\tau'} = \frac{5}{48}\left(\Delta_{\mu\nu,\mu'\nu'}\,\Delta_{\sigma\tau,\sigma'\tau'} + \Delta_{\mu\nu,\sigma'\tau'}\,\Delta_{\sigma\tau,\mu'\nu'}\right)$$
$$-\frac{1}{24}\,\Delta_{\mu\nu,\sigma\tau}\,\Delta_{\mu'\nu',\sigma'\tau'}. \tag{11.39}$$

The double contraction $\{\sigma\tau\} = \{\sigma'\tau'\}$, in the last equation, yields

$$\Delta_{\mu\nu,\lambda\kappa,\sigma\tau}\,\Delta_{\mu'\nu',\lambda\kappa,\sigma\tau} = \frac{7}{12}\,\Delta_{\mu\nu,\mu'\nu'}, \tag{11.40}$$

and consequently

$$\Delta_{\mu\nu,\lambda\kappa,\sigma\tau}\,\Delta_{\mu\nu,\lambda\kappa,\sigma\tau} = \frac{35}{12},$$

which implies $\Delta_{\mu\nu,\lambda\kappa,\sigma\tau}\Delta_{\mu\nu,\lambda\kappa,\sigma\tau} = \Delta_{\mu\nu,\nu\kappa,\kappa\mu}$, see the last equation of (11.38). The following contraction of four isotropic tensors yields the same numerical value, viz.

$$\Delta_{\mu\nu,\beta\gamma,\mu'\nu'}\,\Delta_{\mu'\nu',\alpha\gamma,\sigma\tau}\,\Box_{\sigma\tau,\lambda,\mu\nu}\,\varepsilon_{\beta\lambda\alpha} = \frac{35}{12}. \tag{11.41}$$

Furthermore, by analogy to (11.22), the tensor defined by (11.36) can also be expressed by

$$\Delta_{\mu\nu,\lambda\kappa,\alpha\beta} = \frac{5}{2}\,\Delta^{(3)}_{\mu\nu\sigma,\alpha\beta\tau}\,\Delta_{\lambda\kappa,\sigma\tau}. \tag{11.42}$$

11.4.2 Tensor Product of Second Rank Tensors

For irreducible second rank tensors S and A, the expression (11.35) reduces to

$$\overline{(\mathbf{S}\cdot\mathbf{A})_{\mu\nu}} \equiv \overline{S_{\mu\lambda}A_{\lambda\nu}} = \Delta_{\mu\nu,\lambda\kappa,\mu'\nu'}\,S_{\lambda\kappa}\,A_{\mu'\nu'}. \tag{11.43}$$

Products of the coupling tensor $\Delta_{\mu\nu,\lambda\kappa,\sigma\tau}$ with symmetric traceless dyadic tensors constructed from the components of two vectors $\mathbf{a}$ and $\mathbf{b}$ are listed next:

$$\overline{a_\mu a_\lambda}\,\overline{b_\lambda b_\nu} \equiv \Delta_{\mu\nu,\lambda\kappa,\sigma\tau}\,\overline{a_\lambda a_\kappa}\,\overline{b_\sigma b_\tau}$$
$$= \mathbf{a}\cdot\mathbf{b}\,\overline{a_\mu b_\nu} - \frac{1}{3}\left(b^2\,\overline{a_\mu a_\nu} + a^2\,\overline{b_\mu b_\nu}\right), \tag{11.44}$$

$$\overline{a_\mu b_\lambda}\,\overline{a_\lambda b_\nu} \equiv \Delta_{\mu\nu,\lambda\kappa,\sigma\tau}\,\overline{a_\lambda b_\kappa}\,\overline{a_\sigma b_\tau}$$
$$= \frac{-1}{6}\,\mathbf{a}\cdot\mathbf{b}\,\overline{a_\mu b_\nu} + \frac{1}{4}\left(b^2\,\overline{a_\mu a_\nu} + a^2\,\overline{b_\mu b_\nu}\right), \tag{11.45}$$

$$\overline{a_\mu b_\lambda} \; \overline{a_\lambda a_\nu} \equiv \Delta_{\mu\nu,\lambda\kappa,\sigma\tau} \; \overline{a_\lambda b_\kappa} \; \overline{a_\sigma a_\tau}$$
$$= \frac{1}{6} a^2 \overline{a_\mu b_\nu} + \frac{1}{6} \mathbf{a} \cdot \mathbf{b} \, \overline{a_\mu a_\nu} . \tag{11.46}$$

Putting $\mathbf{b} = \mathbf{a}$, in these equations yields

$$\overline{a_\mu a_\lambda} \; \overline{a_\lambda a_\nu} \equiv \Delta_{\mu\nu,\lambda\kappa,\sigma\tau} \; \overline{a_\lambda a_\kappa} \; \overline{a_\sigma a_\tau} = \frac{1}{3} a^2 \overline{a_\mu a_\nu} . \tag{11.47}$$

Multiplication of the last equation by $\overline{a_\mu a_\nu}$ leads to

$$\overline{a_\mu a_\nu} \; \overline{a_\nu a_\lambda} \; \overline{a_\lambda a_\mu} = \Delta_{\mu\nu,\lambda\kappa,\sigma\tau} \; \overline{a_\mu a_\nu} \; \overline{a_\lambda a_\kappa} \; \overline{a_\sigma a_\tau} = \frac{2}{9} a^6 . \tag{11.48}$$

11.5 Coupling of a Vector with Irreducible Tensors

The product of a vector $\mathbf{b}$ with an irreducible tensor $\mathbf{A}$ of rank ℓ yields a tensor of rank $\ell + 1$ which can be decomposed into an irreducible tensor of rank $\ell + 1$ and terms involving irreducible tensors of ranks ℓ and $\ell - 1$. With the help of isotropic Δ-tensors and the $\square$-tensor, this decomposition reads:

$$b_\mu A_{\mu_1\mu_2\cdots\mu_\ell} = \Delta^{(\ell+1)}_{\mu\mu_1\mu_2\cdots\mu_\ell,\nu\nu_1\nu_2\cdots\nu_\ell} b_\nu A_{\nu_1\nu_2\cdots\nu_\ell}$$
$$- \frac{\ell(2\ell - 1)}{\ell(2\ell + 1) - 1} \square^{(\ell)}_{\mu_1\mu_2\cdots\mu_\ell,\mu,\nu_1\nu_2\cdots\nu_\ell} (\mathbf{b} \times \mathbf{A})_{\nu_1\nu_2\cdots\nu_\ell}$$
$$+ \frac{(2\ell - 1)}{(2\ell + 1)} \Delta^{(\ell)}_{\mu_1\mu_2\cdots\mu_\ell,\nu\,\nu_1\nu_2\cdots\nu_{\ell-1}} (\mathbf{b} \cdot \mathbf{A})_{\nu_1\nu_2\cdots\nu_{\ell-1}} . \tag{11.49}$$

The first term on the right hand side of (11.49) corresponds to

$$\overline{b_\mu A_{\mu_1\mu_2\cdots\mu_\ell}} . \tag{11.50}$$

The cross product and the dot product occurring in the second and third term are given by

$$(\mathbf{b} \times \mathbf{A})_{\nu_1\nu_2\cdots\nu_\ell} = \overline{\varepsilon_{\nu_1\lambda\kappa} b_\lambda A_{\kappa\,\nu_2\cdots\nu_\ell}} , \tag{11.51}$$

and

$$(\mathbf{b} \cdot \mathbf{A})_{\nu_1\nu_2\cdots\nu_{\ell-1}} = b_\lambda A_{\lambda\nu_1\nu_2\cdots\nu_{\ell-1}} . \tag{11.52}$$

For $\ell = 1$, i.e. when $\mathbf{A}$ is a vector, these relations reduce to the expressions given in Chap. 6 for the decomposition of dyadics.

Now, let **A** be the traceless tensor constructed from the components of the vector **b**, viz. $A_{\nu_1\nu_2\cdots\nu_\ell} = \overline{b_{\mu_1}b_{\mu_2}\cdots b_{\mu_\ell}}$. Then, the second term on the right hand side of (11.49), involving the cross product, vanishes. Due to

$$\overline{b_\mu\, \overline{b_\mu b_{\mu_1}b_{\mu_2}\cdots b_{\mu_{\ell-1}}}} = \frac{\ell}{(2\ell-1)}\, b^2\, \overline{b_{\mu_1}b_{\mu_2}\cdots b_{\mu_{\ell-1}}},$$

cf. (10.19), the relation (11.49), with (11.50) and (11.52), reduces to

$$\overline{b_\mu\,\overline{b_{\mu_1}b_{\mu_2}\cdots b_{\mu_\ell}}} = \overline{b_\mu\, b_{\mu_1}b_{\mu_2}\cdots b_{\mu_\ell}} \tag{11.53}$$
$$+ \frac{\ell}{(2\ell+1)}\, b^2\, \Delta^{(\ell)}_{\mu_1\mu_2\cdots\mu_\ell,\mu\,\nu_1\nu_2\cdots\nu_{\ell-1}}\, \overline{b_{\nu_1}b_{\nu_2}\cdots b_{\nu_{\ell-1}}}\,.$$

For $\ell = 1$, this relation reduces to $\overline{b_\mu b_\nu} = \overline{b_\mu b_\nu} + \frac{1}{3}\delta_{\mu\nu}$. The case $\ell = 2$ yields

$$\overline{b_\mu\,\overline{b_\nu b_\lambda}} = \overline{b_\mu b_\nu b_\lambda} + \frac{2}{5}\, b^2\, \Delta_{\nu\lambda,\mu\kappa}\, b_\kappa. \tag{11.54}$$

Multiplication of (11.53) by the components $a_\mu a_{\mu_1}\cdots a_{\mu_\ell}$ of the vector **a**, use of $a_\mu\,\overline{a_\mu a_{\mu_1}a_{\mu_2}\cdots a_{\mu_{\ell-1}}} = \ell/(2\ell-1)a^2\,\overline{a_{\mu_1}a_{\mu_2}\cdots a_{\mu_{\ell-1}}}$, cf. (10.19), and of the abbreviation, cf. (9.10), $\mathscr{P}_\ell(\mathbf{a},\mathbf{b}) = \overline{a_{\mu_1}a_{\mu_2}\cdots a_{\mu_\ell}}\,\overline{b_{\mu_1}b_{\mu_2}\cdots b_{\mu_\ell}}$ yields

$$\mathbf{a}\cdot\mathbf{b}\,\mathscr{P}_\ell(\mathbf{a},\mathbf{b}) = \mathscr{P}_{\ell+1}(\mathbf{a},\mathbf{b}) + \frac{\ell}{(2\ell+1)}\frac{\ell}{(2\ell-1)}a^2 b^2\,\mathscr{P}_{\ell-1}(\mathbf{a},\mathbf{b}). \tag{11.55}$$

Due to $\mathscr{P}_\ell(\mathbf{a},\mathbf{b}) = a^\ell\, b^\ell\, N_\ell\, P_\ell(x)$, with $x = \widehat{\mathbf{a}}\cdot\widehat{\mathbf{b}}$ and $N_\ell = \ell!/(2\ell-1)!!$, cf. (9.11), the (11.55) is equivalent to the recursion relation for the Legendre polynomials

$$x\,P_\ell(x) = \frac{\ell+1}{2\ell+1}\,P_{\ell+1}(x) + \frac{\ell}{2\ell+1}\,P_{\ell-1}(x). \tag{11.56}$$

11.6 Coupling of Second Rank Tensors with Irreducible Tensors

The product of second rank tensor with an irreducible tensor of rank ℓ can be decomposed by analogy to (11.49), where now irreducible tensors of ranks $\ell+2,\,\ell+1,\,\ell,$ $\ell-1,\,\ell-2$ occur. Here, the special case is considered where both the second rank tensor and the ℓ-th rank tensor are constructed from the components of the vector **b**. Then the tensors of ranks $\ell\pm1$ vanish and the desired result can be obtained by a twofold application of (11.53). Thus one finds

$$\overline{b_\nu b_\lambda} \ \overline{b_{\mu_1} b_{\mu_2} \cdots b_{\mu_\ell}} = \overline{b_\nu b_\lambda b_{\mu_1} b_{\mu_2} \cdots b_{\mu_\ell}}$$

$$+ \frac{2\ell}{2\ell+3} b^2 \, \Delta^{(\ell,2,\ell)}_{\mu_1\mu_2\cdots\mu_\ell,\nu\lambda,\nu_1\nu_2\cdots\nu_\ell} \ \overline{b_{\nu_1} b_{\nu_2} \cdots b_{\nu_\ell}} \qquad (11.57)$$

$$+ \frac{\ell(\ell-1)}{(2\ell+1)(2\ell-1)} b^4 \Delta^{(\ell)}_{\mu_1\mu_2\cdots\mu_\ell,\nu\lambda\nu_1\nu_2\cdots\nu_{\ell-2}} \ \overline{b_{\nu_1} b_{\nu_2} \cdots b_{\nu_{\ell-2}}} \, .$$

For $\ell = 1$, the (11.54) is recovered. Of special interest is the case $\ell = 2$. Here (11.57) is equivalent to

$$\overline{b_\mu b_\nu} \ \overline{b_\lambda b_\kappa} = \overline{b_\mu b_\nu b_\lambda b_\kappa} + \frac{4}{7} b^2 \Delta_{\mu\nu,\lambda\kappa,\sigma\tau} \ \overline{b_\sigma b_\tau} + \frac{2}{15} b^4 \Delta_{\mu\nu,\lambda\kappa}. \qquad (11.58)$$

Multiplication of this equation with symmetric tensors $a_{\mu\nu}$ and $c_{\lambda\kappa}$ yields

$$a_{\mu\nu} \overline{b_\mu b_\nu} \ c_{\lambda\kappa} \overline{b_\lambda b_\kappa} = a_{\mu\nu} c_{\lambda\kappa} \overline{b_\mu b_\nu b_\lambda b_\kappa} + \frac{4}{7} b^2 a_{\mu\nu} \overline{b_\nu b_\lambda} \ c_{\lambda\mu} + \frac{2}{15} b^4 a_{\mu\nu} c_{\mu\nu}.$$
$$(11.59)$$

Applications of these relations involving the products of irreducible tensors are found in the following sections.

11.7 Scalar Product of Three Irreducible Tensors

11.7.1 Scalar Invariants

Consider three symmetric irreducible tensors a, b, c, viz. $a_{\mu_1\mu_2\cdots\mu_{\ell_1}}$, $b_{\nu_1\nu_2\cdots\nu_{\ell_2}}$ and $c_{\kappa_1\kappa_2\cdots\kappa_\ell}$. Provided that $\ell = \ell_1 + \ell_2$, their product and total contraction yields the scalar

$$a_{\mu_1\mu_2\cdots\mu_{\ell_1}} \, b_{\nu_1\nu_2\cdots\nu_{\ell_2}} \, c_{\mu_1\mu_2\cdots\mu_{\ell_1}\nu_1\nu_2\cdots\nu_{\ell_2}}.$$

However, using in the product of the tensors a, b two subscripts which are equal and summed over, one also obtains a scalar, when $\ell = \ell_1 + \ell_2 - 2$ applies and the appropriate contractions are performed. In this case one has

$$a_{\mu_1\cdots\mu_{\ell_1-1}\lambda} \, b_{\nu_1\cdots\nu_{\ell_2-1}\lambda} \, c_{\mu_1\mu_2\cdots\mu_{\ell_1-1}\nu_1\nu_2\cdots\nu_{\ell_2-1}}.$$

Obviously, this can be generalized to form scalars from the product of three tensors of ranks ℓ_1, ℓ_2 and ℓ with

$$\ell = \ell_1 + \ell_2, \ \ell_1 + \ell_2 - 2, \ldots, |\ell_1 - \ell_2|.$$

All these triple products of three tensors are invariant under a rotation of the coordinate system.

Now let the tensors a, b, c be constructed from the components of unit vectors **a**, **b** and **c**, viz.

$$a_{\mu_1\mu_2\cdots\mu_{\ell_1}} = \overline{a_{\mu_1}a_{\mu_1}\cdots a_{\mu_{\ell_1}}},$$

$$b_{\nu_1\nu_2\cdots\nu_{\ell_2}} = \overline{b_{\nu_1}b_{\nu_1}\cdots b_{\nu_{\ell_2}}},$$

$$c_{\kappa_1\kappa_2\cdots\kappa_\ell} = \overline{c_{\kappa_1}c_{\kappa_2}\cdots c_{\kappa_\ell}}.$$

Then the scalar quantity

$$\mathscr{P}^{(\ell_1,\ell_2,\ell)}(\mathbf{a},\mathbf{b},\mathbf{c}), \quad \ell = \ell_1 + \ell_2, \; \ell_1 + \ell_2 - 2, \ldots, |\ell_1 - \ell_2|, \tag{11.60}$$

constructed in the manner just described, defines a *generalized Legendre polynomial*, apart from a numerical factor. For $\ell_1 = \ell$, $\ell_2 = 0$, one has $\mathscr{P}^{(\ell,0,\ell)} \sim P_\ell$, where P_ℓ is a Legendre polynomial, cf. Sect. 9.3. For the cases $\ell_1 = 0$, $\ell_2 = \ell$, and $\ell_1 = \ell_2$, $\ell = 0$, the generalization also reduces to ordinary Legendre polynomials. An explicit example for $\ell = \ell_1 + \ell_2$ is

$$\mathscr{P}^{(\ell_1,\ell_2,\ell_1+\ell_2)}(\mathbf{a},\mathbf{b},\mathbf{c}) = a_{\mu_1..\mu_{\ell_1}}\, b_{\nu_1..\nu_{\ell_2}}\, c_{\mu_1..\mu_{\ell_1}\nu_1..\nu_{\ell_2}}$$

$$= a_{\mu_1}..a_{\mu_{\ell_1}}\, b_{\nu_1}..b_{\nu_{\ell_2}}\, \overline{c_{\mu_1}..c_{\mu_{\ell_1}}c_{\nu_1}..c_{\nu_{\ell_2}}}. \tag{11.61}$$

The special case $\ell_1 = \ell_2 = \ell = 2$ is

$$\overline{a_\mu a_\lambda}\, \overline{b_\lambda b_\nu}\, \overline{c_\nu c_\mu}.$$

The interaction potential between non-spherical particles, see e.g. [34], involves scalar invariants of the kind discussed here.

11.7.2 Interaction Potential for Uniaxial Particles

Consider two uniaxial non-spherical particles whose symmetry axes are specified by the unit vectors $\mathbf{u}_1$ and $\mathbf{u}_2$. These particles can be rod-like or disc-like. In many cases, weakly biaxial particles can be treated as effectively uniaxial. The centers of mass of the two particles are at the positions $\mathbf{r}_1$ and $\mathbf{r}_2$, $\mathbf{r} = \mathbf{r}_1 - \mathbf{r}_2$ is their relative position vector. The interaction potential Φ between the particles depends on their orientations, viz. on $\mathbf{u}_1$ and $\mathbf{u}_2$, and on $\mathbf{r} = r\,\hat{\mathbf{r}}$. The angle dependence of Φ can be described by the functions $\mathscr{P}^{(\ell_1,\ell_2,\ell)}$ defined above. Thus the potential function is expressed as the expansion

$$\Phi = \Phi(\mathbf{u}_1, \mathbf{u}_2, \mathbf{r}) = \sideset{}{'}\sum_{\ell} \sum_{\ell_1} \sum_{\ell_2} \phi_{(\ell_1, \ell_2, \ell)}(r) \, \mathscr{P}^{(\ell_1, \ell_2, \ell)}(\mathbf{u}_1, \mathbf{u}_2, \widehat{\mathbf{r}}). \qquad (11.62)$$

The prime in the summation over ℓ indicates that only the "allowed" values are used, as indicated in (11.60), viz. $\ell = \ell_1 + \ell_2, \ell_1 + \ell_2 - 2, \ldots, |\ell_1 - \ell_2|$.

The scalar functions $\phi_{(\ell_1, \ell_2, \ell)}(r)$ characterize the radial dependence of the different angular contributions. The function $\phi_{(0,0,0)}$ is the *spherical symmetric part*, also referred to as *isotropic part* of the potential. The dipole-dipole, dipole-quadrupole and quadrupole-quadrupole interactions considered in Sect. 10.4.5 are examples for the cases (ℓ_1, ℓ_2, ℓ) equal to $(1, 1, 2)$, $(1, 2, 3)$ and $(2, 2, 4)$. The anisotropic part of the interaction between *Janus spheres*, these are spherical particles with two different 'faces', can be modeled, cf. [35], with three terms proportional to $\mathbf{u}_1 \cdot \mathbf{u}_2$, to $\mathbf{u}_1 \cdot \mathbf{r} - \mathbf{u}_2 \cdot \mathbf{r}$ and to $\mathbf{u}_1 \cdot \mathbf{r}\,\mathbf{u}_2 \cdot \mathbf{r} - \frac{1}{3}\mathbf{u}_1 \cdot \mathbf{u}_2$. These correspond to (ℓ_1, ℓ_2, ℓ) equal to $(1, 1, 0)$, $(1, 0, 1) - (0, 1, 1)$ and $(1, 1, 2)$. A non-spherical interaction potential for nematic liquid crystals, introduced in [36], involves the anisotropic terms $\overline{\mathbf{u}_1\mathbf{u}_1} : \overline{\mathbf{u}_2\mathbf{u}_2}$ and $(\overline{\mathbf{u}_1\mathbf{u}_1} + \overline{\mathbf{u}_2\mathbf{u}_2}) : \overline{\mathbf{r}\mathbf{r}}$, corresponding to (ℓ_1, ℓ_2, ℓ) equal to $(2, 2, 0)$ and $(2, 0, 2) + (0, 2, 2)$. This potential is simpler than the commonly used *Gay-Berne potential*, cf. [37].

Chapter 12
Integral Formulae and Distribution Functions

Abstract This chapter is devoted to integral formulae and distribution functions. Firstly, integrals over the unit sphere are considered, in particular, results are presented for integrals of the product of two and more irreducible tensors. Then the orientational distribution function needed for orientational averages and the expansion of the distribution with respect to irreducible tensors are introduced, Applications to the anisotropic dielectric tensor, field-induced orientation of non-spherical particles, Kerr effect, Cotton-Mouton effect, non-linear susceptibility, the orientational entropy and the Fokker-Planck equation governing the orientational dynamics, are discussed. Secondly, averages over velocity distributions are treated, expansions about a global or a local Maxwell distribution are analyzed and applied for kinetic equations. Thirdly, anisotropic pair correlation functions and static structure factors are considered. Examples for two-particle averages are the potential contributions to the energy and to the pressure tensor of a liquid. The shear-flow induced distortion of the pair-correlation is discussed, in particular for a plane Couette flow. The pair correlation for a system with cubic symmetry is described. The chapter is concluded by a derivation of the quantum-mechanical selection rules for electromagnetic radiation using the expansion of wave functions with respect to irreducible Cartesian tensors.

12.1 Integrals Over Unit Sphere

Here, an integral over the unit sphere means a surface integral, as discussed in Sect. 8.2.4. As in (8.31), $d^2\hat{r}$ stands for the surface element of the unit sphere which is equal to $\sin\theta d\theta d\varphi$, when spherical polar coordinates are used. The surface of the unit sphere is recalled: $\int d^2\hat{r} = 4\pi$. Many integrals of interest can be evaluated effectively, without any explicit integration over angles, by taking the isotropy of the sphere and the symmetry properties of the integrands into account, and by using properties of isotropic tensors.

© Springer International Publishing Switzerland 2015

S. Hess, *Tensors for Physics*, Undergraduate Lecture Notes in Physics,

DOI 10.1007/978-3-319-12787-3_12

12.1.1 *Integrals of Products of Two Irreducible Tensors*

Integrals $\int \ldots \mathrm{d}^2\hat{r}$ of products of irreducible tensors $\overline{\hat{r}_{\mu_1}\hat{r}_{\mu_2}\cdots\hat{r}_{\mu_\ell}}$ and $\overline{\hat{r}_{\nu_1}\hat{r}_{\nu_2}\cdots\hat{r}_{\nu_{\ell'}}}$ over the surface of the unit sphere must be proportional to an isotropic tensor, since a sphere possesses no preferential direction. Furthermore, the particular isotropic tensor must have the same subscripts with the same symmetry properties as the integrands. There is no isotropic tensor which is symmetric traceless in both sets of subscripts, unless $\ell' = \ell$ holds true. So the integral must be proportional to the Kronecker delta $\delta_{\ell\ell'}$ and to the $\Delta^{(\ell)}$-tensor. Thus the ansatz

$$\int \overline{\hat{r}_{\mu_1}\hat{r}_{\mu_2}\cdots\hat{r}_{\mu_\ell}} \; \overline{\hat{r}_{\nu_1}\hat{r}_{\nu_2}\cdots\hat{r}_{\nu_{\ell'}}} \, \mathrm{d}^2\hat{r} = C_\ell \, \delta_{\ell\ell'} \, \Delta^{(\ell)}_{\mu_1\mu_2\cdots\mu_\ell, \, \nu_1\nu_2\cdots\nu_\ell}.$$

The coefficient C_ℓ is determined via the total contraction $\{\mu_1\mu_2\cdots\mu_\ell\} = \{\nu_1\nu_2\cdots\nu_\ell\}$. With $\overline{\hat{r}_{\mu_1}\hat{r}_{\mu_2}\cdots\hat{r}_{\mu_\ell}} \; \overline{\hat{r}_{\mu_1}\hat{r}_{\mu_2}\cdots\hat{r}_{\mu_\ell}} = \ell!/((2\ell-1)!!$, cf. Sect. 9.3, $\int \mathrm{d}^2\hat{r} = 4\pi$ and $\Delta^{(\ell)}_{\mu_1\mu_2\cdots\mu_\ell,\mu_1\mu_2\cdots\mu_\ell} = 2\ell+1$, one obtains $C_\ell = 4\pi\,\ell!/((2\ell+1)!!)$, and consequently

$$\frac{1}{4\pi} \int \overline{\hat{r}_{\mu_1}\hat{r}_{\mu_2}\cdots\hat{r}_{\mu_\ell}} \; \overline{\hat{r}_{\nu_1}\hat{r}_{\nu_2}\cdots\hat{r}_{\nu_{\ell'}}} \, \mathrm{d}^2\hat{r} = \frac{\ell!}{(2\ell+1)!!} \, \delta_{\ell\ell'} \, \Delta^{(\ell)}_{\mu_1\mu_2\cdots\mu_\ell, \, \nu_1\nu_2\cdots\nu_\ell}.$$

$$(12.1)$$

Notice, the integral $\frac{1}{4\pi} \int \ldots \mathrm{d}^2\hat{r}$ is an orientational average. The orientational average of corresponding tensors constructed from the vector $\mathbf{r}$ rather than the unit vector $\hat{\mathbf{r}}$ is given by the right hand side of (12.1), multiplied by $r^{2\ell}$.

Some special cases of (12.1) are

$$\int \hat{r}_\mu \mathrm{d}^2\hat{r} = \int \overline{\hat{r}_\mu\hat{r}_\nu} \, \mathrm{d}^2\hat{r} = \int \overline{\hat{r}_\mu\hat{r}_\nu} \, \hat{r}_\lambda \mathrm{d}^2\hat{r} = 0, \qquad (12.2)$$

and

$$\frac{1}{4\pi} \int \hat{r}_\mu\hat{r}_\nu \mathrm{d}^2\hat{r} = \frac{1}{3} \delta_{\mu\nu}, \qquad (12.3)$$

$$\frac{1}{4\pi} \int \overline{\hat{r}_\mu\hat{r}_\nu} \; \overline{\hat{r}_\lambda\hat{r}_\kappa} = \frac{2}{15} \Delta_{\mu\nu,\lambda\kappa}. \qquad (12.4)$$

The numerical factor occurring here can be checked immediately by observing that $\delta_{\mu\mu} = 3$, $\overline{\hat{r}_\mu\hat{r}_\nu}\,\overline{\hat{r}_\mu\hat{r}_\nu} = 2/3$, and $\Delta_{\mu\nu,\mu\nu} = 5$.

In terms of the $Y_{\ldots}$-tensors defined in Sect. 10.1.4, equation (12.1) is equivalent to

$$\frac{1}{4\pi} \int Y_{\mu_1\mu_2\cdots\mu_\ell} \, Y_{\nu_1\nu_2\cdots\nu_{\ell'}} \mathrm{d}^2\hat{r} = \ell! \,(2\ell-1)!! \, \delta_{\ell\ell'} \, \frac{1}{2\ell+1} \, \Delta^{(\ell)}_{\mu_1\mu_2\cdots\mu_\ell, \, \nu_1\nu_2\cdots\nu_\ell}.$$

$$(12.5)$$

The orientational average of the corresponding multipole potential tensors X... rather than the Y...-tensors is given by the right hand side of (12.5), multiplied by $r^{-2(\ell+1)}$.

12.1.2 Multiple Products of Irreducible Tensors

Integrals involving products of more than two irreducible tensors can also be evaluated with the help of symmetry considerations and the contraction of tensors.

For instance, the orientational average of the fourfold product $\widehat{r}_\mu \widehat{r}_\nu \widehat{r}_\lambda \widehat{r}_\kappa$ of components of the unit vector must be proportional to the isotropic fourth rank tensor with the appropriate symmetry, viz. $\delta_{\mu\nu}\delta_{\lambda\kappa} + \delta_{\mu\lambda}\delta_{\nu\kappa} + \delta_{\mu\kappa}\delta_{\lambda\nu}$. By analogy to the consideration given above, one finds

$$\frac{1}{4\pi} \int \widehat{r}_\mu \widehat{r}_\nu \widehat{r}_\lambda \widehat{r}_\kappa \, \mathrm{d}^2\widehat{r} = \frac{1}{15} (\delta_{\mu\nu}\delta_{\lambda\kappa} + \delta_{\mu\lambda}\delta_{\nu\kappa} + \delta_{\mu\kappa}\delta_{\lambda\nu}). \qquad (12.6)$$

To check the numerical factor on the right hand side, put λ equal to κ and compare with (12.3).

The integral of three irreducible second tensors

$$\widehat{r_\mu r_\nu} \ \widehat{r_\lambda r_\kappa} \ \widehat{r_\sigma r_\tau} \ ,$$

over the unit sphere must be proportional to the sixth rank isotropic tensor $\Delta_{\mu\nu,\lambda\kappa,\sigma\tau}$, defined by (11.36). The resulting equation is

$$\frac{1}{4\pi} \int \widehat{r_\mu r_\nu} \ \widehat{r_\lambda r_\kappa} \ \widehat{r_\sigma r_\tau} \, \mathrm{d}^2\widehat{r} = \frac{8}{105} \Delta_{\mu\nu,\lambda\kappa,\sigma\tau}. \qquad (12.7)$$

The numerical factor $8/105$ can be verified in the following exercise.

The integral of four irreducible second tensors

$$\widehat{r_{\mu_1} r_{\nu_1}} \ \widehat{r_{\mu_2} r_{\nu_2}} \ \widehat{r_{\mu_3} r_{\nu_3}} \ \widehat{r_{\mu_4} r_{\nu_4}}$$

can be found in the same spirit. Here, the result must be proportional to an isotropic tensor of rank 8 with the required symmetry properties. There are two expressions of this type, viz.

$$\Delta_{\mu_1\nu_1,\lambda\kappa} \ \Delta_{\mu_2\nu_2,\kappa\sigma} \ \Delta_{\mu_3\nu_3,\sigma\tau} \ \Delta_{\mu_4\nu_4,\tau\lambda} \ ,$$

which is a generalization of the $\Delta^{(2,2,2)}$-tensor, cf. Sect. 11.4, and

$$\Delta_{\mu_1\nu_1,\mu_2\nu_2} \ \Delta_{\mu_3\nu_3,\mu_4\nu_4} + \Delta_{\mu_1\nu_1,\mu_3\nu_3} \ \Delta_{\mu_2\nu_2,\mu_4\nu_4} + \Delta_{\mu_1\nu_1,\mu_4\nu_4} \ \Delta_{\mu_2\nu_2,\mu_3\nu_3} \ ,$$

which is constructed by analogy to the fourth rank tensor occurring on the right hand side of (12.6). Multiplication of these isotropic tensors with $S_{\mu_1\nu_1}S_{\mu_2\nu_2}S_{\mu_3\nu_3}S_{\mu_4\nu_4}$, where S is symmetric traceless second rank tensor, yields a product of the form $S_{\mu\nu}S_{\nu\lambda}S_{\lambda\kappa}S_{\kappa\mu}$, in the first case and $3(S_{\mu\nu}S_{\nu\mu})^2$, in the second case. Both terms are proportional to each other, see (5.52). Thus also the two isotropic tensors of rank 8 are equivalent, apart from a numerical factor. More specifically, one has

$$\Delta_{\mu_1\nu_1,\lambda\kappa}\,\Delta_{\mu_2\nu_2,\kappa\sigma}\,\Delta_{\mu_3\nu_3,\sigma\tau}\,\Delta_{\mu_4\nu_4,\tau\lambda}$$
$$= \frac{1}{6}(\Delta_{\mu_1\nu_1,\mu_2\nu_2}\Delta_{\mu_3\nu_3,\mu_4\nu_4} + \Delta_{\mu_1\nu_1,\mu_3\nu_3}\Delta_{\mu_2\nu_2,\mu_4\nu_4} + \Delta_{\mu_1\nu_1,\mu_4\nu_4}\Delta_{\mu_2\nu_2,\mu_3\nu_3}).$$
$$(12.8)$$

Thus it suffices to consider one of these isotropic tensor, e.g. the second one. Then the desired integral is

$$\frac{1}{4\pi}\int \overline{\widehat{r}_{\mu_1}\widehat{r}_{\nu_1}}\;\overline{\widehat{r}_{\mu_2}\widehat{r}_{\nu_2}}\;\overline{\widehat{r}_{\mu_3}\widehat{r}_{\nu_3}}\;\overline{\widehat{r}_{\mu_4}\widehat{r}_{\nu_4}}\,\mathrm{d}^2\widehat{r}$$
$$= \frac{4}{105}\frac{1}{3}(\Delta_{\mu_1\nu_1,\mu_2\nu_2}\Delta_{\mu_3\nu_3,\mu_4\nu_4} + \Delta_{\mu_1\nu_1,\mu_3\nu_3}\Delta_{\mu_2\nu_2,\mu_4\nu_4} + \Delta_{\mu_1\nu_1,\mu_4\nu_4}\Delta_{\mu_2\nu_2,\mu_3\nu_3}).$$
$$(12.9)$$

Multiplication of this equation with $S_{\mu_1\nu_1}S_{\mu_2\nu_2}S_{\mu_3\nu_3}S_{\mu_4\nu_4}$ implies

$$\frac{1}{4\pi}\int \left(S_{\mu\nu}\,\overline{\widehat{r}_{\mu}\widehat{r}_{\nu}}\right)^4 \mathrm{d}^2\widehat{r} = \frac{4}{105}\,(S_{\mu\nu}\,S_{\mu\nu})^2. \qquad (12.10)$$

12.1 Exercise: Verify the Numerical Factor in (12.7) for the Integral over a Triple Product of Tensors
Hint: Put $\nu = \lambda$, $\kappa = \sigma$, $\tau = \mu$ and use the relevant formulae given in Sect. 11.4.

12.2 Orientational Distribution Function

12.2.1 Orientational Averages

The orientation of a single molecule or particle with uniaxial symmetry can be specified by a unit vector $\mathbf{u}$. When needed, the vector $\mathbf{u}$ can be expressed in terms of the spherical polar angles θ and φ, just as the unit vector $\widehat{\mathbf{r}}$.

The orientation of many of those molecules in a molecular liquid or a nematic liquid crystal or of rod-like particles in a colloidal solution, is characterized by an orientational distribution function $f = f(\mathbf{u})$. With the normalization

$$\int f(\mathbf{u})\mathrm{d}^2u = 1, \qquad (12.11)$$

where d^2u is the surface element on the unit sphere, the average $\langle\psi\rangle$ of an angle dependent quantity $\psi = \psi(\mathbf{u})$, is given by

$$\langle\psi\rangle = \int \psi(\mathbf{u})f(\mathbf{u})d^2u. \tag{12.12}$$

In an isotropic state, where no preferential direction exists, the orientational distribution function f_0 does not depend on $\mathbf{u}$. Due to the normalization condition one has $f_0 = (4\pi)^{-1}$. Averages over this isotropic distribution are denoted by $\langle\cdots\rangle_0$, viz.

$$\langle\psi\rangle_0 = \frac{1}{4\pi}\int \psi(\mathbf{u})d^2u. \tag{12.13}$$

12.2.2 Expansion with Respect to Irreducible Tensors

In general, the orientational distribution function $f(\mathbf{u})$ can be expanded in terms of irreducible tensors constructed from the components of $\mathbf{u}$. It is convenient to use the tensors

$$\phi_{\mu_1\mu_2\cdots\mu_\ell} \equiv \sqrt{\frac{(2\ell+1)!!}{\ell!}}\; \overline{u_{\mu_1}u_{\mu_2}\cdots u_{\mu_\ell}}. \tag{12.14}$$

The basis functions $\phi_{...}$ are orthogonal and normalized according to

$$\frac{1}{4\pi}\int \phi_{\mu_1\mu_2\cdots\mu_\ell}\phi_{\nu_1\nu_2\cdots\nu_{\ell'}}d^2u = \langle\phi_{\mu_1\mu_2\cdots\mu_\ell}\phi_{\nu_1\nu_2\cdots\nu_{\ell'}}\rangle_0 = \delta_{\ell\ell'}\Delta^{(\ell)}_{mu_1\mu_2\cdots\mu_\ell,\nu_1\nu_2\cdots\nu_\ell}, \tag{12.15}$$

cf. (12.1). Then the expansion reads

$$f(\mathbf{u}) = f_0(1+\Phi), \quad f_0 = (4\pi)^{-1}, \quad \Phi = \sum_{\ell=1}^{\infty} a_{\mu_1\mu_2\cdots\mu_\ell}\phi_{\mu_1\mu_2\cdots\mu_\ell}. \tag{12.16}$$

Clearly, Φ is the deviation of f from the isotropic distribution f_0. The expansion coefficients $a_{...}$ are the moments of the distribution function, viz.

$$a_{\mu_1\mu_2\cdots\mu_\ell} = \int \phi_{\mu_1\mu_2\cdots\mu_\ell}f(\mathbf{u})d^2u \equiv \langle\phi_{\mu_1\mu_2\cdots\mu_\ell}\rangle. \tag{12.17}$$

These quantities are referred to as *alignment tensors* or *order parameter tensors*. In general, they may depend on the time t and the position $\mathbf{r}$ in space.

The expansion tenors pertaining to $\ell = 1, 2, 3$ and $\ell = 4$ are

$$\phi_\mu = \sqrt{3}\, u_\mu\,, \quad \phi_{\mu\nu} = \sqrt{\frac{15}{2}}\, \overline{u_\mu u_\nu}\,, \quad \phi_{\mu\nu\lambda} = \frac{1}{2}\sqrt{70}\, \overline{u_\mu u_\nu u_\lambda}\,,$$

$$\phi_{\mu\nu\lambda\kappa} = \frac{3}{4}\sqrt{70}\, \overline{u_\mu u_\nu u_\lambda u_\kappa}\,.$$

The first few terms of $f(\mathbf{u}) = \frac{1}{4\pi}(1 + a_\mu \phi_\mu + a_{\mu\nu}\phi_{\mu\nu} + \ldots)$ of the expansion can also be written as

$$f(\mathbf{u}) = \frac{1}{4\pi}\left(1 + 3\langle u_\mu\rangle u_\mu + \frac{15}{2}\langle \overline{u_\mu u_\nu}\rangle\, \overline{u_\mu u_\nu} + \ldots\right). \tag{12.18}$$

The second rank alignment tensor $a_{\mu\nu} \sim \langle \overline{u_\mu u_\nu}\rangle$ has the same symmetry as the quadrupole moment tensor, cf. (10.29). For this reason the tensor $\langle \overline{u_\mu u_\nu}\rangle$ is also denoted by $Q_{\mu\nu}$ and referred to as Q-tensor, in the liquid crystal literature, [67].

12.2.3 Anisotropic Dielectric Tensor

The anisotropy of the dielectric tensor gives rise to birefringence, cf. Sect. 5.3.4. In fluids of optically anisotropic, uniaxial particles, the symmetric traceless part of the dielectric tensor is proportional to the alignment tensor:

$$\overline{\varepsilon_{\mu\nu}} = \varepsilon_a\, a_{\mu\nu}. \tag{12.19}$$

The proportionality coefficient ε_a depends on the density n of the system and on the difference $\alpha_\parallel - \alpha_\perp$ of the molecular polarizabilities parallel and perpendicular to $\mathbf{u}$, cf. Sect. 5.3.3.

The electric polarization $\mathbf{P}$ is the average of the induced dipole moment, which is proportional to the electric field 'felt' by the particle. In dilute systems, this is equal to the applied field $\mathbf{E}$. Then one has

$$P_\mu = \varepsilon_0\, n\, \langle \alpha_{\mu\nu}\rangle\, E_\nu$$

and consequently

$$\overline{\varepsilon_{\mu\nu}} = n\,(\alpha_\parallel - \alpha_\perp)\langle \overline{u_\mu u_\nu}\rangle\,.$$

In dense systems, the particles 'feel' a local field $\mathbf{E}^{\mathrm{loc}}$, modified by the surrounding particles. The Lorentz field approximation

$$\mathbf{E}^{\mathrm{loc}} = \mathbf{E}^{\mathrm{Lor}} \equiv \mathbf{E} + \frac{1}{3\varepsilon_0}\,\mathbf{P}\,,$$

leads to

$$\overline{\varepsilon_{\mu\nu}} = n \left(\frac{\varepsilon_{\mathrm{iso}}+2}{3}\right)^2 \langle \overline{\alpha_{\mu\nu}} \rangle + \dots,$$

where $\varepsilon_{\mathrm{iso}}$ is the orientationally averaged dielectric coefficient. This result is equivalent to (12.19) with

$$\varepsilon_a = n \left(\frac{\varepsilon_{\mathrm{iso}}+2}{3}\right)^2 (\alpha_\| - \alpha_\perp)\, \zeta_2^{-1}, \qquad (12.20)$$

where $\zeta_2 = \sqrt{\frac{15}{2}}$ is the normalization factor occurring in the definition of the alignment tensor.

12.2.4 Field-Induced Orientation

In the presence of electric or magnetic fields, the energy H of a particle with permanent or induced electric or magnetic dipole moments depends on its orientation relative to the direction of the applied field. For a uniaxial particle, one has $H = H(\mathbf{u})$. In thermal equilibrium, the orientational distribution function $f = f_{\mathrm{eq}}$ is proportional to $\exp[-H/k_B T] = \exp[-\beta H]$:

$$f_{\mathrm{eq}} = Z^{-1}\exp[-\beta H(\mathbf{u})], \quad Z = \int \exp[-\beta H]\mathrm{d}^2 u, \quad \beta = \frac{1}{k_B T}. \qquad (12.21)$$

It is assumed that $\langle H \rangle_0 = 0$ where $\langle \dots \rangle_0 = (4\pi)^{-1}\int \dots \mathrm{d}^2 u$ indicates the unbiased orientational average in an isotropic state. Then the *high temperature expansion* for the state function Z reads

$$Z = (4\pi)\left(1 + \frac{1}{2}\beta^2 \langle H^2\rangle_0 - \frac{1}{6}\beta^3\langle H^3\rangle_0 + \frac{1}{24}\beta^4\langle H^4\rangle_0 \mp \dots\right).$$

First, the interaction of an electric or magnetic dipole moment is considered, which is parallel to $\mathbf{u}$ and subjected to an electric or magnetic field, which is denoted by $\mathbf{F}$. Then H is equal to

$$H = H^{(1)} \equiv H^{\mathrm{dip}} = -d\, F_\mu u_\mu,$$

where d stands for the magnitude of the relevant dipole moment. In this case, $-\beta H$ is written as $\beta_1 F_\mu \phi_\mu$ with $\phi_\mu = \sqrt{3}u_\mu$ and $\beta_1 = \beta d/\sqrt{3}$, and Z, in lowest order in β_1, reduces to

$$Z = (4\pi)\,(1 + \frac{1}{2}\beta_1^2\, F_\mu F_\mu + \dots).$$

Likewise, up to second order in β_1, the distribution function is equal to

$$f_{\text{eq}} = (4\pi)^{-1} \left(1 + \beta_1 \, F_\mu \phi_\mu + \sqrt{\frac{3}{10}} \, \beta_1^2 F_\mu F_\nu \phi_{\mu\nu} + \dots \right). \tag{12.22}$$

In this high temperature approximation, the first two moments are given by

$$a_\mu = \langle \phi_\mu \rangle = \beta_1 F_\mu \,, \quad a_{\mu\nu} = \langle \phi_{\mu\nu} \rangle = \sqrt{\frac{3}{10}} \, \beta_1^2 \, \overline{F_\mu F_\nu} = \sqrt{\frac{3}{10}} \, \overline{a_\mu a_\nu} \,. \tag{12.23}$$

These relations imply

$$\langle \overline{u_\mu u_\nu} \rangle = c^{(2|11)} \, \overline{\langle u_\mu \rangle \langle u_\nu \rangle} \,. \tag{12.24}$$

The numerical factor $c(2|11)$ is equal to $3/5$ in the high temperature approximation. For a very strong orienting field which induces a practically perfect orientation, $c^{(2|11)} = c^{(2|11)}(\beta_1 F)$ approaches 1.

In the case of a dipolar orientation, the distribution and consequently all its moments have uniaxial symmetry. Let $\mathbf{e}$ be a unit vector parallel to the field $\mathbf{F}$, viz. $F_\mu = F e_\mu$, where F is the strength of the field. Then the energy can be written as $H^{\text{dip}} = -dF e_\mu u_\mu = -dFx$, where $x = e_\mu u_{\mu|}$ is the cosine of the angle between $\mathbf{u}$ and the direction of the field. Furthermore, one has $e_\mu \phi_\mu = \sqrt{3} x$ and $-H^{\text{dip}}/k_B T = \sqrt{3} \beta_1 Fx$. The average $\langle x^n \rangle$ is evaluated according to

$$\langle x^n \rangle = \frac{1}{2} Z^{-1} \int_{-1}^{1} x^n \exp\left[\sqrt{3} \, \beta_1 \, F \, x \right] dx \,,$$

with the state function Z given by

$$Z = L\left(\sqrt{3} \, \beta_1 \, F \right), \quad L(z) = \frac{1}{2z} \left(\exp[z] - \exp[-z] \right) = 1 + z^2/6 + z^4/120 + \dots \,. \tag{12.25}$$

Let L_k be the k-derivative of $L(z)$ with respect to z, viz. $L_k = \mathrm{d}^k L/\mathrm{d}z^k$. The average $\langle x^k \rangle$ is then determined by $L_k(z)/L(z)$, with $z = \sqrt{3}\beta_1 F = \beta dF$. The results for the lowest moments are

$$\langle x \rangle = \frac{1 + z + \exp[2z](z-1)}{z \exp[2z] - z} = z/3 - z^3/45 \pm \dots \,, \tag{12.26}$$

$$\langle x^2 \rangle - \frac{1}{3} = \frac{1 + z + \exp[2z](z-1)}{z \exp[2z] - z} - \frac{1}{3} = 2z^2/45 - 4z^4/945 \pm \dots \tag{12.27}$$

For $z \to \infty$, both $\langle x \rangle$ and $\langle x^2 \rangle$ approach 1.

The coefficient $c^{(2|11)}$ defined in (12.24) is given by

$$c^{(2|11)} = \langle P_2(x) \rangle / \langle P_1(x) \rangle^2 = \frac{3}{2} \left(\langle x^2 \rangle - \frac{1}{3} \right) / \langle x \rangle^2. \qquad (12.28)$$

The high-temperature approximation is

$$c^{(2|11)} = 3/5 + 4z^2/175 + 4z^4/2625 + \ldots, \qquad (12.29)$$

for $z \to \infty$ one has $c^{(2|11)} \to 1$.

Next the case is considered where the orientational energy involves the scalar product of the symmetric traceless tensor field $F_{\mu\nu}$ and the tensor $\phi_{\mu\nu} = \sqrt{15/2}$ $\overline{u_\mu u_\mu}$, viz.

$$H = H^{(2)} \sim -F_{\mu\nu} \overline{u_\mu u_\mu}, \quad -H^{(2)}/k_B T = \beta_2 F_{\mu\nu} \phi_{\mu\nu}. \qquad (12.30)$$

Interactions of this type occur for linear molecules with anisotropic polarizability in the presence of an electric field $\mathbf{E}$, cf. Sect. 5.3.3, and for particles with an electric quadrupole moment in the presence of an electric field gradient $\nabla \mathbf{E}$. In the first case, the field tensor $F_{\mu\nu} = \overline{E_\mu E_\nu}$ is uniaxial, in the second case, where $F_{\mu\nu} = \overline{\nabla_\mu E_\nu}$ applies, the field tensor is biaxial, unless the gradient is parallel to the $\mathbf{E}$-field.

The distribution function pertaining to (12.30) is

$$\begin{aligned} f_{eq} &= Z^{-1} \exp[\beta_2 F_{\mu\nu} \phi_{\mu\nu}] \\ &= Z^{-1} \left(1 + \beta_2 F_{\mu\nu} \phi_{\mu\nu} + \frac{1}{2} \beta_2^2 F_{\mu\nu} F_{\lambda\kappa} \phi_{\mu\nu} \phi_{\lambda\kappa} + \ldots \right), \end{aligned}$$

with

$$Z = (4\pi)^{-1} \left(1 + \frac{1}{2} \beta_2^2 F_{\mu\nu} F_{\mu\nu} + z^{(3)} + z^{(4)} + \ldots \right).$$

The third and fourth order terms are

$$z^{(3)} = \frac{1}{6} \beta_2^3 F_{\mu\nu} F_{\lambda\kappa} F_{\sigma\tau} \langle \phi_{\mu\nu} \phi_{\lambda\kappa} \phi_{\sigma\tau} \rangle_0,$$

$$z^{(4)} = \frac{1}{24} \beta_2^4 F_{\mu\nu} F_{\lambda\kappa} F_{\sigma\tau} F_{\alpha\beta} \langle \phi_{\mu\nu} \phi_{\lambda\kappa} \phi_{\sigma\tau} \phi_{\alpha\beta} \rangle_0.$$

The relation (12.7) implies

$$F_{\mu\nu} F_{\lambda\kappa} F_{\sigma\tau} \langle \phi_{\mu\nu} \phi_{\lambda\kappa} \phi_{\sigma\tau} \rangle_0 = \sqrt{30} \frac{2}{7} F_{\mu\nu} F_{\nu\kappa} F_{\kappa\mu}$$

and similarly, from (12.10) follows

$$F_{\mu\nu}F_{\lambda\kappa}F_{\sigma\tau}F_{\alpha\beta}\langle\phi_{\mu\nu}\phi_{\lambda\kappa}\phi_{\sigma\tau}\phi_{\alpha\beta}\rangle_0 = \frac{15}{7}\left(F_{\mu\nu}F_{\mu\nu}\right)^2.$$

Thus up to terms of fourth order in the field tensor, the partition function Z is

$$Z = (4\pi)^{-1}\left(1 + \frac{1}{2}\beta_2^2 F_{\mu\nu}F_{\mu\nu} + \frac{1}{27}\sqrt{30}\beta_2^3 F_{\mu\nu}F_{\nu\kappa}F_{\kappa\mu} + \frac{5}{56}\beta_2^4(F_{\mu\nu}F_{\mu\nu})^2 + \cdots\right).$$

$$(12.31)$$

For H given by (12.30), the equilibrium average $a_{\mu\nu} = \langle\phi_{\mu\nu}\rangle$ can also be evaluated according to

$$a_{\mu\nu} = \beta_2^{-1}Z^{-1}\frac{\partial Z}{\partial F_{\mu\nu}} = \beta_2^{-1}\frac{\partial \ln Z}{\partial F_{\mu\nu}}. \tag{12.32}$$

With the help of (11.59) and $\phi_{\mu\nu\lambda\kappa} = \frac{3}{4}\sqrt{70}\,\overline{u_\mu u_\nu u_\lambda u_\kappa}$, the second order term in the expression for f_{eq} is rewritten as

$$F_{\mu\nu}F_{\lambda\kappa}\,\phi_{\mu\nu}\phi_{\lambda\kappa} = \frac{15}{2}F_{\mu\nu}F_{\lambda\kappa}\left(\frac{3}{4}\sqrt{70}\right)^{-1}\phi_{\mu\nu\lambda\kappa} + \frac{4}{7}\sqrt{\frac{15}{2}}F_{\mu\lambda}F_{\lambda\nu}\phi_{\mu\nu} + F_{\mu\nu}F_{\mu\nu}.$$

Thus in high temperature approximation, up to second order in β_2, the second and fourth rank alignment tensors are given by

$$a_{\mu\nu} = \beta_2 F_{\mu\nu} + \frac{1}{7}\sqrt{30}\,\beta_2^2\,\overline{F_{\mu\lambda}F_{\lambda\nu}}, \quad a_{\mu\nu\lambda\kappa} = \frac{1}{2}\sqrt{\frac{10}{7}}\,\beta_2^2\,\overline{F_{\mu\nu}F_{\lambda\kappa}}. \tag{12.33}$$

A relation similar to (12.24) links the fourth rank alignment tensor with the product of two second rank alignment tensors, viz. $a_{\mu\nu\lambda\kappa} \sim \overline{a_{\mu\nu}a_{\lambda\kappa}}$ and

$$\langle\overline{u_\mu u_\nu u_\lambda u_\kappa}\rangle = c^{(4|22)}\,\overline{\langle u_\mu u_\nu\rangle\langle u_\lambda u_\kappa\rangle}. \tag{12.34}$$

The coefficient $c^{(4|22)}$ depends on the field strength. The high temperature approximation is $c^{(4|22)} = 5/7$.

12.2.5 Kerr Effect, Cotton-Mouton Effect, Non-linear Susceptibility

The birefringence induced by an applied electric field or by a magnetic field are called *Kerr effect*, named after J. Kerr, and *Cotton-Mouton effect*, named after A. Cotton and H. Mouton, who discovered these effects in 1875 and 1907, respectively.

Phenomenologically, the *electro-optic Kerr effect* is described by

$$\overline{\varepsilon_{\mu\nu}} = 2K \, \overline{E_\mu E_\nu} \,, \qquad (12.35)$$

where the *Kerr coefficient K* quantifies this effect. Due to (12.35), the difference $\delta\nu = \nu_\parallel - \nu_\perp$ of the indices of refraction of the optical electric field parallel and perpendicular to the applied electric field $\mathbf{E}$ is determined by $\nu_\parallel^2 - \nu_\perp^2 = 2KE^2 = (\nu_\parallel + \nu_\perp)\delta\nu$. Thus, with average index of refraction $\bar{\nu} = (\nu_\parallel + \nu_\perp)/2$, one has

$$\bar{\nu}\,\delta\nu = K E^2 \,.$$

In the corresponding relations for the Cotton-Mouton effect, the applied electric field $\mathbf{E}$ and K are replaced by the magnetic field and another characteristic coefficient. Sometimes, $K/\bar{\nu}^2$ is referred to as Kerr coefficient. The term 'electro-optic Kerr effect' is used in order to distinguish this effect from another one associated with *Kerr*, viz. the *magneto-optic Kerr effect*.

Several mechanisms contribute to these effects. Firstly, strong fields influence the electronic structure and change the 'shape' of atoms and molecules. Secondly, the field-induced orientation of optically anisotropic particles, in fluids, gives rise to birefringence. While the first contribution is independent of the temperature T, the second one, in general, contains contribution proportional to T^{-1} and T^{-2}, in the low-field and high-temperature limit. The key is the relation (12.19) between the symmetric traceless part of the dielectric tensor and the alignment tensor, viz. $\overline{\varepsilon_{\mu\nu}} = \varepsilon_a a_{\mu\nu}$, for $= \varepsilon_a$ see (12.20). In the high temperature approximation, permanent dipole moments yield a contribution proportional to T^{-2}, cf. (12.23). Induced dipole moments linked with an anisotropic polarizability lead to a contribution to K which is proportional to T^{-1}, as given by the first term on the right hand side of (12.33).

In (12.35) it is understood, that $\mathbf{E}$ is an applied electric field which is to be distinguished from the electric field $\mathbf{E}^{\text{light}}$ of the light. When the optical electric field $\mathbf{E}^{\text{light}}$ is strong enough and denoted by $\mathbf{E}$, the Kerr effect gives rise to a nonlinear susceptibility, cf. (2.59), where the third order contribution to the electric polarization is $\mathbf{P}^{(3)} = \varepsilon_0 \chi^{(3)}_{\mu\nu\lambda\kappa} E_\nu E_\lambda E_\kappa$ with the susceptibility tensor $\chi^{(3)}_{\mu\nu\lambda\kappa} = 2K \Delta_{\mu\nu\lambda\kappa}$.

12.2.6 Orientational Entropy

The entropy of a system in an ordered state is lower than that in an isotropic state. The entropy, per particle, associated with the orientation, characterized by the orientational distribution function $f(\mathbf{u})$, is determined by the Boltzmann like expression $-k_B \int f \ln f \, d^2 u$. This should be compared with the corresponding expression for the isotropic distribution f_0, viz. $s_0 = -k_B \int f_0 \ln f_0 d^2 u = -k_B \int f \ln f_0 d^2 u$. The second equality follows from the fact that $\ln f_0$ is a constant and that the normalization imposes $\int f d^2 u = \int f_0 d^2 u = 1$. Thus the difference between the entropy, per particle, in the ordered and the isotropic state is

$$s_a = -k_B \int f \ln(f/f_0) d^2u = -k_B \langle \ln(f/f_0) \rangle. \tag{12.36}$$

Use of $f = f_0(1 + \Phi)$, cf. (12.16) yields

$$s_a = -k_B \langle (1 + \Phi) \ln(1 + \Phi) \rangle_0. \tag{12.37}$$

With the help of the power series expansion

$$(1 + x) \ln(1 + x) = x + \sum_{n=2}^{\infty} (-1)^n \frac{1}{n(n-1)} x^n,$$

and due to $\langle \Phi \rangle_0 = 0$, the entropy (12.37) is equivalent to

$$s_a = -k_B \sum_{n=2}^{\infty} (-1)^n \frac{1}{n(n-1)} \langle \Phi^n \rangle_0. \tag{12.38}$$

The first few terms in this series are

$$s_a = -k_B \left(\frac{1}{2} \langle \Phi^2 \rangle_0 - \frac{1}{6} \langle \Phi^3 \rangle_0 + \frac{1}{12} \langle \Phi^4 \rangle_0 \pm \dots \right). \tag{12.39}$$

In lowest order in Φ, the orthogonality and the normalization (12.15) of the expansion tensors imply

$$s_a = -k_B \left(\frac{1}{2} \sum_{\ell=1}^{\infty} a_{\mu_1 \mu_2 \cdots \mu_\ell} a_{\mu_1 \mu_2 \cdots \mu_\ell} \pm \dots \right). \tag{12.40}$$

The dots stand for terms associated with third and higher powers in Φ. Examples for the role of higher order terms are discussed in the Exercise 15.1.

12.2.7 Fokker-Planck Equation for the Orientational Distribution

In the absence of any orientating torque and for a spatially homogeneous system, the distribution function for the orientation of (effectively) uniaxial particles immersed in a fluid obeys the dynamic equation, frequently called orientational *Fokker-Planck equation*,

$$\frac{\partial f(\mathbf{u})}{\partial t} - \nu_0 \mathscr{L}_\mu \mathscr{L}_\mu f(\mathbf{u}) = 0. \tag{12.41}$$

Here $\mathscr{L}_\mu \mathscr{L}_\mu$, with

$$\mathscr{L}_\mu = \varepsilon_{\mu\nu\lambda} u_\nu \frac{\partial}{\partial u_\lambda} ,$$

is the Laplace operator on the unit sphere. The properties of $\mathscr{L}$ as used here are equivalent to those discussed in Sect. 7.6 where the unit vector $\hat{\mathbf{r}}$ parallel to the position vector $\mathbf{r}$ occurred instead of the unit vector which is parallel to the figure axis of a non-spherical particle. The relaxation frequency coefficient $\nu_0 > 0$ has the dimension one over time. The (12.41) essentially describes a diffusional motion on the unit sphere. For this reason, the coefficient ν_0 is referred to as orientational diffusion coefficient.

By analogy to (11.31), the irreducible tensors $\phi_{\mu_1\mu_2\cdots\mu_\ell}$ defined by (12.14), are eigenfunctions of the orientational Laplace operator with the eigenvalues $-\ell(\ell+1)$, viz.

$$\mathscr{L}_\lambda \mathscr{L}_\lambda \, \phi_{\mu_1\mu_2\cdots\mu_\ell} = -\ell(\ell+1)\, \phi_{\mu_1\mu_2\cdots\mu_\ell}. \qquad (12.42)$$

Multiplication of the dynamic equation (12.41) by $\phi_{\mu_1\mu_2\cdots\mu_\ell}$, subsequent integration over d^2u, use of the expansion (12.16) for the distribution function and of the ortho-normalization (12.1) yields the relaxation equations

$$\frac{da_{\mu_1\mu_2\cdots\mu_\ell}}{dt} + \nu_\ell \, a_{\mu_1\mu_2\cdots\mu_\ell} = 0, \quad \ell \geq 1, \qquad (12.43)$$

for the tensorial moments $a_{\mu_1\mu_2\cdots\mu_\ell}$. The relaxation coefficients ν_ℓ are given by

$$\nu_\ell = \ell(\ell+1)\,\nu_0. \qquad (12.44)$$

Equation (12.43) implies an exponential relaxation

$$a_{\mu_1\mu_2\cdots\mu_\ell}(t) = \exp(-t/\tau_\ell)\, a_{\mu_1\mu_2\cdots\mu_\ell}(0), \quad t > 0,$$

with the relaxation time $\tau_\ell = \nu_\ell^{-1}$. Clearly, higher moments pertaining to larger values of ℓ relax faster. In the long time limit all moments approach zero and the distribution is isotropic, corresponding to the equilibrium state of an isotropic fluid.

Some Historical Remarks

A dynamic equation for an orientational distribution function as discussed here was first introduced by Adriaan Fokker in his thesis in 1913 and published 1914 [43]. So the (12.41) should actually be called *Fokker equation*. How got Planck involved? In 1917, Max Planck was asked by colleagues to explain the work of Fokker. So at a meeting of the Prussian Academy of Science, in 1917, he presented what he was inspired to, viz. a dynamic equation for the velocity distribution function of a Brownian particle immersed in a liquid [44]. So his equation might be called *Planck equation*. However, it is common practice to refer to both types of these equations and generalizations thereof as *Fokker-Planck equation*.

In the presence of an external torque **T**, which can be derived from a potential function $H = H(\mathbf{u})$ according to

$$T_\lambda = -\mathscr{L}_\lambda H(\mathbf{u}),\qquad(12.45)$$

the orientational Fokker-Planck equation contains an additional term. More specifically, the kinetic equation now reads

$$\frac{\partial f(\mathbf{u})}{\partial t} - v_0\,\mathscr{L}_\mu\left(\mathscr{L}_\mu f(\mathbf{u}) + \beta f(\mathbf{u})\,\mathscr{L}_\mu H(\mathbf{u})\right) = 0\,,\quad \beta = \frac{1}{k_\mathrm{B}T}.\qquad(12.46)$$

In this case, the stationary solution of the Fokker-Planck equation is the equilibrium distribution $f = f_\mathrm{eq}$ which is proportional to $\exp[-\beta H] = \exp[-H/k_\mathrm{B}T]$, cf. Sect. 12.2.4.

Multiplication of the (12.46) by a function $\psi = \psi(\mathbf{u})$, integration over d^2u and integrations by part leads to the *moment equation* for the time change of the average $\langle\psi\rangle$:

$$\frac{\mathrm{d}}{\mathrm{d}t}\langle\psi\rangle - v_0\left(\langle\mathscr{L}_\mu\mathscr{L}_\mu\psi\rangle - \beta\,\langle(\mathscr{L}_\mu\psi)(\mathscr{L}_\mu H)\rangle\right) = 0.\qquad(12.47)$$

Examples for such moment equations, however for the case $H = 0$, are the relaxation (12.43). For $H \neq 0$, the relaxation equation for the ℓth rank tensor $a_{\mu_1\mu_2\cdots\mu_\ell}$ is coupled with tensors of different ranks, in general. In particular, when $H(\mathbf{u})$ involves the first and second rank tensors u_μ and $\overline{u_\mu u_\nu}$, as considered in Sect. 12.2.4, the tenor of rank ℓ is coupled with tensors of ranks $\ell\pm1$ and $\ell\pm2$, respectively, quite in analogy to the quantum mechanical selection rules for electric dipole and quadrupole radiation, cf. Sect. 12.5. Applications of a generalized Fokker-Planck equation to the dynamics of liquid crystals is discussed in Sect. 16.4.4.

12.2 Exercise: Prove that the Fokker-Planck Equation Implies an Increase of the Orientational Entropy with Increasing Time
Hint: The time change of an orientational average is $\mathrm{d}\langle\psi\rangle/\mathrm{d}t = \int \partial(\psi f)/\partial t\,\mathrm{d}^2u$, use the expression (12.36) for the orientational entropy.

12.3 Averages Over Velocity Distributions

Atoms and molecules never sit still. In equilibrium, they move in all directions with equal probability and their average speed is higher, at higher temperatures. In non-equilibrium situations, preferential direction can be favored. The velocity distribution function characterizes this behavior. The distribution is isotropic in equilibrium and, in general, anisotropic in non-equilibrium. Here the relevant tools needed in the

Kinetic Theory of Gases are presented. Physical phenomena associated with the velocity distribution, both for an equilibrium situation and for non-equilibrium, in particular for transport processes, are e.g. discussed in [40], see also [13].

12.3.1 Integrals Over the Maxwell Distribution

Let $\mathbf{v}$ be the velocity of an atom or molecule in a gas or liquid. The distribution of the different velocities is characterized by the velocity distribution function, also denoted by $f = f(\mathbf{v})$, which is conventionally normalized such that

$$\int f(\mathbf{v}) d^3 v = n. \tag{12.48}$$

Here n is the number density. The average $\langle \psi \rangle$ of a function $\psi(\mathbf{v})$ is given by

$$\langle \psi \rangle = \frac{1}{n} \int \psi(\mathbf{v}) f(\mathbf{v}) d^3 v. \tag{12.49}$$

In thermal equilibrium, at temperatures and densities, where quantum effects play no role, f is equal to the Maxwell distribution f_0

$$f_0(\mathbf{v}) \equiv n_0 \left(\frac{m}{2\pi k_B T_0} \right)^{3/2} \exp\left(-\frac{m v^2}{2 k_B T_0} \right), \tag{12.50}$$

where m is the mass of a particle. The constant density $n_0 = N/V$, of the N particles confined to the volume V and the constant temperature T_0 characterize the absolute equilibrium state. The Boltzmann constant is denoted by k_B. It is convenient to introduce a dimensionless velocity variable $\mathbf{V}$ via

$$V^2 = \frac{m v^2}{2 k_B T_0}, \tag{12.51}$$

which implies

$$\mathbf{v} = \sqrt{2} c_0 \mathbf{V}, \quad c_0 = \sqrt{k_B T_0 / m}, \tag{12.52}$$

and to use the velocity distribution $F = F(\mathbf{V})$, linked with $f(\mathbf{v})$, such that

$$f(\mathbf{v}) d^3 v = F(\mathbf{V}) d^3 V.$$

Instead of (12.49), averages are then evaluated according to

$$\langle \psi \rangle = \frac{1}{n} \int \psi(\mathbf{V}) F(\mathbf{V}) d^3 V. \tag{12.53}$$

In thermal equilibrium, F is equal to the absolute Maxwellian F_0, which is the Gaussian function

$$F_0(\mathbf{V}) \equiv n_0 \, \pi^{-3/2} \, \exp(-V^2). \tag{12.54}$$

Averages evaluated with this Maxwell velocity distribution function are denoted by $\langle \ldots \rangle_0$, viz.

$$\langle \psi \rangle_0 \equiv \pi^{-3/2} \int \psi(\mathbf{V}) \, \exp(-V^2) \mathrm{d}^3 V = \frac{1}{n_0} \int \psi(\mathbf{V}) \, F_0(\mathbf{V}) \mathrm{d}^3 V. \tag{12.55}$$

The equilibrium average of powers of the magnitude V of the dimensionless velocity $\mathbf{V}$ are

$$\langle V^n \rangle_0 = 2 \, \pi^{-1/2} \, \Gamma \left(\frac{n+3}{2} \right), \tag{12.56}$$

where n is a positive integer number and $\Gamma(x)$ is the gamma-function, with the property $\Gamma(n+1) = n!$. For the first few even powers, (12.56) implies $\langle 1 \rangle_0 = 1$, and

$$\langle V^2 \rangle_0 = \frac{3}{2}, \quad \langle V^4 \rangle_0 = \frac{15}{4}, \quad \langle V^6 \rangle_0 = \frac{105}{8}, \quad \langle V^8 \rangle_0 = \frac{945}{16}. \tag{12.57}$$

Clearly, the Maxwell distribution is isotropic. Thus equilibrium averages of the irreducible tensors $\overline{V_{\mu_1} V_{\mu_2} \cdots V_{\mu_\ell}}$ vanish:

$$\langle \overline{V_{\mu_1} V_{\mu_2} \cdots V_{\mu_\ell}} \rangle_0 = 0. \tag{12.58}$$

This is not the case for products of irreducible tensors of the same rank. In particular, from (12.1) and (12.56), with $2\pi^{-1/2} \Gamma(\frac{2\ell+3}{2}) = \frac{(2\ell+1)!!}{2^\ell}$, follows

$$\langle \overline{V_{\mu_1} V_{\mu_2} \cdots V_{\mu_\ell}} \; \overline{V_{\nu_1} V_{\nu_2} \cdots V_{\nu_{\ell'}}} \rangle_0 = \frac{\ell!}{2^\ell} \delta_{\ell\ell'} \, \Delta^{(\ell)}_{\mu_1\mu_2\cdots\mu_\ell, \nu_1\nu_2\cdots\nu_\ell}. \tag{12.59}$$

The special cases $\ell = \ell' = 1, 2$ correspond to

$$\langle V_\mu V_\nu \rangle_0 = \frac{1}{2} \delta_{\mu\nu}, \quad \langle \overline{V_\mu V_\nu} \; \overline{V_\lambda V_\kappa} \rangle_0 = \frac{1}{2} \Delta_{\mu\nu,\lambda\kappa}. \tag{12.60}$$

12.3.2 Expansion About an Absolute Maxwell Distribution

Clearly, the Maxwell distribution is isotropic. In a non-equilibrium situation, however, this is not the case. In general, the velocity distribution is anisotropic. The directional properties of the velocity distribution can be characterized by irreducible tensors constructed from the components of the velocity variable **V**. However, additional scalar polynomial functions, depending on V^2, are needed the characterize the full dependence of f on the magnitude and direction of the velocity. These scalar functions are the *Sonine polynomials* [12, 13], which are closely related to the *associated Laguerre polynomials*. The orthogonal expansion functions are, cf. [13, 41],

$$\phi_{\mu_1\mu_2\cdots\mu_\ell}^{(s+1)} \sim (-1)^{s+1} S_{(\ell+1/2)}^{(s)}(V^2) \, \overline{V_{\mu_1} V_{\mu_2} \cdots V_{\mu_\ell}}, \qquad (12.61)$$

where the label s characterizes the different Sonine polynomials. The Sonine polynomials are defined by

$$S_{(k)}^{(s)}(x) = (s!)^{-1} \frac{\partial^s}{\partial z^s} (1-z)^{-(1+k)} \exp[zx/(z-1)], \quad z \to 0. \qquad (12.62)$$

The first few of these polynomials are $S^{(0)} = 1$ and

$$S_{(k)}^{(1)}(x) = k+1-x, \quad S_{(k)}^{(2)}(x) = (1/2)(k+1)(k+2) - (k+2)x + (1/2)\, x^2. \quad (12.63)$$

The functions (12.61) are orthogonal and normalized according to

$$\langle \phi_{\mu_1\mu_2\cdots\mu_\ell}^{(s)} \, \phi_{\nu_1\nu_2\cdots\nu_{\ell'}}^{(s')} \rangle_0 = \delta_{ss'} \, \delta_{\ell\ell'} \, \Delta_{\mu_1\mu_2\cdots\mu_\ell, \nu_1\nu_2\cdots\nu_\ell}^{(\ell)}. \qquad (12.64)$$

The velocity distribution is written as

$$F(\mathbf{V}) = F_0(\mathbf{V})\,(1+\Phi), \qquad (12.65)$$

where $\Phi(t, \mathbf{r}, \mathbf{V})$ characterizes the deviation of F from the Maxwellian F_0. The expansion of Φ reads

$$\Phi = \sum_{\ell=0}^{\infty} \sum_{s=1}^{\infty} a_{\mu_1\mu_2\cdots\mu_\ell}^{(s)} \, \phi_{\mu_1\mu_2\cdots\mu_\ell}^{(s)}. \qquad (12.66)$$

The expansion coefficients

$$a_{\mu_1\mu_2\cdots\mu_\ell}^{(s)} = \langle \phi_{\mu_1\mu_2\cdots\mu_\ell}^{(s)} \rangle \qquad (12.67)$$

are the moments of the velocity distribution.

Just the first few terms of the expansion are of importance in most applications, so only the first few expansion functions are listed here. The first three scalar expansion functions are,

$$\Phi^{(1)} = 1, \quad \Phi^{(2)} = \sqrt{\frac{2}{3}} \left(V^2 - \frac{3}{2}\right), \quad \Phi^{(3)} = \sqrt{\frac{2}{15}} \left(V^4 - 5\,V^2 + \frac{15}{4}\right).$$

$$(12.68)$$

These functions are orthogonal and normalized according to $\langle \Phi^{(s)} \Phi^{(s')} \rangle_0 = \delta_{ss'}$. The first two vectorial expansion functions are

$$\Phi_\mu^{(1)} = \sqrt{2}\, V_\mu, \quad \Phi_\mu^{(2)} = \frac{2}{\sqrt{5}} \left(V^2 - \frac{5}{2} \right) V_\mu. \tag{12.69}$$

Here the normalization corresponds to $\langle \Phi_\mu^{(s)} \Phi_\nu^{(s')} \rangle_0 = \delta_{ss'} \delta_{\mu\nu}$. The first second rank tensorial expansion functions are

$$\Phi_{\mu\nu}^{(1)} = \sqrt{2}\, \overline{V_\mu V_\nu}, \quad \Phi_{\mu\nu}^{(2)} = \frac{2}{\sqrt{7}} \left(V^2 - \frac{7}{2} \right) \overline{V_\mu V_\nu}. \tag{12.70}$$

For comparison with [13] notice that $s - 1$ corresponds to the label r used by Waldmann, which is the number of zero-values of the Sonine polynomials involved.

The first two scalar and first vectorial moments are essentially the relative deviation of the number density n from the constant density n_0, the relative deviation of the temperature T from T_0, and the average flow velocity $\langle \mathbf{v} \rangle$, viz.

$$a^{(1)} = n/n_0 - 1, \quad a^{(2)} = \sqrt{\frac{3}{2}}\, (T/T_0 - 1), \quad a_\mu^{(1)} = c_0^{-1} \langle v_\mu \rangle, \quad c_0 = (k_B T_0/m)^{1/2}. \tag{12.71}$$

In pure gases, these are associated with conserved quantities. The number density, thus $a^{(1)}$, is the only conserved quantity for colloidal particles in liquids. The moments $a_\mu^{(2)}$ and $a_{\mu\nu}^{(1)}$ are proportional to the parts of the heat flux and of the friction pressure tensor which are associated with the translational motion. The specific relations are stated in Sect. 12.3.4.

12.3.3 Kinetic Equations, Flow Term

The time change $\partial f/\partial t$ of the velocity distribution $f = f(t, \mathbf{r}, \mathbf{v})$ contains a contribution due to the translational motion of the particles, viz. $v_\lambda \nabla_\lambda f$ which is referred to as *flow term*. The corresponding expression in the *kinetic equation* for the dimensionless distribution function $F = F(t, \mathbf{r}, \mathbf{V})$ is

$$\frac{\partial}{\partial t} F + \sqrt{2}\, c_0\, V_\lambda \nabla_\lambda F - \left(\frac{\delta F}{\delta t} \right)_{..} = 0. \tag{12.72}$$

Here $\left(\frac{\delta F}{\delta t} \right)_{..}$ stands either for the collision term $\left(\frac{\delta F}{\delta t} \right)_{\text{coll}}$ of the Boltzmann equation or for the damping term $\left(\frac{\delta F}{\delta t} \right)_{\text{FP}}$ of Planck's version of the Fokker-Planck equation. In any case, these terms guarantee the approach to equilibrium. The Fokker-Planck expression, used for the dynamics of colloidal Brownian particles, is

$$-\left(\frac{\delta F}{\delta t}\right)_{\text{FP}} = F_M \, \Omega_{\text{FP}}(\Phi) = -\frac{1}{2} \, v_0 \, \frac{\partial}{\partial V_\kappa} \left(F_M \frac{\partial}{\partial V_\kappa} \Phi\right). \tag{12.73}$$

Here $v_0 > 0$ is the translational relaxation frequency. For particles with a radius R larger than the molecules of the surrounding liquid, the hydrodynamic result for the Stokes friction force (10.58) applies. Then one has $v_0 = m^{-1} 6\pi \eta R$. The quantity $\Omega_{\text{FP}}(\Phi)$, defined by (12.73), is referred to as the Fokker-Planck relaxation operator. The expansion tensors used here are eigenfunctions of this operator.

The implications of the flow term for the moment equations apply to both the Boltzmann and the Fokker-Planck equations. After an insertion of the expansion (12.66) into the flow term of (12.72), the multiplication of the expansion tensors of rank ℓ, e.g. $\phi_{\nu_1 \nu_2 \cdots \nu_{\ell'}}^{(s'+1)}$, with the vector V_λ yields irreducible tensors of ranks $\ell' - 1$ and $\ell' + 1$, see Sect. 11.5. Multiplication of (12.72) by the function $\phi_{\mu_1 \mu_2 \cdots \mu_\ell}^{(s)}$, subsequent integration over $d^3 V$ and the use of the relevant orthogonality relations leads to

$$\frac{\partial}{\partial t} a_{\mu_1 \mu_2 \cdots \mu_\ell}^{(s)} + \sum_{s'} c(\ell s | \ell - 1 s') \, \nabla_\lambda a_{\lambda \mu_2 \cdots \mu_{\ell-1}}^{(s')}$$

$$+ \sum_{s'} c(\ell s | \ell + 1 s') \, \overline{\nabla_\lambda a_{\mu_1 \mu_2 \cdots \mu_{\ell-1}}^{(s')}} + \cdots = 0. \tag{12.74}$$

The terms which stem from the collision term or from the damping term of the Fokker-Planck equation are indicated by the dots. Due to

$$\langle \phi_{\mu_1 \mu_2 \cdots \mu_\ell}^{(s)} \, V_\mu \, \phi_{\nu_1 \nu_2 \cdots \nu_{\ell-1}}^{(s')} \rangle_0 \sim \Delta_{\mu_1 \mu_2 \cdots \mu_\ell, \mu \nu_1 \nu_2 \cdots \nu_{\ell-1}}^{(\ell)},$$

and (11.7), the flow term coefficients $c(\ldots)$ are determined by

$$c(\ell s | \ell - 1 s') = \sqrt{2} c_0 \, (2\ell + 1)^{-1} \, \langle \phi_{\mu_1 \mu_2 \cdots \mu_\ell}^{(s)} \, V_{\mu \ell} \, \phi_{\mu_1 \mu_2 \cdots \mu_{\ell-1}}^{(s')} \rangle_0. \tag{12.75}$$

The coefficient $c(\ell s | \ell + 1 s')$ can be inferred from $c(\ell - 1 s' | \ell s) = c(\ell s | \ell - 1 s')$ and subsequent replacement $\ell \to \ell + 1$ and the interchange of s and s'. The first few of the coefficients $c(\ldots)$ are

$$c(01 | 1 s') = c_0 \, \delta_{1,s'}, \quad c(11 | 0 s') = c_0 \, \delta_{1,s'} + \sqrt{\frac{2}{3}} \, c_0 \, \delta_{2,s'}, \quad c(11 | 2 s') = \sqrt{2} \, c_0 \, \delta_{1,s'}.$$

For the Fokker-Planck case, the first two of the moment equations are

$$\frac{\partial}{\partial t} a^{(1)} + c_0 \, \nabla_\mu \, a_\mu^{(1)} = 0,$$

$$\frac{\partial}{\partial t} a_\mu^{(1)} + c_0 \nabla_\mu \, a^{(1)} + \sqrt{\frac{2}{3}} \, c_0 \, \nabla_\mu \, a^{(2)} + \sqrt{2} \, c_0 \, \nabla_\nu \, a_{\nu\mu}^{(1)} + v_1 \, a_\mu^{(1)} = 0, \tag{12.76}$$

with the relaxation frequency $\nu_1 = \nu_0$. Clearly, the flow term couples the moment $a_\mu^{(1)}$, which is essentially the average velocity of the Brownian particles, with the scalar $a^{(2)}$ and the second rank tensor $a_{\mu\nu}^{(1)}$. When these terms can be disregarded, the (12.76) and (12.77) are equivalent to the following equations for the number density n and the flux density $\mathbf{j} = n\langle\mathbf{v}\rangle$ of the colloidal particles:

$$\frac{\partial}{\partial t}n + \nabla_\mu j_\mu = 0, \tag{12.77}$$

$$\frac{\partial}{\partial t}j_\mu + c_0^2\,\nabla_\mu n + \nu_1 j_\mu = 0. \tag{12.78}$$

For time changes which are slow compared with the relaxation time $\tau_1 = \nu_1^{-1}$, the second of these equations reduces to $j_\mu = -D\nabla_\mu n$ with the diffusion coefficient

$$D = c_0^2\,\tau_1 = \frac{k_B T}{m\nu_1} = \frac{k_B T}{6\pi\eta R}. \tag{12.79}$$

The third equality is based on the Stokes friction law (10.58).

12.3.4 Expansion About a Local Maxwell Distribution

The expansion of the velocity distribution function about an absolute Maxwellian, as used in [13] and [21] is appropriate for small deviations from equilibrium. In general, it is more advantageous, cf. [41], to expand the velocity distribution of gases about a local Maxwellian, as treated next.

The difference between the velocity of a particle, at position $\mathbf{r}$, and the local average flow velocity $\langle v_\mu\rangle = \langle v_\mu\rangle(\mathbf{r}, t)$, viz.

$$c_\mu = v_\mu - \langle v_\mu\rangle, \tag{12.80}$$

is referred to as the *peculiar velocity* of the particle. The local temperature $T = T(\mathbf{r}, t)$ is linked with the average peculiar kinetic energy via $(m/2)\langle c^2\rangle = (3/2)k_B T$.

The distribution of the peculiar velocities is also denoted by f. It is normalized according to

$$\int f(\mathbf{c})d^3c = n, \tag{12.81}$$

The average $\langle\psi\rangle$ of a function $\psi(\mathbf{c})$ is given by

$$\langle\psi\rangle = \frac{1}{n}\int \psi(\mathbf{c})f(\mathbf{c})d^3c. \tag{12.82}$$

In local thermal equilibrium, at temperatures and densities, where quantum effects play no role, f is equal to the local Maxwell distribution f_M

$$f_M(\mathbf{c}) \equiv n \left(\frac{m}{2\pi k_B T} \right)^{3/2} \exp\left(-\frac{mc^2}{2k_B T} \right), \tag{12.83}$$

where m is the mass of a particle.

It is convenient to introduce a dimensionless velocity variable $\mathbf{V}$, cf. (12.51), now via

$$V_\mu = \sqrt{\frac{m}{2k_B T}} c_\mu = \sqrt{\frac{m}{2k_B T}} (v_\mu - \langle v_\mu \rangle), \tag{12.84}$$

which implies

$$V^2 = \frac{mc^2}{2k_B T}, \tag{12.85}$$

and to use the velocity distribution $F = F(\mathbf{V})$, linked with $f(\mathbf{c})$, such that

$$f(\mathbf{c})d^3 c = F(\mathbf{V})d^3 V \,.$$

Averages are now evaluated according to

$$\langle \psi \rangle = \frac{1}{n} \int \psi(\mathbf{V}) F(\mathbf{V}) d^3 V, \tag{12.86}$$

which is mathematically identical to (12.53).

In thermal equilibrium, F is equal to the local Maxwellian F_M

$$F_M(\mathbf{V}) \equiv n \pi^{-3/2} \exp(-V^2). \tag{12.87}$$

Averages evaluated with this Maxwell velocity distribution function are denoted by $\langle \ldots \rangle_M$, viz.

$$\langle \psi \rangle_M \equiv \pi^{-3/2} \int \psi(\mathbf{V}) \exp(-V^2) d^3 V = \frac{1}{n} \int \psi(\mathbf{V}) F_M(\mathbf{V}) d^3 V. \tag{12.88}$$

The results given above for $\langle \ldots \rangle_0$, in particular (12.56)–(12.60), apply also to the averages $\langle \ldots \rangle_M$, evaluated with the local Maxwell distribution function.

Similar to (12.65), the full distribution function is now written as

$$F(\mathbf{V}) = F_M(\mathbf{V})(1 + \Phi), \tag{12.89}$$

where $\Phi(t, \mathbf{r}, \mathbf{V})$ characterizes the deviation of F from the local Maxwellian F_M. The expansion of Φ is formally similar to (12.66), with one fundamental difference:

now Φ has to be orthogonal to the moments associated with the conserved quantities, with the number density, the kinetic energy and the velocity. This means, the first and second scalar, viz. 1 and $V^2 - 3/2$, as well as the first vectorial expansion function $\mathbf{V}$, must not be included in the expansion. Instead of (12.66), the expansion now reads

$$\Phi = \sum_{s=3}^{\infty} a^{(s)} \phi^{(s)} + \sum_{s=2}^{\infty} a_\mu^{(s)} \phi_\mu^{(s)} + \sum_{\ell=2}^{\infty} \sum_{s=1}^{\infty} a_{\mu_1\mu_2\cdots\mu_\ell}^{(s)} \phi_{\mu_1\mu_2\cdots\mu_\ell}^{(s)}. \tag{12.90}$$

As before, the expansion coefficients

$$a_{\mu_1\mu_2\cdots\mu_\ell}^{(s)} = \langle \phi_{\mu_1\mu_2\cdots\mu_\ell}^{(s)} \rangle \tag{12.91}$$

are the moments of the velocity distribution.

Notice, the variable $\mathbf{V}$ depends on the time t and on the position $\mathbf{r}$ via the temperature and the average velocity. This is a second fundamental point which has to be observed in applications, e.g. see [41].

In many applications, the most relevant terms of the expansion (12.90) involve the translational or kinetic parts q_μ^{kin} and $\overline{p_{\mu\nu}^{\text{kin}}}$ of the heat flux vector and of the symmetric traceless friction pressure tensor. These quantities are given by

$$q_\mu^{\text{kin}} = n \left\langle \left(\frac{m}{2} c^2 - \frac{5}{2} k_B T \right) c_\mu \right\rangle = n k_B T (k_B T/m)^{1/2} \sqrt{2} \left\langle \left(V^2 - \frac{5}{2} \right) V_\mu \right\rangle, \tag{12.92}$$

or, equivalently,

$$q_\mu^{\text{kin}} = n k_B T (k_B T/m)^{1/2} \sqrt{\frac{5}{2}} \langle \phi_\mu \rangle, \tag{12.93}$$

and

$$\overline{p_{\mu\nu}^{\text{kin}}} = nm \langle \overline{c_\mu c_\nu} \rangle = n k_B T \, 2 \langle \overline{V_\mu V_\nu} \rangle = n k_B T \sqrt{2} \langle \phi_{\mu\nu} \rangle. \tag{12.94}$$

The relevant vectorial and tensorial expansion functions occurring here are

$$\phi_\mu \equiv \phi_\mu^{(2)} = \frac{2}{\sqrt{5}} \left(V^2 - \frac{5}{2} \right) V_\mu, \quad \phi_{\mu\nu} \equiv \phi_{\mu\nu}^{(1)} = \sqrt{2} \, \overline{V_\mu V_\nu}. \tag{12.95}$$

In the approximation where only these two expansion functions and the corresponding moments characterize the deviation Φ from the local equilibrium, one has

$$\Phi = \langle \phi_\mu \rangle \phi_\mu + \langle \phi_{\mu\nu} \rangle \phi_{\mu\nu}$$

$$= (n k_B T)^{-1} \left[(k_B T/m)^{-1/2} \sqrt{\frac{2}{5}} \, q_\mu^{\text{kin}} \phi_\mu + \frac{\sqrt{2}}{2} \overline{p_{\mu\nu}^{\text{kin}}} \phi_{\mu\nu} \right]. \tag{12.96}$$

The *thirteen moments approximation* of kinetic gas theory [45] employs just these 3 plus 5 moments describing the heat flux and the friction pressure tensor, in addition to the 5 variables associated with the conserved quantities, viz. the number density n, the temperature T and average flow velocity $\mathbf{v}$.

In addition to the heat conductivity and the viscosity, also a contribution to the symmetric traceless part of the pressure tensor which is proportional to the gradient of the heat flux $\mathbf{q}$ and thus proportional to the second spatial derivative of the temperature T is described by the thirteen moments approximation. This phenomenon, where one has

$$p_{\mu\nu} \sim -\overline{\nabla_\mu q_\nu} \sim \overline{\nabla_\mu \nabla_\nu}\, T,$$

was already predicted by Maxwell, it is referred to as *Maxwell's thermal pressure*. An experimental manifestation of this effect is provided by light-induced velocity selective heating or cooling in gases, cf. [46]. For some applications, however, more than thirteen moments have to be included for the solution of the Boltzmann equation [41, 47].

A side remark: in dilute gases and in the hydrodynamic regime, the heat flux and the viscous friction pressure tensor are independent of the density n. The factor $(nk_\mathrm{B}T)^{-1}$ in (12.96) implies that the deviation Φ from the Maxwell velocity distribution is proportional to n^{-1}, thus it is the larger the smaller the density n is. At a lower density, fewer particles have to 'work' harder to transport the energy and the linear momentum. Maxwell's thermal pressure is also proportional to n^{-1}.

12.3 Exercise: Second Order Contributions of the Kinetic Heat Flux and Friction Pressure Tensor to the Entropy

The 'non-equilibrium' entropy, per particle, associated with the velocity distribution function $f = f_\mathrm{M}(1+\Phi)$, is given by $s = -k_\mathrm{B}\langle \ln(f/f_\mathrm{M})\rangle = -k_\mathrm{B}\langle(1+\Phi)\ln(1+\Phi)\rangle_\mathrm{M}$, where f_M is the local Maxwell distribution and Φ is the deviation of f from f_M. By analogy with (12.39), the contribution up to second order in the deviation is $s = -k_\mathrm{B}\frac{1}{2}\langle\Phi^2\rangle_\mathrm{M}$.

Determine the second order contributions to the entropy associated with heat flux and the symmetric traceless pressure tensor.

12.4 Anisotropic Pair Correlation Function and Static Structure Factor

The particles surrounding any given reference particle in a dense fluid, in a liquids or a colloidal solution, possess a short ranged order which is also referred to as the *structure of a liquid*. This property is characterized by the *pair correlation function* or by the *static structure factor*. For fluids composed of spherical particles, these functions are isotropic, in thermal equilibrium. In non-equilibrium situations, these quantities become anisotropic. The basics and examples for the use of tensors needed to characterize the anisotropic properties are presented here. For more information on the physics of liquids e.g. consult [48–52].

12.4.1 Two-Particle Density, Two-Particle Averages

Consider a fluid of N spherical particles with positions $\mathbf{r}^i$, $i = 1, 2, \ldots N$, contained within a volume V. The two-particle probability density to find one particle at position $\mathbf{r}^a$ and another one at $\mathbf{r}^b$ is the *two-particle density* $n^{(2)}(\mathbf{r}^a, \mathbf{r}^b)$. It is given by

$$n^{(2)}(\mathbf{r}^a, \mathbf{r}^b) = \left\langle \sum_I \sum_{j \neq i} \delta(\mathbf{r}^a - \mathbf{r}^i)\, \delta(\mathbf{r}^b - \mathbf{r}^j) \right\rangle, \tag{12.97}$$

where the bracket $\langle \ldots \rangle$ indicates a N-particle average. It can e.g. be a canonical average or a time-average, but it need not to be specified here. The integral of $n^{(2)}$ over both $\mathbf{r}^a$ and $\mathbf{r}^b$ yields the number of pairs $\{i, j\}$ viz. $N(N-1)$.

Consider a quantity

$$\Psi = \sum_I \sum_{j \neq i} \psi(\mathbf{r}^i, \mathbf{r}^j),$$

where ψ is a function which depends on the position vectors of two particles. Then its average is given by

$$\langle \Psi \rangle = \int \psi(\mathbf{r}_1, \mathbf{r}_2)\, n^{(2)}(\mathbf{r}_1, \mathbf{r}_2) \mathrm{d}^3 r_1 \mathrm{d}^3 r_2, \tag{12.98}$$

where now the integration variables are denoted by $\mathbf{r}_1, \mathbf{r}_2$ rather than $\mathbf{r}^a, \mathbf{r}^b$. When the function ψ depends on the difference between two position vectors only, just like the binary interaction potential, the average $\langle \Psi \rangle$ can be written as

$$\langle \Psi \rangle = \frac{N^2}{V} \int \psi(\mathbf{r})\, g(\mathbf{r}) \mathrm{d}^3 r. \tag{12.99}$$

Here the *pair-correlation function* $g = g(\mathbf{r})$ is defined by

$$g(\mathbf{r}) = \frac{V}{N^2} \int n^{(2)}(\mathbf{r}_2 + \mathbf{r}, \mathbf{r}_2) \mathrm{d}^3 r_2. \tag{12.100}$$

The vector

$$\mathbf{r} = \mathbf{r}_1 - \mathbf{r}_2$$

is not an ordinary position vector but it is the difference vector in the $\{\mathbf{r}_1, \mathbf{r}_2\}$ pair-space. Due to (12.97), g is also given by the N-particle average

$$\frac{N}{V} g(\mathbf{r}) = \frac{1}{N} \left\langle \sum_{I} \sum_{j \neq i} \delta(\mathbf{r} - \mathbf{r}^{ij}) \right\rangle, \quad \mathbf{r}^{ij} = \mathbf{r}^i - \mathbf{r}^j. \tag{12.101}$$

Clearly, $\mathbf{r}^{ij}$ is the difference vector between the position vectors of particles i and j.

Notice that g as given by (12.100) does not depend on the absolute positions of two particles within the volume V, but rather on their relative position vector. The definition (12.100) and consequently (12.101) also apply to spatially inhomogeneous systems. For a spatially homogeneous case, where the two-particle density $n^{(2)} = n^{(2)}(\mathbf{r}_1, \mathbf{r}_2)$ depends on the difference $\mathbf{r} = \mathbf{r}_1 - \mathbf{r}_2$ only, $g(\mathbf{r})$ can also be defined by

$$n^{(2)}(\mathbf{r}) = \left(\frac{N}{V}\right)^2 g(\mathbf{r}) = n^2 g(\mathbf{r}), \tag{12.102}$$

where it is understood that $n = N/V$ is the spatially constant number density. For a 'pure system', i.e. a substance composed of one type of particles, the interchange of the labels 1, 2 of two particles, which implies the replacement of $\mathbf{r}$ by $-\mathbf{r}$, should not make any difference for g. Thus g is an even function of $\mathbf{r}$:

$$g(\mathbf{r}) = g(-\mathbf{r}). \tag{12.103}$$

For particles which cannot penetrate each other due to their short range repulsion one has $g(0) = 0$. On the other hand, particles in isotropic media without long-ranged correlations are uncorrelated when they are separated by distances r large compared with their size. Then $g \to 1$ holds true for $r \to \infty$. Typically, the orientationally averaged part of g has a maximum at a value r which corresponds to the first neighbor distance. In dense systems there are several additional maxima at larger distances, with smaller height, however.

Examples for averages which can be evaluated as integrals over $g(\mathbf{r})$ are the potential contributions to the energy and to the pressure tensor as well as the static structure factor.

12.4.2 Potential Contributions to the Energy and to the Pressure Tensor

Assuming that the total potential energy of the particles is given by the sum of the binary interaction potential $\phi = \phi(\mathbf{r}) = \phi(-\mathbf{r})$, the total potential energy is

$$\Phi = \frac{1}{2} \sum_{i} \sum_{j \neq i} \phi(\mathbf{r}^i - \mathbf{r}^j). \tag{12.104}$$

The factor $\frac{1}{2}$ stems from the fact that the interaction between two particles i and j has to be counted only once whereas the summation over i and j contains the equivalent pairs $\{i, j\}$ and $\{j, i\}$. The average potential energy per particle is $u^{\text{pot}} = \langle \Phi \rangle / N$. The corresponding integral over the pair correlation function is

$$u^{\text{pot}} = \frac{1}{2} n \int \phi(\mathbf{r}) \, g(\mathbf{r}) \mathrm{d}^3 r, \tag{12.105}$$

where $n = N/V$ is the average number density.

The force associated with the binary interaction ϕ is

$$F_\mu = F_\mu(\mathbf{r}) = -\frac{\partial}{\partial r_\mu} \phi(\mathbf{r}).$$

The potential contribution to the pressure tensor $p_{\nu\mu}^{\text{pot}}$ then is given by the integral

$$p_{\nu\mu}^{\text{pot}} = \frac{1}{2} n^2 \int r_\nu F_\mu(\mathbf{r}) \, g(\mathbf{r}) \mathrm{d}^3 r. \tag{12.106}$$

For a substance composed of spherical particles, the force $\mathbf{F}$ is parallel to $\mathbf{r}$, thus the pressure tensor (12.106) is symmetric. It can be decomposed into its isotropic part $p\delta_{\mu\nu}$ and its symmetric irreducible (traceless) part $\overline{p_{\mu\nu}}$, cf. Chap. 6

$$p_{\nu\mu}^{\text{pot}} = p^{\text{pot}} \delta_{\mu\nu} + \overline{p_{\mu\nu}^{\text{pot}}},$$

with

$$p^{\text{pot}} = \frac{1}{6} n^2 \int r_\lambda F_\lambda(\mathbf{r}) \, g(\mathbf{r}) \mathrm{d}^3 r, \quad \overline{p_{\mu\nu}^{\text{pot}}} = \frac{1}{2} n^2 \int \overline{r_\nu F_\mu(\mathbf{r})} \, g(\mathbf{r}) \mathrm{d}^3 r. \tag{12.107}$$

The pressure tensor is the sum of its kinetic and potential contributions.

12.4.3 Static Structure Factor

The static structure factor $S = S(\mathbf{k})$ can be measured in experiments where electromagnetic radiation or particles, like neutrons, are scattered. The scattered intensity is determined by the *form factor*, which reflects the shape of the scatters, and by the static structure factor, which contains the information on the correlations between the positions of the particles. The *scattering wave vector* $\mathbf{k} = \mathbf{k}_{\text{sc}} - \mathbf{k}_{\text{in}}$ is the difference between the wave vectors $\mathbf{k}_{\text{sc}}$ and $\mathbf{k}_{\text{in}}$ of the detected scattered and the incident beam, see Fig. 12.1.

Fig. 12.1 Schematic
scattering geometry with
wave vectors $\mathbf{k}_{in}$ and $\mathbf{k}_{sc}$ for
the incoming and the
scattered beams

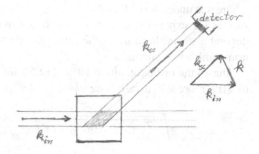

The scattering is elastic, i.e. one has $\mathbf{k}_{sc}^2 = \mathbf{k}_{in}^2$. The quantity $S(\mathbf{k}) - 1$ is essentially the spatial Fourier transform of $g(\mathbf{r}) - 1$, viz.

$$S(\mathbf{k}) = 1 + n \int \exp[-i\,\mathbf{k} \cdot \mathbf{r}]\,(g(\mathbf{r}) - 1)\mathrm{d}^3 r. \qquad (12.108)$$

Due to $g(\mathbf{r}) = g(-\mathbf{r})$, the exponential function in (12.108) can be replaced by $\cos(\mathbf{k} \cdot \mathbf{r})$. At first glance, curves of the isotropic part of S as function of k look similar to those of g as function of r. Typically, there is a first maximum at $k \approx 2\pi/r_{nn}$, where r_{nn} is the nearest neighbor distance associated with the first maximum of g. The behavior of S for k approaching zero is fundamentally different from that of $g(0) = 0$. More specifically, one has $S(0) = \langle(\delta\mathcal{N})^2\rangle/N$, where $\mathcal{N}$ is the number of scatters in the open scattering volume and $\delta\mathcal{N} = \mathcal{N} - \langle\mathcal{N}\rangle$ is its fluctuation. The mean square fluctuation of the number of particles is related to the isothermal compressibility $\kappa_T = n^{-1}(\partial n/\partial p)_T$. More specifically, one has $S(0) = n k_B T \kappa_T$. This quantity is small, but finite, in dense liquids. Close to the critical point, however, $S(0)$ becomes very large, this underlies the critical opalescence.

12.4.4 Expansion of $g(\mathbf{r})$

In thermal equilibrium, the pair correlation function of a fluid composed of spherical particles is isotropic, i.e. it depends on r of the vector $\mathbf{r}$, but not on its direction specified by the unit vector $\hat{\mathbf{r}}$. In general, however, in particular in non-equilibrium situations, g is a function of both r and $\hat{\mathbf{r}}$. The angular dependence, also called directional dependence, of g can be taken into account explicitly by an expansion with respect to irreducible tensors of rank ℓ constructed from the components of the unit vector $\hat{\mathbf{r}}$. Since g is an even function of $\mathbf{r}$, only even values $\ell = 0, 2, 4, \ldots$ occur in the expansion. The first terms are

$$g(\mathbf{r}) = g_s + g_{\mu\nu} \overline{\hat{r}_\mu \hat{r}_\nu} + g_{\mu\nu\lambda\kappa} \overline{\hat{r}_\mu \hat{r}_\nu \hat{r}_\lambda \hat{r}_\kappa} + \ldots, \qquad (12.109)$$

where it is understood, that the "spherical" part g_s of the pair correlation function, as well as the expansion tensors $g_{...}$ are function of the inter-particle distance r. They depend also on the time t when the pair correlation function applies to a general non-equilibrium situation.

Due to the orthogonality relation (12.5), the quantities $g_{...}$ are tensorial moments of $g(\mathbf{r})$, more specifically, one has

$$\frac{1}{4\pi} \int \overline{\widehat{r}_{\mu_1}\widehat{r}_{\mu_2}\cdots\widehat{r}_{\mu_\ell}}\, g(\mathbf{r})\mathrm{d}^2\widehat{r} = \frac{\ell!}{(2\ell+1)!!}\, g_{\mu_1\mu_2\cdots\mu_\ell}. \qquad (12.110)$$

For $\ell = 2$ and $\ell = 4$ this relation implies, see also (12.1),

$$g_{\mu\nu} = \frac{15}{2}\frac{1}{4\pi}\int \overline{\widehat{r}_\mu\widehat{r}_\nu}\, g(\mathbf{r})\mathrm{d}^2\widehat{r}, \qquad (12.111)$$

and

$$g_{\mu\nu\lambda\kappa} = \frac{315}{8}\frac{1}{4\pi}\int \overline{\widehat{r}_\mu\widehat{r}_\nu\widehat{r}_\lambda\widehat{r}_\kappa}\, g(\mathbf{r})\mathrm{d}^2\widehat{r}. \qquad (12.112)$$

Insertion of the expansion (12.109) into (12.107) leads to expressions which involve just the integration over r, but no longer over the angles. These relations are

$$p^{\mathrm{pot}} = \frac{2\pi}{3} n^2 \int r_\lambda F_\lambda\, g_s(r)\, r^2\mathrm{d}r, \qquad \overline{p^{\mathrm{pot}}_{\mu\nu}} = \frac{4\pi}{15} n^2 \int r_\lambda F_\lambda\, g_{\mu\nu}(r)\, r^2\mathrm{d}r. \qquad (12.113)$$

Notice that $r_\lambda F_\lambda = -r\phi'$. Here the prime indicates the differentiation with respect to r.

For a fluid composed of spherical particles, all moments with $\ell \geq 2$ vanish in thermal equilibrium. This, however, is no longer the case in non-equilibrium situations. In particular, certain components of $g_{\mu\nu}$ are non-zero for a viscous flow. This is already the case in the limit of small shear rates. At higher shear rates, also higher moments, e.g. with $\ell = 4$, cf. Sect. 12.4.5 and Exercise 12.4.

The expansion (12.109) is equivalent to an expansion with respect to spherical harmonics $Y_\ell^{(m)} = Y_\ell^{(m)}(\widehat{\mathbf{r}})$, cf. Sect. 9.4.2. The first few terms corresponding to (12.109) are

$$g(\mathbf{r}) = g_s + \sum_{m=-2}^{2} g^{2m}\, Y_2^{(m)} + \sum_{m=-4}^{4} g^{4m}\, Y_4^{(m)} + \dots, \qquad (12.114)$$

where the expansion coefficients are functions of r. The $Y_\ell^{(m)}$ obey the ortho-normalization $\int Y_\ell^{(m)}(Y_{\ell'}^{(m')})^* \mathrm{d}^2\widehat{r} = \delta_{\ell\ell'}\delta_{mm'}$, thus

$$g^{\ell m}(r) = \int (Y_\ell^{(m)}(\widehat{\mathbf{r}})^*\, g(\mathbf{r})\mathrm{d}^2\widehat{r}. \qquad (12.115)$$

It depends on the geometry of applications whether an expansion with respect to spherical or to Cartesian tensors is preferred. In the case of the distortion of the pair correlation function caused by a simple shear flow pertaining to the plane Couette geometry, cf. Fig. 7.6, the Cartesian version (12.109) is more appropriate.

12.4.5 Shear-Flow Induced Distortion of the Pair Correlation

The time change of the pair density $n^{(2)}(\mathbf{r}_1, \mathbf{r}_2)$ involves the flow term

$$\left[v_\mu(\mathbf{r}_1) \frac{\partial}{\partial r_{1\mu}} + v_\mu(\mathbf{r}_2) \frac{\partial}{\partial r_{2\mu}} \right] n^{(2)}(\mathbf{r}_1, \mathbf{r}_2).$$

When $n^{(2)} = n^2 g$, with constant number density n, depends on the difference variable $\mathbf{r} = \mathbf{r}_1 - \mathbf{r}_2$ only, this expression reduces to

$$\left(v_\mu(\mathbf{r}_1) - v_\mu(\mathbf{r}_2) \right) \frac{\partial}{\partial r_\mu} n^2 g(\mathbf{r}).$$

Consider a linear flow profile, cf. Sect. 7.2.2,

$$v_\mu(\mathbf{r}) = r_\nu (\nabla_\nu v_\mu) = r_\nu (\varepsilon_{\nu\mu\lambda} \omega_\lambda + \gamma_{\mu\nu}). \tag{12.116}$$

The *vorticity* ω_λ and the *deformation rate* or *shear rate* tensor $\gamma_{\mu\nu}$ are given by

$$\omega_\lambda = \frac{1}{2} \varepsilon_{\lambda\kappa\tau} \nabla_\kappa v_\tau, \quad \gamma_{\mu\nu} = \overline{\nabla_\nu v_\mu}. \tag{12.117}$$

Here $\nabla_\nu v_\nu = 0$, i.e. a divergence-free flow is assumed.

The flow term in the kinetic equation for g can be split into two contributions involving ω_λ and $\gamma_{\mu\nu}$, which induce a local rotation and deformation, respectively, in pair-space. This equation reads

$$\frac{\partial}{\partial t} g + \omega_\lambda \mathscr{L}_\lambda g + \gamma_{\mu\nu} \mathscr{L}_{\mu\nu} g + \mathscr{D}(g) = 0 \tag{12.118}$$

with the differential operators

$$\mathscr{L}_\lambda = \varepsilon_{\lambda\kappa\tau} r_\kappa \frac{\partial}{\partial r_\tau}, \quad \mathscr{L}_{\mu\nu} = \overline{r_\mu \frac{\partial}{\partial r_\tau}}. \tag{12.119}$$

The vector operator $\mathscr{L}_\lambda$, cf. (7.80), is the generator of the infinitesimal rotation. The symmetric traceless second rank tensor operator $\mathscr{L}_{\mu\nu}$ is associated with an infinitesimal volume conserving deformation. The damping term $\mathscr{D}(g)$ guarantees the approach of g to its equilibrium value $g_{eq} = g_{eq}(r)$ when the flow is switched off. A specific expression for $\mathscr{D}$ was proposed by Kirkwood [53], which is analogous to what Smoluchowski had used for the diffusion in the presence of an external potential. For this reason, the kinetic equation is referred to as *Kirkwood-Smoluchowski equation*. The potential used by Kirkwood is an effective potential ϕ_{eff} determined by the equilibrium pair correlation function g_{eq} according to $\phi_{eff} = -k_B T \ln g_{eq}$. A generalization and applications are discussed in [54]. The simple *relaxation time approximation*

$$\mathscr{D}(g) = \tau^{-1} (g - g_{eq}),$$

suffices to analyze the essential features associated with the shear flow induced distortion of g. Here τ is a structural relaxation time, sometimes also called *Maxwell relaxation time*. With $g = g_{eq} + \delta g$ and $\mathscr{L}_\lambda g_{eq} = 0$, the kinetic equation (12.118) reduces to

$$\frac{\partial}{\partial t} \delta g + \omega_\lambda \mathscr{L}_\lambda \delta g + \gamma_{\mu\nu} \mathscr{L}_{\mu\nu} \delta g + \tau^{-1} \delta g = -\gamma_{\mu\nu} \mathscr{L}_{\mu\nu} (g_{eq} - 1), \qquad (12.120)$$

where $\mathscr{L}_{\mu\nu} g_{eq} = \mathscr{L}_{\mu\nu}(g_{eq} - 1)$ was used.

Next, a stationary situation is considered, where the time derivative of g vanishes. Furthermore, g is written as

$$g = g_{eq} + \delta g^{(1)} + \delta g^{(2)} + \cdots,$$

where $g^{(k)}$ is of the order k in the shear rate. In first order, the kinetic equation (12.120) yields

$$\delta g^{(1)} \equiv -\tau \, \gamma_{\mu\nu} \mathscr{L}_{\mu\nu} \, g_{eq} = -\tau \, \gamma_{\mu\nu} \, \overline{r_\mu r_\nu} \, r^{-1} g'_{eq}, \qquad (12.121)$$

where the prime denotes the differentiation with respect to r. Comparison with (12.109) shows that

$$g_{\mu\nu} = -\tau \, \gamma_{\mu\nu} \, r g'_{eq}, \qquad (12.122)$$

in this approximation.

The contribution of second order in the shear rate is given by

$$\delta g^{(2)} = -\tau \, (\omega_\lambda \mathscr{L}_\lambda + \gamma_{\mu\nu} \mathscr{L}_{\mu\nu}) \delta g^{(1)} = \tau^2 \, (\omega_\lambda \mathscr{L}_\lambda + \gamma_{\mu\nu} \mathscr{L}_{\mu\nu}) \gamma_{\kappa\sigma} \mathscr{L}_{\kappa\sigma} \, g_{eq}.$$

Due to (12.121), the term involving the vorticity yields

$$2 \tau^2 \, \varepsilon_{\mu\kappa\lambda} \, \omega_\lambda \gamma_{\kappa\nu} \, \overline{r_\mu r_\nu} \, r^{-1} g'_{eq}.$$

The term involving the product $\mathscr{L}_{\mu\nu}\mathscr{L}_{\kappa\sigma}$ gives contributions to the isotropic part of g and to its second and fourth rank irreducible parts. More specifically, the term $\gamma_{\mu\nu}\mathscr{L}_{\mu\nu}\gamma_{\kappa\sigma}\mathscr{L}_{\kappa\sigma}\,g_{eq}$ is rewritten and computed as

$$\gamma_{\mu\nu}\,\mathscr{L}_{\mu\nu}\,\gamma_{\kappa\sigma}r_\kappa r_\sigma\,r^{-1}g'_{eq} = 2\gamma_{\mu\nu}\gamma_{\kappa\nu}\,r_\kappa r_\mu\,r^{-1}g'_{eq} + \gamma_{\mu\nu}\,\gamma_{\kappa\sigma}r_\nu r_\mu r_\kappa r_\sigma\,r^{-1}(r^{-1}g'_{eq})'.$$

Due to $r_\kappa r_\mu = \overline{r_\kappa r_\mu} + \frac{1}{3}r^2\delta_{\mu\nu}$ and with the help of relation (11.59), the second order distortion $\delta g^{(2)}$ can be decomposed into the parts $\delta g_\ell^{(2)}$, associated with tensors of ranks $\ell = 0, 2, 4$. The term involving the vorticity contributes to $\delta g_2^{(2)}$ only. Thus one has

$$g = g_{eq} + \delta g^{(1)} + \delta g^{(2)} + \dots\,, \qquad \delta g^{(2)} = \delta g_0^{(2)} + \delta g_2^{(2)} + \delta g_4^{(2)}, \qquad (12.123)$$

with

$$\delta g_0^{(2)} = \tau^2\gamma_{\mu\nu}\gamma_{\mu\nu}\left(\frac{2}{3}rg'_{eq} + \frac{2}{15}r^3(r^{-1}g'_{eq})'\right),$$

$$\delta g_2^{(2)} = 2\tau^2\varepsilon_{\mu\kappa\lambda}\omega_\lambda\gamma_{\kappa\nu}\,\overline{r_\mu r_\nu}\,r^{-1}g'_{eq} + \tau^2\gamma_{\mu\kappa}\gamma_{\kappa\nu}\,\overline{r_\nu r_\mu}\left(2r^{-1}g'_{eq} + \frac{4}{7}r(r^{-1}g'_{eq})'\right),$$

$$\delta g_4^{(2)} = \tau^2\gamma_{\mu\nu}\gamma_{\kappa\sigma}\,\overline{r_\nu r_\mu r_\kappa r_\sigma}\,r^{-1}(r^{-1}g'_{eq})'. \qquad (12.124)$$

Consequences of corresponding relations for a plane Couette flow are presented in [55], see also the Exercise 12.4. Symmetry considerations for this simple flow geometry are discussed next.

The shear-induced distortion of the structure of a liquid or of a colloidal dispersion can be detected in scattering experiments where the static structure factor, cf. Sect. 12.4.8 is analyzed. Direct observations of the distorted pair correlation function is possible in *confocal microscopy* experiments [56] and, in particular in *Non-Equilibrium Molecular Dynamics* (NEMD) computer simulations [57, 58].

12.4.6 Plane Couette Flow Symmetry

A plane Couette flow geometry is considered with the velocity $\mathbf{v} = \mathbf{v}(\mathbf{r})$ in the x-direction and its gradient in the y-direction, cf. (7.28), thus

$$v_\mu = e_\mu^x\, e_\nu^y\, r_\nu = y\, e_\mu^x, \qquad (12.125)$$

with $x = e_\nu^x r_\nu$ and $y = e_\nu^y r_\nu$, where $\mathbf{e}^x$ and $\mathbf{e}^y$ are unit vectors parallel to the x-axis and y-axis, respectively. The unit vector parallel to the z-axis is $\mathbf{e}^z$. The constant shear rate is

$$\gamma = \frac{\partial v_x}{\partial y}.$$

(12.126)

The vorticity and the shear rate tensor, cf. are

$$\omega_\lambda = -\frac{1}{2} \gamma \, e_\lambda^z, \quad \gamma_{\mu\nu} = \gamma \, \overline{e_\nu^y e_\mu^x} = \frac{1}{2} \gamma \, (e_\nu^y e_\mu^x + e_\mu^y e_\nu^x).$$

(12.127)

Here, the first order expression (12.121) implies

$$\delta g^{(1)} = -\tau \, \gamma \, x \, y \, r^{-1} g'_{eq} \, ,$$

and one has

$$g_{\mu\nu} = -\tau \, \gamma \, r g'_{eq} \, \overline{e_\nu^y e_\mu^x} \, ,$$

in this approximation. The more general ansatz

$$g_{\mu\nu} = g_+ \, \overline{e_\nu^y e_\mu^x} + g_- \frac{1}{2}(e_\nu^x e_\mu^x - e_\nu^y e_\mu^y) + g_0 \, \overline{e_\nu^z e_\mu^z}$$

(12.128)

contains all terms which obey the plane Couette symmetry, viz. which are invariant when both $\mathbf{e}^x$ and $\mathbf{e}^y$ are replaced by $-\mathbf{e}^x$ and $-\mathbf{e}^y$. The remaining two terms of the 5 components of the irreducible second rank tensor which, however, do not have this symmetry, are proportional to $\overline{e_\nu^z e_\mu^x}$ and $\overline{e_\nu^z e_\mu^y}$.

The quantities g_+, g_- and g_0 are functions of r. In first order in the shear rate γ, one has $g_+ = -\tau \gamma r^{-1} g'_{eq}$ and $g_- = g_0 = 0$. In second order, $g_- = \gamma \tau g_+$, $g_0 \neq 0$ is found. The calculation is deferred to the next exercise.

12.4 Exercise: Pair Correlation Distorted by a Couette Flow

Compute the functions g_+, g_- and g_0 in first and second order in the shear rate γ, in steady state, from the plane Couette version of the kinetic equation (12.120)

$$\gamma \, y \, \frac{\partial}{\partial x} \delta g + \tau^{-1} \delta g = -\gamma \, y \, \frac{\partial}{\partial x} g_{eq}.$$

Hint: use $y^2 = (x^2 + y^2)/2 - (x^2 - y^2)/2$ and $x^2 + y^2 = r^2 - z^2 = 2r^2/3 - (z^2 - r^2/3)$, furthermore decompose $x^2 y^2 = e_\mu^x e_\nu^x e_\lambda^y e_\kappa^y r_\mu r_\nu r_\lambda r_\kappa$ into its parts associated with tensors of ranks $\ell = 0, 2, 4$ with the help of (9.6). Compare g_- with g_+.

12.4.7 Cubic Symmetry

When a crystal melts, its highly anisotropic structure relaxes to an isotropic state typical for a liquid in equilibrium. For an initial state with cubic symmetry, cubic harmonics are appropriate to characterize the anisotropic pair correlation, as e.g. studied in [25]. The pair correlation function is expanded with respect to the cubic harmonics with full cubic symmetry, cf. Sect. 9.5.2, viz.

$$g(\mathbf{r}) = g^{(0)}(r) + g_4(r)\,K_4(\hat{\mathbf{r}}) + g_6(r)\,K_6(\hat{\mathbf{r}}) + g_8(r)\,K_8(\hat{\mathbf{r}}) + \dots . \qquad (12.129)$$

The functions K_4, K_6, K_8 are defined in (9.33)–(9.35). In general, the radially symmetric part $g^{(0)}$ as well as the partial correlation functions g_4, g_6, g_8 characterizing the anisotropy of $g(\mathbf{r})$ are functions of the time t. Examples, from [25], are shown in Figs. 12.2, 12.3 and 12.4. The 'data' stem from a molecular dynamics computer simulation. Initially the particles are put on body centered cubic (bcc) lattice sites at a temperature and density where the system is fluid. Thus $g^{(0)}$ approaches a pair correlation function typical for a dense liquid, the functions g_4, g_6, $\dots$ associated with the anisotropy decay to zero. The simulation [25] was performed for 1024 particles interacting with a r^{-12} potential cut off appropriately, which is also referred to as 'soft sphere' potential. Periodic boundary conditions and a 'thermostat' were used.

The decay of the cubic anisotropy, in the first coordination shell, is relatively slow compared with the approach of the thermodynamic variables, like the pressure and the energy, to their equilibrium values in the liquid state. This slowing down of the relaxation becomes more pronounced when the initial state is closer to the thermodynamic state where the solid phase is stable.

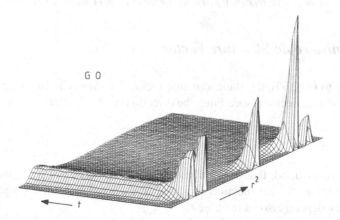

Fig. 12.2 Perspective view of the isotropic part of the pair correlation function. The variable r^2 runs right-upward, the time t runs left. The initial state shows the positions of the first, second, $\dots$ coordination shells in the bcc solid

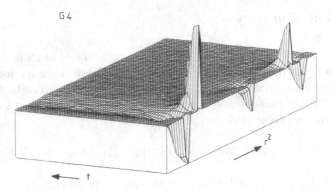

Fig. 12.3 The cubic pair correlation function g_4, variables r^2 and t are as in Fig. 12.2. The sign changes from one coordination shell to next

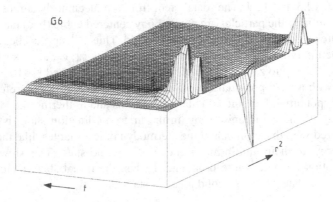

Fig. 12.4 The cubic pair correlation function g_6, variables r^2 and t are as in Fig. 12.2

12.4.8 Anisotropic Structure Factor

By analogy to (12.109), the static structure factor $S = S(\mathbf{k})$ can be expanded with respect to the unit vector $\widehat{\mathbf{k}}$ specifying the direction of the scattering wave vector $\mathbf{k}$:

$$S(\mathbf{k}) = S_{\mathrm{s}} + S_{\mu\nu} \, \overline{\widehat{k}_\mu \widehat{k}_\nu} + S_{\mu\nu\lambda\kappa} \, \overline{\widehat{k}_\mu \widehat{k}_\nu \widehat{k}_\lambda \widehat{k}_\kappa} + \dots, \qquad (12.130)$$

where it is understood, that the isotropic or spherical part S_{s} of S, as well as the expansion tensors $S_{\dots}$ are function of k, viz. of the magnitude of the vector $\mathbf{k}$. In general, they depend also on the time t.

On account of the orthogonality relation (12.5), the quantities $S_{\dots}$ are tensorial moments of $S(\mathbf{k})$, thus

$$\frac{1}{4\pi} \int \overline{\widehat{k}_{\mu_1} \widehat{k}_{\mu_2} \cdots \widehat{k}_{\mu_\ell}} \, S(\mathbf{k}) \mathrm{d}^2 \widehat{k} = \frac{\ell!}{(2\ell + 1)!!} \, S_{\mu_1 \mu_2 \cdots \mu_\ell}. \qquad (12.131)$$

For $\ell = 2$ this relation implies, see also (12.1),

$$S_{\mu\nu} = \frac{15}{2} \frac{1}{4\pi} \int \overline{\widehat{k}_\mu \widehat{k}_\nu} \, S(\mathbf{k}) d^2\widehat{k}, \tag{12.132}$$

the case $\ell = 4$ is analogous to (12.112).

The static structure factor and the pair correlation function are related to each other by a spatial Fourier transformation, see (12.108), so there exist also interrelations between the tensors $S_{\mu_1,\mu_2\cdots\mu_\ell}$ and $g_{\mu_1,\mu_2\cdots\mu_\ell}$. The key for this connection is the *Rayleigh expansion*

$$\exp[-i\,\mathbf{k}\cdot\mathbf{r}] = \sum_{\ell=0}^{\infty}(-i)^\ell\,(2\ell+1)j_\ell(k\,r)\,P_\ell(\widehat{k}\cdot\widehat{r}), \quad j_\ell(k\,r) = \left(\frac{\pi}{2kr}\right)^{\frac{1}{2}} J_{\ell+\frac{1}{2}}(k\,r),$$

$$\tag{12.133}$$

where the $J_{..}$ are *Bessel functions* and the j_ℓ are referred to as *spherical Bessel functions*, [66], sometimes also called Sommerfeld's Bessel functions. The Legendre polynomial P_ℓ is the scalar product of the ℓth rank irreducible tensors constructed from the components of the unit vectors $\widehat{k}$ and $\widehat{r}$, cf. Sect. 9.3. The integral relation analogous to (12.1) leads to

$$\frac{1}{4\pi} \int \overline{\widehat{k}_{\mu_1}\widehat{k}_{\mu_2}\cdots\widehat{k}_{\mu_\ell}} \, P_{\ell'}(\widehat{k}\cdot\widehat{r})d^2\widehat{k} = \delta_{\ell\ell'}\,(2\ell+1)^{-1}\overline{\widehat{r}_{\mu_1}\widehat{r}_{\mu_2}\cdots\widehat{r}_{\mu_\ell}}. \tag{12.134}$$

As a consequence, insertion of (12.108) into (12.131), use of the relations just stated and of $\int \ldots d^3r = \int \ldots d^2r d^2\widehat{r}$ leads to

$$S_{\mu_1\mu_2\cdots\mu_\ell}(k) = (-i)^\ell \int j_\ell(k\,r)\,g_{\mu_1\mu_2\cdots\mu_\ell}(r)\,r^2 dr, \tag{12.135}$$

for $\ell \geq 1$. The interrelation between the spherical parts, corresponding to the case $\ell = 0$, is

$$S_s(k) - 1 = \int j_0(k\,r)\,(g_s(r) - 1)\,r^2 dr.$$

The first few of the spherical Bessel functions are

$$j_0(x) = x^{-1}\sin x, \quad j_1(x) = x^{-1}(x^{-1}\sin x - \cos x)$$
$$j_2(x) = 3x^{-2}(x^{-1}\sin x - x/3 - \cos x).$$

The spherical Bessel functions obey the recursion relation

$$j_{\ell+1}(x) = \frac{\ell}{x}j_\ell - \frac{d}{dx}j_\ell = -x^\ell\frac{d}{dx}(x^{-\ell}j_\ell).$$

The shear flow induced distortion of the pair correlation function implies a corresponding anisotropy of the static structure factor. This anisotropy has been observed in computer simulations [59], in light scattering [60] and in neutron scattering experiments [61].

12.5 Selection Rules for Electromagnetic Radiation

12.5.1 Expansion of the Wave Function

Let $\Psi = \Psi(t, \mathbf{r})$ be the wave function, in spatial representation, which obeys the *Schrödinger equation* for a quantum mechanical single particle problem, e.g. the electron bound by the proton, in the hydrogen atom. The pertaining Hamilton operator is the sum of the operator for the kinetic energy, viz.

$$H_{\text{kin}}^{\text{op}} = -\frac{\hbar^2}{2m} \Delta = -\frac{\hbar^2}{2m} \Delta_{\text{r}} - \frac{\hbar^2}{2m} r^{-2} \mathscr{L}_\mu \mathscr{L}_\mu \,,$$

cf. Sect. 7.6.4, and the potential energy $V = V(\mathbf{r})$. For the radial part Δ_r of the Laplace operator see (7.91). The angular part of the Laplacian involves the differential operator $\mathscr{L}_\mu = \varepsilon_{\mu\lambda\nu} r_\lambda \partial/\partial r_\nu$, cf. (7.80).

To describe the angle dependence of the wave function $\Psi(t, \mathbf{r})$ an expansion can be made with respect to spherical harmonics, as found in text books on Quantum Mechanics, or with respect to Cartesian basis tensors, as presented here. With the help of the normalized basis functions, see also (12.14),

$$\phi_{\mu_1 \cdots \mu_\ell} \equiv \sqrt{\frac{(2\ell+1)!!}{\ell!}} \, \widehat{\overline{r_{\mu_1} \cdots r_{\mu_\ell}}} \,, \tag{12.136}$$

the expansion is written as

$$\Psi(t, \mathbf{r}) = \sum_{\ell=0}^{\infty} c_{\mu_1 \cdots \mu_\ell}(t, r) \, \phi_{\mu_1 \cdots \mu_\ell}(\widehat{\mathbf{r}}). \tag{12.137}$$

In general, the moment tensors $c_{\mu_1 \cdots \mu_\ell}$, depend on the time t and on $r = |\mathbf{r}|$. There is no explicit time dependence when Ψ is a solution of the stationary Schrödinger equation. Being symmetric traceless, the ℓth rank tensor $c_{\mu_1 \cdots \mu_\ell}$ has $2\ell + 1$ independent components, in accord with the same number of m-values of the spherical components.

The normalization condition $\int |\Psi|^2 d^3 r = 1$ implies

$$\sum_{\ell=0}^{\infty} |c_\ell|^2 = 1 , \quad |c_\ell|^2 = 4\pi \int c^*_{\nu_1 \cdots \nu_\ell} c_{\nu_1 \cdots \nu_\ell} r^2 dr. \tag{12.138}$$

In the following, it is assumed that the part

$$\Psi_\ell = c_{\mu_1 \cdots \mu_\ell} \phi_{\mu_1 \cdots \mu_\ell}$$

of the wave function associated with orbital angular momentum ℓ is a solution of the stationary Schrödinger equation $\mathscr{H}\Psi_\ell = E\Psi_\ell$ with a radially symmetric potential V. The function $c_{..}$ then obeys the equation

$$\left[-\frac{\hbar^2}{2m} \Delta_r + \ell(\ell+1) \frac{\hbar^2}{2m} r^{-2} + V(r) \right] c_{\mu_1 \cdots \mu_\ell} = E\, c_{\mu_1 \cdots \mu_\ell} ,$$

and the appropriate boundary and integrability conditions. For a radially symmetric interaction potential, the tensor functions $c_{..}$ are the product of a scalar radial wave function $R_\ell(r)$ and a tensor $C_{\mu_1 \cdots \mu_\ell}$, which is complex, in general., viz.

$$c_{\mu_1 \cdots \mu_\ell} = R_\ell(r)\, C_{\mu_1 \cdots \mu_\ell} .$$

The radial functions $R_\ell(r)$ are characterized by additional quantum numbers, like the main quantum number of the H-atom. Explicit expressions are not needed for the discussion of the selection rules. Assuming $C^*_{\nu_1 \cdots \nu_\ell} C_{\nu_1 \cdots \nu_\ell} = 1$, the normalization condition (12.138) implies

$$\sum_{\ell=0}^{\infty} 4\pi \int |R_\ell(r)|^2 r^2 dr = 1.$$

The orientational properties of a state described by a wave function is determined by its angle dependent part $\phi_\ell(\hat{\mathbf{r}})$. Assuming that both the radial and the angular parts are appropriately normalized, the expectation value of an operator $\mathscr{O} = \mathscr{O}(\mathbf{L})$, which is a function of the angular momentum operator $\mathbf{L}$, is given by

$$\langle \mathscr{O}(\mathbf{L}) \rangle = (4\pi)^{-1} \int \Phi^*_\ell \,\mathscr{O}(\mathbf{L})\, \Phi_\ell d^2 \hat{r}, \quad \Phi_\ell = C_{\mu_1 \cdots \mu_\ell} \phi_{\mu_1 \cdots \mu_\ell}. \tag{12.139}$$

The tensors $C_{..}$ are complex, in general. Examples for $\mathscr{O}$ are the vector polarization $\langle L_\mu \rangle$ and the tensor polarization $\langle \overline{L_\mu L_\nu} \rangle$. Exercise 12.5 deals with an application.

12.5.2 Electric Dipole Transitions

Electromagnetic waves, induce transitions between an 'initial' stationary state 1 and a 'final' state 2 with the energies E_1 and E_2, provided that the radiation has the right frequency $\omega = (E_2 - E_1)/\hbar$. Furthermore, for weak fields, the transition has to be 'allowed' by the *selection rules*. These rules follow from expressions for the transition rate which, in turn, is proportional to the absolute square of a 'matrix element' $\langle \Psi_{\text{final}} | H^{\text{pert}} | \Psi_{\text{initial}} \rangle$, where H^{pert} stands for the time independent part of the 'perturbation Hamiltonian' which characterizes the interaction between the atom and the electromagnetic field. In spatial representation, the matrix element is computed as an integral over space.

The electric dipole transitions are associated with the perturbation Hamiltonian $H^{\text{pert}} = H^{\text{dip}} \equiv -\mathbf{p}^{\text{e}} \cdot \mathbf{E}$ with the electric dipole moment $\mathbf{p}^{\text{e}} = q\mathbf{r}$, where q is the electric charge. The unit vector parallel to the electric field $\mathbf{E}$ is denoted by $\mathbf{e}$. Now Ψ_ℓ, i.e. a state with a well defined magnitude of the angular momentum is chosen as initial state. Then $H^{\text{dip}} | \Psi_{\text{initial}} \rangle$ is proportional to

$$e_\lambda \widehat{r}_\lambda \phi_{v_1 \cdots v_\ell} c_{v_1 \cdots v_\ell}$$
$$= \left(\sqrt{\frac{\ell+1}{2\ell+3}} \phi_{v_1 \cdots v_\ell \lambda} + \sqrt{\frac{\ell}{2\ell+1}} \Delta^{(\ell)}_{v_1 \cdots v_\ell, \lambda \kappa_1 \cdots \kappa_{\ell-1}} \phi_{\kappa_1 \cdots \kappa_{\ell-1}} \right) e_\lambda c_{v_1 \cdots v_\ell} . \quad (12.140)$$

Here the (11.53) was used: the product of a vector with an irreducible tensor of rank ℓ constructed from this vector yield an irreducible tensor of rank $\ell + 1$ and another one of rank $\ell - 1$. This is already the essence of the selection rule for the electric dipole transitions. Multiplication of the expression (12.140) by $\Psi_{\ell'}^*$ and subsequent integration over $d^3 r$ yields non-zero contributions only for

$$\ell' = \ell \pm 1 .$$

The resulting dipole transition matrix elements are

$$(4\pi)^{-1} \int \Psi_{\ell+1}^{ast} e_\lambda r_\lambda \Psi_\ell d^3 r = \sqrt{\frac{\ell+1}{2\ell+3}} e_{v_{\ell+1}} \int r \, c^*_{v_1 \cdots v_{\ell+1}} c_{v_1 \cdots v_\ell} r^2 dr ,$$

$$(4\pi)^{-1} \int \Psi_{\ell-1}^* e_\lambda r_\lambda \Psi_\ell d^3 r = \sqrt{\frac{\ell}{2\ell+1}} e_{v_\ell} \int r \, c^*_{v_1 \cdots v_{\ell-1}} c_{v_1 \cdots v_\ell} r^2 dr .$$

The selection rule determines which angular momentum state can be reached in an 'allowed' transition. The strength of the transition rate is determined by the remaining 'overlap integral' $\int \cdots dr$. The Cartesian indices characterize properties, e.g. the direction of the electric field and the orientational sub-states of the initial and final state.

The electric-field-induced transition from an $\ell = 0$ to a $\ell = 1$ state prepares an orientationally well defined state, depending on the polarization of the incident radiation. According to (12.140) the angle dependent part of the $\ell = 1$ wave function is $\Phi_1(\widehat{\mathbf{r}}) = C_\mu \phi_\mu$, with $C_\mu = e_\mu \equiv E^{-1} E_\mu$, thus

$$\Phi_1 = \Phi_1(\mathbf{r}) = \sqrt{3}\, e_\lambda \widehat{r}_\lambda. \tag{12.141}$$

Clearly, $\mathbf{e}$ is the unit vector parallel to the field. For linearly polarized light, propagating in z-direction, $\mathbf{e} = \mathbf{e}^x$ can be chosen. The field orientation of circular polarized light is described by $\mathbf{e} = 2^{-1/2}(\mathbf{e}^x \pm i\mathbf{e}^y)$. Notice that the wave function is complex, in this case. The vector and tensor polarizations in this exited state are

$$\langle L_\mu \rangle = 0, \quad \langle \overline{L_\mu L_\nu} \rangle = \overline{e_\mu^x e_\nu^x}, \tag{12.142}$$

for the linearly polarized light. For circular polarization, one obtains

$$\langle L_\mu \rangle = e_\mu^z, \quad \langle \overline{L_\mu L_\nu} \rangle = -\overline{e_\mu^z e_\nu^z}. \tag{12.143}$$

The computations leading to these results are deferred to the Exercise 12.5.

12.5 Exercise: Compute the Vector and Tensor Polarization for a $\ell = 1$ State
Hint: use the wave function (12.141) with $e_\mu = e_\mu^x$ and $e_\mu = (e_\mu^x + ie_\mu^y)/\sqrt{2}$ for the linear and circular polarized cases. For the angular momentum operator and its properties see Sect. 7.6.2.

12.5.3 Electric Quadrupole Transitions

The Hamiltonian H^{quad} inducing electric quadrupole transitions is proportional to $r_\lambda r_\kappa k_\kappa E_\lambda$, where $\mathbf{k} = k\widehat{\mathbf{k}}$ is the wave vector of the incident electric field $\mathbf{E} = E\mathbf{e}$. From (11.57) follows, that the application of H^{quad} on the wave function Ψ_ℓ yields three contributions $\Psi_{\ell'}$ with

$$\ell' = \ell, \ell \pm 2.$$

More specifically, one has

$$e_\lambda \widehat{k}_\kappa \, \overline{\widehat{r}_\lambda \widehat{r}_\kappa}\, \phi_{\mu_1 \cdots \mu_\ell} c_{\mu_1 \cdots \mu_\ell}$$
$$= \left(\sqrt{\frac{(\ell+2)(\ell+1)}{(2\ell+5)(2\ell+3)}}\, \phi_{\mu_1 \cdots \mu_\ell \lambda \kappa} + \frac{2\ell}{2\ell+3} \Delta^{(\ell,2,\ell)}_{\mu_1 \mu_2 \cdots \mu_\ell, \lambda\kappa, \nu_1 \nu_2 \cdots \nu_\ell} \phi_{\nu_1 \cdots \nu_\ell} \right.$$
$$\left. + \sqrt{\frac{\ell(\ell-1)}{(2\ell+1)(2\ell-1)}} \Delta^{(\ell)}_{\mu_1 \mu_2 \cdots \mu_\ell, \nu\lambda\, \nu_1 \nu_2 \cdots \nu_{\ell-2}} \phi_{\nu_1 \cdots \nu_{\ell-2}} \right) e_\lambda \widehat{k}_\kappa c_{\mu_1 \cdots \mu_\ell}. \tag{12.144}$$

Multiplication of (12.144) by $\Psi_{\ell'}^*$ and subsequent integration over d^3r yields non-zero contributions only for $\ell' - \ell = 0, \pm 2$. The resulting quadrupole transition matrix elements are

$$(4\pi)^{-1} \int \Psi_{\ell+2}^* e_\lambda \widehat{k_\kappa} \; \overline{\widehat{r_\lambda r_\kappa}} \; \Psi_\ell d^3 r = \sqrt{\frac{(\ell+2)(\ell+1)}{(2\ell+5)(2\ell+3)}} \; \overline{e_{\nu_{\ell+1}} \widehat{k}_{\nu_{\ell+2}}}$$

$$\times \int r^2 c_{\nu_1 \cdots \nu_{\ell+1} \nu_{\ell+2}}^* c_{\nu_1 \cdots \nu_\ell} r^2 dr \,,$$

$$(4\pi)^{-1} \int \Psi_\ell^* e_\lambda \widehat{k_\kappa} \; \overline{\widehat{r_\lambda r_\kappa}} \; \Psi_\ell d^3 r = \frac{2\ell}{2\ell+3} \; \overline{e_\lambda \widehat{k}_\kappa} \int r^2 c_{\lambda \nu_2 \cdots \nu_\ell}^* c_{\kappa \nu_2 \cdots \nu_\ell} r^2 dr \,,$$

$$(4\pi)^{-1} \int \Psi_{\ell-2}^* e_\lambda \widehat{k_\kappa} \; \overline{\widehat{r_\lambda r_\kappa}} \; \Psi_\ell d^3 r = \sqrt{\frac{\ell(\ell+1)}{(2\ell+1)(2\ell-1)}} \; \overline{e_{\nu_{\ell-1}} \widehat{k}_{\nu_\ell}}$$

$$\times \int r^2 c_{\nu_1 \cdots \nu_{\ell-2}}^* c_{\nu_1 \cdots \nu_\ell} r^2 dr \,.$$

The selection rule determines which angular momentum state can be reached in an allowed transition. The strength of the transition rate is determined by the remaining overlap integral $\int \cdots dr$.

The relevant perturbation Hamiltonian for *two-quantum absorption* processes is proportional to $r_\lambda r_\kappa E_\kappa E_\lambda$, thus of second order in the radiation field $\mathbf{E}$. Here the quadrupole selection rules $\ell' - \ell = 0, \pm 2$ apply as well.

Chapter 13
Spin Operators

Abstract Spin operators are introduced in this chapter. The spin 1/2 and 1 are looked upon explicitly. Projectors into magnetic sub-states and irreducible spin tensors are defined. Spin traces of multiple products of these tensors and their role for the expansion of density operators and the evaluation of averages are elucidated. The last section deals with the rotational angular momenta of linear molecules, in particular with tensor operators. One application is the anisotropic dielectric tensor of a gas of rotating molecules.
The orbital angular momentum of particles is linked with their linear momentum. Most elementary particles, like electrons, protons, neutrons, neutrinos posses an intrinsic angular momentum, conventionally called 'spin', which is not caused by their translational motion. Here properties of spin operators are discussed and rules are presented for tensors constructed from the Cartesian components of the spin operators. Furthermore, tensor operators associated with the rotational motion of linear molecules are treated.

13.1 Spin Commutation Relations

13.1.1 Spin Operators and Spin Matrices

The spin operator of a particle with spin s, in units of $\hbar$, is denoted by **s**. The components of this operator can be represented by hermitian $(2s + 1) \times (2s + 1)$ matrices. For $s = \frac{1}{2}$, e.g. these are the two-by-two Pauli matrices. The quantity s is a positive integer or halve-integer number, viz. $s = \frac{1}{2}$, or $s = 1$, or $s = \frac{3}{2}$, or etc.

The Cartesian components of the spin operator obey the angular momentum commutation relations, analogous to those of the dimensionless orbital angular momentum, cf. Sect. 7.6.2. The spin commutation relations

$$s_\mu s_\nu - s_\nu s_\mu \equiv [s_\mu, s_\nu]_- = i\, \varepsilon_{\mu\nu\lambda}\, s_\lambda. \qquad (13.1)$$

© Springer International Publishing Switzerland 2015
S. Hess, *Tensors for Physics*, Undergraduate Lecture Notes in Physics,
DOI 10.1007/978-3-319-12787-3_13

Here, $[.., ..]_-$ indicates the commutator. Much as (7.87) implies the relation (7.88), the spin commutation relation is equivalent to

$$\varepsilon_{\lambda\mu\nu}\, s_\mu\, s_\nu = i\, s_\lambda, \tag{13.2}$$

or

$$\mathbf{s} \times \mathbf{s} = i\, \mathbf{s}. \tag{13.3}$$

Clearly, the components of the quantum mechanical angular momentum do not commute, in contradistinction to the components of the classical angular momentum for which the corresponding cross product vanishes.

The scalar product of two spin s vector operators is

$$\mathbf{s} \cdot \mathbf{s} = s_\mu\, s_\mu = s\,(s+1)\,\mathbf{1}. \tag{13.4}$$

Here $\mathbf{1}$ stands for the unit operator in the spin space. In matrix representation, this is just the $(2s+1) \times (2s+1)$ unit matrix. This unit operator is omitted frequently, when no danger of confusion exists.

13.1.2 Spin 1/2 and Spin 1 Matrices

The spin matrices for $s = 1/2$, the spin matrices s_x, s_y, s_z are

$$\frac{1}{2}\begin{pmatrix} 0 & 1 \\ 1 & 0 \end{pmatrix}, \quad \frac{i}{2}\begin{pmatrix} 0 & -1 \\ 1 & 0 \end{pmatrix}, \quad \frac{1}{2}\begin{pmatrix} 1 & 0 \\ 0 & -1 \end{pmatrix}. \tag{13.5}$$

Apart from the factor 1/2, these are the Pauli matrices σ_x, σ_y, σ_z.

For $s = 1$, the spin matrices s_x, s_y, s_z are

$$\frac{1}{\sqrt{2}}\begin{pmatrix} 0 & 1 & 0 \\ 1 & 0 & 1 \\ 0 & 1 & 0 \end{pmatrix}, \quad \frac{i}{\sqrt{2}}\begin{pmatrix} 0 & -1 & 0 \\ 1 & 0 & -1 \\ 0 & 1 & 0 \end{pmatrix}, \quad \begin{pmatrix} 1 & 0 & 0 \\ 0 & 0 & 0 \\ 0 & 0 & -1 \end{pmatrix}. \tag{13.6}$$

13.1 Exercise: Verify the Normalization for the Spin 1 Matrices

Compute explicitly $s_x^2 + s_y^2 + s_z^2$ for the spin matrices (13.6) in order to check the normalization relation (13.4).

13.2 Magnetic Sub-states

13.2.1 Magnetic Quantum Numbers and Hamilton Cayley

The spin operator possesses sub-states, which are eigenstates of one of its Cartesian components. Let this preferential direction be parallel to the unit vector $\mathbf{h}$. The symbol $\mathbf{h}$ alludes to the direction of a magnetic field $\mathbf{H}$. Frequently, the z-direction of a coordinate system is chosen parallel to $\mathbf{h}$. The eigenvalues are referred to as *magnetic quantum numbers* since they determine the strength of the interaction of a magnetic moment, which is parallel or anti-parallel to the spin operator $\mathbf{s}$, in the presence of a magnetic field. More specifically, the energy of a magnetic moment $\mathbf{m}$ in the presence of a magnetic field $\mathbf{B} = B\mathbf{h}$ is $-\mathbf{m} \cdot \mathbf{B}$. With the magnetic moment given by $\mathbf{m} = \gamma \hbar \mathbf{s}$, where γ is the *gyromagnetic ratio*, the corresponding Hamilton operator is $H^{\mathrm{mag}} = -\hbar \gamma B\mathbf{h} \cdot \mathbf{s}$.

The eigenvalues of $\mathbf{h} \cdot \mathbf{s}$ are denoted by m. These magnetic quantum numbers m are of relevance, even when no magnetic field is applied. Due to the *Richtungs-Quantelung*, the allowed values for m are

$$-s, -s+1, \ldots, s-1, s,$$

for a spin s. The m are integer or halve-integer numbers, depending on whether s is an integer or a halve-integer number. Altogether, there are $2s + 1$ magnetic quantum numbers and magnetic sub-states. Clearly, the smallest non-zero s is $s = \frac{1}{2}$.

The *Hamilton-Cayley* relation for the magnetic sub-states is

$$\prod_{m=-s}^{s} (\mathbf{h} \cdot \mathbf{s} - m) = \mathbf{0}. \tag{13.7}$$

This is a polynomial of degree $2s + 1$, in $\mathbf{h} \cdot \mathbf{s}$. Thus $(\mathbf{h} \cdot \mathbf{s})^{(2s+1)}$ is equal to a linear combination of lower powers of $\mathbf{h} \cdot \mathbf{s}$. The same applies for $(\mathbf{h} \cdot \mathbf{s})^p$, when the power p is larger than $2s + 1$. The Hamilton-Cayley relation also implies, that symmetric traceless tensors of rank ℓ, constructed from the Cartesian components of the spin operator, are non-zero only up to rank $\ell = 2s$, see the following section.

13.2.2 Projection Operators into Magnetic Sub-states

The projection operator $P^{(m)}$ into the sub-state with the eigenvalue m is defined via

$$P^{(m)} \mathbf{h} \cdot \mathbf{s} = m P^{(m)}. \tag{13.8}$$

These projectors have the properties

$$P^{(m)} P^{(m')} = \delta_{mm'} P^{(m)}, \quad \sum_{m=-s}^{s} P^{(m)} = 1. \tag{13.9}$$

Thus they are orthogonal, idempotent as any projector, and they form a complete set.

The projectors can be expressed in terms of powers $\mathbf{h} \cdot \mathbf{s}$, analogous to the Hamilton-Cayley relation. In particular, one has

$$P^{(m)} = \prod_{m' \neq m} \frac{(\mathbf{h} \cdot \mathbf{s} - m')}{m - m'}. \tag{13.10}$$

Clearly, in the product, the magnetic quantum numbers run over all allowed values, except m. The highest power of $\mathbf{h} \cdot \mathbf{s}$ occurring in (13.10) is $2s$.

For spin $s = \frac{1}{2}$, the projection operators are

$$P^{(1/2)} = \frac{1}{2} + \mathbf{h} \cdot \mathbf{s}, \quad P^{(-1/2)} = \frac{1}{2} - \mathbf{h} \cdot \mathbf{s}. \tag{13.11}$$

It is understood, that additive numbers, like the $\frac{1}{2}$ here, have to be multiplied by the appropriate unit matrix, when the spin operators are represented by matrices.

For $s = 1$, the projection operators are

$$P^{(\pm 1)} = \frac{1}{2} \mathbf{h} \cdot \mathbf{s} (1 \pm \mathbf{h} \cdot \mathbf{s}), \quad P^{(0)} = (1 - \mathbf{h} \cdot \mathbf{s})(1 + \mathbf{h} \cdot \mathbf{s}). \tag{13.12}$$

13.3 Irreducible Spin Tensors

13.3.1 Defintions and Examples

The ℓ-rank irreducible tensor constructed from the components of the spin operator $\mathbf{s}$ is the symmetric traceless tensor

$$\overline{s_{\mu_1} s_{\mu_2} \cdots s_{\mu_\ell}} = \Delta^{(\ell)}_{\mu_1 \mu_2 \cdots \mu_\ell, \nu_1 \nu_2 \cdots \nu_\ell} s_{\nu_1} s_{\nu_2} \cdots s_{\nu_\ell}. \tag{13.13}$$

Here the symmetrization matters. This is in contradistinction to tensors constructed from vectors whose components commute.

The second rank irreducible tensor is explicitly given by

$$\overline{s_\mu s_\nu} = \frac{1}{2}(s_\mu s_\nu + s_\mu s_\nu) - \frac{1}{3} s(s+1)\delta_{\mu\nu}. \tag{13.14}$$

As stated before, for a spin s, the irreducible tensors of ranks $\ell \geq 2s + 1$ vanish. For $\ell = 2s + 1$, the proof is indicated as follows. Multiplication of the left-hand side of the Hamilton-Cayley relation (13.7) by the irreducible tensor $\overline{h_{\mu_1} h_{\mu_2} \cdots h_{\mu_{(2s+1)}}}$ and subsequent integration over $d^2 h$ picks out the highest order term

$$(\mathbf{h} \cdot \mathbf{s})^{(2s+1)} = h_{\nu_1} s_{\nu_1} h_{\nu_2} s_{\nu_2} \cdots h_{\nu_{2s+1}} s_{\nu_{(2s+1)}}$$

in the product, because terms of lower order in $\mathbf{h}$ give no contribution in the integral. The only non-vanishing term is equivalent to

$$\int \overline{h_{\mu_1} h_{\mu_2} \cdots h_{\mu_{(2s+1)}}} \, h_{\nu_1} h_{\nu_2} \cdots h_{\nu_{(2s+1)}} d^2 h \, s_{\nu_1} s_{\nu_2} \cdots s_{\nu_{(2s+1)}}.$$

By analogy to (12.1), the integral is equal to the isotropic tensor $\Delta^{(2s+1)}_{\ldots\ldots}$, apart from numerical factors. Multiplication of this tensor with the product of the Cartesian components of $\mathbf{s}$, cf. (13.13) yields the irreducible spin tensor of rank $2s + 1$. Thus the Hamilton-Cayley relation implies

$$\overline{s_{\mu_1} s_{\mu_2} \cdots s_{\mu_{(2s+1)}}} = 0. \tag{13.15}$$

For particles with spin s, the existing irreducible spin-tensors are of ranks $2s$ or smaller.

13.2 Exercise: Verify a Relation Peculiar for Spin 1/2

For spin $s = 1/2$, the peculiar relation

$$s_\mu s_\nu = \frac{-i}{2} \varepsilon_{\mu\nu\lambda} s_\lambda + \frac{1}{4} \delta_{\mu\nu}$$

holds true. To prove it, start from $\overline{s_\mu s_\nu} = 0$, for $s = 1/2$, and use the commutation relation.

13.3.2 Commutation Relation for Spin Tensors

The angular momentum commutation relation (13.1) for spin vectors leads to a generalization for irreducible spin tensors, which reads

$$[\overline{s_{\mu_1} s_{\mu_2} \cdots s_{\mu_\ell}}, s_\lambda]_- = i\,\ell\,\square^{(\ell)}_{\mu_1\mu_2\cdots\mu_\ell,\lambda,\nu_1\nu_2\cdots\nu_\ell} \overline{s_{\nu_1} s_{\nu_2} \cdots s_{\nu_\ell}}. \tag{13.16}$$

For $\ell = 1$, this relation is identical to (13.1). The case $\ell = 2$ corresponds to

$$[\overline{s_\mu s_\nu}, s_\lambda]_- = 2\,i\,\square_{\mu\nu,\lambda,\alpha\beta} \overline{s_\alpha s_\beta}. \tag{13.17}$$

In the *Heisenberg picture*, the time dependence of an operator $\mathscr{O}$ is governed by the commutator with the relevant Hamilton operator $\mathscr{H}$:

$$\frac{d\mathscr{O}}{dt} = \frac{i}{\hbar} [\mathscr{H}, \mathscr{O}]_-. \tag{13.18}$$

For particles with a magnetic moment μs_λ, in the presence of a magnetic field $B_\lambda = B h_\lambda$, where **h** is unit vector, one has $\mathscr{H} = -\mu B h_\lambda s_\lambda$. Then the commutation relation (13.16) for the ℓ-th rank spin tensor implies

$$\frac{d}{dt} \overline{s_{\mu_1} s_{\mu_2} \cdots s_{\mu_\ell}} = -\ell \, \omega_B \, h_\lambda \, \square^{(\ell)}_{\mu_1 \mu_2 \cdots \mu_\ell, \lambda, \nu_1 \nu_2 \cdots \nu_\ell} \overline{s_{\nu_1} s_{\nu_2} \cdots s_{\nu_\ell}}, \tag{13.19}$$

with the precession frequency $\omega_B = \mu B / \hbar$.

The commutation relation for two second rank spin tensors is

$$[\overline{s_\mu s_\nu}, \overline{s_\kappa s_\lambda}]_- = i \left\{ \square_{\mu\nu,\lambda,\alpha\beta} (s_\kappa \overline{s_\alpha s_\beta} + \overline{s_\alpha s_\beta} \, s_\kappa) \right.$$
$$\left. + \square_{\mu\nu,\kappa,\alpha\beta} (s_\lambda \overline{s_\alpha s_\beta} + \overline{s_\alpha s_\beta} \, s_\lambda) \right\}. \tag{13.20}$$

Commutators of this type occur in applications, when the Hamilton operator involves the second rank spin tensor, as in the case of a nucleus with an electric quadrupole moment, e.g. the deuteron, in the presence of an electric field.

13.3.3 Scalar Products

The scalar product or total contraction of two irreducible spin tensors of rank ℓ is given by

$$\overline{s_{\mu_1} s_{\mu_2} \cdots s_{\mu_\ell}} \; \overline{s_{\mu_1} s_{\mu_2} \cdots s_{\mu_\ell}} = \frac{\ell!}{(2\ell - 1)!!} S_0^2 \, S_1^2 \cdots S_{\ell-1}^2. \tag{13.21}$$

The factor $N_\ell = \frac{\ell!}{(2\ell-1)!!}$ occurs in the corresponding expression for the contraction of irreducible tensors constructed from vectors with commuting components, cf. (9.10) and (9.11). The factors

$$S_k^2 = s(s+1) - \frac{k}{2}\left(\frac{k}{2} + 1\right), \tag{13.22}$$

reflect quantum mechanical features of the spin. Notice, one has $S_k^2 = 0$ for $k = 2s$, and the norm of the irreducible spin tensor of rank $2s + 1$ is zero, in accord with (13.15). On the other hand, for $s \gg 1$ and $\ell \ll s$, the product $S_0^2 S_1^2 \cdots S_{\ell-1}^2$ approaches its 'classical' value $s^{2\ell}$.

For $\ell = 2$, (13.21) corresponds to

$$\overline{s_\mu s_\nu}\;\overline{s_\mu s_\nu} = \frac{2}{3}\,S_0^2\,S_1^2, \quad S_0^2 = s\,(s+1), \quad S_1^2 = S_0^2 - \frac{3}{4}. \tag{13.23}$$

Clearly, one has $S_1^2 = 0$ for $s = 1/2$.

Some relations, where the contraction is just over one subscript, also follow from the properties of the spin operator, viz.

$$s_\lambda\,s_\mu s_\lambda = (S_0^2 - 1)\,s_\mu, \quad s_\lambda\,\overline{s_\mu s_\lambda} = \overline{s_\mu s_\lambda}\,s_\lambda = \frac{2}{3}\,S_1^2\,s_\mu, \tag{13.24}$$

$$s_\lambda\,\overline{s_\mu s_\nu}\,s_\lambda = (S_0^2 - 3)\,\overline{s_\mu s_\nu}, \tag{13.25}$$

$$\overline{s_\mu s_\lambda}\;\overline{s_\lambda s_\nu} = \frac{1}{3}\left(S_0^2 - \frac{2}{3}\right)\overline{s_\mu s_\nu} + \frac{i}{2}\varepsilon_{\nu\lambda\kappa}\,\overline{s_\mu s_\lambda}\,s_\kappa + \frac{i}{3}S_1^2\varepsilon_{\mu\nu\kappa}\,s_\kappa + \frac{2}{9}S_0^2 S_1^2 \delta_{\mu\nu}. \tag{13.26}$$

13.4 Spin Traces

13.4.1 Traces of Products of Spin Tensors

In the following, spin traces are denoted by the symbol tr. When spin operators are represented by matrices, the tr-operation corresponds to the standard summation over diagonal elements. For a spin s, the trace of the relevant unit matrix is $2s + 1$, i.e. it is equal to the number of magnetic sub-states. The expression

$$\frac{1}{2s + 1}\,\mathrm{tr}\{\ldots\}$$

is equivalent to an orientational average. For classical variables, as discussed in Chap. 12 , e.g. in connection with the integration over the unit sphere, cf. Sect. 12.1, or over the directions of the velocity, cf. Sect. 12.3.1, a continuum of directions is possible. In contradistinction, the spin allows a discrete set of directions only. As demonstrated above, results of classical orientational averages are obtained without performing explicit integrations over angle variables, when symmetry properties are employed. In the same spirit, the traces of products of spin operators are found, in the following, without using an explicit matrix representation. The trace operation is a rotationally invariant process. Consequently, isotropic tensors come into play again.

By analogy to symmetry arguments which lead to (12.1), the trace of the product of two irreducible spin tensors is given by

$$\frac{1}{2s+1}\,\mathrm{tr}\{\overbrace{s_{\mu_1}s_{\mu_2}\cdots s_{\mu_\ell}}\,\overbrace{s_{\nu_1}s_{\nu_2}\cdots s_{\nu_{\ell'}}}\}$$

$$= \frac{\ell!}{(2\ell+1)!!}\,S_0^2 S_1^2 \cdots S_{l-1}^2 \delta_{\ell\ell'}\,\Delta^{(\ell)}_{\mu_1\mu_2\cdots\mu_\ell,\nu_1\nu_2\cdots\nu_\ell}. \qquad (13.27)$$

For S_k^2 see (13.22). As a consequence, one has

$$\mathrm{tr}\{\overbrace{s_{\mu_1}s_{\mu_2}\cdots s_{\mu_\ell}}\} = 0, \qquad (13.28)$$

for $\ell \geq 1$.

Special cases of (13.27) are, for $\ell = \ell' = 1$,

$$\frac{1}{2s+1}\,\mathrm{tr}\{s_\mu s_\nu\} = \frac{1}{3}\,S_0^2\,\delta_{\mu\nu}, \qquad (13.29)$$

and for $\ell = \ell' = 2$,

$$\frac{1}{2s+1}\,\mathrm{tr}\{\overbrace{s_\mu s_\nu}\,\overbrace{s_\lambda s_\kappa}\} = \frac{2}{15}\,S_0^2 S_1^2\,\Delta_{\mu\nu,\lambda\kappa}. \qquad (13.30)$$

13.4.2 Triple Products of Spin Tensors

No classical analogue exists for the trace of the product of the spin vector components and two irreducible spin tensors of rank ℓ. Due to symmetry considerations, the trace must be proportional to the $\square^{(\ell)}$-tensor. The result is

$$\frac{1}{2s+1}\,\mathrm{tr}\{\overbrace{s_{\mu_1}s_{\mu_2}\cdots s_{\mu_\ell}}\,s_\lambda\,\overbrace{s_{\nu_1}s_{\nu_2}\cdots s_{\nu_\ell}}\}$$

$$= i\frac{\ell}{2}\frac{\ell!}{(2\ell+1)!!}\,S_0^2 S_1^2 \cdots S_{l-1}^2\,\square^{(\ell)}_{\mu_1\mu_2\cdots\mu_\ell,\lambda,\nu_1\nu_2\cdots\nu_\ell}. \qquad (13.31)$$

Special cases of this expression for $\ell = 1$ and $\ell = 2$ are

$$\frac{1}{2s+1}\,\mathrm{tr}\{s_\mu\,s_\lambda\,s_\nu\} = \frac{i}{6}\,S_0^2\,\varepsilon_{\mu\lambda\nu}, \qquad (13.32)$$

and

$$\frac{1}{2s+1}\,\mathrm{tr}\{\overbrace{s_\mu s_\nu}\,s_\lambda\,\overbrace{s_\kappa s_\sigma}\} = \frac{2\,i}{15}\,S_0^2 S_1^2\,\square_{\mu\nu,\lambda,\kappa\sigma}. \qquad (13.33)$$

13.4.3 Multiple Products of Spin Tensors

The trace of the fourfold product of the spin is similar to the classical expression (12.6). Yet it differs due to the non-commutativity of the spin components:

$$\frac{1}{2s+1}\,\mathrm{tr}\{s_\mu s_\nu s_\lambda s_\kappa\} = \frac{S_0^2}{30}\left[2\,S_2^2\,\delta_{\mu\lambda}\delta_{\nu\kappa} + (2\,S_0^2 + 1)\,(\delta_{\mu\nu}\delta_{\lambda\kappa} + \delta_{\mu\kappa}\delta_{\nu\lambda})\right].$$

(13.34)

To check the numerical factor on the right hand side, put λ equal to κ and compare with (13.29).

The trace of the triple product of second rank spin tensors, analogous to (12.7), is

$$\frac{1}{2s+1}\,\mathrm{tr}\{\overline{s_\mu s_\nu}\;\overline{s_\lambda s_\kappa}\;\overline{s_\sigma s_\tau}\} = \frac{8}{105}\,S_0^2 S_1^2 S_3^2\,\Delta_{\mu\nu,\lambda\kappa,\sigma\tau}.$$

(13.35)

Two trace formulas are listed which involve fourfold products of second rank spin tensors, contracted such that the results are scalars:

$$\frac{1}{2s+1}\,\mathrm{tr}\{\overline{s_\mu s_\nu}\;\overline{s_\lambda s_\kappa}\;\overline{s_\mu s_\nu}\;\overline{s_\lambda s_\kappa}\} = S_0^2 S_1^2\left(7 + \frac{4}{9}S_0^4 - \frac{13}{3}S_0^2\right),$$

(13.36)

$$\frac{1}{2s+1}\,\mathrm{tr}\{\overline{s_\mu s_\nu}\,[\,\overline{s_\lambda s_\kappa}\,,\,\overline{s_\mu s_\nu}\,]_-\,\overline{s_\lambda s_\kappa}\} = S_0^2 S_1^2\left(4S_0^2 - 7\right).$$

(13.37)

13.5 Density Operator

13.5.1 Spin Averages

The orientation of the spins, in an ensemble, is described by the *spin density operator* ρ. Just as the spin operator of a particle with spin s, it can be represented by a hermitian $(2s+1) \times (2s+1)$ matrix. For this reason, the ρ is also called *spin density matrix*. In many applications, there is no need for an explicit matrix notation. Alternatively, $\rho = \rho(\mathbf{s})$ is considered as a function of the spin operator and its algebraic properties, as given above, are used. In the following, ρ is normalized to 1, viz. it obeys the condition $\mathrm{tr}\{\rho\} = 1$.

The average $\langle\Psi\rangle$ of a function $\Psi = \Psi(\mathbf{s})$, assumed to be a polynomial of the spin operator $\mathbf{s}$, is determined by

$$\langle\Psi\rangle = \mathrm{tr}\{\Psi(\mathbf{s})\,\rho(\mathbf{s})\},\quad \mathrm{tr}\{\rho(\mathbf{s})\} = 1.$$

(13.38)

The density operator can be expressed as a linear combination of the projection operators $P^{(m)}$, introduced in Sect. 13.2.2, viz.

$$\rho = \sum_{m=-s}^{s} c_m \, P^{(m)}, \quad c_m = \langle P^{(m)} \rangle, \quad \sum_{m=-s}^{s} c_m = 1. \qquad (13.39)$$

Notice that $\mathrm{tr}\{P^{(m)}\} = 1$. The coefficients c_m determine the relative weight of the magnetic sub-state m. For a system of N particles, where $N^{(m)}$ particles are in the sub-state m, one has

$$c_m = \frac{N^{(m)}}{N}, \quad N = \sum_{m=-s}^{s} N^{(m)}. \qquad (13.40)$$

The (13.39) is analogous to the representation of a vector as a linear combination of basis vectors. The projectors $P^{(m)}$ play the role of the basis vectors, the coefficients c_m are the relevant components.

An ensemble of spins is said to be *unpolarized*, when all magnetic sub-states occur equally, i.e. when $c_m = 1/(2s + 1)$, for all m. This means ρ does not depend on $\mathbf{s}$, it is just proportional to the unit operator $\mathbf{1}$. Due to the normalization condition, the density operator ρ_0 for an unpolarized state is

$$\rho_0 = \frac{1}{2s+1} \mathbf{1}. \qquad (13.41)$$

Averages in the unpolarized state are denoted by $\langle \ldots \rangle_0$. The trace formulas presented above are such averages. Averages in a partially polarized state are discussed in the next section.

13.5.2 Expansion of the Spin Density Operator

The deviation of the quantum mechanical spin density operator ρ from its isotropic or unpolarized state can be expanded with respect to irreducible spin tensors. This is similar to the description of the deviation from isotropy of the orientational distribution function $f(\mathbf{u})$ of Sect. 12.2. There, the expansion is with respect to irreducible tensors of rank ℓ, constructed from the components of the classical vector $\mathbf{u}$. In the classical case, in principle, tensors of all ranks, from $\ell = 1$ up to $\ell = \infty$ are needed for a complete characterization, cf. (12.6). For spin s, on the other hand, irreducible tensors of ranks $\ell > \ell_{max} \equiv 2s$ vanish. Thus in the spin case, the expansion runs from $\ell = 1$ up to $\ell = 2s$. Here the expansion is formulated as

$$\rho(\mathbf{s}) = \rho_0 \left(1 + \Phi\right), \quad \rho_0 = (2s + 1)^{-1}, \quad \Phi = \sum_{\ell=1}^{2s} b_{\mu_1\mu_2\cdots\mu_\ell} \overline{s_{\mu_1} s_{\mu_2} \cdots s_{\mu_\ell}}.$$

$$(13.42)$$

Clearly, Φ is the deviation of ρ from the isotropic density operator ρ_0. The expansion coefficients $b_{...}$ are the moments of the density. Computation of the average of the ℓ-th rank spin tensor with ρ given by (13.42) and use of the trace formula (13.27) yields

$$\langle \overline{s_{\mu_1} s_{\mu_2} \cdots s_{\mu_\ell}} \rangle \equiv \mathrm{tr}\{\overline{s_{\mu_1} s_{\mu_2} \cdots s_{\mu_\ell}}\, \rho(\mathbf{s})\} = \frac{\ell!}{(2\ell+1)} S_0^2 S_1^2 \cdots S_{\ell-1}^2 \, b_{\mu_1\mu_2\cdots\mu_\ell}.$$

$$(13.43)$$

The *tensor polarization* $P_{\mu_1\mu_2\cdots\mu_\ell}$ of rank ℓ is defined by

$$P_{\mu_1\mu_2\cdots\mu_\ell} \equiv s^{-\ell} \langle \overline{s_{\mu_1} s_{\mu_2} \cdots s_{\mu_\ell}} \rangle, \quad \ell \geq 1. \qquad (13.44)$$

In terms of these tensor polarizations, the expansion (13.42) is equivalent to

$$\rho(\mathbf{s}) = \rho_0 \left[1 + \sum_{\ell=1}^{2s} \frac{s^\ell (2\ell+1)!!}{\ell! \, S_0^2 S_1^2 \cdots S_{\ell-1}^2} P_{\mu_1\mu_2\cdots\mu_\ell} \overline{s_{\mu_1} s_{\mu_2} \cdots s_{\mu_\ell}} \right]. \qquad (13.45)$$

For $\ell = 1$, the vector P_μ occurring here is called *vector polarization*. It is the only type of polarization possible for particles with spin $s = 1/2$. Of course, particles with a larger spin may also have a vector polarization. Frequently, the term tensor polarization is used for $P_{\mu\nu}$, the case corresponding to $\ell = 2$. Particles with spin $s = 1$, or with a higher spin, can have this type of tensor polarization. The special cases $s = 1/2$ and $s = 1$ are discussed next.

13.5.3 Density Operator for Spin 1/2 and Spin 1

Electrons, protons and neutrons have spin 1/2, For them, the spin density operator reads

$$\rho(\mathbf{s}) = \frac{1}{2} [1 + 2 P_\mu s_\mu]. \qquad (13.46)$$

It is understood, that additive numbers, like the $\frac{1}{2}$ here, have to be multiplied by the appropriate unit matrix, when the spin operators are represented by matrices.

Let $N^{(1/2)}$ and $N^{(-1/2)}$ be the number of particles in the magnetic substates $m = \pm 1/2$, $N = N^{(1/2)} + N^{(-1/2)}$ is the total number of particles. The relative numbers $c_{\pm 1/2} = N^{(\pm 1/2)}/N$ are determined by the averages $\langle P^{(\pm 1/2)} \rangle$ of the projection operators given in (13.11). Thus the relative difference of the occupation numbers is a measure for the degree of polarization of spin $1/2$ particles. It is determined by

$$c_{1/2} - c_{-1/2} = \frac{N^{(1/2)} - N^{(-1/2)}}{N^{(1/2)} + N^{(-1/2)}} = \langle 2h_\nu s_\nu \rangle = h_\mu P_\mu. \tag{13.47}$$

To obtain the last equality, the spin density (13.46) and the trace formula (13.29) are used. By definition, the coefficients c_m are positive and bounded by 1. Thus one has $-1 \le \mathbf{h} \cdot \mathbf{P} \le 1$. A state with $\mathbf{h} \cdot \mathbf{P} = \pm 1$, associated with $c_{1/2} = 1, c_{-1/2} = 0$ and $c_{1/2} = 0, c_{-1/2} = 1$, respectively, is completely polarized, with respect to the preferential direction $\mathbf{h}$. The cases in between the complete polarizations correspond to a partially polarized state.

For spin $s = 1$, the expression (13.45) reduces to

$$\rho(\mathbf{s}) = \frac{1}{3} \left[1 + \frac{3}{2} P_\mu s_\mu + 3 P_{\mu\nu} \overline{s_\mu s_\nu} \right]. \tag{13.48}$$

To link the vector polarization P_μ and the tensor polarization $P_{\mu\nu}$ with the relative occupation numbers c_1, c_0, c_{-1}, spin $s = 1$ projection operators (13.12) are employed. Notice that $\mathsf{P}^{(1)} - \mathsf{P}^{(-1)} = \mathbf{h} \cdot \mathbf{s}$, furthermore $\mathsf{P}^{(1)} + \mathsf{P}^{(-1)} = (\mathbf{h} \cdot \mathbf{s})^2$, and $\mathsf{P}^{(0)} = 1 - (\mathbf{h} \cdot \mathbf{s})^2$. Thus one obtains

$$c_1 - c_{-1} = \frac{N^{(1)} - N^{(-1)}}{N} = h_\nu \langle s_\nu \rangle = h_\mu P_\mu, \tag{13.49}$$

and

$$c_1 + c_{-1} - 2 c_0 = \frac{N^{(1)} + N^{(-1)} - 2 N^{(0)}}{N} = 3 \overline{h_\mu h_\nu} \langle \overline{s_\mu s_\nu} \rangle = 3 \overline{h_\mu h_\nu} \, P_{\mu\nu}. \tag{13.50}$$

13.3 Exercise: Compute the Tensor Polarization for Spin 1

Compute explicitly the relation (13.50) between relative occupation numbers and the tensor polarization, for spin $s = 1$.

13.6 Rotational Angular Momentum of Linear Molecules, Tensor Operators

13.6.1 Basics and Notation

Within reasonable approximation, the rotation of a linear molecule like H_2, N_2 or CO_2, can be treated as the rotational motion of a stiff *linear rotator*. The unit vector parallel to the axis of the rotor is denoted by $\mathbf{u}$. The rotational angular momentum is perpendicular to $\mathbf{u}$. This also applies to the rotational momentum operator $\hbar\mathbf{J}$, thus one has $\mathbf{J} \cdot \mathbf{u} = 0$. The eigenvalues of $\mathbf{J} \cdot \mathbf{J}$ are $j(j+1)$ where j can have the integer values $0, 1, 2, 3, \ldots$ for hetero-nuclear molecules like HD or CO. For homonuclear molecules, j must be even or odd, depending on the nuclear spins, e.g. one has $j = 0, 2, 4, \ldots$ for para-hydrogen, where the total spin of the two protons is zero, and $j = 1, 3, \ldots$ for ortho-hydrogen the total nuclear spin is 1, in units of $\hbar$. Magnetic quantum numbers m, again in units of $\hbar$, assume the values $-j, -j+1$, $\ldots j - 1, j$. When the z-axis is identified with the quantization direction, one has, in 'bra-ket' notation,

$$J_z |jm\rangle = m |jm\rangle, \quad \mathbf{J} \cdot \mathbf{J} |jm\rangle = j(j+1) |jm\rangle.$$

Here $|jm\rangle$ indicates the quantum mechanical state vector. Despite of this name, the quantity $|jm\rangle$ is not a vector in the sense of being a tensor of rank 1, as defined in Sect. 2.5 . The components of $\mathbf{J}$ obey the angular momentum commutation relations

$$J_\mu J_\nu - J_\nu J_\mu \equiv [J_\mu, J_\nu]_- = i\,\varepsilon_{\mu\nu\lambda}\, J_\lambda. \tag{13.51}$$

In contradistinction to the spin operators $\mathbf{s}$, cf. 13.1, the magnitude of the rotational angular momentum is not fixed and it cannot have half-integer eigenvalues.

13.6.2 Projection into Rotational Eigenstates, Traces

A state with a fixed eigenvalue j is obtained with the help of the projection operator

$$\mathsf{P}^j = \sum_{m=-j}^{j} |jm\rangle\langle jm|. \tag{13.52}$$

Applications may require the projection of an observable $\mathcal{O} = \mathcal{O}(\mathbf{u})$ depending on $\mathbf{u}$, into rotational eigenstates. This means, expressions of the form

$$\mathcal{O}^{jj'} \equiv \mathsf{P}^j\, \mathcal{O}(\mathbf{u})\, \mathsf{P}^{j'}, \tag{13.53}$$

are needed. The cases $j = j'$ and $j \neq j'$ are often called 'diagonal' and 'non-diagonal', without mentioning that these terms just refer to the rotational quantum numbers and not to the magnetic quantum numbers. Examples for O are: the electric dipole moment parallel to $\mathbf{u}$, the electric quadrupole moment or the anisotropic part of a molecular polarizability tensor, proportional to $\overline{\mathbf{u}\mathbf{u}}$. The pertaining applications are dipole and quadrupole radiation, birefringence and light scattering.

The diagonal part of $\mathcal{O}(\mathbf{u})$ can be expressed as a function of the angular momentum operator, viz.

$$\sum_j \mathsf{P}^j \, \mathcal{O}(\mathbf{u}) \, \mathsf{P}^{j'} = \mathcal{O}(\mathbf{J}). \tag{13.54}$$

Examples are presented next, the case $j \neq j'$ is treated in Sect. 13.6.6.

The trace operation Tr involves the summation over the magnetic quantum numbers, just like the trace operation tr for a spin, and an additional summation over the rotational quantum numbers of the diagonal elements of an operator $\mathcal{O}$,

$$\mathrm{Tr}\{\mathcal{O}\} = \sum_j \sum_{m=-j}^{j} \langle jm|\mathcal{O}|jm\rangle = \sum_j \mathrm{tr}\{\mathcal{O}^{jj}\}. \tag{13.55}$$

An unbiased orientational average of an operator corresponds to $(2j+1)^{-1}\,\mathrm{tr}\{..\}$. The formulas for the traces tr of spins **s** also apply for the rotational angular momentum **J**.

13.6.3 Diagonal Operators

Observables $\mathcal{O}(\mathbf{u})$ which are even functions of $\mathbf{u}$ possess a part $\mathcal{O}(\mathbf{u})$ which is diagonal in j. As an instructive example, the second rank irreducible tensor $\overline{u_\mu u_\nu}$ is considered. By symmetry, it should be proportional to the symmetric traceless tensor constructed from the components of **J**. Thus the ansatz

$$(\overline{u_\mu u_\nu})^{jj} = c\,\mathsf{P}^j\,\overline{J_\mu J_\nu}$$

is made. To determine the proportionality coefficient c, multiply this equation by $\overline{J_\mu J_\nu}$ and use $\mathbf{J} \cdot \mathbf{u} = 0$. The left hand side of the equation yields

$$\overline{J_\mu J_\nu}\,\mathsf{P}^j\,\overline{u_\mu u_\nu}\,\mathsf{P}^j = -\frac{1}{3}\,j(j+1)\,\mathsf{P}^j.$$

The right hand side is found to be

$$\overline{J_\mu J_\nu}\,\overline{J_\mu J_\nu}\,\mathsf{P}^j = \frac{2}{3}\,j_0^2\,j_1^2\,\mathsf{P}^j,$$

by analogy to (13.23). One obtains $c = -\frac{1}{2}j_1^{-2}$. The abbreviations

$$j_0^2 = j(j+1), \quad j_1^2 = j_0^2 - \frac{3}{4}$$

are used. Thus the diagonal part $(\overline{u_{m\mu}u_\nu})^{\text{diag}}$ of $\overline{u_\mu u_\nu}$ is given by

$$(\overline{u_\mu u_\nu})^{\text{diag}} = -\frac{1}{2}\left(J^2 - \frac{3}{4}\right)^{-1}\overline{J_\mu J_\nu}. \qquad (13.56)$$

The anisotropic part of the molecular polarizability tensor $\alpha_{\mu\nu}$ is proportional to $\overline{u_\mu u_\nu}$, cf. Sect. 5.3.3 . The average of the diagonal part of this tensor is closely related to the *birefringence* and the *depolarized Rayleigh light scattering* in gases of rotating molecules, [22].

13.6.4 Diagonal Density Operator, Averages

The part of the density operator of a gas of rotating linear molecules, which is diagonal with respect to the rotational quantum numbers, can be considered as a distribution function $\rho = \rho(\mathbf{J})$ depending on the operator $\mathbf{J}$. The average $\langle\Psi\rangle$ of a function $\Psi = \Psi(\mathbf{J})$ is computed according to

$$\langle\Psi\rangle = \text{Tr}\{\Psi(\mathbf{J})\,\rho(\mathbf{J})\} = \sum_j \text{tr}\{\mathsf{P}^j\Psi(\mathbf{J})\,\rho(\mathbf{J})\}, \qquad (13.57)$$

where it is understood that ρ is normalized, viz. $\text{Tr}\{\rho(\mathbf{J})\} = 1$.

In thermal equilibrium, and in the absence of orienting fields, the molecules of a gas have a random orientation of their rotational angular momenta $\mathbf{J}$. The square of the angular momentum J^2 is distributed with the canonical weight factor $\exp[-\mathscr{H}/k_BT]$. The Hamiltonian of a linear rotator with the moment of inertia θ is

$$\mathscr{H} = \frac{1}{2\theta}\hbar^2 J^2.$$

The equilibrium distribution operator is

$$\rho_{\text{eq}} = Z^{-1}\exp[-\mathscr{H}/k_BT] = Z^{-1}\exp\left[-\frac{\hbar^2 J^2}{2\theta\,k_BT}\right], \qquad (13.58)$$

with the 'state sum' Z given by

$$Z = \text{Tr}\,\exp[-\mathscr{H}/k_BT] = \sum_j (2j+1)\,\exp\left[-\frac{\hbar^2 j(j+1)}{2\theta\,k_BT}\right]. \qquad (13.59)$$

Notice that

$$\mathrm{tr}\{\mathsf{P}^j\} = 2\,j + 1.$$

The summation over j has to be taken with the allowed values, which may be all integers, all even or all odd integers, depending on the symmetry and the nuclear spin of the molecules.

By analogy to (12.65) and (13.42), the density operator is written as

$$\rho(\mathbf{J}) = \rho_{\mathrm{eq}}\,(1 + \Phi),\tag{13.60}$$

where the quantity $\Phi = \Phi(t, \mathbf{J}))$ specifies the deviation from the equilibrium. Analogous to the expansion of the orientational and the velocity distribution functions, as well as the spin density operator, it is possible to expand Φ with respect to orthogonal basis tensors which here depend on $\mathbf{J}$. The first term of the expansion, and the most relevant one in some applications, involves the second rank *tensor polarization* proportional to $\overline{\langle J_\mu J_\nu \rangle}$. More specifically, the tensor operator

$$\phi_{\mu\nu}^{\mathrm{T}} = \sqrt{\frac{15}{2}}\, c_0^{-1}\, c(J^2) \left[J^2 \left(J^2 - \frac{3}{4} \right) \right]^{-1/2} \overline{J_\mu J_\nu},\tag{13.61}$$

is introduced, where the superscript T stands for Tensor. It is normalized according to

$$\langle \phi_{\mu\nu}\, \phi_{\mu'\nu'} \rangle_{\mathrm{eq}} = \Delta_{\mu\nu,\mu'\nu'}.$$

Here

$$\langle \ldots \rangle_{\mathrm{eq}} = \mathrm{Tr}\{\ldots \rho_{\mathrm{eq}}\}\tag{13.62}$$

indicates the average evaluated with the isotropic equilibrium density operator ρ_{eq}. The weight function $c(J^2)$ can be chosen appropriately. The normalization requires that $c_0^2 = \langle c^2 \rangle_{\mathrm{eq}}$. The average

$$a_{\mu\nu}^{\mathrm{T}} = \langle \phi_{\mu\nu}^{\mathrm{T}} \rangle\tag{13.63}$$

is referred to as *tensor polarization*. The choice $c = 1$ in (13.61) means that the basis tensor is essentially constructed from a unit vector parallel to $\mathbf{J}$. The case $c(J^2) = [J^2 (J^2 - \frac{3}{4})]^{1/2}$ implies an expansion tensor operator which is similar to the expansion function $\overline{V_\mu V_\nu}$ used for the velocity distribution, just with the classical velocity $\mathbf{V}$ replaced by the angular momentum operator $\mathbf{J}$.

13.6.5 Anisotropic Dielectric Tensor of a Gas of Rotating Molecules

The anisotropic part $\overline{\varepsilon_{\mu\nu}}$ of the dielectric tensor, cf. Sect. 5.3.4, of a gas of rotating linear molecules is related to the average of the anisotropic part of the molecular polarizability tensor, cf. Sect. 5.3.3, by

$$\overline{\varepsilon_{\mu\nu}} = n\,(\alpha_\| - \alpha_\perp)\,\langle(\overline{u_\mu u_\nu})^{\text{diag}}\rangle = -\frac{1}{2}n\,(\alpha_\| - \alpha_\perp)\left\langle\left(J^2 - \frac{3}{4}\right)^{-1}\overline{J_\mu J_\nu}\right\rangle. \tag{13.64}$$

Here $\alpha_\|$ and $\alpha_\perp$ are the polarizability for an electric field parallel and perpendicular to $\mathbf{u}$, respectively, and n is the number density of the gas. The relation (13.56) was used to obtain the second equality in (13.64).

The density operator needed for the evaluation of the averages is given by

$$\rho = \rho_{\text{eq}}\,(1 + a_{\mu\nu}^{\mathrm{T}}\,\phi_{\mu\nu}^{\mathrm{T}} + \cdots),$$

where the dots stand for terms involving higher rank tensors, for $a_{\mu\nu}^{\mathrm{T}}$ and $\phi_{\mu\nu}^{\mathrm{T}}$ see (13.61) and (13.63). The resulting average needed for the dielectric tensor is

$$\overline{\varepsilon_{\mu\nu}} = \varepsilon_a^{\mathrm{T}}\,a_{\mu\nu}^{\mathrm{T}},\quad \varepsilon_a^{\mathrm{T}} = -\frac{1}{2}n\,(\alpha_\| - \alpha_\perp)\sqrt{\frac{2}{15}}\,\xi_{\mathrm{T}},$$

$$\xi_{\mathrm{T}} = \langle c(J^2)^2\rangle_{\text{eq}}^{-1}\left\langle c(J^2)\sqrt{\frac{J^2}{J^2 - \frac{3}{4}}}\right\rangle_{\text{eq}}. \tag{13.65}$$

For $c = 1$ and rotational states $j = 4$ and higher, the factor ξ_{T} approaches 1.

The interrelation (13.65) between the anisotropic part of the dielectric tensor and the tensor polarization $\mathbf{a}^{\mathrm{T}}$ plays a key role in the kinetic theory for the depolarized Rayleigh scattering and the flow birefringence of molecular gases [17, 22, 62–64].

13.6.6 Non-diagonal Tensor Operators

Spherical tensor operators are defined by

$$T_{\ell m}^{jj'} = \sum_{m'}\sum_{m''}(-1)^{j-m''}\,(j'm', j - m''|\ell m)\,|jm''\rangle\langle j'm'|, \tag{13.66}$$

where the symbol $(.., ..|..)$ indicates a Clebsch-Gordan coefficient These coefficients govern the coupling of two angular momentum states $|j_1 m_1\rangle$ and $|j_2 m_2\rangle$ to a state $|jm\rangle$ according to, cf. [65],

$$\sum_{m_1} \sum_{m_2} |j_1 m_1\rangle\, |j_2 m_2\rangle\, (j_1 m_1, j_2 m_2 | jm) = |jm\rangle.$$

The nondiagonal elements of the spherical harmonic $Y_\ell^{(m)}(\mathbf{u})$ are related to the spherical tensor operators by

$$\left[Y_\ell^{(m)}(\mathbf{u})\right]^{jj'} = \mathrm{P}^j Y_\ell^{(m)}(\mathbf{u})\mathrm{P}^{j'} = \sqrt{\frac{2j+1}{4\pi}}\,(j\,0, \ell\,0 | j'\,0)\, T_{\ell m}^{jj'}. \qquad (13.67)$$

By analogy to (13.67), the operator form of the Cartesian tensor $\overline{u_{\mu_1} \cdots u_{\mu_\ell}}$ is

$$(\overline{u_{\mu_1} \cdots u_{\mu_\ell}})^{jj'} = \sqrt{\frac{\ell!}{(2\ell+1)!!}}\,\sqrt{2j+1}\,(j\,0, \ell\,0 | j'\,0)\, T_{\mu_1 \cdots \mu_\ell}^{jj'}. \qquad (13.68)$$

The tensor operators have the properties:

(i) The hermitian adjoint of $T_{\mu_1 \cdots \mu_\ell}^{jj'}$ is

$$(T_{\mu_1 \cdots \mu_\ell}^{jj'})^\dagger = (-1)^{j-j'}\, T_{\mu_1 \cdots \mu_\ell}^{j'j}. \qquad (13.69)$$

(ii) Orthogonality and normalization

$$\mathrm{tr}\{T_{\mu_1 \cdots \mu_\ell}^{jj'}\, (T_{\nu_1 \cdots \nu_{\ell'}}^{jj'})^\dagger\} = \delta_{\ell\ell'}\, \Delta_{\mu_1 \cdots \mu_\ell, \nu_1 \cdots \nu_\ell}^{(\ell)}. \qquad (13.70)$$

The theoretical description of the *rotational Raman scattering* involves the elements of the anisotropic molecular polarizability tensor which are non-diagonal with respect to the rotational quantum number, cf. (13.53). The relevant operators are

$$\mathrm{P}^j\, \overline{u_\mu u_\nu}\, \mathrm{P}^{j'} = (\overline{u_\mu u_\nu})^{jj'},$$

with $j' = j \pm 2$. In particular, one has

$$(\overline{u_\mu u_\nu})^{jj\pm 2} = \sqrt{\frac{2}{15}}\,\sqrt{2j+1}\,(j\,0, 2\,0 | j \pm 2\,0)\, T_{\mu\nu}^{jj\pm 2}. \qquad (13.71)$$

The Clesch-Gordan coefficients are

$$(j\,0, 2\,0 | j + 2\,0) = \sqrt{\frac{3}{2}}\,\sqrt{\frac{(j+1)(j+2)}{(2j+1)(2j+3)}},$$

$$(j\,0, 2\,0 | j - 2\,0) = \sqrt{\frac{3}{2}}\,\sqrt{\frac{j(j-1)}{(2j+1)(2j-1)}}.$$

Both coefficients approach $\frac{1}{2}\sqrt{\frac{3}{2}}$ for large values of j.

The diagonal tensor operator $T^{jj}_{\mu_1\cdots\mu_\ell}$ is essentially the hermitian tensor operator $\mathsf{P}^j\,\overline{J_{\mu_1}\cdots J_{\mu_\ell}}$, viz.

$$T^{jj}_{\mu_1\cdots\mu_\ell} = \mathsf{P}^j\,\overline{J_{\mu_1}\cdots J_{\mu_\ell}}\left(\frac{\ell!}{(2\ell+1)!!}\right)^{-1/2}(2j+1)^{-1/2}(j_0\,j_1\cdots j_{\ell-1})^{-1}.$$

$$(13.72)$$

The quantities j_k are analogousl to the S_k, defined in (13.22), i.e.

$$j_k^2 = j\,(j+1) - \frac{k}{2}\left(\frac{k}{2}+1\right).$$

The second rank tensor is

$$T^{jj}_{\mu\nu} = \left(\frac{15}{2}\right)^{1/2}(2j+1)^{-1/2}(j_0\,j_1)^{-1}\,\mathsf{P}^j\,\overline{J_\mu J_\nu}\,.$$

$$(13.73)$$

Apart from the factor $(2j+1)^{-1/2}$, the tensor operator $T^{jj}_{\mu\nu}$ is equal to $\mathsf{P}^j\phi^T_{\mu\nu}$, as defined in (13.61), with $c=1$.

Chapter 14
Rotation of Tensors

Abstract This chapter is concerned with the active rotation of tensors. Firstly, infinitesimal and finite rotations of vectors are described by second rank rotation tensors, the connection with spherical components is pointed out. Secondly, the rotation of second rank tensors is treated with the help of fourth rank projection tensors. The fourth rank rotation tensor is a linear combination of these projectors. The scheme is generalized to the rotation of tensors of rank $\ell > 2$. Thirdly, the projection tensors are applied to the solution of tensor equations. An example deals with the effect of a magnetic field on the electrical conductivity.

The *active rotation of a tensor*, to be considered here, has to be distinguished from the rotation of the coordinate system, see Sect. 2.4. The passive rotation of the coordinate system must not affect the physics. The rotation of a tensor, on the other hand, describes physical changes. The rotation of vectors is discussed before second and higher rank tensors are treated. This section follows and generalizes the material given in the appendix of [42].

14.1 Rotation of Vectors

14.1.1 Infinitesimal and Finite Rotation

The rotation of a vector **a** by the infinitesimal angle $\delta\varphi$ about an axis, which is parallel to the axial unit vector **h**, generates a vector $\mathbf{a}'$, whose components are given by

$$a'_\mu = a_\mu + \delta\varphi \, \varepsilon_{\mu\lambda\nu} h_\lambda a_\nu = a_\mu + \delta\varphi \, H_{\mu\nu} a_\nu = (\delta_{\mu\nu} + \delta\varphi \, H_{\mu\nu}) a_\nu. \quad (14.1)$$

The antisymmetric second rank tensor H is defined by

$$H_{\mu\nu} \equiv \varepsilon_{\mu\lambda\nu} h_\lambda. \quad (14.2)$$

© Springer International Publishing Switzerland 2015

S. Hess, *Tensors for Physics*, Undergraduate Lecture Notes in Physics,
DOI 10.1007/978-3-319-12787-3_14

The rotation by an finite angle $\varphi = n\delta\varphi$ is given by $(1 + \delta\varphi H)^n$. With $\delta\varphi = \varphi/n$, the limit $n \to \infty$ leads to

$$a'_\mu = (\exp[\varphi\,H])_{\mu\nu}\,a_\nu \equiv R_{\mu\nu}(\varphi)\,a_\nu. \tag{14.3}$$

In principal, the rotation tensor R can be expressed in terms of the power series

$$R_{\mu\nu}(\varphi) = \delta_{\mu\nu} + \varphi\,H_{\mu\nu} + \frac{1}{2}\varphi^2 H_{\mu\kappa}\,H_{\kappa\nu} + \ldots. \tag{14.4}$$

Due to the special properties of H, to be discussed next, R can be represented in a more compact form.

14.1.2 Hamilton Cayley and Projection Tensors

Due to $H_{\mu\kappa}\,H_{\kappa\nu} = h_\mu h_\nu - \delta_{\mu\nu}$ and $h_\sigma\,H_{\sigma\nu} = 0$, the tensor H obeys the relation

$$H^3 + H = 0. \tag{14.5}$$

This corresponds to a Hamilton-Cayley equation for H with the eigenvalues $i\,m$, where $m = 0, \pm1$. Second rank projection tensors $P^{(m)}$ are defined by

$$P^{(m)} = \prod_{m'\neq m} \frac{H - im'\,1}{im - im'}, \quad m, m' = 0, \pm1. \tag{14.6}$$

In (14.6), the symbol **1** stands for the second rank unit tensor, viz. for $\delta_{\mu\nu}$. These projectors are explicitly given by

$$P^{(0)}_{\mu\nu} = h_\mu h_\nu, \quad P^{(\pm1)}_{\mu\nu} = \frac{1}{2}\left(\delta_{\mu\nu} - h_\mu h_\nu \mp i\,\varepsilon_{\mu\lambda\nu}\,h_\lambda\right). \tag{14.7}$$

The projection tensors possess the following properties:

$$P^{(m)}_{\mu\kappa}\,P^{(m')}_{\kappa\nu} = \delta_{mm'}\,P^{(m)}_{\mu\nu}, \tag{14.8}$$

$$(P^{(m)}_{\mu\nu})^* = P^{(-m)}_{\mu\nu} = P^{(m)}_{\nu\mu}, \tag{14.9}$$

$$\sum_{m=-1}^{1} P^{(m)}_{\mu\nu} = \delta_{\mu\nu}, \quad P^{(m)}_{\mu\mu} = 1. \tag{14.10}$$

The eigenvalue equation

$$P^{(m)}_{\mu\kappa}\,H_{\kappa\nu} = H_{\mu\kappa}\,P^{(m)}_{\kappa\nu} = i\,m\,P^{(m)}_{\mu\nu}, \tag{14.11}$$

reflects that the $P_{\mu\nu}^{(m)}$ are 'eigen-tensors' of the tensor $H_{\mu\nu}$. On the other hand, $H_{\mu\nu}$ can be represented as a linear combination of the projection tensors, viz.

$$H_{\mu\nu} = \sum_{m=-1}^{1} i\, m\, P_{\mu\nu}^{(m)}. \tag{14.12}$$

Some additional formulas involving second rank projection operators are:

$$\overline{h_\nu\, h_\mu a_\nu} = \left(\frac{2}{3}P_{\mu\nu}^{(0)} + \frac{1}{2}P_{\mu\nu}^{(1)} + \frac{1}{2}P_{\mu\nu}^{(-1)}\right) a_\nu, \tag{14.13}$$

$$\overline{h_\nu h_\kappa\, h_\mu b_{\nu\kappa}} = \left(\frac{3}{5}P_{\mu\nu}^{(0)} + \frac{8}{15}P_{\mu\nu}^{(1)} + \frac{8}{15}P_{\mu\nu}^{(-1)}\right) b_{\nu\kappa}\, h_\kappa, \tag{14.14}$$

where a_ν and $b_{\mu\nu}$ are a vector and a second rank tensor.

14.1.3 Rotation Tensor for Vectors

With the help of the projection tensors, the rotation tensor for vectors can now be expressed as

$$R_{\mu\nu}(\varphi) = \sum_{m=-1}^{1} P_{\mu\kappa}^{(m)}(\exp[\varphi\, H])_{\kappa\nu} = \sum_{m=-1}^{1} \exp[i\, m\, \varphi]\, P_{\mu\nu}^{(m)}. \tag{14.15}$$

Decomposition into real and imaginary parts leads to

$$R_{\mu\nu}(\varphi) = P_{\mu\nu}^{(0)} + \cos\varphi \left(P_{\mu\nu}^{(1)} + P_{\mu\nu}^{(-1)}\right) + \sin\varphi\, i \left(P_{\mu\nu}^{(1)} - P_{\mu\nu}^{(-1)}\right). \tag{14.16}$$

Notice that

$$P_{\mu\nu}^{(0)} = h_\mu h_\nu \equiv P_{\mu\nu}^{\|}, \quad P_{\mu\nu}^{(1)} + P_{\mu\nu}^{(-1)} = \delta_{\mu\nu} - h_\mu h_\nu \equiv P_{\mu\nu}^{\perp}, \tag{14.17}$$

correspond to the projection tensors onto the direction parallel and perpendicular to **h**, denoted by $P^{\|}$ and $P^{\perp}$, respectively. Furthermore, one has

$$i \left(P_{\mu\nu}^{(1)} - P_{\mu\nu}^{(-1)}\right) = H_{\mu\nu}. \tag{14.18}$$

Thus the rotation tensor also reads

$$R_{\mu\nu}(\varphi) = h_\mu h_\nu + \sin\varphi\, \varepsilon_{\mu\lambda\nu}\, h_\lambda + \cos\varphi\, (\delta_{\mu\nu} - h_\mu h_\nu). \tag{14.19}$$

In hindsight, this result is not unexpected for the rotation of a vector. However, the
formal considerations presented here are suitable for a generalization to the rotation
of tensors.

The orthogonal transformation matrix U for the rotation of the coordinate system,
as introduced in Sect. 2.4.2, is related to R by $U_{\mu\nu}(\varphi) = R_{\mu\nu}(-\varphi)$. To compare with
(2.41), choose $\mathbf{h}$ parallel to the 3-axis, also referred to as to the z-axis.

14.1 Exercise: Scalar Product of two Rotated Vectors
Let $\tilde{a}_\mu = R_{\mu\nu}(\varphi)a_\nu$ and $\tilde{b}_\mu = R_{\mu\kappa}(\varphi)a_\kappa$ be the Cartesian components of the vectors
$\mathbf{a}$ and $\mathbf{b}$ which have been rotated by the same angle φ about the same axis. Prove that
the scalar products $\tilde{\mathbf{a}} \cdot \tilde{\mathbf{b}}$ is equal to $\mathbf{a} \cdot \mathbf{b}$.

14.1.4 Connection with Spherical Components

The complex basis vectors $\mathbf{e}^{(m)}$, introduced by (9.16) and employed with the definition of spherical components, cf. (9.18) and (9.19), are eigenvectors of the projection
tensors, provided that $\mathbf{h}$ is chosen parallel to the unit vector $\mathbf{e}^{(z)}$, more specifically:

$$P_{\mu\nu}^{(m)} e_\nu^{(m')} = \delta_{mm'} e_\mu^{(m)}, \quad m, m' = -1, 0, 1. \tag{14.20}$$

Thus, due to (9.18), application the projector on a vector $\mathbf{a}$ yields

$$P_{\mu\nu}^{(m)} a_\nu = a^{(m)} e_\mu^{(m)}, \tag{14.21}$$

where $a^{(m)}$ is a spherical component of this vector. Furthermore, the projection tensor
can be expressed by

$$P_{\mu\nu}^{(m)} = \left(e_\mu^{(m)}\right)^* e_\nu^{(m)}, \tag{14.22}$$

when one chooses $\mathbf{h} = \mathbf{e}^{(z)}$.

14.2 Rotation of Second Rank Tensors

14.2.1 Infinitesimal Rotation

Let $A_{\mu\nu} = a_\mu a_\nu$ be a second rank tensor composed of the components of the vector
$\mathbf{a}$. The infinitesimal rotation by the angle $\delta\varphi$ about an axis parallel to $\mathbf{h}$, as described
by (14.1) with (14.2) implies that the rotated tensor $A'_{\mu\nu}$ is

$$A'_{\mu\nu} = A_{\mu\nu} + \delta\varphi\,(H_{\mu\mu'}\,\delta_{\nu\nu'} + H_{\nu\nu'}\,\delta_{\mu\mu'})A_{\mu'\nu'} = A_{\mu\nu} + \delta\varphi\,\mathscr{H}_{\mu\nu,\mu'\nu'}\,A_{\mu'\nu'}. \tag{14.23}$$

The fourth rank tensor $\mathscr{H}$ is defined by

$$\mathscr{H}_{\mu\nu,\mu'\nu'} \equiv \varepsilon_{\mu\lambda\mu'}\, h_\lambda\, \delta_{\nu\nu'} + \varepsilon_{\nu\lambda\nu'}\, h_\lambda\, \delta_{\mu\mu'} = H_{\mu\mu'}\, \delta_{\nu\nu'} + H_{\nu\nu'}\, \delta_{\mu\mu'}. \tag{14.24}$$

The infinitesimal rotation of a symmetric tensor S is also described by $\mathscr{H}$, viz.

$$S'_{\mu\nu} = S_{\mu\nu} + \delta\varphi\, \mathscr{H}_{\mu\nu,\mu'\nu'}\, S_{\mu'\nu'}. \tag{14.25}$$

In the following, it is assumed that S is also traceless, i.e. it is an irreducible second rank tensor: $S_{\mu\nu} = \overline{S_{\mu\nu}}$. The rotated tensor $S'_{\mu\nu}$ is also irreducible. Then one has

$$\mathscr{H}_{\mu\nu,\mu'\nu'}\, S_{\mu'\nu'} = \overline{\mathscr{H}_{\mu\nu,\mu'\nu'}\, S_{\mu'\nu'}},$$

and $\mathscr{H}$ is equivalent to

$$\mathscr{H}_{\mu\nu,\mu'\nu'} = 2\, h_\lambda\, \square_{\mu\nu,\lambda,\mu'\nu'}, \tag{14.26}$$

for $\square_{...}$ see (11.19).

14.2.2 Fourth Rank Projection Tensors

Fourth rank projection tensors $\mathscr{P}^{(m_1,m_2)}$ are defined via products of the second rank projectors (14.7):

$$\mathscr{P}^{(m_1,m_2)}_{\mu\nu,\mu'\nu'} = P^{(m_1)}_{\mu\mu'}\, P^{(m_2)}_{\nu\nu'}. \tag{14.27}$$

The fourth rank projectors have the property

$$\mathscr{P}^{(m_1,m_2)}_{\mu\nu,\lambda\kappa}\, \mathscr{P}^{(m'_1,m'_2)}_{\lambda\kappa,\mu'\nu'} = \mathscr{P}^{(m_1,m_2)}_{\mu\nu,\mu'\nu'}\, \delta_{(m_1 m'_1)}\, \delta_{(m_2 m'_2)}. \tag{14.28}$$

The relation

$$\mathscr{H}_{\mu\nu,\mu'\nu'} S_{\mu'\nu'} = \sum_{m_1=-1}^{1} \sum_{m_2=-1}^{1} i\,(m_1+m_2)\, \mathscr{P}^{(m_1,m_2)}_{\mu\nu,\mu'\nu'}\, S_{\mu'\nu'}$$

follows the definition of $\mathscr{H}$ and the properties of H. Insertion of $\mathscr{P}$ into this equation and use of (14.28) leads to

$$\mathscr{H}_{\mu\nu,\lambda\kappa}\, \mathscr{P}^{(m_1,m_2)}_{\lambda\kappa,\mu'\nu'} = i\,(m_1+m_2)\, \mathscr{P}^{(m_1,m_2)}_{\mu\nu,\mu'\nu'}. \tag{14.29}$$

Notice that $m_1 + m_2$ assumes the five values $m = -2, -1, 0, 1, 2$. The fourth rank tensor obeys the eigenvalue equation

$$\prod_{m=-2}^{2} (\mathcal{H} - i\,m\,\mathbf{1}) = 0, \tag{14.30}$$

with these five eigenvalues for m. The corresponding eigen-tensors are

$$\mathscr{P}^{(m)}_{\mu\nu,\mu'\nu'} = \sum_{m_1=-1}^{1} \sum_{m_2=-1}^{1} P^{(m_1)}_{\mu\mu'} P^{(m_2)}_{\nu\nu'} \delta(m, m_1 + m_2), \tag{14.31}$$

where the $\delta(m, m_1 + m_2) = 1$ for $m = m_1 + m_2$, and $\delta(m, m_1 + m_2) = 0$, for $m \neq m_1 + m_2$. In terms of these projectors, the spectral decomposition of $\mathcal{H}$ reads

$$\mathcal{H}_{\mu\nu,\mu'\nu'} = \sum_{m=-2}^{2} i\,m\,\mathscr{P}^{(m)}_{\mu\nu,\mu'\nu'}. \tag{14.32}$$

14.2.3 Fourth Rank Rotation Tensor

By analogy to the rotation of a vector, cf. (14.3), the rotation of a second rank tensor by the finite angle φ is given by

$$A'_{\mu\nu} = (\exp[\varphi\,\mathcal{H}])_{\mu\nu,\mu'\nu'} A_{\mu'\nu'} \equiv \mathscr{R}_{\mu\nu,\mu'\nu'}(\varphi)\, A_{\mu'\nu'}, \tag{14.33}$$

with the fourth rank rotation tensor

$$\mathscr{R}_{\mu\nu,\mu'\nu'}(\varphi) = \sum_{m=-2}^{2} \exp[i\,m]\, \mathscr{P}^{(m)}_{\mu\nu,\mu'\nu'}. \tag{14.34}$$

Decomposition into real and imaginary parts yields, by analogy to (14.16)

$$\mathscr{R}_{\mu\nu,\mu'\nu'}(\varphi) = \mathscr{P}^{(0)}_{\mu\nu,\mu'\nu'} + \sum_{m=1}^{2} \left[\cos(m\varphi) \left(\mathscr{P}^{(m)}_{\mu\nu,\mu'\nu'} + \mathscr{P}^{(-m)}_{\mu\nu,\mu'\nu'} \right) \right.$$
$$\left. + \sin(m\varphi)i \left(\mathscr{P}^{(m)}_{\mu\nu,\mu'\nu'} - \mathscr{P}^{(-m)}_{\mu\nu,\mu'\nu'} \right) \right]. \tag{14.35}$$

The comparison of the formulas for the rotation of second rank tensor with those for the rotation of a vector indicates how the general case of a rotation of a tensor of rank ℓ can be treated.

14.3 Rotation of Tensors of Rank ℓ

The obvious generalization of the generator for the infinitesimal rotation of second rank tensors, viz. (14.24), to tensors of rank ℓ, is the 2ℓth rank tensor $\mathscr{H}^{(\ell)}$ defined by

$$\mathscr{H}^{(\ell)}_{\mu_1\mu_2\cdots\mu_{(\ell)},\nu_1\nu_2\cdots\nu_\ell} \tag{14.36}$$

$$\equiv \varepsilon_{\mu_1\lambda\nu_1} h_\lambda \delta_{\mu_2\nu_2} \cdots \delta_{\mu_\ell\nu_\ell} + \cdots + \delta_{\mu_1\nu_1} \cdots \delta_{\mu_{\ell-1}\nu_{\ell-1}} \varepsilon_{\mu_\ell\lambda\nu_\ell} h_\lambda$$

$$= H_{\mu_1\nu_1} \delta_{\mu_2\nu_2} \cdots \delta_{\mu_\ell\nu_\ell} + \delta_{\mu_1\nu_1} H_{\mu_2\nu_2} \cdots \delta_{\mu_\ell\nu_\ell} + \cdots + \delta_{\mu_1\nu_1} \cdots \delta_{\mu_{\ell-1}\nu_{\ell-1}} H_{\mu_\ell\nu_\ell}.$$

The infinitesimal rotation of a symmetric tensor S is also described by $\mathscr{H}^{(\ell)}$, viz.

$$S'_{\mu_1\mu_2\cdots\mu_\ell} = S_{\mu_1\mu_2\cdots\mu_\ell} + \delta\varphi\, \mathscr{H}^{(\ell)}_{\mu_1\mu_2\cdots\mu_{(\ell)},\nu_1\nu_2\cdots\nu_\ell} S_{\nu_1\nu_2\cdots\nu_\ell}. \tag{14.37}$$

In the following, it is assumed that S is also traceless, i.e. it is an irreducible tensor. The rotated tensor S' is also irreducible. Then one has

$$\mathscr{H}^{(\ell)} \odot S = \overline{\mathscr{H}^{(\ell)} \odot S},$$

where the symbol $\odot$ indicates the ℓ-fold contraction as occurring in the equations above, and $\mathscr{H}^{(\ell)}$ is equivalent to

$$\mathscr{H}^{(\ell)}_{\mu_1\mu_2\cdots\mu_{(\ell)},\nu_1\nu_2\cdots\nu_\ell} = \ell\, h_\lambda\, \square^{(\ell)}_{\mu_1\mu_2\cdots\mu_{(\ell)},\lambda,\nu_1\nu_2\cdots\nu_\ell}, \tag{14.38}$$

for $\square_{...}$ see (11.16).

Projection tensors $\mathscr{P}^{(m)}$, where $m = -\ell, -\ell+1, \ldots, 0, \ldots, \ell-1, \ell$, are defined by analogy to (14.31). These projectors are eigen-tensors of $\mathscr{H}^{(\ell)}$, viz.

$$\mathscr{H}^{(\ell)} \odot \mathscr{P}^{(m)} = \mathscr{P}^{(m)} \odot \mathscr{H}^{(\ell)} = i\, m\, \mathscr{P}^{(m)}. \tag{14.39}$$

In terms of these projectors, the spectral decomposition of $\mathscr{H}^{(\ell)}$ reads

$$\mathscr{H}^{(\ell)}_{\mu_1\mu_2\cdots\mu_{(\ell)},\nu_1\nu_2\cdots\nu_\ell} = \sum_{m=-\ell}^{\ell} i\, m\, \mathscr{P}^{(m)}_{\mu_1\mu_2\cdots\mu_{(\ell)},\nu_1\nu_2\cdots\nu_\ell}. \tag{14.40}$$

The obvious generalizations of equations (14.34) and (14.35) describing the rotation of second rank tensors to those of ℓth rank tensors is

$$\mathscr{R}^{(\ell)}_{\mu_1\mu_2\cdots\mu_{(\ell)},\nu_1\nu_2\cdots\nu_\ell}(\varphi) = \sum_{m=-\ell}^{\ell} \exp[i\, m]\, \mathscr{P}^{(m)}_{\mu_1\mu_2\cdots\mu_{(\ell)},\nu_1\nu_2\cdots\nu_\ell}, \tag{14.41}$$

and

$$\mathscr{R}^{(\ell)}_{\cdots,\cdots}(\varphi) = \mathscr{P}^{(0)}_{\cdots,\cdots} + \sum_{m=1}^{\ell}\left[(\cos(m\varphi)\left(\mathscr{P}^{(m)}_{\cdots,\cdots} + \mathscr{P}^{(-m)}_{\cdots,\cdots}\right)\right.$$

$$\left. + \sin(m\varphi)i\left(\mathscr{P}^{(m)}_{\cdots,\cdots} - \mathscr{P}^{(-m)}_{\cdots,\cdots}\right)\right]. \tag{14.42}$$

The projection operators also allow the solution of tensor equations, as discussed next.

14.4 Solution of Tensor Equations

14.4.1 Inversion of Linear Equations

Let $a_{\mu_1\cdot\mu_\ell}$ and $b_{\mu_1\cdot\mu_\ell}$ be irreducible tensors of rank ℓ which obey the rotation-like linear relation

$$a_{\mu_1\mu_2\cdots\mu_\ell} + \varphi\,\mathscr{H}^{(\ell)}_{\mu_1\mu_2\cdots\mu_{(\ell)},\,\nu_1\nu_2\cdots\nu_\ell}a_{\nu_1\nu_2\cdots\nu_\ell} = c_0\,b_{\mu_1\mu_2\cdots\mu_\ell}, \tag{14.43}$$

where c_0 is a given coefficient. With the properties of the projectors $\mathscr{P}^{(m)}$ given above, this equation is inverted for $a_{\mu_1\cdot\mu_\ell}$ according to

$$a_{\mu_1\mu_2\cdots\mu_\ell} = \sum_{m=-\ell}^{\ell} c^{(m)}\,\mathscr{P}^{(m)}_{\mu_1\mu_2\cdots\mu_{(\ell)},\nu_1\nu_2\cdots\nu_\ell}\,b_{\mu_1\mu_2\cdots\mu_\ell}, \tag{14.44}$$

with

$$c^{(m)} = c_0\,(1 + m\,i\,\varphi)^{-1}. \tag{14.45}$$

When a more general linear relation between two tensors is cast into the form

$$\sum_{m=-\ell}^{\ell} c^{(m)}\,\mathscr{P}^{(m)}_{\mu_1\mu_2\cdots\mu_{(\ell)},\nu_1\nu_2\cdots\nu_\ell}\,a_{\mu_1\mu_2\cdots\mu_\ell} = b_{\mu_1\mu_2\cdots\mu_\ell}, \tag{14.46}$$

with given coefficients $c^{(m)}$, the inversion of this equation reads

$$a_{\mu_1\mu_2\cdots\mu_\ell} = \sum_{m=-\ell}^{\ell} \left(c^{(m)}\right)^{-1}\mathscr{P}^{(m)}_{\mu_1\mu_2\cdots\mu_{(\ell)},\nu_1\nu_2\cdots\nu_\ell}\,b_{\mu_1\mu_2\cdots\mu_\ell}, \tag{14.47}$$

A simple application, for $\ell = 1$, is the computation of the electrical conductivity in the presence of a magnetic field, as discussed next. The case of the fourth rank viscosity tensor of a fluid in the presence of a magnetic field, is treated in Sect. 16.3.2.

14.4.2 Effect of a Magnetic Field on the Electrical Conductivity

In a stationary situation, the linear relation between the electric flux density $\mathbf{j}$ and an applied electric field $\mathbf{E}$ is described by

$$j_\mu = \sigma_{\mu\nu} E_\nu, \tag{14.48}$$

where $\sigma_{\mu\nu}$ is the electrical conductivity tensor. For the isotropic case, where $\sigma_{\mu\nu} \sim \delta_{\mu\nu}$, this corresponds to the local formulation of Ohm's law. The influence of a magnetic field $\mathbf{B} = B\mathbf{h}$, with $\mathbf{h} \cdot \mathbf{h} = 1$, on the conductivity is analyzed next for a simple model. Consider the case of single carriers with mass m, charge e, number density n and an average velocity $\mathbf{v}$, then the flux density is $\mathbf{j} = ne\mathbf{v}$. The velocity is assumed to obey the damped equation of motion

$$m\dot{\mathbf{v}} = e(\mathbf{E} + \mathbf{v} \times \mathbf{B}) - m\,\tau^{-1}\,\mathbf{v},$$

where τ is a relaxation time. For a stationary situation, the time derivative $\dot{\mathbf{v}}$ vanishes and the equation above for $\mathbf{v}$ reduces to an expression of the type (14.43), just for $\ell = 1$, viz.

$$v_\mu + \varphi H_{\mu\nu} v_\nu = c_0 E_\mu, \tag{14.49}$$

with $\varphi = e\,B\,\tau/m$ and $c_0 = e\,\tau/m$. The solution of this equation for $\mathbf{v}$, cf. (14.44), is

$$v_\mu = \frac{e\tau}{m} \sum_{k=-1}^{1} (1 + ik\,\varphi)^{-1} P_{\mu\nu}^{(k)} E_\nu. \tag{14.50}$$

Thus the dc-conductivity tensor is

$$\sigma_{\mu\nu} = \frac{ne^2\tau}{m} \sum_{k=-1}^{1} \left(1 + ik\frac{eB\tau}{m}\right)^{-1} P_{\mu\nu}^{(k)}. \tag{14.51}$$

This result is equivalent to

$$\sigma_{\mu\nu} = \sigma^{\parallel} h_\mu h_\nu + \sigma^{\perp}(\delta_{\mu\nu} - h_\mu h_\nu) + \sigma^{\text{trans}} \varepsilon_{\mu\lambda\nu} h_\lambda, \tag{14.52}$$

with the longitudinal, perpendicular and transverse conductivity coefficients determined by

$$\sigma^{\parallel} = \sigma_0 \equiv n\,e^2\,\tau/m, \quad \sigma^{\perp} = \sigma_0\,(1+\varphi^2)^{-1}, \quad \sigma^{\text{trans}} = \sigma_0\,\varphi\,(1+\varphi^2)^{-1}. \quad (14.53)$$

The magnetic field is an axial vector, the same applies for $\mathbf{h}$. The constitutive law (14.48), with the conductivity tensor given here conserves parity. Notice that the conductivity tensor has the symmetry property

$$\sigma_{\mu\nu}(\mathbf{h}) = \sigma_{\nu\mu}(-\mathbf{h}). \quad (14.54)$$

Since $\tau > 0$, one has $\sigma^{\parallel} > 0$ and $\sigma^{\perp} > 0$. The relation of the associated longitudinal and perpendicular parts of the current density with the electric field violate time-reversal invariance, typical for an irreversible process. The transverse coefficient which underlies the Hall-effect, is of reversible character, the coefficient σ^{trans} may have either sign.

14.5 Additional Formulas Involving Projectors

The application of the fourth rank projection tensor on the symmetric traceless tensor $a_{\mu\nu}$ is explicitly given by

$$\mathscr{P}^{(0)}_{\mu\nu,\mu'\nu'}\,a_{\mu'\nu'} = \frac{3}{2}\,\overline{h_\mu h_\nu}\,h_{\mu'}h_{\nu'}a_{\mu'\nu'}, \quad (14.55)$$

$$\mathscr{P}^{(\pm 1)}_{\mu\nu,\mu'\nu'}\,a_{\mu'\nu'} = \frac{1}{2}\,(h_\mu h_\kappa a_{\kappa\nu} + h_\nu h_\kappa a_{\kappa\mu}) - h_\mu h_\nu\,a_{\mu'\nu'}\,h_{\mu'}h_{\nu'}$$
$$\mp \frac{i}{2}\,(h_\mu H_{\nu\tau}a_{\tau\kappa}h_\kappa + h_\nu H_{\mu\tau}a_{\tau\kappa}h_\kappa),$$

$$\mathscr{P}^{(\pm 2)}_{\mu\nu,\mu'\nu'}\,a_{\mu'\nu'} = \frac{1}{2}\,a_{\mu\nu} - \overline{h_\mu h_\kappa a_{\kappa\nu}} + \frac{1}{4}\,\overline{h_\mu h_\nu}\,h_{\mu'}h_{\nu'}a_{\mu'\nu'} \quad (14.56)$$
$$\mp \frac{i}{4}\,(H_{\mu\tau}a_{\tau\nu} - h_\mu H_{\nu\tau}a_{\tau\kappa}h_\kappa + H_{\nu\tau}a_{\tau\mu} - h_\nu H_{\mu\tau}a_{\tau\kappa}h_\kappa).$$

Multiplication of the expressions above by the symmetric traceless tensor $b_{\mu\nu}$ leads to

$$b_{\mu\nu}\mathscr{P}^{(0)}_{\mu\nu,\mu'\nu'}\,a_{\mu'\nu'} = \frac{3}{2}\,(h_\mu b_{\mu\nu}h_\nu)\,(h_{\mu'}a_{\mu'\nu'}h_{\nu'}), \quad (14.57)$$

$$b_{\mu\nu}\mathscr{P}^{(\pm 1)}_{\mu\nu,\mu'\nu'}\,a_{\mu'\nu'} = h_\mu b_{\mu\nu}a_{\nu\kappa}h_\kappa - (h_\mu b_{\mu\nu}h_\nu)\,(h_{\mu'}a_{\mu'\nu'}h_{\nu'}) \mp i\,h_\mu b_{\mu\nu} H_{\nu\tau}a_{\tau\kappa}h_\kappa,$$

$$b_{\mu\nu} \mathscr{P}^{(\pm 2)}_{\mu\nu,\mu'\nu'} a_{\mu'\nu'} = \frac{1}{2} b_{\mu\nu} a_{\mu\nu} - h_\mu b_{\mu\nu} a_{\nu\kappa} h_\kappa + \frac{1}{4} (h_\mu b_{\mu\nu} h_\nu)(h_{\mu'} a_{\mu'\nu'} h_{\nu'})$$

$$\mp \frac{i}{2} [b_{\mu\nu} H_{\nu\tau} a_{\tau\mu} - h_\mu b_{\mu\nu} H_{\nu\tau} a_{\tau\kappa} h_\kappa]. \tag{14.58}$$

Now let the tensor **a** be constructed from the components of the unit vectors **e**, viz. $a_{\mu\nu} = \overline{e_\mu e_\nu}$. Then the real and imaginary parts of (14.55) and (14.56) are

$$\mathscr{P}^{(0)}_{\mu\nu,\mu'\nu'} \overline{e_{\mu'} e_{\nu'}} = \frac{3}{2} \overline{h_\mu h_\nu} \left[(\mathbf{h} \cdot \mathbf{e})^2 - \frac{1}{3} \right], \tag{14.59}$$

$$(\mathscr{P}^{(1)}_{\mu\nu,\mu'\nu'} + \mathscr{P}^{(-1)}_{\mu\nu,\mu'\nu'}) \overline{e_{\mu'} e_{\nu'}} = \frac{1}{2} (\overline{h_\mu e_\nu} + \overline{h_\nu e_\mu})(\mathbf{h} \cdot \mathbf{e}) - \overline{h_\mu h_\nu} (\mathbf{h} \cdot \mathbf{e})^2,$$

$$(\mathscr{P}^{(2)}_{\mu\nu,\mu'\nu'} + \mathscr{P}^{(-2)}_{\mu\nu,\mu'\nu'}) \overline{e_{\mu'} e_{\nu'}} = \overline{e_\mu e_\nu} - 2 \overline{h_\mu e_\nu} (\mathbf{h} \cdot \mathbf{e}) + \frac{1}{2} \overline{h_\mu h_\nu} [1 + (\mathbf{h} \cdot \mathbf{e})^2],$$

$$i \left(\mathscr{P}^{(1)}_{\mu\nu,\mu'\nu'} - \mathscr{P}^{(-1)}_{\mu\nu,\mu'\nu'} \right) \overline{e_{\mu'} e_{\nu'}} = [h_\mu (\mathbf{h} \times \mathbf{e})_\nu + h_\nu (\mathbf{h} \times \mathbf{e})_\mu](\mathbf{h} \cdot \mathbf{e}),$$

$$i \left(\mathscr{P}^{(2)}_{\mu\nu,\mu'\nu'} - \mathscr{P}^{(-2)}_{\mu\nu,\mu'\nu'} \right) \overline{e_{\mu'} e_{\nu'}} = \frac{1}{2} [(\mathbf{h} \times \mathbf{e})_\mu e_\nu + (\mathbf{h} \times \mathbf{e})_\nu e_\mu]$$

$$- \frac{1}{2} [h_\mu (\mathbf{h} \times \mathbf{e})_\nu + h_\nu (\mathbf{h} \times \mathbf{e})_\mu] (\mathbf{h} \cdot \mathbf{e}). \tag{14.60}$$

The cross product $(\mathbf{h} \times \mathbf{e})$ stems from $H_{\mu\tau} e_\tau = (\mathbf{h} \times \mathbf{e})_\mu$. For $\mathbf{e} = \mathbf{h}$, all terms on the right hand side of (14.60) and in the second and third equations of (14.59) vanish. This is obvious since a rotation about an axis parallel to **e** does not change the direction of **e**. For **e** perpendicular to **h**, the equations involving $\mathscr{P}^{(\pm 1)}$ yield zero, the remaining equations reduce to

$$\mathscr{P}^{(0)}_{\mu\nu,\mu'\nu'} \overline{e_{\mu'} e_{\nu'}} = -\frac{1}{2} \overline{h_\mu h_\nu}, \tag{14.61}$$

$$\left(\mathscr{P}^{(2)}_{\mu\nu,\mu'\nu'} + \mathscr{P}^{(-2)}_{\mu\nu,\mu'\nu'} \right) \overline{e_{\mu'} e_{\nu'}} = \overline{e_\mu e_\nu} + \frac{1}{2} \overline{h_\mu h_\nu},$$

$$i \left(\mathscr{P}^{(2)}_{\mu\nu,\mu'\nu'} - \mathscr{P}^{(-2)}_{\mu\nu,\mu'\nu'} \right) \overline{e_{\mu'} e_{\nu'}} = \frac{1}{2} [(\mathbf{h} \times \mathbf{e})_\mu e_\nu + (\mathbf{h} \times \mathbf{e})_\nu e_\mu].$$

Next, the special case $a_{\mu\nu} = \overline{e_\mu u_\nu}$ is considered, where the unit vectors are perpendicular to each other. Then (14.55) and (14.56) lead to

$$\mathscr{P}^{(0)}_{\mu\nu,\mu'\nu'} \overline{e_{\mu'} u_{\nu'}} = \frac{3}{2} \overline{h_\mu h_\nu} (\mathbf{h} \cdot \mathbf{e})(\mathbf{h} \cdot \mathbf{u}), \tag{14.62}$$

$$\left(\mathscr{P}^{(1)}_{\mu\nu,\mu'\nu'} + \mathscr{P}^{(-1)}_{\mu\nu,\mu'\nu'} \right) \overline{e_{\mu'} u_{\nu'}} = \frac{1}{4} [(\overline{h_\mu u_\nu} + \overline{h_\nu u_\mu})(\mathbf{h} \cdot \mathbf{e}) + (\overline{h_\mu e_\nu} + \overline{h_\nu e_\mu})(\mathbf{h} \cdot \mathbf{u})]$$

$$- 2 \overline{h_\mu h_\nu} (\mathbf{h} \cdot \mathbf{e}) (\mathbf{h} \cdot \mathbf{u}),$$

$$\left(\mathscr{P}^{(2)}_{\mu\nu,\mu'\nu'} + \mathscr{P}^{(-2)}_{\mu\nu,\mu'\nu'}\right) \overline{e_{\mu'}u_{\nu'}} = \overline{e_\mu u_\nu} + \frac{1}{2}\overline{h_\mu h_\nu}\,(\mathbf{h}\cdot\mathbf{e})(\mathbf{h}\cdot\mathbf{u})$$
$$-[\overline{h_\mu e_\nu}\,(\mathbf{h}\cdot\mathbf{u}) + \overline{h_\mu u_\nu}\,(\mathbf{h}\cdot\mathbf{e})],$$

$$2i\left(\mathscr{P}^{(1)}_{\mu\nu,\mu'\nu'} - \mathscr{P}^{(-1)}_{\mu\nu,\mu'\nu'}\right) \overline{e_{\mu'}u_{\nu'}} = [h_\mu(\mathbf{h}\times\mathbf{e})_\nu + h_\nu(\mathbf{h}\times\mathbf{e})_\mu]\,(\mathbf{h}\cdot\mathbf{u}) \quad (14.63)$$
$$+[h_\mu(\mathbf{h}\times\mathbf{u})_\nu + h_\nu(\mathbf{h}\times\mathbf{u})_\mu]\,(\mathbf{h}\cdot\mathbf{e}),$$

$$4i\left(\mathscr{P}^{(2)}_{\mu\nu,\mu'\nu'} - \mathscr{P}^{(-2)}_{\mu\nu,\mu'\nu'}\right) \overline{e_{\mu'}u_{\nu'}} = \{u_\mu(\mathbf{h}\times\mathbf{e})_\nu + u_\nu(\mathbf{h}\times\mathbf{e})_\mu$$
$$+ e_\mu(\mathbf{h}\times\mathbf{u})_\nu + e_\nu(\mathbf{h}\times\mathbf{u})_\mu\}$$
$$- \{[h_\mu(\mathbf{h}\times\mathbf{e})_\nu + h_\nu(\mathbf{h}\times\mathbf{e})_\mu]\,(\mathbf{h}\cdot\mathbf{u})$$
$$+ [h_\mu(\mathbf{h}\times\mathbf{u})_\nu + h_\nu(\mathbf{h}\times\mathbf{u})_\mu]\,(\mathbf{h}\cdot\mathbf{e})\}.$$

Direct application of the equation (14.31) defining the fourth rank projectors in terms of the second rank tensors and use of symbolic notation, leads to

$$\mathscr{P}^{(\pm1)} : (\mathbf{e}\mathbf{u}+\mathbf{u}\mathbf{e}) = \frac{1}{2}\,[\mathbf{e}^\|\mathbf{u}^\perp + \mathbf{u}^\|\mathbf{e}^\perp + \mathbf{u}^\perp\mathbf{e}^\| + \mathbf{e}^\perp\mathbf{u}^\|] \qquad (14.64)$$
$$\mp\frac{i}{2}\,[\mathbf{e}^\|\mathbf{u}^{\mathrm{tr}} + \mathbf{u}^\|\mathbf{e}^{\mathrm{tr}} + \mathbf{u}^{\mathrm{tr}}\mathbf{e}^\| + \mathbf{e}^{\mathrm{tr}}\mathbf{u}^\|],$$

$$\mathscr{P}^{(\pm2)} : (\mathbf{e}\mathbf{u}+\mathbf{u}\mathbf{e}) = \frac{1}{4}\,[\mathbf{e}^\perp\mathbf{u}^\perp + \mathbf{u}^\perp\mathbf{e}^\perp - \mathbf{u}^{\mathrm{tr}}\mathbf{e}^{\mathrm{tr}} - \mathbf{e}^{\mathrm{tr}}\mathbf{u}^{\mathrm{tr}}] \qquad (14.65)$$
$$\mp\frac{i}{4}\,[\mathbf{e}^\perp\mathbf{u}^{\mathrm{tr}} + \mathbf{u}^\perp\mathbf{e}^{\mathrm{tr}} + \mathbf{u}^{\mathrm{tr}}\mathbf{e}^\perp + \mathbf{e}^{\mathrm{tr}}\mathbf{u}^\perp].$$

Here $\mathbf{e}$ and $\mathbf{u}$ are two arbitrary unit vectors which, in special cases, may be parallel or perpendicular to each other. The parts of a vector $\mathbf{e}$ which are parallel, perpendicular and transverse with respect to $\mathbf{h}$, are defined by

$$\mathbf{e}^\| = P^\| \cdot \mathbf{e}, \quad \mathbf{e}^\perp = P^\perp \cdot \mathbf{e}, \quad \mathbf{e}^{\mathrm{tr}} = H \cdot \mathbf{e} = \mathbf{h}\times\mathbf{e}.$$

Due to $\mathscr{P}^{(\pm2)}_{\mu\nu,\mu'\nu'}\delta_{\mu'\nu'} = 0$, the expressions $(\mathbf{e}\mathbf{u}+\mathbf{u}\mathbf{e})$ in the equations above may be replaced by $2\,\overline{\mathbf{e}\mathbf{u}}$. Furthermore, the resulting dyadics in (14.64) and (14.65) are automatically traceless. This is not the case for the corresponding expression involving the projector $\mathscr{P}^{(0)} = P^{(0)}P^{(0)} + P^{(1)}P^{(-1)} + P^{(-1)}P^{(1)}$. Here one has $\mathscr{P}^{(0)}_{\mu\nu,\mu'\nu'}\delta_{\mu'\nu'} = P^{(0)}_{\mu\nu} + P^{(1)}_{\mu\nu} + P^{(-1)}_{\mu\nu} = \delta_{\mu\nu}$. Thus one obtains

$$\mathscr{P}^{(0)}_{\mu\nu,\mu'\nu'}\,2\,\overline{e_{\mu'}u_{\nu'}} = e^\|_\mu u^\|_\nu + e^\|_\nu u^\|_\mu \qquad (14.66)$$
$$+\frac{1}{2}\,[e^\perp_\mu u^\perp_\nu + u^\perp_\mu e^\perp_\nu + u^{\mathrm{tr}}_\mu e^{\mathrm{tr}}_\nu + e^{\mathrm{tr}}_\mu u^{\mathrm{tr}}_\nu] - \frac{2}{3}\,\mathbf{e}\cdot\mathbf{u}\,\delta_{\mu\nu}.$$

This expression is also traceless, notice that $e^{\mathrm{tr}}_\mu u^{\mathrm{tr}}_\mu = e^\perp_\mu u^\perp_\mu$ and $e^\parallel_\mu u^\parallel_\mu + e^\perp_\mu u^\perp_\mu = \mathbf{e}\cdot\mathbf{u}$. The application of the fourth rank projectors onto a symmetric traceless tensor yields a symmetric traceless tensor. By symmetry,

$$\mathscr{P}^{(0)}_{\mu\nu,\mu'\nu'}\, 2\,\overline{e_{\mu'}u_{\nu'}} = c\,\overline{h_\mu h_\nu}$$

is expected, with a proportionality factor c. Multiplication of this equation by $h_\mu h_\nu$ and use of (14.66) yields $c = \frac{3}{2}[2\mathbf{h}\cdot\mathbf{eh}\cdot\mathbf{u} - \frac{2}{3}\mathbf{e}\cdot\mathbf{u}]$. This is in accord with

$$\mathscr{P}^{(0)}_{\mu\nu,\mu'\nu'} = \frac{3}{2}\,\overline{h_\mu h_\nu}\;\overline{h_{\mu'}h_{\nu'}}\,,$$

as already implied by (14.55).

Application of the fourth rank projector $\mathscr{P}^{(\mathrm{m})}$ on $\overline{h_\mu h_\kappa}\,a_{\kappa\nu}$, where $a_{\mu\nu}$ is an irreducible second rank tensor, yields

$$\mathscr{P}^{(\mathrm{m})}_{\mu\nu,\mu'\nu'}\,\overline{h_{\mu'}h_\kappa}\,a_{\kappa\nu'} = \left(\frac{1}{3} - \frac{m^2}{6}\right)\,\mathscr{P}^{(\mathrm{m})}_{\mu\nu,\mu'\nu'}\,a_{\mu'\nu'}. \tag{14.67}$$

Furthermore,

$$h_{\mu_2}\cdots h_{\mu_\ell}\,\mathscr{H}^{(\ell)}_{\mu_1\mu_2\cdots\mu_{(\ell)},\nu_1\nu_2\cdots\nu_\ell}a_{\nu_1\nu_2\cdots\nu_\ell} = i\left(P^{(1)}_{\mu_1\mu'_1} - P^{(-1)}_{\mu_1\mu'_1}\right)A_{\mu'_1}, \tag{14.68}$$

with

$$A_{\mu'_1} = h_{\mu'_2}\cdots h_{\mu'_\ell}\,a_{\mu'_1\mu'_2\cdots\mu'_\ell}, \tag{14.69}$$

where $a_{\mu'_1\mu'_2\cdots\mu'_\ell}$ is an irreducible ℓth rank tensor.

Chapter 15
Liquid Crystals and Other
Anisotropic Fluids

Abstract This chapter deals with equilibrium properties of liquid crystals and other anisotropic fluids. After some remarks on nematic, cholesteric and smectic liquid crystals and blue phases, the second rank alignment tensor is introduced as the relevant order parameter for the nematic state. Theories for the phase transition isotropic-nematic are presented. The orientational elastic behavior of nematics and cholesterics is firstly described by the director elasticity involving the Frank coefficients and then by the alignment tensor elasticity theory. Systems with cubic and with tetrahedral symmetry, referred to as cubatics and tetradics, are characterized by fourth and third rank order parameter tensors. Some examples for the energetic coupling of order parameter tensors of equal and of different ranks are considered.

Crystalline solids are anisotropic. Fluids in thermal equilibrium, on the other hand, are commonly looked upon as isotropic substances. However, fluids can also become anisotropic, be it through the application of external fields or by a spontaneous phase transition into an state with orientational order. *Liquid crystals* are the most prominent *anisotropic fluids*. Here the emphasis is on *nematic liquid crystals*. Properties of some other anisotropic fluids are discussed briefly. In any case, tensors are the tools needed to characterize the anisotropy of these substances. This section is devoted to equilibrium properties. Non-equilibrium phenomena are treated in Sect. 16.4 and Chap. 17.

A standard publication for the physics of liquid crystals is the classic book of de Gennes from 1973, a revised and extended second edition appeared in 1993 [67]. A good introduction is [68], for phase types and structures see also [69]. An extensive survey of the literature up to 1980 is given in [70]. Concepts and experiments are discussed in [71]. Classic papers on liquid crystals are reproduced and commented in [72]. Optical experiments and their theoretical foundation are treated in [73] for equilibrium and non-equilibrium properties of complex fluids, in particular for polymers and liquid crystals.

© Springer International Publishing Switzerland 2015
S. Hess, *Tensors for Physics*, Undergraduate Lecture Notes in Physics,
DOI 10.1007/978-3-319-12787-3_15

15.1 Remarks on Nomenclature and Notations

Liquid crystals are substances which can flow, and thus are liquid-like and, at the same time, exhibit anisotropic properties which are typical for crystalline solids. Around 1880, Otto Lehmann coined the words "flowing crystal" and "liquid crystal" to refer to same types of *anisotropic fluids*. Some contemporary scientists opposed to his ideas that fluid and crystalline properties can occur in a homogeneous substance and that various liquid-crystalline states are phases in the sense of thermodynamics, just as *gas, isotropic liquid and crystalline solid*. Nowadays the notion *liquid crystal* is well accepted, in particular "LCD", i.e. "Liquid Crystal Display" became a household word.

Liquid crystals are composed of non-spherical particles which have an orientational degree of freedom. Prototypes for effectively axisymmetric or uniaxial particles have a rod-like or a disc-like shape. They are referred to as *prolate* and *oblate* particles.

The main types of liquid crystals are called *nematic*, *cholesteric* and *smectic* liquid crystals.

The terms *thermotropic liquid crystal* and *lyotropic liquid crystal* are used to indicate that the change of the temperature or of the concentration in a solution drives the phase transition from an isotropic liquid to a liquid crystalline state.

15.1.1 Nematic and Cholesteric Phases, Blue Phases

In *nematics*, the main axis of the particles have a long range preferential order whereas the positions of their centers of mass have no long range order. Usually, it is understood that nematics are composed of prolate particles, otherwise the notion "discotic nematic" is used for fluids composed of oblate particles. Symmetry considerations, however, apply to both types of nematics: they are characterized by an *order parameter* which is a symmetric traceless second rank tensor.

In ordinary nematics, the phase has uniaxial symmetry, even when its constituents are biaxial particles. Fluids with an overall biaxial symmetry, but without any long ranged positional order are referred to as *biaxial nematics*. One has to distinguish between the symmetry of the particles and the symmetry of the phase. In principle, it is possible to have a biaxial phase composed of uniaxial particles and a uniaxial phase composed of biaxial particles. Typically, however, the rare biaxial nematic phase is found for substances composed of biaxial particles.

A cartoon of the orientation of 'particles' in the nematic phase is shown in Fig. 15.1. Although 'up and down' or 'head and tail' can be distinguished for each particle, the average orientation has *head-tail symmetry*.

A *cholesteric liquid crystal* is essentially a nematic, where the preferential direction of the phase is twisted in space, with a twist axis pointing in a certain direction. Cholesterics possess a spontaneously formed helix, which is geometrically similar to the steps of a spiral staircase. The twist axis is also called *helical axis*.

Fig. 15.1 Cartoon of the orientation of molecules in the nematic phase, as shown in [74]

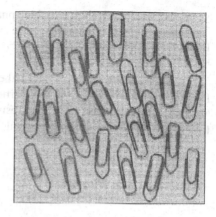

Fig. 15.2 Schematic double-twist configuration

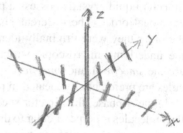

When the local helical axis points in the radial directions perpendicular to another fixed axis, the director configuration is referred to as *double twist structure*, cf. Fig. 15.2. There the short lines indicate the director, the thin horizontal lines mark the directions of two of the helical axes which are orthogonal to the cylinder axis. The molecular arrangement in a double-twist cylinder with a diameter determined such that the twist from the center of the cylinder axis is about 45°, can be more stable than the single-twist configuration of an ordinary cholesteric state. In larger volumes 3D supra-molecular structures are spontaneously formed in the *blue phases*. These structures contain defects where three orthogonal double-twist cylinders touch each other. The 3D arrangement of the defects determines the symmetry of the blue phase. The phases referred to as 'BP1 and BP2 have cubic symmetry of fcc and bcc type. Blue phases with icosahedral symmetry and with an irregular structure also exist.

The *blue phases* are found in chiral substances between the 'ordinary' cholesteric and the isotropic liquid phase. The name comes from the blue shine observed in *cholesteryl benzoate*, as first reported by Reinitzer in 1888. He sent this substance to Lehmann who studied the spontaneous birefringence as function of the temperature. Lehman noticed that the 'blue phase' is optically isotropic, in contradistinction to the optically anisotropic cholesteric phase which Lehmann then called 'liquid crystal', a name he had previously used for substances now called *superionic conductors*. Decades later it was recognized that there is not one blue phase but several types of blue phases which occur in a very narrow temperature intervals. For over hundred years, the research dealing with blue phases was considered as a rather exotic topic. This changed in 2005 when substances with a wide temperature range of blue-phase liquid crystalline state were found [75]. In the meantime, a blue phase LC display has been developed.

Colloidal particles are localized at the defects of liquid crystals and they may form periodic structures [76]. Colloidal particles immersed in a blue phase liquid crystal stabilize the blue phase [77]. The resulting 3D periodic structure has the optical properties of ordinary colloidal crystals [78] but is mechanically more stable.

15.1.2 Smectic Phases

Smectic liquid crystals possess a partial positional ordering, in addition to an orientational order. The different phases are labelled by the letters $A, B, C, \ldots$ in the order they were originally identified as distinct thermodynamic phases, before the underlying microscopic structure was identified. The most prominent cases are the *smectic A* and *smectic C* phases, where the centers of mass of the molecules are preferentially located in planes.

In the A-phase, the director specifying the average direction of the long axes of the molecules is perpendicular to the planes, in the C-phase, it is tilted with respect to the planes. A cartoon of the orientation of 'particles' in the smectic A and C phases is shown in Fig. 15.3. The tilt angle ϑ which distinguishes the C from the A phase can be used as an order parameter, e.g. see [67]. Alternatively, cf. [79], the transition $A \rightarrow C$ can be looked upon as a spontaneous shear displacement $\mathbf{u}(\mathbf{r})$ of the centers of mass of the molecules where $\mathbf{u}$ and its gradient are parallel and perpendicular, respectively, to the director $\mathbf{n}$. With $\mathbf{u}$ in x-direction and its gradient in y-direction, the deformation $\partial u_x / \partial y$ is equal to $\tan \vartheta$. In the *ferro-electric* smectic C^* liquid crystals, the electric polarization is proportional to the axial vector associated with the antisymmetric part of the deformation tensor $\nabla_\mu u_\nu$, [79].

The *smectic B* phase is similar to the A-phase but with an additional hexagonal short range order of neighbor molecules within a plane. The *smectic D* phase has a 3D cubic order, rather than the layered structure typical for smectics. For chiral and ferro-electric smectics, as well as columnar phases, and *banana phases* see, e.g. [67–69].

Fig. 15.3 Cartoon of the orientation of molecules in the smectic A (*left*) and smectic C (*right*) phases, as shown in [74]

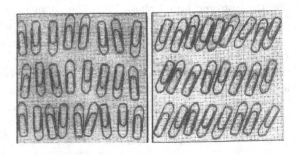

15.2 Isotropic ↔ Nematic Phase Transition

15.2.1 Order Parameter Tensor

The existence of a non-zero second rank alignment tensor in thermal equilibrium, distinguishes the nematic phase of a liquid crystal from its isotropic liquid state. The order parameter tensor can be introduced phenomenologically via the anisotropic, i.e. symmetric traceless part of the electric or magnetic susceptibility tensor or, and this is preferred here, microscopically as an average over the orientational distribution function. As in Sect. 12.2.2, particles with a symmetry axis parallel to the unit vector **u** are considered. In most nematics, the orientational distribution $f = f(\mathbf{u})$ does not depend on the sign of **u**, even when the particles do have a polar character. The property $f(\mathbf{u}) = f(-\mathbf{u})$ is referred to as *head-tail symmetry*. The lowest moment which distinguishes an anisotropic distribution from an isotropic one is the *second rank alignment tensor*, cf. (12.14) and (12.17),

$$a_{\mu\nu} = \langle \phi_{\mu\nu} \rangle, \quad \phi_{\mu\nu} = \zeta_2 \, \overline{u_\mu u_\nu}, \quad \zeta_2 = \sqrt{\frac{15}{2}}. \tag{15.1}$$

The bracket indicates the average over the orientational distribution, viz.

$$\langle \ldots \rangle = \int \ldots f(\mathbf{u}) \mathrm{d}^2 u.$$

The quadrupole moment tensor, cf. (10.29) has the same symmetry as the second rank alignment tensor $a_{\mu\nu} \sim \langle \overline{u_\mu u_\nu} \rangle$. Therefore, the tensor $\langle \overline{u_\mu u_\nu} \rangle$ is also denoted by $Q_{\mu\nu}$ and called Q-tensor, [67].

The general properties of symmetric second rank tensors discussed in Chap. 5 apply to the traceless tensor $a_{\mu\nu}$ defined in (15.1). In particular, in a principal axes frame with the principal axes parallel to the mutually perpendicular unit vectors $\mathbf{e}^{(i)}$, $i = 1, 2, 3$, the tensor is expressed as,

$$a_{\mu\nu} = \sqrt{\frac{3}{2}} a_0 \, \overline{e_\mu^{(3)} e_\nu^{(3)}} + \frac{\sqrt{2}}{2} a_1 \left(e_\mu^{(1)} e_\nu^{(1)} - e_\mu^{(2)} e_\nu^{(2)} \right). \tag{15.2}$$

The factors in front of the quantities a_0 and a_1 have been chosen such that the magnitude of the alignment tensor, viz. the second scalar invariant $I_2 = a_{\mu\nu} a_{\mu\nu}$ is equal to $a_0^2 + a_1^2$. For a comparison with the relations given in Sect. 5.2.4, notice that the quantities corresponding to $\bar{S}$, s and q of (5.11) are 0, $\sqrt{\frac{3}{2}} a_0$ and $a_1/\sqrt{2}$. By analogy to (5.13), the principal values $a^{(i)}$, $i = 1, 2, 3$ of the alignment tensor are

$$a^{(1)} = -\frac{1}{\sqrt{6}} a_0 + \frac{1}{\sqrt{2}} a_1, \quad a^{(2)} = -\frac{1}{\sqrt{6}} a_0 - \frac{1}{\sqrt{2}} a_1, \quad a^{(3)} = \sqrt{\frac{2}{3}} a_0. \tag{15.3}$$

The alignment tensor is uniaxial for $a_1 = 0$, with its symmetry axis parallel to $\mathbf{e}^{(3)}$. The tensor is planar biaxial for $a_0 = 0$. For this reason, a_0 and a_1 are referred to as *uniaxial order parameter* and *biaxial order parameter*. Notice, however, that the alignment tensor is also uniaxial when $a_1 = \pm\sqrt{3}a_0$ holds true. In these cases, the symmetry axis is parallel to $\mathbf{e}^{(2)}$ and $\mathbf{e}^{(1)}$, respectively.

In terms of a_0 and a_1, the third scalar invariant $I_3 \sim a_{\mu\nu}a_{\nu\kappa}a_{\kappa\mu}$, which is essentially the determinant, cf. (5.44), is determined by

$$I_3 = \sqrt{6}\,a_{\mu\nu}a_{\nu\kappa}a_{\kappa\mu} = a_0\,(a_0^2 - 3\,a_1^2). \tag{15.4}$$

The factor $\sqrt{6}$, which was not included in (5.44), is inserted here for convenience. The biaxiality parameter b, cf. Sect. 5.5.2, is now given by

$$b^2 = 1 - I_3^2/I_2^3. \tag{15.5}$$

With

$$a_0 = a\cos\alpha, \quad a_1 = a\sin\alpha, \tag{15.6}$$

where a determines the magnitude of the alignment and the angle α is a measure for the biaxiality, the scalar invariants and the biaxiality parameter are given by

$$I_2 = a_0^2 + a_1^2 = a^2, \quad I_3 = a^3\,\cos\alpha\,(\cos^2\alpha - 3\sin^2\alpha) = a^3\cos 3\alpha, \quad b = \sin 3\alpha. \tag{15.7}$$

The argument 3α reflect the fact that the roles of the three principal axes can be interchanged without changing the physics described.

Ordinary nematic liquid crystals are uniaxial in thermal equilibrium and when no distortions are imposed. Then one has $\alpha = 0$, $a_1 = 0$ and $a_0 = a$. Furthermore, the unit vector parallel to the space-fixed symmetry direction is denoted by $\mathbf{n}$, rather than $\mathbf{e}^{(3)}$, and called *director*. The alignment tensor is written as

$$a_{\mu\nu} = \sqrt{\frac{3}{2}}\,a\,\overline{n_\mu n_\nu}. \tag{15.8}$$

Due to

$$n_\mu n_\nu a_{\mu\nu} = \sqrt{\frac{2}{3}}\,a,$$

and (15.1), the order parameter a is determined by

$$a = \sqrt{\frac{3}{2}}\,\varsigma_2\,\langle\overline{u_\mu u_\nu}\rangle\,\overline{n_\mu n_\nu} = \sqrt{5}\,S_2, \quad S_2 \equiv \langle P_2(\mathbf{u}\cdot\mathbf{n})\rangle, \tag{15.9}$$

where $P_2(x) = \frac{3}{2}(x^2 - \frac{1}{3})$ is the second Legendre polynomial. Frequently, the quantity S_2 is denoted by S and referred to as *Maier-Saupe order parameter*.

Theoretical approaches to the *phase transition isotropic ↔ nematic* are discussed next.

15.2.2 Landau-de Gennes Theory

The *Landau-de Gennes theory* for the phase transition isotropic ↔ nematic is based on finding the minimum of a free energy $\mathcal{F}_a$ associated with the alignment. This free energy is written as

$$\mathcal{F}_a = N \, k_B T \, \Phi, \tag{15.10}$$

where N is the number of particles, k_B is the Boltzmann constant, T is the temperature and Φ is a dimensionless thermodynamic potential function which depends on the scalar invariants I_2, I_3 of the alignment tensor. In the Landau de Gennes theory the ansatz

$$\Phi = \Phi^{\mathrm{LdG}} \equiv \frac{1}{2} A I_2 - \frac{1}{3} B I_3 + \frac{1}{4} C I_2^2, \quad A = A_0\left(1 - \frac{T^*}{T}\right), \quad A_0, B, C > 0, \tag{15.11}$$

or explicitly,

$$\Phi^{\mathrm{LdG}} = \frac{1}{2} A \, a_{\mu\nu} a_{\nu\mu} - \frac{1}{3} B \sqrt{6} \, a_{\mu\nu} a_{\nu\kappa} a_{\kappa\mu} + \frac{1}{4} C \, (a_{\mu\nu} a_{\nu\mu})^2, \tag{15.12}$$

is made. The phenomenological coefficients $A_0, B, C > 0$ are assumed to be practically constant in the vicinity of the phase transition, and T^* is a pseudo-critical temperature, which is somewhat below the isotropic-nematic transition temperature T_{ni}.

The equilibrium value of the alignment is inferred from the minimum of the free energy, which in turn, is obtained by putting the first derivative of the potential Φ with respect to the alignment tensor equal to zero. To compute the derivative $\partial \Phi / \partial \mathbf{a}$, replace $\mathbf{a}$ in $\Phi(\mathbf{a})$ by $\mathbf{a} + \delta \mathbf{a}$ where $\delta \mathbf{a}$ is a small distortion. Then find the factor in the term of $\delta\Phi = \Phi(\mathbf{a} + \delta\mathbf{a}) - \Phi(\mathbf{a})$ which is linear in $\delta\mathbf{a}$. For the present case, use of

$$\delta\Phi = \left(\frac{\partial\Phi}{\partial I_2}\frac{\partial I_2}{\partial a_{\mu\nu}} + \frac{\partial\Phi}{\partial I_3}\frac{\partial I_3}{\partial a_{\mu\nu}}\right)\delta a_{\mu\nu},$$

with

$$\frac{\partial I_2}{\partial a_{\mu\nu}}\delta a_{\mu\nu} = 2 a_{\mu\nu}\delta a_{\mu\nu},$$

$$\frac{\partial I_3}{\partial a_{\mu\nu}}\delta a_{\mu\nu} = 3\sqrt{6} a_{\nu\kappa} a_{\kappa\mu}\delta a_{\mu\nu} = 3\sqrt{6}\,\overline{a_{\nu\kappa} a_{\kappa\mu}}\,\delta a_{\mu\nu},$$

or direct computation from (15.11) leads to

$$\Phi_{\mu\nu}^{\text{LdG}} \equiv \frac{\partial \Phi^{\text{LdG}}}{\partial a_{\mu\nu}} = A\, a_{\mu\nu} - B\sqrt{6}\, \overline{a_{\mu\kappa} a_{\kappa\nu}} + C a_{\mu\nu}\, a_{\lambda\kappa} a_{\lambda\kappa}. \qquad (15.13)$$

Before the equilibrium condition $\Phi_{\mu\nu}^{\text{LdG}} = 0$ is discussed further, the Landau de Gennes potential (15.11) is expressed in terms of the variables a and α as introduced in the previous Sect. 15.2.1. The result is

$$\Phi^{\text{LdG}} = \frac{1}{2}Aa^2 - \frac{1}{3}Ba^3 \cos 3\alpha + \frac{1}{4}Ca^4. \qquad (15.14)$$

The conditions for an extremum of this function are

$$\frac{\partial \Phi^{\text{LdG}}}{\partial \alpha} = Ba^3 \sin 3\alpha = 0,$$

and

$$\frac{\partial \Phi^{\text{LdG}}}{\partial a} = aA - Ba^2 \cos 3\alpha + Ca^3 = 0.$$

The first of these conditions implies that the biaxiality parameter $b = \sin 3\alpha$ vanishes. Thus the equilibrium state is uniaxial and $\alpha = 0$ is used. Then the second condition is

$$a\,(A - Ba + Ca^2) = 0. \qquad (15.15)$$

The solutions are $a = 0$ and, provided that $B^2 \geq AC$, also

$$a = a_{1,2} = \frac{B}{2C} \pm \frac{1}{2C}\sqrt{B^2 - 4AC}.$$

The case $a = 0$ corresponds to an isotropic state. In an uniaxially ordered state one has $a \neq 0$. Notice that $A = A(T)$ is a function of the temperature T. At the isotropic-nematic coexistence temperature T_{ni}, the potential has a minimum at the value of a where $\Phi^{\text{LdG}}(a) = 0$ holds true. With $A_{\text{ni}} = A(T_{\text{ni}}) = A_0(1 - T^*/T_{\text{ni}})$, one has

$$\frac{1}{2}A_{\text{ni}} - \frac{1}{3}Ba + \frac{1}{4}Ca^2 = 0. \qquad (15.16)$$

Since

$$A_{\text{ni}} - Ba + Ca^2 = 0,$$

has to be obeyed at equilibrium, these relations imply

$$a = a_{\text{ni}} \equiv \frac{2B}{3C}, \quad A_{\text{ni}} = A_0(1 - T^*/T_{\text{ni}}) = \frac{1}{3}Ba_{\text{ni}} = \frac{2B^2}{9C}. \tag{15.17}$$

Typical thermotropic nematic liquid crystals have an order parameter S_2 of about 0.4, at the transition temperature. This corresponds to $a_{\text{ni}} = 0.9 \approx 1$. The relative difference between the transition temperature T_{ni} and T^*, viz.

$$\delta_{\text{ni}} = (T_{\text{ni}} - T^*)/T_{\text{ni}} = 2B^2/(9A_0C) = a_{\text{ni}}^2 C/(2B_0), \tag{15.18}$$

is of the order 10^{-2}. The Exercise 15.1 provides a derivation of the potential function (15.12) and it yields specific values for the coefficients A_0, B, C. In the literature, variables referring to the nematic-isotropic phase transition are also labelled with the letter "K", rather than "ni", like T_K or a_K instead of T_{ni} and a_{ni}. The letter "K" stems from "Klärpunkt", meaning "clearing point". The reason is: polycrystalline liquid crystals are turbid and they become clear in the isotropic phase.

It is convenient to introduce the scaled alignment tensor $a_{\mu\nu}^*$, a reduced potential Φ^* and a reduced relative temperature ϑ via

$$a_{\mu\nu} = a_{\text{ni}} a_{\mu\nu}^*, \quad \Phi = a_{\text{ni}}^2 A_{\text{ni}} \Phi^*, \tag{15.19}$$

$$\vartheta = A(T)/A_{\text{ni}} = (1 - T^*/T)/(1 - T^*/T_{\text{ni}}) = (T_{\text{ni}}/T)(T - T^*)/(T_{\text{ni}} - T^*). \tag{15.20}$$

Then the resulting expressions for the scaled Landau de Gennes potential

$$(\Phi^{\text{LdG}})^* = \frac{1}{2}\vartheta\, a_{\mu\nu}^* a_{\nu\mu}^* - \sqrt{6}\, a_{\mu\nu}^* a_{\nu\kappa}^* a_{\kappa\mu}^* + \frac{1}{2}(a_{\mu\nu}^* a_{\nu\mu}^*)^2, \tag{15.21}$$

is universal in the sense that the original coefficients A_0, B, C no longer show up explicitly. With $a = a_{\text{ni}}a^*$, the scaled expression corresponding to (15.14) is

$$(\Phi^{\text{LdG}})^* = \frac{1}{2}\vartheta\,(a^*)^2 - (a^*)^3 + \frac{1}{2}(a^*)^4. \tag{15.22}$$

The transition temperature T_{ni} and the temperature T^* correspond to $\vartheta = 1$ and $\vartheta = 0$, respectively. The relation corresponding to (15.15) is

$$a^*(\vartheta - 3a^* + 2(a^*)^2) = 0.$$

The resulting equilibrium value of the order parameter in the nematic phase is

$$a_{\text{eq}}^* = \frac{3}{4} + \frac{1}{4}\sqrt{9 - 8\vartheta}, \quad \vartheta \leq \frac{9}{8}. \tag{15.23}$$

Clearly, $a_{eq}^* = 1$ for $\vartheta = 1$. The nematic state is metastable in the range $1 < \vartheta < \frac{9}{8}$ of the reduced temperature. The isotropic state corresponding to $a = 0$ is metastable for $0 < \vartheta < 1$.

The scaled variables can be denoted by the original symbols without the star, when no confusion arises. Notice that the scaled potential function (15.21) corresponds to (15.12) with $A = \vartheta$ and the universal coefficients $B = 3$, $C = 2$.

By definition, the Maier-Saupe order parameter $S_2 = \langle P_2 \rangle$ lies within the range $-\frac{1}{2} \leq S_2 \leq 1$. Consequently the order parameter $a = \sqrt{5}S_2$ is bounded by $-\sqrt{5}/2 \approx 1.12$ and $\sqrt{5} \approx 2.24$. The bounds for the corresponding scaled variable $a^* = a/a_{ni}$ involve the factor $1/a_{ni}$. For typical thermotropic nematics,

$$-1.25 < a^* < 2.5.$$

is the range of the scaled order parameter. The Landau-de Gennes free energy is well suited to study the isotropic state and the nematic phase in the vicinity of the transition temperature. The bounds on the order parameter just discussed, however, are not taken care of. An amended version of a Landau-de Gennes type potential function which implies an upper bound of the magnitude of the order parameter was considered in [86].

15.1 Exercise: Derivation of the Landau-de Gennes Potential

In general, the free energy F is related to the internal energy U and the entropy S by $F = U - TS$. Thus the contributions to these thermodynamic functions which are associated with the alignment obey the relation

$$\mathscr{F}_a = \mathscr{U}_a - T\,\mathscr{S}_a.$$

Assume that the relevant internal energy is equal to

$$\mathscr{U}_a = -N\frac{1}{2}\varepsilon\, a_{\mu\nu}a_{\mu\nu},$$

where $\varepsilon > 0$ is a characteristic energy, per particle, associated with the alignment. It is related to the temperature T^* by $k_B T^* = \varepsilon/A_0$. Furthermore, approximate the entropy by the single particle contribution

$$\mathscr{S}_a = -N\,k_B\,\langle \ln(f/f_0)\rangle_0,$$

cf. Sect. 12.2.6, where the entropy per particle s_a was considered. Notice that $\mathscr{S}_a = Ns_a$. Use $f = f_0(1 + a_{\mu\nu}\phi_{\mu\nu})$ and (12.39) to compute the entropy and consequently the free energy up to fourth order in the alignment tensor. Compare with the expression (15.12) to infer A_0, B, C. Finally, use these values to calculate a_{ni} and $\delta = (T_{ni} - T^*)/T_{ni}$, cf. (15.17) and (15.18).

15.2.3 Maier-Saupe Mean Field Theory

Although Maier and Saupe [80] followed a different line of reasoning, the essence of their theory for the isotropic-nematic phase transition is based on a mean field approach. More specifically, it is assumed that a molecule feels an orienting internal field caused by the orientation of its neighbors which, in turn, is proportional to the second rank alignment tensor. By analogy with (12.30), the Hamilton function for the orientational interaction is determined by

$$H = H^{(MS)} \sim -a_{\mu\nu} \overline{u_\mu u_\mu}, \quad -H^{(MS)}/k_B T = \beta_{MS} a_{\mu\nu} \phi_{\mu\nu}, \quad \beta_{MS} = T^*/T.$$
(15.24)

The equilibrium distribution function is proportional to $\exp[\beta_{MS} a_{\mu\nu} \phi_{\mu\nu}]$. Thus $a_{\mu\nu} = \langle \phi_{\mu\nu} \rangle$ evaluated with this distribution leads to a nonlinear equation for the alignment tensor, from which the phase transition behavior can be inferred.

For a uniaxial alignment, and this is the case treated by Maier and Saupe, one has $a_{\mu\nu} = \sqrt{3/2}\,\overline{n_\mu n_\nu}$ with $a_{\mu\nu} a_{\mu\nu} = a^2$ and $a = \sqrt{5}\langle P_2(\mathbf{n} \cdot \mathbf{n}) \rangle$. Then the Hamilton function reduces to

$$H^{(MS)} = -\sqrt{5}\,k_B T^* a\, P_2(\mathbf{n} \cdot \mathbf{u}) = -5 k_B T^* S\, P_2(\mathbf{n} \cdot \mathbf{u}), \quad S = \langle P_2(\mathbf{n} \cdot \mathbf{u}) \rangle. \quad (15.25)$$

The self-consistency relation determining the equilibrium values of the order parameter is

$$a = \mathscr{J}(a), \quad \mathscr{J}(a) = Z(a)^{-1}\sqrt{5} \int_0^1 P_2(x) \exp[(T^*/T)\sqrt{5}\,a\, P_2(x)]\,dx,$$

$$Z(a) = \int_0^1 \exp[(T^*/T)\sqrt{5}\,a\, P_2(x)]\,dx. \quad (15.26)$$

In terms of $\tilde{a} = (T^*/T)a$ and $F(\tilde{a}) = \mathscr{J}((T/T^*)\tilde{a})$, the relation (15.26) is equivalent to

$$\frac{T}{T^*}\tilde{a} = F(\tilde{a}),$$

where now $F(\tilde{a})$ is a function which does not depend on T. The intersection of the straight lines, cf. Fig. 15.4, with the curve yields the self-consistent value for $\tilde{a} = (T^*/T)a$.

The equilibrium value for the order parameter a can be plotted as function of T/T^* by a parametric plot of $F(x)/x$ via $F(x)$, with x instead of $\tilde{a}$, in the appropriate range, see Fig. 15.5. The pertaining Gibbs free energy has a minimum with the value 0 at $T = T_{ni} = 1.099\,T^* \approx 1.1\,T^*$. The order parameter, at the coexistence temperature is $a_{ni} = 0.98 \approx 1$, corresponding to the value 0.44 for the Maier-Saupe order parameter S at the phase transition temperature T_{ni}. The right end of the curve in

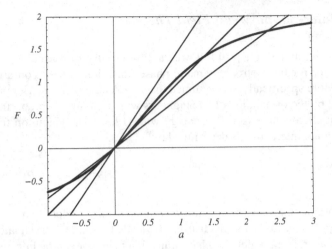

Fig. 15.4 The Maier-Saupe graphical solution for the order parameter. The *three straight lines* are for the temperatures $T/T^* = 1.5$, 1.0, and 0.75, from *left* to *right*

Fig. 15.5 The order parameter $a = \sqrt{5}\, S$ versus the temperature T in units of T^*. The *dashed part* of the curve corresponds to unstable states

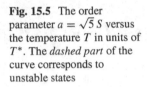
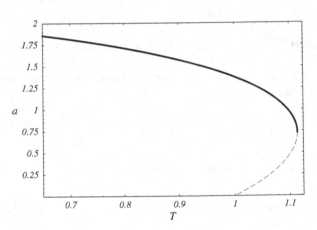

Fig. 15.5 lies at the slightly higher temperature $T_{ni}^* = 1.114\, T^*$. For this temperature, the straight line in Fig. 15.4 just touches the curve rather than intersecting it. At this point, the order parameter a is equal to 0.724 corresponding to $S \approx 0.33$.

15.3   Elastic Behavior of Nematics

The elastic behavior of nematic liquid crystals and their great sensitivity to an applied electric field play an essential role for the operation of liquid crystal displays (LCD). The speed of switching a LCD is influenced by viscous properties, which are discussed in Sect. 16.4.1.

15.3.1 Director Elasticity, Frank Coefficients

Standard nematic liquid crystals, in thermal equilibrium, are uniaxial, cf. (15.8), (15.9), and their order parameter $S = S_2$ is constant. The local director $\mathbf{n}$, however, depends on the spatial position $\mathbf{r}$. The spatial variation, in general, is influenced by boundary conditions and by orienting magnetic or electric fields. The functional dependence $\mathbf{n} = \mathbf{n(r)}$ of the director field is governed by an equation, which follows from a variational principle for the relevant free energy density f_{elast}. The pertaining free energy $F_{\text{elast}} = \int f_{\text{elast}} d^3 r$ is the spatial integral of the energy density. The "elasticity" of the director field discussed here is of a different character as compared with the elastic behavior of solids under deformations, cf. Sect. 16.2. The standard ansatz for the free energy density associated with the 'elasticity' of the director field is

$$f_{\text{elast}} = \frac{1}{2}\left[K_1 \left(\nabla \cdot \mathbf{n}\right)^2 + K_2 \left(\mathbf{n} \cdot \left(\nabla \times \mathbf{n}\right)\right)^2 + K_3 \left(\mathbf{n} \times \nabla \times \mathbf{n}\right)^2\right], \qquad (15.27)$$

with the *Frank elasticity coefficients* K_1, K_2, K_3. The distortions of the director field described by the divergence $\nabla \cdot \mathbf{n}$, by a rotation parallel to the director $\mathbf{n} \cdot (\nabla \times \mathbf{n})$, and by a rotation perpendicular to $\mathbf{n}$, viz. $\mathbf{n} \times \nabla \times \mathbf{n}$), are referred to as *splay, twist,* and *bend deformations*, as indicated in the sketch Fig. 15.6.

The undistorted, spatially homogeneous state has the lower free energy, provided that $K_i > 0$, $i = 1, 2, 3$ holds true. Expressions like (15.27) were first introduced by Oseen and Zocher, later refined by Frank [82]. In the literature, the coefficient K_i are called *Frank-Oseen elasticity*, or mostly, *Frank-elasticity* coefficients.

The director $\mathbf{n} = \mathbf{n(r)}$ is a unit vector, thus $\mathbf{n} \cdot \mathbf{n} = 1$, and consequently one has

$$n_\nu \, \nabla_\lambda \, n_\nu = 0. \qquad (15.28)$$

In Cartesian component notation, the free energy density (15.27) reads

$$f_{\text{elast}} = \frac{1}{2}K_1 \left(\nabla_\mu n_\mu\right)\left(\nabla_\nu n_\nu\right) \qquad (15.29)$$

$$+ \frac{1}{2}K_2 \left[\left(\nabla_\nu n_\mu\right)\left(\nabla_\nu n_\mu\right) - \left(\nabla_\nu n_\mu\right)\left(\nabla_\mu n_\nu\right) - n_\nu \left(\nabla_\nu n_\lambda\right) n_\mu \left(\nabla_\mu n_\lambda\right)\right]$$

$$+ \frac{1}{2}K_3 \, n_\nu \left(\nabla_\nu n_\lambda\right) n_\mu \left(\nabla_\mu n_\lambda\right).$$

Fig. 15.6 The splay, twist and bend deformation of a director field

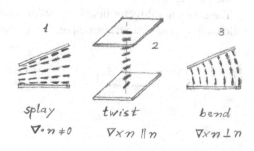

splay   twist   bend
$\nabla \cdot n \neq 0$ $\nabla \times n \parallel n$ $\nabla \times n \perp n$

For the derivation of this expression from (15.27), the condition (15.28) equations
(4.8), (4.10) for the product of two epsilon-tensors have been used. With help of the
relation

$$\nabla_\mu (n_\nu \nabla_\nu n_\mu - n_\mu \nabla_\nu n_\nu) = (\nabla_\nu n_\mu)(\nabla_\mu n_\nu) - (\nabla_\mu n_\mu)(\nabla_\nu n_\nu),$$

Equation (15.29) can be cast into the form

$$f_{\text{elast}} = \frac{1}{2} K_2 (\nabla_\nu n_\mu)(\nabla_\nu n_\mu) + \frac{1}{2}(K_1 - K_2) (\nabla_\mu n_\mu) (\nabla_\nu n_\nu) \qquad (15.30)$$
$$+ \frac{1}{2}(K_3 - K_2) n_\nu (\nabla_\nu n_\lambda) n_\mu (\nabla_\mu n_\lambda) - \frac{1}{2} K_2 \nabla_\mu (n_\nu \nabla_\nu n_\mu - n_\mu \nabla_\nu n_\nu).$$

The last term, being a total spatial derivative, contributes at the surface only, when
the free energy density is integrated over a volume. For typical low molecular weight
nematic liquid crystal, one has $K_2 < K_1 < K_3$. Some qualitative features of the
nematic elasticity can be treated theoretically in the "isotropic" approximation $K_1 =
K_2 = K_3 = K$. Then (15.30), with the surface term disregarded, reduces to

$$f_{\text{elast}} = f_{\text{elast}}^{\text{iso}} \equiv \frac{1}{2} K (\nabla_\nu n_\mu)(\nabla_\nu n_\mu). \qquad (15.31)$$

The Frank elasticity coefficients have the dimension of *energy density times length
squared*. On the other hand, ordinary elastic coefficients, like the shear modulus, as
treated in Sect. 16.2, have the dimension of an *energy density* or equivalently, of a
pressure.

   In the presence of external electric or magnetic fields, the free energy density contains additional contributions. For substances without permanent dipole moments,
but with anisotropic electric and magnetic susceptibilities $\overline{\chi_{\mu\nu}^{\text{el}}} = \chi_a^{\text{el}} \overline{n_\mu n_\nu}$ and
$\overline{\chi_{\mu\nu}^{\text{mag}}} = \chi_a^{\text{mag}} \overline{n_\mu n_\nu}$, cf. (5.34), one has

$$f_{\text{field}} = -\frac{1}{2}(\varepsilon_0 \chi_a^{\text{el}} E_\mu E_\nu + \mu_0^{-1} \chi_a^{\text{mag}} B_\mu B_\nu) \overline{n_\mu n_\nu} = -\frac{1}{2} F_{\mu\nu} \overline{n_\mu n_\nu}, \qquad (15.32)$$

where $\chi_a = \chi_\| - \chi_\perp$ is the difference between the relevant susceptibilities parallel
and perpendicular to the director $\mathbf{n}$. The symmetric traceless field tensor $F_{\mu\nu}$ is
defined by (15.32).

   The stationary director field is the solution of a spatial differential equation which
follows from a variational principle, viz. the spatial integral

$$F = \int f \mathrm{d}^3 r = \int (f_{\text{elast}} + f_{\text{field}}) \mathrm{d}^3 r$$

has to be a minimum. This implies $\delta F = \int f(\mathbf{n} + \delta\mathbf{n})\mathrm{d}^3r - \int f(\mathbf{n})\mathrm{d}^3r = 0$. The change $\delta\mathbf{n}$ must conserve the length of the director, thus $\mathbf{n} \cdot \delta\mathbf{n} = 0$ has to hold true. In the expression for f_{elast}, an integration by parts removes the spatial derivative acting on $\delta\mathbf{n}$. For the simple case of the isotropic elastic energy (15.31), the resulting differential equation is

$$-K \, \nabla_\nu \nabla_\nu n_\mu - F_{\mu\nu} n_\nu = 0.$$

The constraint $\nabla_\nu n_\nu = 1$ can be taken care of by a cross product of this equation with $\mathbf{n}$,

$$\varepsilon_{\lambda\kappa\mu} n_\kappa (K \, \Delta n_\mu + F_{\mu\nu} n_\nu) = 0. \tag{15.33}$$

This is essentially a torque balance, cf. (5.34). Of course, the differential equation has to be supplemented by boundary conditions. An example for the director field is shown in Fig. 7.7. The defect seen in the lower left corner of the right figure with a 'half integer winding number' is typical for a tensor field. The winding number counts the number of 360° turns of the director when one follows its direction on a closed path, for 360°, around the defect. The head-tail symmetry of the director allows half integer winding numbers, viz. the turn of the director for 180° only. Defects with integer winding number, the only ones allowed for a vector field, are also possible for a director field.

15.3.2 The Cholesteric Helix

In *cholesteric liquid crystals*, the director field has a screw-like spatial behavior. The cholesteric phase is essentially a spontaneously twisted nematic state with a characteristic pitch P of the helix. For cholesterics, the twist-part of the free energy density (15.27), involving the elasticity coefficient K_2, contains a term linear in the spatial derivative,

$$f_{\text{elast}}^{\text{chol}} = \frac{1}{2}\left[K_1 \, (\nabla \cdot \mathbf{n})^2 + K_2 \, [\mathbf{n} \cdot (\nabla \times \mathbf{n}) + q_0]^2 + K_3 \, (\mathbf{n} \times \nabla \times \mathbf{n})^2 \right]. \tag{15.34}$$

Conservation of parity requires that the quantity q_0 is a pseudo scalar. It is non-zero only for substances containing chiral particles or which possess a helical short range structure.

Let the spatial dependence of $\mathbf{n}$ be of the twist type

$$\mathbf{n} = \cos\alpha \, \mathbf{e}^x + \sin\alpha \, \mathbf{e}^y, \quad \alpha = \alpha(z).$$

Since $\nabla_\nu n_\mu = \alpha' e_\nu^z(-\sin\alpha e_\mu^x + \cos\alpha e_\mu^y)$, where α' stands for $d\alpha/dz$, one has $n_\lambda \varepsilon_{\lambda\nu\mu} \nabla_\nu n_\mu = -\alpha'$, and (15.34) reduces to

$$f_{\text{elast}}^{\text{chol}} = \frac{1}{2} K_2 (\alpha' - q_0)^2.$$

The free energy density has a minimum when $\alpha = q_0 z$. The quantity $P = 2\pi/q_0$ is referred to as the *pitch* of the helix. Due to the fact that $\mathbf{n}$ and $-\mathbf{n}$ are physically equivalent, the periodicity of the cholesteric helix is π/q_0.

15.3.3 Alignment Tensor Elasticity

The alignment tensor approach used for the description of the phase transition isotropic $\leftrightarrow$ nematic, cf. Sect. 15.2.2, can be generalized to spatially inhomogeneous systems. This allows to treat the elastic behavior and to derive expressions for the Frank elasticity coefficients which are closer to a microscopic interpretation. The free energy $\mathscr{F}_a$ is written as an integral over a free energy density f_a associated with the alignment and its spatial derivatives, viz. $\mathscr{F}_a = \int f_a \mathrm{d}^3 r$ with

$$f_a = (\rho/m) k_B T \, \Phi + f_a^{\text{inhom}}.$$

Here ρ/m is the number density, of a fluid with the mass density ρ, composed of particles with mass m. As in Sect. 15.2.2, Φ stands for the dimensionless free energy functional depending on the alignment, e.g. the Landau de Gennes expression (15.11), (15.12) involving the second rank tensor $\mathbf{a}$, which now depends on the position $\mathbf{r}$. The additional contribution f_a^{inhom} characterizes the 'energy cost' associated with spatial derivatives of the alignment.

First, the ansatz (15.12) is used with

$$f_a^{\text{inhom}} = (\rho/m)\,\varepsilon_0\,\xi_0^2 \left[\frac{1}{2}\sigma_1(\nabla_\nu a_{\nu\mu})(\nabla_\lambda a_{\lambda\mu}) + \frac{1}{2}\sigma_2(\nabla_\lambda a_{\nu\mu})(\nabla_\lambda a_{\nu\mu}) \right], \quad (15.35)$$

where ε_0 and ξ_0 are a reference energy and a reference length. The energy scale can be associated with the transition temperature T_{ni}, viz. $\varepsilon_0 = k_B T_{\text{ni}}$. The length scale is of the order of a molecular length and can be linked with the average inter-particle distance according to $\xi_0^3 = (\rho/m)^{-1}$. The dimensionless characteristic coefficients σ_1 and σ_2 can be expressed in terms of integrals involving the anisotropic interaction potential and the pair correlation function [83]. Here, these quantities are treated as phenomenological coefficients.

In equilibrium, and in the absence of external orienting fields, the order parameter tensor is uniaxial, cf. (15.8), thus one has $a_{\mu\nu} = \sqrt{\frac{3}{2}} a_{\text{eq}} \overline{n_\mu n_\nu}$, where the order parameter $a_{\text{eq}} = \sqrt{5} S_{\text{eq}}$ is essentially the equilibrium value of the Maier-Saupe order parameter. Assuming that S_{eq} is constant, the free energy (15.35) reduces to the form (15.34) with the Frank elasticity coefficients given by

$$K_1 = K_3 = K_0 \left(\frac{1}{2}k_1 + k_2 \right), \quad K_2 = K_0 k_2, \quad (15.36)$$

$$K_0 = (\rho/m)\,\varepsilon_0\,\xi_0^2 = k_B T_{\text{ni}}\,\xi_0^{-1}, \quad k_{1,2} = 3\,a_{\text{eq}}^2\,\sigma_{1,2} = 15\,S_{\text{eq}}^2\,\sigma_{1,2}.$$

Clearly, for $k_1 = 0$, all three elasticity coefficients are equal, corresponding the isotropic case (15.31). On the other hand, to obtain the full anisotropy of the elastic energy with three different values for the Frank coefficients, the ansatz (15.35) has to be extended. An approach discussed in [83] is the inclusion of the fourth rank alignment tensor $a_{\mu\nu\lambda\kappa}$ in the theoretical description. This means, instead of (15.12), the potential

$$\Phi = \Phi^{\text{LdG}} - \frac{\sqrt{70}}{6} D\, a_{\mu\nu} a_{\lambda\kappa} a_{\mu\nu\lambda\kappa} + \frac{1}{2} E_0\, a_{\mu\nu\lambda\kappa} a_{\mu\nu\lambda\kappa},$$

is used with two additional coefficients D and E_0. In equilibrium, this implies

$$a_{\mu\nu\lambda\kappa} = \frac{\sqrt{70}}{6} \frac{D}{E_0} \overline{a_{\mu\nu} a_{\lambda\kappa}}.$$

As a side remark, from a lowest order expansion of the entropy associated with the single particle distribution function follows $A_0 = E_0 = 1$, $B = \sqrt{5}/7$, $C = 5/7$, see the Exercise 15.1, but also $D = 3/7$.

Furthermore, additional terms are included in the expression for the free energy density involving the spatial derivatives of the type

$$(\nabla_\lambda a_{\mu\nu})(\nabla_\kappa a_{\mu\nu\lambda\kappa}), \quad (\nabla_\kappa a_{\mu\nu\lambda\kappa})(\nabla_\tau a_{\mu\nu\lambda\tau}), \quad (\nabla_\tau a_{\mu\nu\lambda\kappa})(\nabla_\tau a_{\mu\nu\lambda\kappa}).$$

The mixed term with the product of the spatial derivatives of the second and fourth rank tensors provides the desired full anisotropy. The experimentally observed temperature dependence of the elasticity coefficients of ten liquid crystals can be fitted rather well when all terms are included in f_a^{inhom}, for details see the 1982 article of [83]. An alternative approach for the computation of the elasticity coefficients is presented in [84]. There a local perfect order is assumed which can be treated by an affine transformation model. A microscopic method for calculations of the twist elasticity coefficient K_2 is derived and tested in [85].

The variational principle applied to a free energy density $f = f(a_{\mu\nu}, \nabla_\lambda a_{\mu\nu}, \nabla_\lambda a_{\lambda\mu})$ leads to the differential equation

$$\frac{\partial f}{\partial a_{\mu\nu}} - \nabla_\lambda \frac{\partial f}{\partial \nabla_\lambda a_{\mu\nu}} - \nabla_\mu \frac{\partial f}{\partial \nabla_\lambda a_{\lambda\nu}} = 0. \tag{15.37}$$

In the simple case corresponding to the isotropic elasticity, this equation is

$$\Phi_{\mu\nu} - F_{\mu\nu} - \xi^2 \Delta a_{\mu\nu} = 0. \tag{15.38}$$

Here $\Phi_{\mu\nu}$ is the derivative of the potential Φ, with respect to $a_{\mu\nu}$, e.g. the Landau de Gennes expression (15.13), the tensor $F_{\mu\nu}$ characterizes the influence of an orienting electric or magnetic field, and the length ξ is linked with the quantities occurring in (15.35) via $\xi^2 = \frac{\varepsilon_0}{k_B T} \xi_0^2 \sigma_2$.

The local orientation of liquid crystals, as observed optically via its birefringence, i.e. between crossed polarizer and analyzer, may show defects. In the theoretical description based on the director field, the defects are treated as mathematical singularities. The alignment tensor theory, like (15.38) is closer to the physical reality. It takes into account, that the defect is not a point, but rather a spatial region, where the alignment tensor is no longer uniaxial, as assumed by the director theory. Furthermore, the magnitude of the order parameters are spatially dependent and in the core of a defect, the alignment can even vanish, i.e. locally, within a small volume, the fluid is isotropic [87]. For a specific example, the comparison between the alignment tensor theory and the director description is presented in [88]. An example for the alignment tensor field, in the vicinity of a point, which would be treated as a defect in a director description, is shown in Fig. 7.8. Notice that the sides of the bricks are the eigenvalues of a second rank tensor which is the sum of the alignment tensor and a constant isotropic tensor, chosen such that all eigenvalues are positive.

In *cholesterics* and *blue phase liquid crystals* an additional term linear in the spatial derivative of the alignment tensor has to be included in the free energy density (15.35). The contribution to the free energy density associated with the chirality is

$$f_a^{\text{chol}} = \frac{1}{2} \, (\rho/m) \, \varepsilon_0 \, \xi_0 \, \sigma^{\text{ch}} \, \varepsilon_{\nu\lambda\kappa} \, a_{\mu\nu} \, \nabla_\lambda \, a_{\kappa\mu}, \qquad (15.39)$$

where the coefficient σ^{ch} is a pseudo-scalar. For a uniaxial alignment $a_{\mu\nu} = \sqrt{3/2} a_{\text{eq}} \overline{n_\mu n_\nu}$ with a spatially constant order parameter $a_{\text{eq}} = \sqrt{5}S$, the expression (15.39) reduces to

$$f_a^{\text{chol}} = \frac{1}{2} \, (\rho/m) \frac{15}{2} S^2 \, \varepsilon_0 \, \xi_0 \, \sigma^{\text{ch}} n_\nu \varepsilon_{\nu\lambda\kappa} \, \nabla_\lambda \, n_\kappa.$$

Comparison with (15.34) shows that the coefficients q_0 and σ^{ch} characterizing the chiral behavior are linked according to $2K_2 q_0 = (\rho/m)\frac{15}{2}S^2\varepsilon_0\xi_0\sigma^{\text{ch}}$, for K_2 see (15.36).

The general structure of the free energy constructed from the spatial derivatives of the alignment tensor up to second order and of all orders in the second rank alignment tensor, with special emphasis on chiral terms, was studied in [89].

15.4 Cubatics and Tetradics

Some anisotropic fluids are composed of particles or have a local structures which cannot be described by a tensor of rank two, but where higher rank tensors, e.g. tensors of rank three or four are needed. These substances are referred to as *tetradics* and *cubatics*. Here, as in [92, 93], the term 'tetradic' is meant as an abbreviation for 'tetrahedratic', indicating a tetrahedral symmetry. Due to the closer resemblance of the theoretical treatment to that of the second rank case, cubatics are discussed first.

15.4.1 Cubic Order Parameter

Consider a reference particle in a dense liquid or solid and let $\mathbf{u}$ be a unit vector pointing to a nearest neighbor. The fourth rank irreducible tensor

$$a_{\mu\nu\lambda\kappa} = \zeta \, \overline{\langle u_\mu u_\nu u_\lambda u_\kappa \rangle} \tag{15.40}$$

is the lowest rank order parameter tensor which distinguishes a state with local cubic symmetry from an isotropic state. The numerical factor ζ can be chosen conveniently. The bracket $\langle \ldots \rangle$ indicates an average evaluated with an orientational distribution function $f(\mathbf{u})$, just as in Sect. 12.2.1. To indicate that here $\mathbf{u}$ does not specify the direction of a particle but rather the relative positions of particle neighbors, the term *bond orientational order* is used.

When the order parameter tensor has the full cubic symmetry, as in cubic crystals, and the coordinate axes are chosen parallel to the symmetry axes, the order tensor is proportional to the fourth rank cubic tensor defined in Sect. 9.5.1:

$$a_{\mu\nu\lambda\kappa} = \sqrt{\frac{5}{6}} a H_{\mu\nu\lambda\kappa}^{(4)},$$

$$H_{\mu\nu\lambda\kappa}^{(4)} = \sum_{i=1}^{3} e_\mu^{(i)} e_\nu^{(i)} e_\lambda^{(i)} e_\kappa^{(i)} - \frac{1}{5}(\delta_{\mu\nu}\delta_{\lambda\kappa} + \delta_{\mu\lambda}\delta_{\nu\kappa} + \delta_{\mu\kappa}\delta_{\nu\lambda}). \tag{15.41}$$

The $\mathbf{e}^{(i)}$, with $i = 1, 2, 3$ are unit vectors parallel to the cubic symmetry axes. Notice that $H_{\mu\nu\lambda\kappa}^{(4)} H_{\mu\nu\lambda\kappa}^{(4)} = 6/5$ and consequently $a_{\mu\nu\lambda\kappa} a_{\mu\nu\lambda\kappa} = a^2$. The order parameter a is essentially the average of a cubic harmonic, cf. (9.3.2),

$$a = \sqrt{\frac{5}{6}} \zeta \langle H_4 \rangle, \quad H_4 = H_{\mu\nu\lambda\kappa}^{(4)} \overline{u_\mu u_\nu u_\lambda u_\kappa} = u_1^4 + u_2^4 + u_3^4 - \frac{3}{5}. \tag{15.42}$$

The choice $\zeta = \sqrt{6/5}$ implies $a = \langle H_4 \rangle$. On the other hand, with $\zeta = (9!!)/(4!)$, one has $a = \langle K_4 \rangle$, where $K_4 \equiv \frac{5}{4}\sqrt{21} H_4$ has the normalization $(4\pi)^{-1} \int K_4^2 d^2 u = 1$, cf. (9.33). In this case, the expansion of the distribution function reads $f(\mathbf{u}) = (4\pi)^{-1}(1 + a K_4 + \ldots)$. The dots $\ldots$ stand for components of the fourth rank tensor which do not have the full cubic symmetry and for terms involving tensors of higher ranks $\ell = 6, 8, \ldots$

15.4.2 Landau Theory for the Isotropic-Cubatic Phase Transition

A phenomenological theory for the phase transition of an isotropic state to one with cubic symmetry can be made in analogy to the isotropic-nematic transition, cf. Sect. 15.2.2. Such an approach was first proposed independently, in [90] and [91] for cubic crystals. Notice, however, that the long range positional order typical for crystalline solids is not treated explicitly in this theory which focuses on the bond orientational order. To stress this point, the ordered state as treated here, is referred to as "cubatic", be it an ordered fluid like a smectic D liquid crystal, a fluid containing oriented cubic particles, or a true cubic crystal.

By analogy with the Landau-de Gennes theory for nematics, cf. Sect. 15.2.2, a dimensionless free energy potential is formulated:

$$\Phi = \Phi^{L} \equiv \frac{1}{2} A a_{\mu\nu\lambda\kappa} a_{\mu\nu\lambda\kappa} - \frac{1}{3}\sqrt{30} B a_{\mu\nu\lambda\kappa} a_{\lambda\kappa\sigma\tau} a_{\sigma\tau\mu\nu} + \frac{1}{4} C (a_{\mu\nu\lambda\kappa} a_{\mu\nu\lambda\kappa})^{2},$$

(15.43)

with

$$A = A_0 \left(1 - \frac{T^*}{T} \right), \quad A_0, C > 0, \quad \frac{2}{9} B^2 < A_0 C.$$

This ansatz is motivated as follows. The specific entropy is the sum of s_0 for the isotropic state and a contribution s_a associated with the bond orientational order. It is assumed that s_a is given by $s_a = \frac{k_B}{m} [\ldots]$, where $[\ldots]$ is equal to Φ^L as given by (15.43) but with A_0 instead of $A = A(T)$. Similarly, the specific volume $\rho^{-1} = \rho_0^{-1} + \rho_a^{-1}$ and the specific internal energy $u = u_0 + u_a$ are made up from isotropic parts and contributions linked with the order. The standard Gibbs relation $ds_0 = T^{-1}(du_0 + P d\rho_0^{-1})$, where P is the hydrostatic pressure, then leads to

$$ds = T^{-1}(du + P d\rho^{-1}) - \frac{k_B}{m} \frac{\partial \Phi}{\partial a_{\mu\nu\lambda\kappa}},$$

(15.44)

with the potential defined by

$$-\frac{k_B}{m} \Phi = s_a - T^{-1}(u_a + P \rho_a^{-1}).$$

(15.45)

The plausible assumption that the ordered state has a smaller energy and a smaller specific volume, characterized by ε and v_a, respectively, viz. $u_a = -\frac{1}{2}\varepsilon a_{\mu\nu\lambda\kappa} a_{\mu\nu\lambda\kappa}$ and $\rho_a^{-1} = -\frac{1}{2} v_a a_{\mu\nu\lambda\kappa} a_{\mu\nu\lambda\kappa}$, leads to the expression (15.43) with

$$A = A_0 \left[1 - T^{-1} \frac{m}{A_0 k_B} (\varepsilon + P v_a) \right] = A_0 \left[1 - \frac{T^*}{T} \right],$$

$$T^* = T^*(P) = \frac{m}{A_0 k_B} (\varepsilon + P v_a).$$

(15.46)

By analogy with the description of the alignment tensor elasticity, cf. Sect. 15.3.3, the potential function

$$\Phi = \Phi^L + \frac{1}{2}\xi_0^2 (\nabla_\tau a_{\mu\nu\lambda\kappa})(\nabla_\tau a_{\mu\nu\lambda\kappa}), \tag{15.47}$$

is used for a spatially inhomogeneous situation. The pertaining equilibrium state obeys the relation $\Phi_{\mu\nu\lambda\kappa} \equiv \frac{\partial \Phi}{\partial a_{\mu\nu\lambda\kappa}} = 0$, with

$$
\begin{aligned}
\Phi_{\mu\nu\lambda\kappa} &= A\,a_{\mu\nu\lambda\kappa} - \sqrt{30}\,B\,\overline{a_{\mu\nu\sigma\tau}a_{\sigma\tau\lambda\kappa}} \\
&\quad + C a_{\mu\nu\lambda\kappa}\,(a_{\mu'\nu'\lambda'\kappa'}a_{\mu'\nu'\lambda'\kappa'}) - \xi_0^2\,\Delta\,a_{\mu\nu\lambda\kappa}.
\end{aligned} \tag{15.48}
$$

When the order parameter tensor has the full cubic symmetry as described by (15.41), the potential function reduces to

$$\Phi = \frac{1}{2}A\,a^2 - \frac{1}{3}Ba^3 + \frac{1}{4}Ca^4 + \frac{1}{2}\xi_0^2\,(\nabla_\sigma a)(\nabla_\sigma a).$$

The pertaining equilibrium condition is

$$\frac{\partial \Phi}{\partial a} = Aa - Ba^2 + Ca^3 - \xi_0^2\,\Delta\,a. \tag{15.49}$$

For the spatially homogeneous situation, the equilibrium condition $a(A - Ba + Ca^2) = 0$ is equal to that one discussed for nematics, cf. Sect. 15.2.2. In particular, the equilibrium transition between the isotropic and cubic phases occurs at the temperature T_s, where $A(T_s) = 2B^2/(9C)$. There the order parameter is $a_s = 2B/(3C)$. At the temperature T^*, where $A(T^*) = 0$, one has $a(T^*) = B/C$. The sign of the equilibrium value of the order parameter $a_{\mathrm{eq}} = \frac{1}{2}BC^{-1}(1 + \sqrt{1 - 4AC/B^2})$, with $T < T_s$, is determined by the sign of B. The cubic order parameter for particles in the first coordination shell of *simple cubic* crystal is positive, that one of *bcc* and *fcc* crystals is negative.

15.2 Exercise: Compute the Cubic Order Parameter $\langle H_4 \rangle$ for Systems with Simple Cubic, bcc and fcc Symmetry
Hint: The coordinates of one the nearest neighbors, in the first coordination shells, are $(1, 0, 0)$ for simple cubic, $(1, 1, 1)/\sqrt{3}$ for bcc and $(1, 1, 0)/\sqrt{2}$ for fcc. Use symmetry arguments!

15.4.3 Order Parameter Tensor for Regular Tetrahedra

Consider a fluid composed of regular tetrahedra or of practically spherical particles which have first coordination shell with tetrahedral symmetry. Let $\mathbf{u}^i$, $i = 1, 2, 3, 4$

Fig. 15.7 Tetrahedron
embedded within a cube.
The *lines* connecting the
center with the four corners
of the tetrahedron show the
directions of the vectors $\mathbf{u}^i$

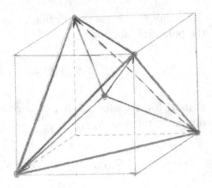

be four unit vectors pointing from the center to the corners of the tetrahedron. In this
case, the orientational order is specified by a third rank tensor $\mathscr{T}_{\mu\nu\lambda}$ defined by

$$\mathscr{T}_{\mu\nu\lambda} = \zeta_3 \left\langle \sum_{i=1}^{4} u^i_\mu u^i_\nu u^i_\lambda \right\rangle \sim \left\langle \overline{e^x_\mu e^y_\nu e^z_\lambda} \right\rangle, \tag{15.50}$$

where ζ_3 is a numerical factor which can be chosen conveniently. The corners of the
tetrahedron can be placed on the corners of a cube, cf. Fig. 15.7. The second relation
in (15.50) involves the body fixed unit vectors $\mathbf{e}^x$, $\mathbf{e}^y$ and $\mathbf{e}^z$ which are parallel to the
axes of this cube.

The tetradic third rank order parameter tensor $\mathscr{T}$ has negative parity, just like a
first rank dipolar order. This is in contradistinction to the second and fourth rank
order parameter tensors, which have positive parity. As first pointed out in [94], the
third rank order parameter is needed for the description of 'banana phases' of liquid
crystals which are composed of particles with a bent core.

15.5   Energetic Coupling of Order Parameter Tensors

Sometimes, more than one order parameter tensor is needed to describe the properties
of a substance and the relevant phenomena. In general, the pertaining free energy
contains coupling terms, whose structure depends on the ranks of the tensors involved.
The case of second and fourth rank tensors was already discussed in Sect. 15.3.3.
Three other examples, viz. the coupling of two second rank tensors, of a second rank
tensor with a vector and with a third rank tensor are presented here.

15.5.1   Two Second Rank Tensors

Let a and b be two symmetric traceless second rank tensors which describe the
orientational properties of a substance. Examples are the alignment tensor associated
with side groups and with the backbone of a side-chain polymer, as studied in [95, 96],

or the molecular alignment and the anisotropy of the pair-correlation function, in the first coordination shell. The dimensionless free energy $\Phi = \Phi(\mathbf{a}, \mathbf{b})$ is written as

$$\Phi = \Phi^a + \Phi^b + \Phi^{ab}, \quad \Phi^{ab} = c_1\, a_{\mu\nu}\, b_{\mu\nu} + \sqrt{6}\, c_2\, a_{\mu\kappa}\, a_{\kappa\nu}\, b_{\mu\nu}, \qquad (15.51)$$

where $\Phi^a = \Phi^a(\mathbf{a})$ and $\Phi^b = \Phi^b(\mathbf{b})$ are e.g. expressions of Landau-de Gennes type with coefficients A_a, B_a, C_a and A_b, B_b, C_b and $\Phi^{ab} = \Phi^{ab}(\mathbf{a}, \mathbf{b})$, with the coefficients c_1, c_2 characterizes the coupling between the two tensors. A term $\sim a_{\mu\kappa}\, b_{\kappa\nu}\, b_{\mu\nu}$ is possible, but disregarded here for simplicity. The derivatives of the potential with respect to the tensors are

$$\frac{\partial \Phi}{\partial a_{\mu\nu}} = \Phi^a_{\mu\nu} + c_1 b_{\mu\nu} + c_2\, 2\, \sqrt{6}\, \overline{a_{\mu\kappa} b_{\kappa\nu}}\,,$$

$$\frac{\partial \Phi}{\partial b_{\mu\nu}} = \Phi^b_{\mu\nu} + c_1 a_{\mu\nu} + c_2\, \sqrt{6}\, \overline{a_{\mu\kappa} a_{\kappa\nu}}\,. \qquad (15.52)$$

In thermal equilibrium, both expressions are zero. Then it is possible to determine $\mathbf{b}$ as function of $\mathbf{a}$ from the second equation of (15.52) and to insert it into the first equation of (15.52). This yields a derivative of a Landau-de Gennes potential, cf. (15.13), with renormalized coefficients A, B, C.

For the special case $\Phi^b = \frac{1}{2} A_b b_{\mu\nu} b_{\mu\nu}$ with $A_b = 1$ due to an appropriate choice of the normalization of $\mathbf{b}$, the result is

$$A = A_a - c_1^2, \quad B = B_a + 3 c_1 c_2, \quad C = A_a - 2 c_2^2. \qquad (15.53)$$

The derivation is deferred to the following exercise.

The renormalized coefficients A and C are smaller than A_a and C_a, irrespective of the sign of c_1 and c_2. The coefficient B is larger or smaller than B_a depending on whether the coupling coefficients c_1 and c_2 have equal or opposite sign.

The treatment of relaxation processes and other non-equilibrium phenomena of the kind presented in Chap. 17 is based on differential equations which contain the derivatives (15.52) of the relevant potential function, e.g. see [96].

15.3 Exercise: Renormalization of Landau-de Gennes Coefficients

Consider the special case where $\Phi^b = \frac{1}{2} A_b b_{\mu\nu} b_{\mu\nu}$, for simplicity put $A_b = 1$. Determine $b_{\mu\nu}$ from $\frac{\partial \Phi}{\partial b_{\mu\nu}} = 0$ with the help of the second equation of (15.52). Insert this expression into the first equation of (15.52) to obtain a derivative of a Landau-de Gennes potential with coefficients A, B, C which differ from the original coefficients A_a, B_a, C_a due to the coupling between the tensors.

Hint: use relation (5.51) for $\mathbf{a}$, viz. $\overline{a_{\mu\kappa} a_{\kappa\lambda} a_{\lambda\nu}} = \frac{1}{2} a_{\mu\nu} a_{\lambda\kappa} a_{\kappa\lambda}$.

15.5.2 Second-Rank Tensor and Vector

Let $\mathbf{d} \sim \langle \mathbf{e} \rangle$ and $\mathbf{a} \sim \langle \overline{\mathbf{u}\mathbf{u}} \rangle >$ be a polar vector and a symmetric traceless second rank alignment tensor which describe the orientational properties of a substance. Here $\mathbf{e}$ is a unit vector parallel to a molecular electric dipole moment which need not be parallel to $\mathbf{u}$. The electric polarization $\mathbf{P}$ is proportional to $\mathbf{d}$. The dimensionless free energy $\Phi = \Phi(\mathbf{d}, \mathbf{a})$ is written as

$$\Phi = \Phi^{\mathrm{d}} + \Phi^{\mathrm{a}} + \Phi^{\mathrm{da}}, \quad \Phi^{\mathrm{da}} = -c_1 d_\mu \nabla_\nu a_{\nu\mu} + \frac{1}{2} c_2 d_\mu d_\nu \, a_{\mu\nu}, \quad (15.54)$$

where $\Phi^{\mathrm{a}} = \Phi^{\mathrm{a}}(\mathbf{a})$ is a Landau-de Gennes potential function, $\Phi^{\mathrm{d}} = \Phi^{\mathrm{d}}(\mathbf{d})$ is a similar expression for the vector $\mathbf{d}$, and $\Phi^{\mathrm{da}} = \Phi^{\mathrm{da}}(\mathbf{d}, \mathbf{a})$, with the coefficients c_1, c_2 characterizes the coupling between the vector and the tensor. Terms of higher order are possible, but not included here, for simplicity. A scalar linear in both $\mathbf{d}$ and $\mathbf{a}$ must involve an additional vector, here it is the nabla-vector. The coupling coefficient c_1 is a true scalar when $\mathbf{d}$ is a polar vector. The corresponding expression for an axial vector must contain a coefficient c_1 which is a pseudo-scalar in order to conserve parity. The coefficient c_2 is a true scalar, in any case.

The derivatives of the potential with respect to the vector and to the tensor are

$$\frac{\partial \Phi}{\partial d_\mu} = \Phi^{\mathrm{d}}_\mu - c_1 \nabla_\nu a_{\nu\mu} + c_2 d_\nu \, a_{\nu\mu}, \quad \frac{\partial \Phi}{\partial a_{\mu\nu}} = \Phi^{\mathrm{a}}_{\mu\nu} + c_1 \overline{\nabla_\mu d_\nu} + \frac{1}{2} c_2 \, \overline{d_\mu d_\nu} \, .$$
$$(15.55)$$

In thermal equilibrium, these derivatives are equal to zero. For the special case where $\Phi^{\mathrm{d}} = \frac{1}{2} d_\mu d_\mu$ applies, one obtains

$$d_\mu + c_2 d_\nu \, a_{\nu\mu} = c_1 \nabla_\nu \, a_{\nu\mu}. \quad (15.56)$$

This relation underlies the *flexo-electric effect*, viz. an electric polarization $\mathbf{P}$ caused by spatial derivatives of the director field $\mathbf{n}$, in nematic liquid crystals. The phenomenological description of this effect is [67]

$$\mathbf{P} = e_1 \, \mathbf{n} \nabla \cdot \mathbf{n} + e_3 \, (\nabla \times \mathbf{n}) \times \mathbf{n},$$

which, due to $n_\nu \nabla_\mu n_\nu = 0$, is equivalent to

$$P_\mu = e_1 \, n_\mu \, \nabla_\nu n_\nu + e_3 \, n_\nu \nabla_\nu n_\mu. \quad (15.57)$$

The phenomenological coefficients e_1 and e_3 characterize the electric polarization caused by splay and by bend deformations, cf. Sect. 15.3.1. For the uniaxial alignment $a_{\mu\nu} = \sqrt{\frac{3}{2}} a_{\mathrm{eq}} \overline{n_\mu n_\nu}$ with the equilibrium order parameter a_{eq}, an expression of the form (15.57) is obtained from (15.56) with the help of the relation $P_\mu = P^{\mathrm{ref}} d_\mu$. Here P^{ref} is a reference value for the electric polarization which is proportional to the number density and to the permanent electric dipole moment of the molecules. The result is derived in the following exercise. Both flexo-electric coefficients are proportional to P^{ref} and to c_1, c_2.

The description of non-equilibrium phenomena dealing with the dynamics of the second rank tensor a, as presented in Chap. 17, can be extended to include the coupling with an electric dipole moment, cf. [97]. The differential equations contain the derivatives (15.55) of the potential function. Flexo-electric effects in cholesteric liquid crystals are treated in [98].

15.4 Exercise: Flexo-electric Coefficients

Start from (15.56) for the vector d_μ, use $a_{\mu\nu} = \sqrt{32}a_{\text{eq}}\,\overline{n_\mu n_\nu}$ and $P_\mu = P^{\text{ref}}d_\mu$ in order to derive an expression of the form (15.57) and express the flexo-electric coefficients e_1 and e_3 to c_1, c_2 and $a_{\text{eq}} = \sqrt{5}S$, where S is the Maier-Saupe order parameter. Furthermore, compute the contribution to electric polarization which is proportional to the spatial derivative of $a_{\text{eq}} = \sqrt{5}S$.

Hint: treat the components of **P** parallel and perpendicular to **n** separately.

15.5.3 Second- and Third-Rank Tensors

By analogy to the coupling between a second rank tensor with a vector, as treated in Sect. 15.5.2, the dimensionless free energy $\Phi = \Phi(a, \mathscr{T})$ underlying the coupling between the second rank tensor a and third rank tensor $\mathscr{T}$, cf. Sect. 15.4.3, is written as

$$\Phi = \Phi^{a} + \Phi^{\mathscr{T}} + \Phi^{a\mathscr{T}}, \quad \Phi^{a\mathscr{T}} = -c_1\,\mathscr{T}_{\mu\nu\lambda}\nabla_\mu a_{\nu\lambda} + \frac{1}{2}c_2\,\mathscr{T}_{\mu\kappa\lambda}\,\mathscr{T}_{\kappa\lambda\nu}\,a_{\mu\nu},$$

(15.58)

where $\Phi^{a} = \Phi^{a}(a)$ is a Landau-de Gennes potential function, $\Phi^{\mathscr{T}} = \Phi^{\mathscr{T}}(\mathscr{T})$ is a similar expression for the third rank tensor $\mathscr{T}$, and $\Phi^{a\mathscr{T}}$, with the coefficients c_1, c_2, characterizes the coupling between the second and third rank tensors. Terms of higher order are possible, but not included here, for simplicity. A scalar linear in both a and $\mathscr{T}$ must involve an additional vector, here it is the nabla-vector. The coupling coefficient c_1 is a true scalar when $\mathscr{T}$ has negative parity. The coefficient c_2 is a true scalar, in any case.

The derivatives of the potential with respect to the second and third rank tensors are

$$\frac{\partial\Phi}{\partial a_{\mu\nu}} = \Phi^{a}_{\mu\nu} + c_1\nabla_\lambda\overline{\mathscr{T}_{\lambda\mu\nu}} + \frac{1}{2}c_2\,\overline{\mathscr{T}_{\mu\lambda\kappa}\mathscr{T}_{\lambda\kappa\nu}}\,,$$

$$\frac{\partial\Phi}{\partial\mathscr{T}_{\mu\nu\lambda}} = \Phi^{\mathscr{T}}_{\mu\nu\lambda} - c_1\,\overline{\nabla_\mu a_{\nu\lambda}} + c_2\,\overline{\mathscr{T}_{\mu\nu\kappa}\,a_{\kappa\lambda}}\,.$$

(15.59)

In thermal equilibrium, these derivatives are equal to zero. In connection with the treatment of non-equilibrium phenomena, as considered in Chap. 17, the governing differential equations contain these derivatives of the potential function. The coupling between the friction pressure tensor and the tetrahedral order parameter tensor as discussed in [93] is described analogous to (15.58). Combined tetrahedral and nematic order in liquid crystals and the consequences for chirality are discussed in [99].

Chapter 16
Constitutive Relations

Abstract In this chapter is devoted to constitutive laws describing equilibrium and non-equilibrium properties in anisotropic media. Firstly, general principles, viz. the Curie principle, energy requirements, positive entropy production and Onsager-Casimir symmetry relations of irreversible thermodynamics are introduced. Secondly, phenomenological considerations and microscopic expressions are presented for the elasticity coefficients describing linear elastic deformations of solids, with emphasis on isotropic and cubic symmetries. Thirdly, the anisotropy of the viscous behavior and non-equilibrium alignment phenomena are studied for various types of fluids. The influence of magnetic and electric fields are analyzed for plane Couette and plane Poiseuille flows. Results of the kinetic theory are presented for the Senftleben-Beenakker effect of the viscosity. Consequences of angular momentum conservation are pointed out for the antisymmetric part of the pressure tensor. The flow birefringence in liquids and in gases of rotating molecules is treated as well as heat-flow birefringence in gases. The phenomenological description of visco-elasticity and of non-linear viscous behavior is discussed. Vorticity-free flow geometries are considered. The fourth part of the chapter deals with the viscosity and alignment in nematic liquid crystals. Viscosity coefficients are introduced which are needed to characterize the anisotropy of the viscosity in an oriented liquid crystal as well as in a free flow. Flow alignment and tumbling are considered. Model computations are presented for the viscosity coefficients as well as the application of a generalized Fokker-Planck equation for the non-equilibrium alignment. A unified theory for the isotropic and nematic phases is introduced and limiting cases are discussed. Equations governing the dynamics of the alignment in spatially inhomogeneous systems are formulated.

In addition to the general laws of physics, special relations are needed for the treatment of physical phenomena in specific substances. These *constitutive relations* involve *material coefficients*. Examples already encountered are relations between the electric polarization and the electric field, between the electric current density and the electric field or between the friction pressure tensor and the velocity gradient. The constitutive relations have to obey certain rules. These, as well as examples for and applications of constitutive relations are presented here.

© Springer International Publishing Switzerland 2015
S. Hess, *Tensors for Physics*, Undergraduate Lecture Notes in Physics,
DOI 10.1007/978-3-319-12787-3_16

16.1 General Principles

16.1.1 Curie Principle

Constitutive relations are laws of physics where, in general, tensors of rank ℓ are linked with tensors of rank k via equations like (2.51), viz.

$$b_{\mu_1\mu_2...\mu_\ell} = C_{\mu_1\mu_2...\mu_\ell \, \nu_1\nu_2...\nu_k} \, a_{\nu_1\nu_2...\nu_k}. \tag{16.1}$$

Here $\mathbf{C}$ is a tensor of rank $\ell + k$.

At about the same time, when Woldemar Voigt invented the notion 'tensor' and presented many applications in physics, Pierre Curie [2] formulated the principle which bears his name.

In our words, the *Curie Principle* says:

> the coefficient tensor $\mathbf{C}$ has to be in accordance with
> the symmetry of the physical system.

The statement can also be reverted: when both the tensors $\mathbf{a}$ and $\mathbf{b}$ are known, the coefficient tensor $\mathbf{C}$ reflects or reveals the symmetry of the system. The symmetry of the underlying physics should not be confused with the symmetry of tensors with respect to an interchange of indices, although there may be a close interrelation. When a microscopic physics model exists for the relation under study, the symmetry properties are usually 'obvious'. In many applications, however, the quantities $\mathbf{a}$ and $\mathbf{b}$ are just phenomenologically defined macroscopic observables. Even when a microscopic picture of the mechanisms underlying a constitutive relation like (16.1) are not known, symmetry considerations provide information on the coefficient tensor $\mathbf{C}$. In particular, the number of independent elements needed to quantify the $\mathbf{C}$ tensor is reduced by symmetry considerations.

Symmetry is closely associated with permanent or induced anisotropies. Examples for the latter are applied electric or magnetic fields, the gradients of these fields, as well as the normal on a bounding surface which also imposes a preferential direction. Preferential orientations in liquid crystals and the structure of crystalline solids are examples for anisotropies which are 'permanent' as long they are not partially destroyed by irreversible processes or by *symmetry breaking* nonlinear phenomena.

Further information on $\mathbf{C}$ is provided by *parity* and *time reversal* arguments, cf. Sects. 2.6 and 2.8. As a reminder: let P_a, P_b and P_C be the parities of the tensors occurring in (16.1). When one has

$$P_b = P_C \, P_a,$$

cf. (2.57), the parity is not violated by the relation (16.1). Usually, the parities of $\mathbf{a}$ and $\mathbf{b}$ are given by their physical meaning. Since the square of the parity value is

1, *parity invariance* of the relation (16.1) requires that the parity of the coefficient tensor **C** has to be

$$P_C = P_a \, P_b.$$

The application of symmetry and parity considerations for the nonlinear electric susceptibility is treated in Exercise 16.1.

Arguments similar to those used for the parity apply to the time reversal behavior. Let T_a, T_b and T_C, with $T^2 = 1$, specify the time reversal behavior of the tensors involved. When time reversal invariance holds true, one has

$$T_b = T_C \, T_a.$$

It is assumed that T_a and T_b are given by the physical meaning of a and b. For time reversal invariance to be valid, the condition

$$T_C = T_a \, T_b, \tag{16.2}$$

has to be fulfilled. When

$$T_C = -T_a \, T_b, \tag{16.3}$$

applies, the relation (16.1) breaks the time reversal invariance, a feature typical for *irreversible processes*. In some applications, the tensor **C** may not have a unique time reversal behavior because it contains contributions which are of reversible and others which are of irreversible character. Of course, this suffices to violate the time reversible invariance.

Energy considerations may require that certain elements of the tensor **C** have to be positive. The *second law of thermodynamics* also imposes conditions on the sign of coefficients describing irreversible processes. Examples are the Ohm law for electrical conduction, cf. Sect. 14.4.2 and the viscosity treated in Sect. 16.3.

16.1 Exercise: Nonlinear Electric Susceptibility in a Polar Material

In a medium without hysteresis, the electric polarization **P** can be expanded in powers of the electric field **E**, cf. (2.59), thus

$$P_\mu = \varepsilon_0 \left(\chi^{(1)}_{\mu\nu} \, E_\nu + \chi^{(2)}_{\mu\nu\lambda} \, E_\nu E_\lambda + \ldots \right).$$

The second rank tensor $\chi^{(1)}_{\mu\nu} \equiv \chi_{\mu\nu}$ characterizes the linear susceptibility. The third rank tensor $\chi^{(2)}_{\mu\nu\lambda}$ describes the next higher order contributions to **P**. Consider a material whose isotropy is broken by a polar unit vector **d**. Formulate the expressions for these tensors which are in accord with the symmetry and with parity conservation. Consider the cases **E** parallel and perpendicular to **d**.

16.1.2 Energy Principle

Consider a constitutive relation (16.1) where $\ell = k$ applies and where the scalar product

$$b_{\mu_1...\mu_\ell} a_{\mu_1...\mu_\ell}$$

is proportional to a contribution to the energy, which has to be positive. From

$$b_{\mu_1...\mu_\ell} a_{\mu_1...\mu_\ell} = a_{\mu_1...\mu_\ell} C_{\mu_1...\mu_\ell \, \nu_1...\nu_\ell} \, a_{\nu_1...\nu_\ell} > 0, \tag{16.4}$$

follows: the part of the tensor **C** which is symmetric under the interchange of the front and back set of ℓ indices, viz.

$$C^{\mathrm{sym}}_{\mu_1...\mu_\ell \nu_1...\nu_\ell} = \frac{1}{2} \left(C_{\mu_1...\mu_\ell \, \nu_1...\nu_\ell} + C_{\nu_1...\nu_\ell \, \mu_1...\mu_\ell} \right)$$

is positive definite. Notice, the part of the tensor **C** which is antisymmetric under the interchange is not necessarily zero.

As an example the dielectric tensor $\varepsilon_{\mu\nu}$ of a linear medium is considered, where the relation

$$D_\mu = \varepsilon_0 \, \varepsilon_{\mu\nu} \, E_\nu$$

applies. In this case, the energy density is $u^{\mathrm{el}} = \frac{1}{2} D_\mu E_\mu$, cf. (8.118), and consequently

$$u^{\mathrm{el}} = \frac{1}{2} E_\mu \, \varepsilon_{\mu\nu} \, E_\nu = \frac{1}{4} \left(\varepsilon_{\mu\nu} + \varepsilon_{\nu\mu} \right) E_\mu E_\nu \geq 0.$$

In general, in particular in the presence of an external or an internal magnetic field, the dielectric tensor possess an antisymmetric part $\varepsilon^{\mathrm{asym}}_{\mu\nu} = \frac{1}{2}(\varepsilon_{\mu\nu} - \varepsilon_{\nu\mu})$. The condition $u^{\mathrm{el}} > 0$ poses a condition on the symmetric part of the dielectric tensor, viz. $\varepsilon^{\mathrm{sym}}_{\mu\nu} = \frac{1}{2}(\varepsilon_{\mu\nu} + \varepsilon_{\nu\mu})$ has to be positive definite.

16.1.3 Irreversible Thermodynamics, Onsager Symmetry Principle

The density of the entropy production caused by irreversible processes is given by expressions of the type [108]

$$\left(\frac{\delta s}{\delta t} \right)_{\mathrm{irrev}} = - \left(J^{(1)}_{\mu_1...\mu_{\ell_1}} F^{(1)}_{\mu_1...\mu_{\ell_1}} + J^{(2)}_{\mu_1...\mu_{\ell_2}} F^{(2)}_{\mu_1...\mu_{\ell_2}} \right), \tag{16.5}$$

where the tensors $\mathbf{J}^{(..)}$ and $\mathbf{F}^{(..)}$ are referred to as *thermodynamic fluxes* and *thermodynamic forces*, respectively. Examples for such fluxes are the heat flux and the friction pressure tensor, the pertaining "forces" are the temperature gradient and velocity gradient tensor. Also a non-equilibrium alignment and its time derivative is such a force-flux pair, cf. Sect. 16.4.5. Typically, the pertaining forces and fluxes have opposite time reversal behavior, i.e. $T_J = -T_F$. Here two force-flux pairs are considered. The reduction to just one such pair is obvious. The generalization to more than two pairs can be formulated along the lines presented here.

In *irreversible thermodynamics*, the linear ansatz

$$
\begin{aligned}
-J^{(1)}_{\mu_1...\mu_{\ell_1}} &= C^{(11)}_{\mu_1...\mu_{\ell_1}\,\nu_1...\nu_{\ell_1}}\, F^{(1)}_{\nu_1...\nu_{\ell_1}} + C^{(12)}_{\mu_1...\mu_{\ell_1}\,\nu_1...\nu_{\ell_2}}\, F^{(2)}_{\nu_1...\nu_{\ell_2}} \\
-J^{(2)}_{\mu_1...\mu_{\ell_2}} &= C^{(21)}_{\mu_1...\mu_{\ell_2}\,\nu_1...\nu_{\ell_1}}\, F^{(1)}_{\nu_1...\nu_{\ell_1}} + C^{(22)}_{\mu_1...\mu_{\ell_2}\,\nu_1...\nu_{\ell_2}}\, F^{(2)}_{\nu_1...\nu_{\ell_2}},
\end{aligned}
\tag{16.6}
$$

is made for the constitutive relations. Positive entropy production requires

$$
\left(\frac{\delta s}{\delta t}\right)_{\text{irrev}} > 0,
\tag{16.7}
$$

and consequently, in symbolic notation,

$$
\mathbf{F}^{(1)} \cdot \mathbf{C}^{(11)} \cdot \mathbf{F}^{(1)} + \mathbf{F}^{(2)} \cdot \mathbf{C}^{(22)} \cdot \mathbf{F}^{(2)} + \mathbf{F}^{(1)} \cdot \mathbf{C}^{(12)} \cdot \mathbf{F}^{(2)} + \mathbf{F}^{(2)} \cdot \mathbf{C}^{(21)} \cdot \mathbf{F}^{(1)} > 0.
\tag{16.8}
$$

This imposes conditions on the coefficient tensors, in particular, the parts of the diagonal tensors $\mathbf{C}^{(11)}$ and $\mathbf{C}^{(22)}$, which are symmetric under the exchange of the front and back indices, have to be positive definite. Furthermore, the magnitude of the non-diagonal tensors $\mathbf{C}^{(12)}$ and $\mathbf{C}^{(21)}$ is bounded by the diagonal ones.

Now it is assumed that $\ell_1 = \ell_2 = \ell$ and that both forces $\mathbf{F}^{(1)}$ and $\mathbf{F}^{(2)}$ have the same behavior under time reversal, in obvious notation $T_{F1} = T_{F2}$, and both fluxes $\mathbf{J}^{(1)}$ and $\mathbf{J}^{(2)}$ have the opposite behavior, viz $T_{J1} = -T_{F1}$, $T_{J2} = -T_{F2}$. Then the *Onsager symmetry relation*, also called *reciprocal relations* [109]

$$
\mathbf{C}^{(12)} = \mathbf{C}^{(21)}
\tag{16.9}
$$

holds true. This symmetry relation for coefficients governing irreversible macroscopic behavior is based on the time reversal invariance of the underlying microscopic dynamics, For $\ell_1 \neq \ell_2$ still a symmetry relation like (16.9) applies, but the back and front indices have to be transposed on one side. Examples for the applications of Onsager symmetry relation are e.g. given in Sects. 16.3.6 and 16.4.5.

When the two forces have opposite time reversal behavior, i.e. when one has $T_{F1} = -T_{F2}$, the *Onsager-Casimir symmetry relation*

$$
\mathbf{C}^{(12)} = -\mathbf{C}^{(21)}
\tag{16.10}
$$

applies, instead of (16.9).

16.2 Elasticity

A property typical for a solid body is its elastic response to a small deformation. Of course, 'small' is relative, and really small for brittle substances. On the other hand, an elastic stick can be bent to a considerable amount. The elastic behavior is described by a linear constitutive relation between the stress tensor and the deformation tensor. The fourth rank elasticity tensor characterizes the elastic properties of specific solids.

16.2.1 Elastic Deformation of a Solid, Stress Tensor

Let $\mathbf{r}$ be the position vector to a volume element within a solid body. When this solid is subjected to a small deformation, this volume element is displaced to the position $\mathbf{r}' = \mathbf{r} + \mathbf{u}(\mathbf{r})$. Now consider two neighboring points which are separated by $d\mathbf{r}$ in the undeformed state. After the deformation, the difference vector between these two points is $d\mathbf{r}' = d\mathbf{r} + d\mathbf{u}$. The difference in the displacement is $du_\mu = dr_\nu \nabla_\nu u_\mu + \ldots$ where the higher order terms, indicated by the dots, can be disregarded for neighboring points. The distance squared, between these two points is $dr^2 = dr_\mu dr_\mu$ in the undeformed state and

$$(dr')^2 = dr'_\mu dr'_\mu = (dr_\mu + dr_\nu \nabla_\nu u_\mu)(dr_\mu + dr_\kappa \nabla_\kappa u_\mu),$$

in the deformed state. Thus one has

$$(dr')^2 = (\delta_{\mu\nu} + 2u_{\nu\mu})dr_\mu dr_\nu, \quad u_{\nu\mu} = \frac{1}{2}\left[\nabla_\nu u_\mu + \nabla_\mu u_\nu + (\nabla_\mu u_\lambda)(\nabla_\nu u_\lambda)\right],$$

$$(16.11)$$

where $u_{\nu\mu} = u_{\mu\nu}$ is the *deformation tensor*. Like any symmetric tensor, $u_{\nu\mu}$ can be diagonalized. With the principal values of the deformation tensor denoted by $u^{(i)}$, $i = 1, 2, 3$, relation (16.11) is equivalent to

$$(dr')^2 = \left(1 + 2u^{(1)}\right)dr_1^2 + \left(1 + 2u^{(2)}\right)dr_2^2 + \left(1 + 2u^{(3)}\right)dr_3^2,$$

in the principal axes system. The relative deformation-induced change of the length, along the principal direction i, is $((dr)'_i - dr_i)/dr_i = \sqrt{1 + 2u^{(i)}} - 1 \approx u^{(i)}$. Accordingly, the volume dV' in the deformed state is related to the original volume dV by

$$dV' = \sqrt{1 + 2u^{(1)}}\sqrt{1 + 2u^{(2)}}\sqrt{1 + 2u^{(3)}}dV \approx \left(1 + u^{(1)} + u^{(2)} + u^{(3)}\right)dV.$$

Thus the relative change $\delta V/dV$ of the volume

$$\delta V/dV = (dV' - dV)/dV = u^{(1)} + u^{(2)} + u^{(3)} = u_{\lambda\lambda}, \qquad (16.12)$$

is determined by the trace of the deformation tensor. The symmetric traceless part is associated with volume conserving squeeze, stretch or shear deformations.

In linear approximation to be used in the following, the deformation tensor defined by (16.11), reduces to

$$u_{\nu\mu} = \frac{1}{2}\left(\nabla_\nu u_\mu + \nabla_\mu u_\nu\right). \tag{16.13}$$

By definition, the *deformation tensor*, which is also called *strain tensor*, is symmetric: $u_{\nu\mu} = u_{\mu\nu}$. An antisymmetric part of $u_{\nu\mu}$ would induce an infinitesimal rotation but not a deformation.

A deformation of a solid causes a stress. Apart from the sign, the stress tensor $\sigma_{\mu\nu}$ is essentially the deviation of the pressure tensor $p_{\mu\nu}$ from its value in the undeformed state. The elastic properties of a solid are expressed in terms of the fourth rank elasticity tensor, which relates the stress tensor to the deformation tensor, viz.

$$-(p_{\mu\nu} - P\delta_{\mu\nu}) \equiv \sigma_{\mu\nu} = G_{\mu\nu,\lambda\kappa}\, u_{\lambda\kappa}. \tag{16.14}$$

In this linear constitutive relation, which is essentially *Hooke's law*, $\sigma_{\mu\nu} = \sigma_{\nu\mu}$ is presupposed. This assumption is common practice in solid state mechanics. The symmetry of the stress tensor holds true for substances composed of particles with a spherical symmetric interaction potential. In general, however, the the pressure tensor and consequently also the stress tensor can possess an antisymmetric part, cf. Sect. 16.3.5.

The deformation costs energy, this means

$$\sigma_{\mu\nu}\, u_{\mu\nu} = u_{\mu\nu}\, G_{\mu\nu,\lambda\kappa}\, u_{\lambda\kappa} > 0, \tag{16.15}$$

i.e. the elasticity tensor is positive definite, in a thermodynamically stable state. The symmetric tensors $\sigma_{\mu\nu}$ and $u_{\mu\nu}$ have 6 independent components, thus the elasticity tensor has 36 components, some of which can be zero. In accord with the Curie principle, the number of independent components of the elasticity tensor is considerably smaller, for certain types of the symmetry of the undeformed solid. In fact, just two coefficients suffice to characterize the linear elastic properties of an isotropic solid. For cubic symmetry, three different material coefficients are needed. These cases are discussed in Sect. 16.2.4. Group theoretical methods for the general crystal symmetries are found in text books devoted to Solid State Physics, in particular to the Mechanics of Solids.

16.2.2 Voigt Coefficients

A notation introduced by Voigt replaces the information contained in the fourth rank elasticity tensor $G_{\mu\nu,\lambda\kappa}$ by a six by six elasticity coefficient matrix c_{ij}. To this purpose, an appropriate Cartesian coordinate system is chosen and the components of $\sigma_{\mu\nu}$ and $u_{\mu\nu}$ are related to the components of σ_i and $u_I, i = 1, 2, .., 6$ according to

$$
\begin{aligned}
&\sigma_1 = \sigma_{xx}, \quad \sigma_2 = \sigma_{yy}, \quad \sigma_3 = \sigma_{zz}, \\
&\sigma_4 = \sigma_{yz} = \sigma_{zy}, \quad \sigma_5 = \sigma_{xz} = \sigma_{zx}, \quad \sigma_6 = \sigma_{xy} = \sigma_{yx} \\
&u_1 = u_{xx}, \quad u_2 = u_{yy}, \quad u_3 = u_{zz}, \\
&u_4 = u_{yz} + u_{zy}, \quad u_5 = u_{xz} + u_{zx}, \quad u_6 = u_{xy} + u_{yx}.
\end{aligned} \qquad (16.16)
$$

In this notation, the linear relation (16.14) between the stress and the strain tensors reads

$$
\sigma_i = \sum_{j=1}^{6} c_{ij}\, u_j. \qquad (16.17)
$$

The matrix c_{ij} of the Voigt elasticity coefficients is symmetric, $c_{ij} = c_{ji}$. The connection with the fourth rank elasticity tensor follows from (16.16), e.g. $c_{11} = G_{xx\,xx}$, $c_{12} = G_{xx\,yy}$, $c_{44} = 2G_{yz\,yz}$.

The coordinate axes are chosen such that they match the crystallographic axes. The choice is obvious for a *cubic system*. In the case of a *hexagonal system*, the z-axis is put parallel to the sixfold symmetry axis. There are crystallographic conventions for the general case.

16.2.3 Isotropic Systems

The trace of the deformation tensor is essentially the relative volume change $\delta V / V = u_{\lambda\lambda}$. The symmetric traceless part $\overline{u_{\mu\nu}}$ describes a squeeze or shear deformation. In an isotropic system, the trace of the stress tensor is linked with the trace of the strain tensor, $\sigma_{\lambda\lambda} \sim u_{\lambda\lambda}$. Similarly, the symmetric traceless parts of these tensors are proportional to each other, $\overline{\sigma_{\mu\nu}} \sim \overline{u_{\mu\nu}}$. Thus two material coefficients only occur in the linear elastic relation. These are the *bulk modulus* B and the *shear modulus* G.

The ansatz for an isotropic elastic solid is

$$
\sigma_{\mu\nu} = B u_{\lambda\lambda}\delta_{\mu\nu} + 2G\, \overline{u_{\mu\nu}}. \qquad (16.18)
$$

In this case, the fourth rank elasticity tensor is given

$$
G_{\mu\nu,\lambda\kappa} = B\delta_{\mu\nu}\delta_{\lambda\kappa} + 2G\, \Delta_{\mu\nu,\lambda\kappa}. \qquad (16.19)
$$

Mechanical stability requires $B > 0$ and $G > 0$. The moduli B and G are related to the tensor by

$$
B = \frac{1}{9}\, G_{\mu\mu,\lambda\lambda}, \quad G = \frac{1}{10}\, \Delta_{\mu\nu,\lambda\kappa}\, G_{\mu\nu,\lambda\kappa}. \qquad (16.20)
$$

The Voigt coefficients are

$$c_{11} = c_{22} = c_{33} = B + \frac{4}{3}G, \quad c_{12} = c_{23} = c_{31} = B - \frac{2}{3}G, \quad c_{44} = c_{55} = c_{66} = G,$$

other coefficients, like c_{14} or c_{45} are equal to zero. In terms of the c-coefficients, the bulk and shear moduli ate given by

$$B = \frac{1}{3}(c_{11} + 2c_{21}), \quad G = \frac{1}{5}(c_{11} - c_{21} + 3c_{44}).$$

Hooke's law (15.19) can be inverted to express the deformation in terms of the stress tensor. The relative volume change is given by

$$u_{\lambda\lambda} = \frac{1}{3B}\,\sigma_{\lambda\lambda}.$$

The full strain tensor obeys the relation

$$u_{\mu\nu} = \frac{1}{9B}\,\sigma_{\lambda\lambda}\,\delta_{\mu\nu} + \frac{1}{2G}\,\overline{\sigma_{\mu\nu}}\,. \tag{16.21}$$

A simple application is a homogeneous deformation of a body, e.g. the elongation or compression of a brick-shaped solid by a force F_z stretching or squeezing it along the z-direction. Than one has $\sigma_{zz} = F_z/A = k_z$, where A is the area of the face normal to the z-direction. In this case, all non-diagonal components of the strain tensor vanish and the diagonal ones are given by

$$u_{zz} = \frac{1}{3}\left(\frac{1}{3B} + \frac{1}{G}\right)k_z = \frac{1}{E}\,k_z, \quad u_{xx} = u_{yy} = -\frac{1}{3}\left(\frac{1}{2G} - \frac{1}{3B}\right)k_z = -\sigma\,u_{zz}.$$

Here E is the Young elastic modulus and σ is the contraction number. These material properties are related to the bulk and shear moduli by

$$E = \frac{9BG}{3B + G}, \quad 2\sigma = \frac{3B - 2G}{3B + G}. \tag{16.22}$$

For a practically incompressible substance, where $B \gg G$ applies, these expressions reduce to $E = 3G$ and $\sigma = 1/2$.

16.2.4 Cubic System

For a system with cubic symmetry, the elastic tensor, cf. the Hooke's law (16.14), is

$$G_{\mu\nu,\lambda\kappa} = B\delta_{\mu\nu}\delta_{\lambda\kappa} + 2G\,\Delta_{\mu\nu,\lambda\kappa} + 2G_c\,H^{(4)}_{\mu\nu\lambda\kappa}. \tag{16.23}$$

In addition to the bulk and shear moduli B and G, which already occur for isotropic systems, a third modulus G_c, specifically associated with the cubic symmetry, is needed here. The fourth rank tensor, cf. Sect. 9.5.1,

$$H_{\mu\nu\lambda\kappa}^{(4)} \equiv \sum_{i=1}^{3} \overline{e_\mu^{(i)} e_\nu^{(i)} e_\lambda^{(i)} e_\kappa^{(i)}} = \sum_{i=1}^{3} e_\mu^{(i)} e_\nu^{(i)} e_\lambda^{(i)} e_\kappa^{(i)} - \frac{1}{5}(\delta_{\mu\nu}\delta_{\lambda\kappa} + \delta_{\mu\lambda}\delta_{\nu\kappa} + \delta_{\mu\kappa}\delta_{\nu\lambda}),$$

reflects the full cubic symmetry. The unit vectors $\mathbf{e}^{(i)}$ are identified with the unit vectors $\mathbf{e}^x$, $\mathbf{e}^y$ and $\mathbf{e}^z$ parallel to the coordinate axes. Due to $H_{\mu\nu\lambda\kappa}^{(4)} \delta_{\mu\nu}\delta_{\lambda\kappa} = 0$, $H_{\mu\nu\lambda\kappa}^{(4)} \Delta_{\mu\nu,\lambda\kappa} = 0$, and $H_{\mu\nu\lambda\kappa}^{(4)} H_{\mu\nu\lambda\kappa}^{(4)} = \frac{6}{5}$, multiplication of (16.23) by the cubic tensor $H_{\mu\nu\lambda\kappa}^{(4)}$ yields

$$G_c = \frac{5}{12} H_{\mu\nu\lambda\kappa}^{(4)} G_{\mu\nu,\lambda\kappa}. \tag{16.24}$$

For the cubic symmetry, the Voigt coefficients are

$$c_{11} = B + \frac{4}{3}G + \frac{4}{5}G_c, \quad c_{12} = B - \frac{2}{3}G - \frac{2}{5}G_c, \quad c_{44} = G - \frac{2}{5}G_c. \tag{16.25}$$

By symmetry, one has $c_{11} = c_{22} = c_{33}, c_{12} = c_{23} = c_{31}$, and $c_{44} = c_{55} = c_{66}$. Other coefficients, like c_{14} or c_{45} are equal to zero. The coefficient $c_{66} = c_{44}$ is the shear modulus for a displacement $\mathbf{u}$ in the x-direction with its gradient in the y-direction. The shear modulus for a deformation in the xy-plane rotated by an angle of $45°$ from these directions is

$$\tilde{c}_{66} = \tilde{c}_{44} = G + \frac{3}{5}G_c. \tag{16.26}$$

In contradistinction to B and G, the cubic coefficient may have either sign. In fact, it is negative for body centered (bcc) and face centered (fcc) cubic crystals, while it is positive for simple cubic (sc) crystals. Mechanical stability requires that both shear moduli c_{44} and $\tilde{c}_{44}$ be positive. This sets lower and upper bounds on G_c:

$$-\frac{5}{3}G < G_c < \frac{5}{2}G. \tag{16.27}$$

In terms of the c-coefficients, the bulk and shear moduli are given by

$$B = \frac{1}{3}(c_{11} + 2c_{12}), \quad G = \frac{1}{5}(c_{11} - c_{12} + 3c_{44}), \quad G_c = \frac{1}{2}(c_{11} - c_{12} - 2c_{44}). \tag{16.28}$$

16.2.5 Microscopic Expressions for Elasticity Coefficients

Consider a system of N particles, located at the positions $\mathbf{r}^i$, $i = 1, 2, \ldots N$ within a volume V. The interaction potential is denoted by $\Phi = \Phi(\mathbf{r}^1, \mathbf{r}^2, \ldots, \mathbf{r}^N)$. When the potential is pairwise additive, one has $\Phi = \sum_{i<j} \phi^{ij} = \sum_{i<j} \phi(\mathbf{r}^{ij})$, where $\mathbf{r}^{ij} = \mathbf{r}^i - \mathbf{r}^j$, and $\phi = \phi(\mathbf{r})$ is the binary interaction potential. In thermal equilibrium, at the temperature T, the configurational part of the free energy F^{pot} is given by

$$\beta F^{\mathrm{pot}} = -\ln Z_{\mathrm{pot}}, \quad Z_{\mathrm{pot}} = \int \exp[-\beta\Phi]\, dr^{\{N\}}, \quad \beta = \frac{1}{k_b T},$$

where Z_{pot} is the configurational partition integral, $dr^{\{N\}}$ is the $3N$-dimensional volume element. The difference between the free energy, where the position vectors are displaced according to $r_\nu^i \to r_\nu^i + r_\mu^i u_{\mu\nu}$ and the original free energy, $\delta F^{\mathrm{pot}} = V p_{\mu\nu}^{\mathrm{pot}} u_{\mu\nu}$ yields the expression

$$V p_{\mu\nu}^{\mathrm{pot}} = \langle \Phi_{\mu\nu} \rangle, \quad \Phi_{\mu\nu} = \sum_i r_\mu^i \partial_\nu^i \Phi = -\sum_i r_\mu^I F_\nu^i, \quad \partial_\nu^i = \frac{\partial}{\partial r_\nu^i}, \tag{16.29}$$

for the potential contribution to the pressure tensor. Here F_ν^i is the force acting on particle i. In the absence of external forces, the total force vanishes: $\sum_i F_\nu^i = 0$. The bracket $\langle \cdots \rangle$ indicates the configurational canonical average

$$\langle \ldots \rangle = Z_{\mathrm{pot}}^{-1} \int \ldots \exp[-\beta\Phi]\, dr^{\{N\}}.$$

The change of the pressure tensor (16.29) under a deformation $r_\lambda^j \to r_\lambda^j + r_\kappa^j u_{\kappa\lambda}$ yields a relation between the potential contribution $\sigma_{\mu\nu}$ of the stress tensor and the deformation tensor, and consequently a microscopic expression for the elastic moduli. Starting from

$$\sigma_{\mu\nu} = -\left(p_{\mu\nu}^{\mathrm{pot,def}} - p_{\mu\nu}^{\mathrm{pot,0}}\right) = -\delta p_{\mu\nu}^{\mathrm{pot}} = \delta\left(V^{-1}\langle\Phi_{\mu\nu}\rangle\right),$$

where $p_{\mu\nu}^{\mathrm{pot,def}}$ and $p_{\mu\nu}^{\mathrm{pot,0}}$ are the pressure tensors in the strained and in the unstrained states, and using $\sigma_{\mu\nu} = G_{\mu\nu,\lambda\kappa}\, u_{\lambda\kappa}$, one obtains

$$VG_{\mu\nu,\lambda\kappa} = \langle\Phi_{\mu\nu,\lambda\kappa}\rangle_0 + V p_{\mathrm{pot}}\, \delta_{\mu\nu}, \delta_{\lambda\kappa} - \beta\left(\langle\Phi_{\mu\nu}\Phi_{\lambda\kappa}\rangle_0 - \langle\Phi_{\mu\nu}\rangle_0\langle\Phi_{\lambda\kappa}\rangle_0\right). \tag{16.30}$$

The subscript 0 in $\langle\ldots\rangle_0$ indicates the average in the undeformed state. The first term viz.

$$\Phi_{\mu\nu,\lambda\kappa} = \sum_j \sum_i r_\mu^i \partial_\nu^i r_\lambda^j \partial_\kappa^j \Phi \tag{16.31}$$

represents the deformation-induced variation of $\Phi_{\mu\nu}$. The second term on the right hand side of (16.30) stems from the volume change δV, since $\delta V/V = u_{\lambda\lambda}$ and $-\langle\Phi_{\mu\nu}\rangle_0 = Vp_{\text{pot}}\delta_{\mu\nu}$, where p_{pot} is the potential contribution to the pressure P.

For pairwise additive interaction, (16.31) reduces to

$$\Phi_{\mu\nu,\lambda\kappa} = \sum_{i<j}\phi^{ij}_{\mu\nu,\lambda\kappa}, \quad \phi^{ij}_{\mu\nu,\lambda\kappa} = \phi_{\mu\nu,\lambda\kappa}(\mathbf{r}^{ij}), \quad \phi_{\mu\nu,\lambda\kappa}(\mathbf{r}) = r_\mu\partial_\nu\, r_\lambda\partial_\kappa\,\phi(\mathbf{r}).$$

$$(16.32)$$

As before, the abbreviation $\mathbf{r}^{ij} = \mathbf{r}^i - \mathbf{r}^j$ is used, and $\phi(\mathbf{r})$ is the pair potential. For spherical particles where $\phi = \phi(r)$, with $r = |\mathbf{r}|$, holds true, one has

$$\phi_{\mu\nu} = r_\mu r_\nu\phi', \quad \phi_{\mu\nu,\lambda\kappa}(\mathbf{r}) = (r_\mu r_\kappa\delta_{\nu\lambda} + r_\nu r_\lambda\delta_{\mu\kappa})r^{-1}\phi' + r_\mu r_\nu r_\lambda r_\kappa r^{-1}(r^{-1}\phi')'.$$

$$(16.33)$$

The prime denotes the differentiation with respect to r.

The *Born-Green* expression for the elastic modulus tensor [100, 101], corresponds to the first and second terms of (16.30), when the total interaction potential is the sum of pair potentials. The remaining terms with the factor β are referred to as *fluctuation* contributions. In solids, the fluctuation parts of the elastic properties are small at low temperatures [102]. They are of crucial importance, however, for the fact that the *low frequency shear modulus* of a liquid vanishes whereas it has a finite value for a solid [103]. This underlies the fundamental difference in the mechanical behavior of a solid and a liquid.

The Born-Green part of the orientationaly averaged shear modulus is

$$G^{\text{BG}} = \frac{1}{15\,V}\left\langle\sum_{i>j}\left(r^3(r^{-1}\phi')'\right)^{ij}\right\rangle_0 - p_{\text{pot}} = \frac{1}{15\,V}\left\langle\sum_{i>j}\left(r^{-2}(r^4\phi')'\right)^{ij}\right\rangle_0.$$

$$(16.34)$$

The shear modulus G^{BG} can be expressed in terms of an integral over the pair correlation function g according to

$$G^{\text{BG}} = \frac{1}{30}\,n^2\,k_{\text{B}}T\int r^{-2}\,(r^4\phi')'\,g(\mathbf{r})\mathrm{d}^3r. \tag{16.35}$$

Here $n = N/V$ is the number density. The quantity G^{BG} is also called *high frequency shear modulus*. It is not only well defined in solid state, but also for liquids, where it reflects a rigidity, observable on a short time scale only. The pair correlation function can be written as $g(r) = \chi(r)\exp[-\beta\phi(r)]$. The quantity χ approaches 1 for small densities, corresponding to a dilute gas. Notice that G^{BG} is not zero even in this limit. The total shear modulus $G = G^{\text{BG}} + G^{\text{fluct}}$, on the other hand, vanishes in the liquid state and the more in a gas. This cancellation of G^{BG} by the fluctuation part G^{fluct} is remarkable, since the computation of G^{fluct} involves not only two-particle, but also three- and four-particle correlations. Let G^{pair} be the approximation for $G^{\text{BG}} + G^{\text{fluct}}$ in a fluid state and for low densities, where three- and four-particle correlations can be disregarded. This quantity can be expressed as an integral over the pair correlation function $g(r) = \chi(r)\exp[-\beta\phi(r)]$ which assumes the form

$$G^{\text{pair}} = \frac{1}{30} n^2 k_B T \int r^2 \chi(r)' \, (\exp[-\beta\phi(r)])' \, d^3r. \qquad (16.36)$$

In contradistinction to G^{BG}, the shear modulus G^{pair} vanishes in the small density limit where $\chi(r)' = 0$ applies.

For pairwise additive interaction, the Born-Green contributions to the bulk modulus B and to the cubic shear modulus G_c are

$$B^{\text{BG}} = \frac{5}{3} G^{\text{BG}} + 2p_{\text{pot}}, \qquad (16.37)$$

and

$$G_c^{\text{BG}} = \frac{5}{12V} \left\langle \sum_{i>j} \left(H^{(4)}(\mathbf{r}) \, r^{-1}(r^{-1}\phi')' \right)^{ij} \right\rangle_0, \quad H^{(4)}(\mathbf{r}) = x^4 + y^4 + z^4 - \frac{3}{5}r^4,$$
$$(16.38)$$

where $H^{(4)}$ is a cubic harmonic of order 4, with full cubic symmetry, cf. Sect. 9.5.2.

The fluctuation contributions to the elastic moduli are given by

$$VB^{\text{fluct}} = -\beta \left(\langle \Phi_{\text{iso}}^2 \rangle_0 - (\langle \Phi_{\text{iso}} \rangle_0)^2 \right), \qquad (16.39)$$

$$VG^{\text{fluct}} = -\frac{1}{5}\beta \left(3\langle \Phi_+^2 \rangle_0 + 2\langle \Phi_-^2 \rangle_0 \right), \quad VG_c^{\text{fluct}} = -\beta \left(\langle \Phi_+^2 \rangle_0 - \langle \Phi_-^2 \rangle_0 \right).$$

Here the abbreviations

$$\Phi_{\text{iso}} = \frac{1}{3}\sum_{i<j}(r\phi')^{ij}, \quad \Phi_+ = \sum_{i<j}(xyr^{-1}\phi')^{ij}, \quad \Phi_- = \frac{1}{2}\sum_{i<j}\left((x^2-y^2)r^{-1}\phi'\right)^{ij},$$
$$(16.40)$$

are used. The elastic moduli are the sum of the Born-Green and fluctuation contributions, e.g. $G = G^{\text{BG}} + G^{\text{fluct}}$.

The total shear modulus tensor also contains the kinetic contribution $nk_B T\delta_{\mu\nu}$ $\delta_{\lambda\kappa}$. This does not affect the shear moduli, but the total Voigt coefficients c_{11}^{total} and c_{12}^{total} are related the coefficients c_{11} and c_{12} used here, and to the moduli c_{11}^0 and c_{12}^0, where the pressure $P = nk_B T + p_{\text{pot}}$ is zero, by

$$c_{11}^{\text{total}} = c_{11} + nk_B T = c_{11}^0 - P, \quad c_{12}^{\text{total}} = c_{12} + nk_B T = c_{12}^0 + P.$$

For zero pressure and at temperatures, where the fluctuation contributions to the elasticity coefficients B and G are negligible, (16.37) implies the *Cauchy relation*

$$3B = 5G. \tag{16.41}$$

Upon the assumptions just mentioned, the Cauchy relation holds true for solids composed of spherical particles interacting with any pairwise additive potential, then the ratio G/B is equal to $3/5 = 0.6$. Experimental values for this ratio are smaller, e.g. one has $G/B \approx 0.5$ for the metals copper, nickel, iron, $G/B \approx 0.3$ for silver, and $G/B \approx 0.2$ for gold. The deviations from the Cauchy relation are closely associated with many particle interactions. An efficient way to include the relevant many particle interactions is provided by the *embedded atom method* [104, 105]. The basic idea of this method is stressed in [106]: the interaction of two particles is influenced by the density of the other particles within a well defined vicinity involving twenty to fifty particles. When this local density is too high or too low, compared with a prescribed density, the two particles under consideration feel an extra repulsion or attraction, respectively. Use of this method does not increase significantly the time needed in molecular dynamics computer simulations. A variant of the embedded atom method, realistic enough to model the elastic properties and simple enough to allow extended non-equilibrium molecular dynamics simulations of the visco-plastic behavior of metals, is presented in [107].

16.3 Viscosity and Non-equilibrium Alignment Phenomena

While the elasticity of solids is an equilibrium property, the viscous flow behavior of fluids is a typical non-equilibrium phenomenon. In both cases, symmetry considerations and the use of tensors play an important role. In this section, the viscosity in simple and in molecular fluids, the influence of external fields on the viscosity, as well as flow birefringence, heat-flow birefringence and visco-elasticity are treated.

16.3.1 General Remarks, Simple Fluids

In thermal equilibrium, the pressure tensor of a fluid is isotropic and it is given by $p_{\nu\mu} = P\delta_{\mu\nu}$, where P is the hydrostatic pressure. The part of the pressure tensor linked with non-equilibrium is decomposed into its irreducible isotropic, symmetric traceless and antisymmetric parts, cf. (7.53), according to

$$p_{\nu\mu} - P\delta_{\mu\nu} = \tilde{p}\,\delta_{\mu\nu} + \overline{p_{\nu\mu}} + \frac{1}{2}\varepsilon_{\nu\mu\lambda}\,p_\lambda,$$

with $p_\lambda = \varepsilon_{\lambda\alpha\beta} p_{\alpha\beta}$. The contributions to the entropy production $(\frac{\delta s}{\delta t})^{dv}_{irrev}$ associated with the flow velocity $\mathbf{v} = \mathbf{v}(t, \mathbf{r})$ follows from the local Gibbs relation linking the entropy density with the internal energy density and the time change of the macroscopic kinetic energy which, in turn, is governed by the local conservation law of the linear momentum, cf. Sect. 7.4.3. The resulting expression is

$$\frac{\rho}{m} T \left(\frac{\delta s}{\delta t}\right)^{dv}_{irrev} = -(p_{\nu\mu} - P\delta_{\mu\nu})\nabla_\nu v_\mu,$$

and consequently, after decomposition into the irreducible parts,

$$\frac{\rho}{m} T \left(\frac{\delta s}{\delta t}\right)_{irrev} = -\left(\tilde{p}\, \nabla_\lambda v_\lambda + \overline{p_{\nu\mu}}\, \overline{\nabla_\nu v_\mu} + \omega_\lambda p_\lambda\right), \quad \omega_\lambda = \frac{1}{2}\varepsilon_{\lambda\nu\mu}\nabla_\nu v_\mu.$$
(16.42)

The first and second terms are force-flux pairs involving tensors of ranks $\ell = 0$ and $\ell = 2$. The discussion of the case $\ell = 1$ associated with the antisymmetric part of the pressure tensor and the vorticity ω, is postponed to Sect. 16.3.5.

First, the attention is focussed on the symmetric pressure tensor. This is the case in *simple fluids*, where the pressure tensor is symmetric, on account of its symmetric kinetic and potential constituents, cf. (12.94) and (12.107). It also applies to more complex molecular fluids when the hydrodynamic processes described by the constitutive relations are slow compared on the time scale over which the antisymmetric part of the pressure tensor relaxes to zero.

In an isotropic fluid and in the absence of any external fields, the constitutive laws governing the viscous behavior are, cf. (7.55)

$$\tilde{p} = -\eta_V\, \nabla_\lambda v_\lambda, \quad \overline{p_{\nu\mu}} = -2\eta\, \overline{\nabla_\nu v_\mu}.$$

This ansatz is in accord with the Curie principle and it obeys the condition of positive entropy production when both the shear viscosity η and the volume viscosity η_V are non-negative.

The general scheme describing the viscous behavior of a fluid with s symmetric pressure tensor is

$$\overline{p_{\mu\nu}} = -2\eta_{\mu\nu\mu'\nu'}\, \overline{\nabla_{\mu'} v_{\nu'}} - \zeta^{(20)}_{\mu\nu}\, \nabla_\lambda v_\lambda,$$

$$\tilde{p} = -\zeta^{(02)}_{\mu\nu}\, \overline{\nabla_\mu v_\nu} - \eta_V\, \nabla_\lambda v_\lambda.$$
(16.43)

Here, $\eta_{\mu\nu\mu'\nu'}$ is the fourth rank shear viscosity tensor, η_V is the volume viscosity and the symmetric traceless coupling tensors $\zeta^{(..)}$ obey the Onsager relation

$$\zeta^{(20)}_{\mu\nu} = \zeta^{(02)}_{\mu\nu}.$$
(16.44)

Positive entropy production requires that

$$\eta^{\text{sym}}_{\mu\nu\mu'\nu'} = \frac{1}{2}(\eta_{\mu\nu\mu'\nu'} + \eta_{\mu'\nu'\mu\nu})$$

is positive definite, and that $\eta_V \geq 0$, as stated before. For the isotropic fluid without external fields, the shear viscosity tensor is proportional to the isotropic fourth rank tensor $\Delta_{\mu\nu,\mu'\nu'}$, thus one has

$$\eta_{\mu\nu\mu'\nu'} = \eta\, \Delta_{\mu\nu,\mu'\nu'}, \quad \eta > 0,$$

and $\zeta^{(20)}_{\mu\nu} = \zeta^{(02)}_{\mu\nu} = 0$, in this case. Next, the more specific expressions for the viscosity tensors are discussed for applied magnetic and electric fields.

16.3.2 Influence of Magnetic and Electric Fields

A magnetic field influences the viscosity via the Lorentz force, when the fluid contains mobile electric charges, or via the precession of magnetic moments in electrically neutral fluids. Examples for the latter substances are ferro-fluids, i.e. colloidal solutions containing particles with permanent or induced magnetic moments [110–112], as well as gases of rotating molecules [17]. The influence of orienting fields on the viscous behavior of liquid crystals deserves a separate discussion in Sect. 16.4.1.

Application of an electric field **E** on a fluid containing particles with permanent or induced electric dipole moments also renders the viscosity anisotropic. The resulting geometric symmetries are alike. The parity of the **B** and **E** fields, however, are different. This implies that terms of odd power in **E** violate parity invariance and are identical to zero, unless one is dealing with chiral substances.

Consider first an isotropic fluid which is subjected to a magnetic field $\mathbf{B} = B\mathbf{h}$, where **h** is a constant unit vector. The viscosity coefficients have to be in accord with this uniaxial symmetry. The obvious ansatz for the coupling tensors occurring in (16.43) is

$$\zeta^{(20)}_{\mu\nu} = \zeta^{(02)}_{\mu\nu} = \zeta\, \overline{h_\mu h_\nu}\,, \tag{16.45}$$

with a scalar phenomenological coefficient $\zeta = \zeta(B)$, which is an even function of B.

There are multiple ways to construct the fourth rank shear viscosity tensors in accord with symmetry of the physical situation. First, the fourth rank projection tensors of Sect. 14.2.2 are employed as basis tensors. By analogy to the construction of the rotation tensors, cf. Sect. 14.2.3, the viscosity tensor is written as, cf. [42],

$$\eta_{\mu\nu\mu'\nu'} = \sum_{m=-2}^{2} \eta^{(m)}\, \mathscr{P}^{(m)}_{\mu\nu,\mu'\nu'}. \tag{16.46}$$

The 5 complex viscosity coefficients $\eta^{(m)}$ have the properties $\eta^{(m)} = (\eta^{(-m)})^*$ and $\eta^{(0)}$, as well as the real parts of $\eta^{(\pm 1)}$ and $\eta^{(\pm 2)}$ are positive.

An alternative, but equivalent representation with real viscosity coefficients is

$$\eta_{\mu\nu,\mu'\nu'} = \eta^{(0)} \mathscr{P}^{(0)}_{\mu\nu,\mu'\nu'} \tag{16.47}$$

$$+ \sum_{m=1}^{2} \left[\eta^{(m+)} \left(\mathscr{P}^{(m)}_{\mu\nu,\mu'\nu'} + \mathscr{P}^{(-m)}_{\mu\nu,\mu'\nu'} \right) + \eta^{(m-)} i \left(\mathscr{P}^{(m)}_{\mu\nu,\mu'\nu'} - \mathscr{P}^{(-m)}_{\mu\nu,\mu'\nu'} \right) \right].$$

The coefficients $\eta^{(m+)} = (\eta^{(m)} + \eta^{(-m)})/2$ and $\eta^{(m-)} = (\eta^{(m)} - \eta^{(-m)})/2i$ are the real and imaginary parts of the coefficients $\eta^{(m)}$. The three non-negative coefficients $\eta^{(0)}$, $\eta^{(1+)}$ and $\eta^{(2+)}$ are even functions of B. The two coefficients $\eta^{(1-)}$ and $\eta^{(2-)}$ may have either sign and they are odd functions of B. The latter two coefficients are also referred to as *transverse viscosity coefficients*.

The shear viscosity tensor, as given by (16.46) or (16.47) obeys the symmetry property

$$\eta_{\mu\nu\mu'\nu'}(\mathbf{h}) = \eta_{\mu'\nu'\mu\nu}(-\mathbf{h}). \tag{16.48}$$

The de Groot-Mazur viscosities η_i^{dGM} of [108] are related to the coefficients $\eta^{(m)}$ by

$$\eta^{(0)} = \eta_1^{\mathrm{dGM}}, \quad \eta^{(1)} = \eta_2^{\mathrm{dGM}} + i\eta_5^{\mathrm{dGM}}, \quad \eta^{(2)} = 2\eta_2^{\mathrm{dGM}} - \eta_1^{\mathrm{dGM}} i - \eta_4^{\mathrm{dGM}}. \tag{16.49}$$

For an electric field $\mathbf{E}$ acting on an electrically neutral fluid containing particles with permanent or induced electric dipole moments the ansatz (16.47) can be used with the axial vector $\mathbf{h}$ replaced by a polar unit vector parallel to the electric field, but the Hall-effect like coefficients $\eta^{(1-)}$ and $\eta^{(2-)}$ are zero, due to parity conservation. In fluids containing chiral particles, however, these coefficients can be non-zero. A pseudo-scalar, characterizing the chirality, occurring as a factor in the Hall-effect like terms, ensures that the parity is still conserved.

16.3.3 Plane Couette and Plane Poiseuille Flow

To elucidate the meaning of the viscosities introduced in (16.46) and (16.47), a simple plane Couette flow is considered first. Again, the velocity is in the x-direction and its gradient in the y-direction. Then the velocity gradient tensor $\overline{\nabla_\mu v_\nu} = \gamma \, \overline{e_\mu^{(y)} e_\nu^{(x)}}$ with the shear rate $\gamma = \partial v_x / \partial y$, is a tensor with the same symmetry as considered in (14.62) and (14.63). An effective shear viscosity $\eta^{\mathrm{Couette}}(\mathbf{h})$, which depends on the direction of $\mathbf{h}$, is defined by

$$P_{yx} = -\eta^{\mathrm{Couette}}(\mathbf{h})\gamma, \quad \eta^{\mathrm{Couette}}(\mathbf{h}) = 2e_\mu^{(y)} e_\nu^{(x)} \eta_{\mu\nu\mu'\nu'}(\mathbf{h}) e_{\mu'}^{(y)} e_{\nu'}^{(x)}. \tag{16.50}$$

Its interrelation with the viscosity coefficients occurring in (16.47) can be inferred with the help of (14.62). The result is

$$\eta^{\text{Couette}}(\mathbf{h}) = 3h_x^2 h_y^2 \, \eta^{(0)} + \left[h_x^2 + h_y^2 - 4h_x^2 h_y^2\right] \eta^{(1+)} + \left[1 + h_x^2 h_y^2 - h_x^2 - h_y^2\right] \eta^{(2+)}. \tag{16.51}$$

The Couette viscosities for $\mathbf{h}$ parallel to the flow velocity, to its gradient and its vorticity, viz. the cases $\mathbf{h} = \mathbf{e}^x, \mathbf{e}^y, \mathbf{e}^z$ are denoted by $\eta_1, \eta_2,$ and η_3, respectively. From (16.51) follows

$$\eta_1 = \eta^{(1+)}, \quad \eta_2 = \eta^{(1+)}, \quad \eta_3 = \eta^{(2+)}. \tag{16.52}$$

The effective viscosity for the field parallel to the bisector between the x- and y-direction, viz. for $h_x^2 = h_y^2 = 1/2$ is denoted by η_{45}, referring to the 45° direction. Here (16.51) implies

$$\eta_{45} = \left(3\eta^{(0)} + \eta^{(2+)}\right)/4.$$

In the liquid crystal literature, the coefficients η_1, η_2, η_3 are called *Miesowicz viscosities* and 4 times the difference between the viscosity η_{45} and one half of the sum of η_1 and η_2, viz.

$$\eta_{12} \equiv 4\eta_{45} - 2(\eta_1 + \eta_2) = 3\eta^{(0)} + \eta^{(2+)} - 4\eta^{(1+)}. \tag{16.53}$$

is called *Helfrich viscosity*. The four effective viscosity coefficients linked with the Couette flow geometry, η_1, η_2, η_3 and η_{12} suffice to characterize the anisotropy of the shear viscosity.

The anisotropic viscosity tensor also gives rise to *normal pressure differences*, e.g. $p_{xx} - p_{yy}$, as well as to transverse components like p_{yz}. The meaning of these terms is elucidated for a *plane Poiseuille flow*.

Consider a plane Poiseuille flow in x-direction between two fixed flat plates which are perpendicular to the y-direction. The geometry is akin to that of the plane Couette, in as much as $\overline{\nabla_\mu v_\nu} = \gamma \, \overline{e_\mu^y e_\nu^x}$. The shear rate $\gamma = \frac{\partial v_x}{\partial y}$ is now, however, not constant, but a linear function of y, such that $\frac{\partial \gamma}{\partial y} = \frac{\partial^2 v_x}{\partial y^2} = \text{const}$.

For a stationary flow and in the absence of external accelerating forces, the linear momentum balance (7.52) implies

$$\nabla_\nu p_{\nu\mu} = \nabla_\mu \delta P + k_\mu = 0, \quad k_\mu = \nabla_\nu p_{\nu\mu}^{\text{fric}}. \tag{16.54}$$

Here δP is the flow-induced change of the hydrodynamic pressure and k_μ is the force density associated with the friction pressure tensor $p_{\nu\mu}^{\text{fric}} = \overline{p_{\nu\mu}} + \cdots$, where the dots stand for term involving the scalar part $\tilde{p}$ and the antisymmetric part of the tensor. When the two latter terms are zero or can be neglected, cf. (7.53), one has

$$\nabla_\mu \delta P = -k_\mu = 2\nabla_\nu \eta_{\mu\nu\mu'\nu'} \overline{\nabla_{\mu'} v_{\nu'}} . \tag{16.55}$$

For the geometry considered here, this relation is equal to

$$\nabla_\mu \delta P = -k_\mu = \gamma' e_\nu^y \eta_{\mu\nu\mu'\nu'} \overline{e_{\mu'}^y e_{\nu'}^x} , \tag{16.56}$$

where $\partial\gamma/\partial y = \gamma'$ is the derivative of the shear rate. The ratio between the *longitudinal pressure gradient* $e_\mu^x \nabla_\mu \delta P$ and γ' defines the effective shear viscosity $\eta^{\text{Pois}} = \eta^{\text{Pois}}(\mathbf{h})$, is equal to the effective viscosity for the Couette geometry (16.51), viz.

$$e_\mu^x \nabla_\mu \delta P / \gamma' = e_\mu^x e_\nu^y \, \eta_{\mu\nu\mu'\nu'} \, \overline{e_{\mu'}^y e_{\nu'}^x} = \eta^{\text{Pois}} = \eta^{\text{Couette}}(\mathbf{h}).$$

Similarly, the effective *normal viscosity* and *transverse viscosity* coefficients η^{norm} and η^{trans} can be defined via the *normal pressure gradient* $e_\mu^y \nabla_\mu \delta P$ and the *transverse pressure gradient* $-e_\mu^z \nabla_\mu \delta P$. These relations are

$$e_\mu^y \nabla_\mu \delta P = \gamma' e_\mu^y e_\nu^y \eta_{\mu\nu\mu'\nu'} \overline{e_{\mu'}^y e_{\nu'}^x} = \gamma' \eta^{\text{norm}}, \tag{16.57}$$

$$\eta^{\text{norm}} = h_x h_y \left[\eta^{(0)} \left(3h_y^2 - 1 \right) + \eta^{(1+)} \left(2 - 4h_y^2 \right) + \eta^{(2+)} \left(h_y^2 - 1 \right) \right]$$
$$+ h_y \left[2\eta^{(1-)} h_y^2 + \eta^{(2-)} \left(1 - h_y^2 \right) \right],$$

$$e_\mu^z \nabla_\mu \delta P = \gamma' e_\mu^z e_\nu^y \eta_{\mu\nu\mu'\nu'} \overline{e_{\mu'}^y e_{\nu'}^x} = \gamma' \eta^{\text{trans}}, \tag{16.58}$$

$$\eta^{\text{trans}} = h_x h_z \left[3\eta^{(0)} h_y^2 + \eta^{(1+)} \left(1 - 4h_y^2 \right) + \eta^{(2+)} \left(h_y^2 - 1 \right) \right]$$
$$+ h_z \left[\eta^{(1-)} \left(1 - 2h_y^2 \right) + \eta^{(2-)} \left(h_y^2 - 1 \right) \right].$$

In contradistinction to the longitudinal viscosity, which is positive, the normal and transverse effective viscosity coefficients may have either sign. The contributions to η^{norm} and η^{trans}, which involve the coefficients $\eta^{(1-)}$, $\eta^{(2-)}$, change sign when $\mathbf{h}$ is replaced by $-\mathbf{h}$.

As above, in connection with the effective shear viscosity, the labels $1, 2, 3$ are used for the field parallel to the x-, y- and z-directions. From (16.57) to (16.58) follows

$$\eta_1^{\text{norm}} = 0, \quad \eta_2^{\text{norm}} = 2\eta^{(1-)}, \quad \eta_3^{\text{norm}} = 0,$$
$$\eta_1^{\text{trans}} = 0, \quad \eta_2^{\text{trans}} = 0, \quad \eta_3^{\text{norm}} = \eta^{(1-)} - \eta^{(2-)}.$$

When the field points in the direction of the bisector between the x- and y-axes, viz. for $\mathbf{h} = (\mathbf{e}^x + \mathbf{e}^y)/\sqrt{2}$, as in (16.51), the (16.57) implies

$$2\eta_{45}^{\text{norm}} = \eta^{(0)} - \eta^{(2+)} + \left(2\eta^{(1-)} - \eta^{(2-)}\right)/\sqrt{2}.$$

The corresponding expression for the transverse pressure gradient with $\mathbf{h}$ parallel to the bisector between the x- and z-axes, now for $\mathbf{h} = (\mathbf{e}^x + \mathbf{e}^z)/\sqrt{2}$, is denoted by $\eta_{45'}^{\text{trans}}$ and given by

$$2\eta_{45'}^{\text{trans}} = \frac{3}{2}\eta^{(0)} + \eta^{(1+)} - \eta^{(2+)} + \sqrt{2}\left(\eta^{(1-)} - \eta^{(2-)}\right).$$

Effective viscosities for other directions of the field, e.g. for $\mathbf{h} = (\mathbf{e}^x + \mathbf{e}^y + \mathbf{e}^z)/\sqrt{3}$ can be inferred from (16.57) to (16.58).

16.3.4 Senftleben-Beenakker Effect of the Viscosity

The influence of a magnetic field on the transport properties of electrically neutral gases is referred to as *Senftleben-Beenakker effect*. This phenomenon was first noticed around 1930 for the paramagnetic gases O_2 and NO [113]. About thirty years later, Beenakker and coworkers [114] demonstrated that the influence of a magnetic field also occurs in diamagnetic gases like N_2. In fact, the effect is typical for all gases composed of rotating molecules [17], which have a rotational magnetic moment of the order of the nuclear magneton μ_N. The field-induced change of the transport properties is small, but relative changes can be detected with a high sensitivity.

The Senftleben-Beenakker effect of the viscosity is mainly due to the collisional coupling between the kinetic part of friction the pressure tensor $\overline{p_{\mu\nu}}$ and the tensor polarization $a_{\mu\nu}^T$, see Sect. 13.6.4. The equations governing the dynamics of these tenors can be derived from a kinetic equation referred to as *Waldmann-Snider equation* [17, 115]. It is a generalized Boltzmann equation for the distribution function operator $f = f(t, \mathbf{r}, \mathbf{p}, \mathbf{J})$, where the the position $\mathbf{r}$ and the linear momentum $\mathbf{p}$ are treated as classical variables, the internal angular momentum $\mathbf{J}$ is a quantum mechanical operator, as discussed in Sect. 13.6. Furthermore, the collision processes are treated quantum mechanically. In the presence of the magnetic field $\mathbf{B} = B\mathbf{h}$, the kinetic equation contains a commutator of the distribution operator f with the relevant Hamilton operator $H^B = -g_{\text{rot}}\mu_N \mathbf{J} \cdot \mathbf{B}$, where g_{rot} the a gyromagnetic factor specific for particular molecules. The quantity $\omega_B = (g_{\text{rot}}\mu_N B)/\hbar$ is the frequency with which the internal angular momentum precesses about the applied field.

Next, for simplicity of notation, $\overline{p_{\mu\nu}} \equiv \overline{p_{\mu\nu}}^{\text{kin}}$ and $a_{\mu\nu} \equiv a_{\mu\nu}^T$ are used. Furthermore, the ideal pressure of a gas with the number density n and the temperature T is denoted by $p_0 = nk_B T$. The resulting *transport-relaxation equations* are

$$\frac{\partial}{\partial t}\overline{p_{\mu\nu}} + 2\,p_0\,\overline{\nabla_\mu v_\nu} + v_p\,\overline{p_{\mu\nu}} + v_{pa}\,\sqrt{2}\,p_0\,a_{\mu\nu} = 0, \quad (16.59)$$

$$\frac{\partial}{\partial t}a_{\mu\nu} - \omega_B\,H_{\mu\nu,\mu'\nu'}\,a_{\mu'\nu'} + \left(\sqrt{2}p_0\right)^{-1}v_{ap}\,\overline{p_{\mu\nu}} + v_a\,a_{\mu\nu} = 0.$$

The fourth rank tensor $H_{\mu\nu,\mu'\nu'}$, defined by (14.26), emerges from the computation of the commutator $h_\lambda[J_\lambda, \overline{J_{\mu'}J_{\nu'}}]_-$ by analogy to (13.17), see also (13.19). The collision frequencies $v_{..}$ can be expressed in terms of collision integrals which involve the scattering amplitude in a binary way. The non-diagonal coefficients obey the Onsager symmetry relation $v_{ap} = v_{pa}$ and the inequalities $v_a > 0$, $v_p > 0$, $v_a v_p > v_{ap}^2$.

For a stationary situation and in the absence of the magnetic field, the equations (16.59) imply

$$\overline{p_{\mu\nu}} = -2\eta\,\overline{\nabla_\mu v_\nu}, \quad \eta = \eta_{iso}\,(1 - A_{pa})^{-1}, \quad \eta_{iso} = \frac{p_0}{v_p}, \quad A_{pa} = \frac{v_{pa}v_{ap}}{v_p\,v_a}.$$
$$(16.60)$$

The viscosity η is larger than η_{iso} which would be the value of the viscosity for an absolutely isotropic state where $a_{\mu\nu} = 0$.

For a stationary situation, with a magnetic field present, the solution of the coupled equations (16.59) with the methods discussed in Sect. 14.4, yields a viscosity tensor of the form (16.47) with the coefficients given as functions of $\varphi_a = \omega_B/v_a$ by

$$\eta^{(0)} = \eta, \quad \eta^{(m+)} - \eta = -\eta A_{pa}\frac{(m\varphi_a)^2}{1 + (m\varphi_a)^2}, \quad \eta^{(m-)} = -\eta A_{pa}\frac{m\varphi_a}{1 + (m\varphi_a)^2},$$
$$(16.61)$$

for $m = 1, 2$. Clearly, here the coefficient $\eta^{(0)}$ is not affected by the magnetic field. The even coefficients $\eta^{(1+)}$ and $\eta^{(2+)}$ decrease with increasing field strength from the zero field value η to the value η_{iso}. The ratio A_{pa} of the relaxation frequencies determines the magnitude of the relative change of the viscosity. The odd coefficients $\eta^{(1-)}$ and $\eta^{(2-)}$ vanish both for weak and for very strong magnetic fields and they have an extremum at $m\varphi_a = 1$, i.e., where the precession frequency $m\omega_B$ is equal to the collision frequency $v_a = \tau_a^{-1}$. The relaxation time τ_a is of the order of the average time between two collisions, which is the longer, the lower the pressure p_0 is. Due to $v_a \sim p_0$, one has $\varphi_a = \omega_B/v_a \sim B/p_0$. Thus the smaller magnetic moment of diamagnetic gases, compared with paramagnetic ones, is compensated by smaller pressures p_0, while the coefficients η and η_{iso} are independent of p_0, in gases at moderate pressures, say between 10^{-3} and 10 times the ambient pressure at room temperature.

Some historical remarks with a personal touch: The occurrence of Hall-effect like transverse terms $\eta^{(1-)}$ and $\eta^{(2-)}$ is surprising for a fluid without free electric charges. In fact, after the publication of the first measurements with paramagnetic gases, Max von Laue discussed the tensorial behavior of the viscosity, following the symmetry arguments of W. Voigt for the shear modulus and he claimed that Hall-effect like terms should not exist in this case. The transverse effects for transport in molecular gases were first treated theoretically in 1964 in the diploma thesis of the present author, in

Erlangen, Germany and, at about the same time independently by F.R. McCourt in his PhD work in Vancouver, Canada. Both advisers of the young scientists, viz. L. Waldmann and R.F. Snider approved the results of the calculations but considered them not worth being published since such effects cannot be measured. However, less than two years later, experiments were performed successfully.

16.3.5 Angular Momentum Conservation, Antisymmetric Pressure and Angular Velocity

Consider a fluid composed of particles with a rotational degree of freedom. Let the fluid have an average rotational velocity $\mathbf{w}$. The pertaining average internal angular momentum is denoted by $\mathbf{J}$, and $\mathbf{J} = \theta \mathbf{w}$ is assumed, with an effective moment of inertia θ. In the absence of torques due to external fields, the total angular momentum, i.e. the sum of the orbital angular momentum $\ell_\lambda = m\varepsilon_{\lambda\nu\mu} r_\nu v_\mu$, associated with average flow velocity $\mathbf{v}$ and the internal angular momentum obey a local conservation equation. Here m is the mass of a particle, ρ/m is the number density. From the local conservation of the linear momentum, cf. (7.54), follows

$$(\rho/m) \frac{d\ell_\lambda}{dt} + \varepsilon_{\lambda\kappa\mu} \nabla_\nu(r_\kappa p_{\nu\mu}) = \varepsilon_{\lambda\nu\mu} p_{\nu\mu} = p_\lambda,$$

where p_λ is the axial vector associated with the antisymmetric part of the pressure tensor. The corresponding equation for the internal angular momentum reads

$$\frac{\rho}{m} \frac{dJ_\lambda}{dt} + \ldots = -p_\lambda, \tag{16.62}$$

where the dots on the left hand side indicate gradient terms linked with the flux of internal angular momentum. The opposite sign of p_λ in the equations for ℓ_λ and J_λ guarantees the conservation of the total angular momentum. The change of the rotational energy $\mathbf{w} \cdot d\mathbf{J}/dt$, taken into account in the energy balance, leads to an extra term $w_\lambda p_\lambda$ in the entropy production (16.42). Thus the part of the entropy production involving axial vectors is

$$\frac{\rho}{m} T \left(\frac{\delta s}{\delta t}\right)^{\text{axvec}}_{\text{irrev}} = -p_\lambda(\omega_\lambda - w_\lambda). \tag{16.63}$$

The ansatz

$$p_\lambda = -2\,\eta_{\text{rot}}\,(\omega_\lambda - w_\lambda), \quad \eta_{\text{rot}} > 0, \tag{16.64}$$

is made where the *rotational viscosity* η_{rot} is a phenomenological coefficient. As a consequence, for a spatially homogeneous system, the average internal angular momentum obeys the relaxation equation

$$\frac{dJ_\lambda}{dt} + \tau_{\text{rot}}^{-1} (J_\lambda - \theta\omega_\lambda) = 0, \quad \tau_{\text{rot}}^{-1} = 2\frac{m}{\rho\theta}\eta_{\text{rot}}, \tag{16.65}$$

with the *rotational relaxation time* τ_{rot}. In the absence of external torques, the angular momentum **J** relaxes to $\theta\omega$, in a time span long compared with τ_{rot}. Then the average rotational velocity **w** matches the vorticity ω and the antisymmetric part of the pressure tensor vanishes.

Side Remarks:
(i) Spin Particles

An equation of the type (16.65) which relates, in a stationary situation, the average internal angular momentum with the vorticity, also applies for particles with spin, even for electrons. The *Barnett effect*, viz. the electron spin polarization and the ensuing magnetization caused by the rotation of a metal, is an experimental evidence for this phenomenon [117]. Here the question arises: what is the relevant moment of inertia θ in this case? Heuristic considerations [21] and a derivation from a generalized quantum mechanical Boltzmann equation with a nonlocal collision term [116] show: θ is determined by the mass of the electron times its thermal de Broglie wave length squared.

(ii) Polymer Coils

Let m be the mass and $\mathbf{r}^i$, $i = 1, \ldots N$ be the position vectors of the monomers of a polymer chain molecule. Its angular momentum is $\mathbf{L} = \sum_i m\mathbf{r}^i \times \dot{\mathbf{r}}^i$. In a liquid, the polymer molecule forms a coil which is spherical, on average, when the system is in thermal equilibrium. When the liquid is flowing, the polymer coil is stretched and it undergoes nonuniform rotations. A remarkable, though approximate, relation between the average angular velocity **w** and the shape of the polymer coil was put forward by Cerf [118]. The shape of the coil is expressed in terms of the radius of gyration tensor $G_{\mu\nu}$, cf. Sect. 5.3.2. More specifically, the long time average $\langle \mathbf{L} \rangle$ of the angular momentum is assumed to be equal to $m\langle \sum_i \mathbf{r}^i \times \mathbf{v}(\mathbf{r}^i) \rangle$ where **v** is the flow velocity field of the liquid. For a plane Couette flow with the velocity in x- and its gradient in y-direction, this relation corresponds to

$$\langle L_z \rangle = -\gamma m \langle G_{yy} \rangle,$$

where γ is the shear rate. The resulting average rotational velocity w_z is obtained by dividing $\langle L_z \rangle$ by the effective moment of inertia $m(\langle G_{xx} \rangle + \langle G_{yy} \rangle)$, thus

$$w_z = -\gamma \frac{\langle G_{yy} \rangle}{\langle G_{xx} \rangle + \langle G_{yy} \rangle} = \omega_z \left(1 - \frac{\langle G_{xx} \rangle - \langle G_{yy} \rangle}{\langle G_{xx} \rangle + \langle G_{yy} \rangle} \right). \tag{16.66}$$

For a weak flow, where the polymer coil retains its effectively spherical shape, one has $\langle G_{xx} \rangle = \langle G_{yy} \rangle$ and consequently $w_z = -\frac{1}{2}\gamma = \omega_z$, i.e. the average angular velocity matches the vorticity. For larger shear rates, the coil is stretched in the flow direction such that $\langle G_{xx} \rangle$ is larger than $\langle G_{yy} \rangle$, then the average rotational velocity is smaller than the vorticity. Non-Equilibrium Molecular Dynamics (NEMD) computer simulations presented in [119] show that the relation (16.66) is obeyed rather well. Simple models for the test of (16.66) are studied in [120, 121].

16.3.6 Flow Birefringence

A fluid composed of non-spherical, i.e. optically anisotropic particles, has optical isotropic properties in its isotropic phase. A viscous flow, however, causes a partial orientation of the particles. As a consequence, the symmetric traceless part of the dielectric tensor becomes non-zero. The resulting double refraction or birefringence is called *flow birefringence* or *streaming double refraction*. This effect was looked for, first observed and described by Maxwell [122], it is also referred to as *Maxwell effect*. The phenomenological ansatz is

$$\overline{\varepsilon_{\mu\nu}} = 2M\,\varepsilon_{\text{iso}}\,\eta\,\overline{\nabla_\mu v_\nu} = -2\beta\,\overline{\nabla_\mu v_\nu}\,. \qquad (16.67)$$

Here M is the *Maxwell coefficient*, ε_{iso} and η are the orientationally averaged dielectric coefficient and the viscosity. The flow birefringence coefficient β is related to the Maxwell coefficient by $\beta = -M\varepsilon_{\text{iso}}\eta$. For a plane Couette flow, with the velocity in x-direction and its gradient in y-direction, two of the principal axes of the dielectric tensor are parallel to the unit vectors $\mathbf{e}^{(1,2)} = (\mathbf{e}^x \pm \mathbf{e}^y)/\sqrt{2}$, the third axis is parallel to $\mathbf{e}^z$.

The relation (16.67) holds true in the absence of additional external fields and for small shear rates. In general, flow birefringence is described by a fourth rank tensor, analogous to the shear viscosity tensor (16.43). The constitutive relation for flow birefringence is

$$\overline{\varepsilon_{\mu\nu}} = -2\,\beta_{\mu\nu\mu'\nu'}\,\overline{\nabla_{\mu'} v_{\nu'}}\,. \qquad (16.68)$$

The simple case (16.67) corresponds to $\beta_{\mu\nu\mu'\nu'} = \beta\Delta_{\mu\nu,\mu'\nu'}$.

The symmetric traceless part of the dielectric tensor of molecular liquids and colloidal dispersions is proportional to the alignment tensor, viz. $\overline{\varepsilon_{\mu\nu}} = \varepsilon_a a_{\mu\nu}$, for ε_a see (12.20). Here flow birefringence is due to the shear flow induced alignment which results from a coupling between the alignment tensor and the friction pressure tensor. Point of departure for a treatment within the framework of irreversible thermodynamics is an expression for the contribution of the alignment to the entropy. In lowest order, this contribution is proportional to $-a_{\mu\nu}a_{\mu\nu}$, see Sect. 12.2.6. The time change of this expression is proportional to $-a_{\mu\nu}\mathrm{d}a_{\mu\nu}/\mathrm{d}t$. For the time change of the alignment tensor the educated guess

$$\frac{\mathrm{d}a_{\mu\nu}}{\mathrm{d}t} - 2\,\overline{\varepsilon_{\mu\lambda\kappa}\omega_\lambda a_{\kappa\nu}} = \left(\frac{\delta a_{\mu\nu}}{\delta t}\right)_{\text{irrev}} \qquad (16.69)$$

is made. The term involving the vorticity describes the time change due to a rotation with an average angular velocity equal to the vorticity. This holds true, when the anti-symmetric part of the pressure tensor vanishes, see the previous section. The operator $\frac{d}{dt} - 2\omega\times$ is referred to as *co-rotational time derivative*. The term $(\frac{\delta a_{\mu\nu}}{\delta t})_{\text{irrev}}$ is the time change due to irreversible processes which occurs in the entropy production. The part associated with second rank tensors is now

$$\frac{\rho}{m} T \left(\frac{\delta s}{\delta t}\right)_{\text{irrev}}^{(2)} = -\left[\overline{p_{\nu\mu}}\ \overline{\nabla_\nu v_\mu} + \frac{\rho}{m} k_B T\ a_{\mu\nu} \left(\frac{\delta a_{\mu\nu}}{\delta t}\right)_{\text{irrev}}\right]. \qquad (16.70)$$

With $a_{\mu\nu}$ and $(\sqrt{2}\frac{\rho}{m}k_B T)^{-1} \overline{p_{\nu\mu}}$ chosen as fluxes, and $(\frac{\delta a_{\mu\nu}}{\delta t})_{\text{irrev}}$ and $\sqrt{2}\ \overline{\nabla_\nu v_\mu}$ as forces, constitutive laws for the second rank tensors are

$$- a_{\mu\nu} = \tau_a \left(\frac{\delta a_{\mu\nu}}{\delta t}\right)_{\text{irrev}} + \tau_{ap} \sqrt{2}\ \overline{\nabla_\nu v_\mu}, \qquad (16.71)$$

$$-\left(\sqrt{2}\frac{\rho}{m} k_B T\right)^{-1} \overline{p_{\nu\mu}} = \tau_{pa} \left(\frac{\delta a_{\mu\nu}}{\delta t}\right)_{\text{irrev}} + \tau_p \sqrt{2}\ \overline{\nabla_\nu v_\mu}.$$

Here the quantities $\tau_{..}$ are relaxation time coefficients where the subscripts a and p refer to "alignment" and "pressure". The non-diagonal coefficients obey the Onsager symmetry relation

$$\tau_{ap} = \tau_{pa}. \qquad (16.72)$$

Positive entropy production is guaranteed by the inequalities

$$\tau_a > 0, \quad \tau_p > 0, \quad \tau_a \tau_p > \tau_{ap}^2. \qquad (16.73)$$

Use of the first of the (16.71) in (16.69) yields the inhomogeneous relaxation equation

$$\frac{da_{\mu\nu}}{dt} - 2\ \overline{\varepsilon_{\mu\lambda\kappa}\omega_\lambda a_{\kappa\nu}} + \tau_a^{-1} a_{\mu\nu} = -\tau_a^{-1} \tau_{ap} \sqrt{2}\ \overline{\nabla_\nu v_\mu}. \qquad (16.74)$$

The term involving the vorticity gives rise to effects nonlinear in the shear rate. When these are disregarded, and for the stationary case, where the time derivative vanishes, (16.74) leads to flow alignment

$$a_{\mu\nu} = -\tau_{ap} \sqrt{2}\ \overline{\nabla_\nu v_\mu}, \qquad (16.75)$$

and consequently. The flow birefringence coefficient is given by

$$\beta = -\frac{\varepsilon_a}{\sqrt{2}} \tau_{ap}. \qquad (16.76)$$

Clearly, the coupling coefficient τ_{ap} is essential for the flow birefringence.

To study the effect of the vorticity on the flow birefringence, a plane Couette geometry is considered, with the flow velocity in x-direction and its gradient in y-direction, cf. (7.28), Fig. 7.6, and Sect. 12.4.6. In this case, the shear rate tensor and the vorticity are given by $\gamma_{\mu\nu} = \overline{\nabla_\nu v_\mu} = \gamma\, \overline{e_\nu^x e_\mu^y}$ and $\omega_\lambda = -\frac{1}{2}\gamma e_\lambda^z$, where $\gamma = \partial v_x/\partial y$ is the shear rate and $\mathbf{e}^i$, $i = x, y, z$ are the unit vectors parallel to the coordinate axes. Insertion of the symmetry adapted ansatz

$$a_{\mu\nu} = a_1\, (e_\nu^x e_\mu^y - e_\nu^y e_\mu^x)/\sqrt{2} + a_2\, \sqrt{2}\, \overline{e_\nu^x e_\mu^y}$$

into the inhomogeneous relaxation equation (16.74) leads to coupled equations for the coefficients a_1 and a_2. These equations are

$$\dot{a}_1 - \gamma\, a_2 + \tau_a^{-1}\, a_1 = 0, \tag{16.77}$$
$$\dot{a}_2 + \gamma\, a_1 + \tau_a^{-1}\, a_2 = -\tau_{ap}\, \tau_a^{-1}\, \gamma.$$

For a stationary situation, one obtains $a_1 = \gamma\tau_a$ and $a_2 = -\gamma\tau_{ap}(1 + \gamma^2\tau_a^2)^{-1}$. In the small shear rate limit where $\gamma\tau_a \ll 1$ applies, this result for a_2 corresponds to (16.75), for the geometry considered here. Due to $a_1 = a\cos 2\varphi$, $a_2 = a\sin 2\varphi$, where $a^2 = a_1^2 + a_2^2$, the stationary solutions are equivalent to

$$a = |\tau_{ap}|\,\frac{\gamma}{\sqrt{1 + \gamma^2\,\tau_a^2}}, \quad \tan 2\,\varphi = \frac{1}{\gamma\tau_a}. \tag{16.78}$$

For large shear rates, the magnitude a of the alignment saturates at the value $|\tau_{ap}|\tau_a^{-1}$. The angle φ, in the present context referred to as flow angle, indicates the directions of the principal axes of the alignment tensor, within the xy-plane. More specifically, one of these axes encloses the angle φ with the x-direction, the other one the angle $\varphi + 90°$, the third principal direction is parallel to the z-axis. In the small shear rate limit, one has $\varphi = 45°$. At higher shear rates, this principal axis approaches the flow direction.

The results (16.77) and (16.78) pertain to (16.74) where terms nonlinear in the shear rate enter only via the co-rotational time derivative. In general, other nonlinearities occur in the dynamic equation for the alignment tensor. These are, e.g. an additional term proportional to $\gamma_{\mu\kappa}a_{\kappa\nu}$ and terms nonlinear in the alignment tensor, as encountered in connection with the phase transition isotropic-nematic. The relevant equations and their consequences are discussed in Sects. 16.4.4, 16.4.5, and 17.3.

For *gases of rotating molecules*, cf. (13.65), one has $\overline{\varepsilon_{\mu\nu}} = \varepsilon_a^T a_{\mu\nu}^T$, where $a_{\mu\nu}^T$ is the tensor polarization associated with the rotational angular momenta. In this case not only a theoretical treatment analogous to that one above is possible, but a kinetic theory approach based on the *Waldmann-Snider equation*, a generalized quantum mechanical Boltzmann equation. The resulting inhomogeneous relaxation equations (16.59) are similar to the ansatz (16.71), just with the role of forces and fluxes exchanged. This means the relaxation time "matrix" formed by the $\tau_{..}$ coefficients is the reciprocal of the relaxation frequency $\nu_{..}$ matrix. Here, reciprocal relaxation time coefficients have a microscopic interpretation since they are expressed in terms of collision integrals involving the binary scattering amplitude [17, 62, 64]. To be more specific, consider a stationary situation in the absence of a **B** field. Then the second equation of (16.59) implies

$$a_{\mu\nu} = -\left(\sqrt{2}\,p_0\right)^{-1} v_a^{-1}\, \nu_{ap}\, \overline{p_{\mu\nu}}$$

and, with $\overline{p_{\mu\nu}} = -2\eta\,\overline{\nabla_\mu v_\nu}$, cf. (16.60), one obtains

$$a_{\mu\nu} = 2\eta \left(\sqrt{2}\,p_0\right)^{-1} v_a^{-1}\, \nu_{ap}\, \overline{\nabla_\mu v_\nu}.$$

Thus in this case the flow birefringence coefficient is given by [62]

$$\beta = -\frac{\varepsilon_a^T}{\sqrt{2}}\,\tau_{ap}, \quad \tau_{ap} = -\nu_{ap}\,(\nu_p\,\nu_a)^{-1}\,(1-A_{pa})^{-1}. \tag{16.79}$$

This relation is similar to (16.76), but with ε_a^T instead of ε_a and the coupling coefficient τ_{ap} is expressed in terms of the relaxation frequencies $\nu_{..}$. Whereas the Maxwell effect in colloidal dispersions, molecular liquids and polymeric fluids [123] has been studied experimentally for over a century, the flow birefringence in molecular gases was first measured by F. Baas in 1971 [124], see also [17, 125].

The flow birefringence is the manifestation of a *cross effect*: a viscous flow causes an alignment. There is a *reciprocal effect*: a non-equilibrium alignment gives rise to an extra contribution $\overline{p_{\nu\mu}^a}$ to the symmetric traceless pressure tensor [126]. The alignment, in turn, influences the flow properties. This back-coupling underlies the influence of a magnetic field on the viscosity in molecular gases as discussed above, and the nonlinear flow behavior in molecular liquids and colloidal dispersions of non-spherical particles, as treated in Sect. 16.3.9.

16.2 Exercise: Acoustic Birefringence

Sound waves cause an alignment of non-spherical particles in fluids. The ensuing birefringence is called *acoustic birefringence*. Use (16.74) to compute the sound-induced alignment tensor for the velocity field $\mathbf{v} = v_0 k^{-1}\mathbf{k}\cos(\mathbf{k}\cdot\mathbf{r} - \omega t)$ where $\mathbf{k}$ and ω are the wave vector and the frequency of the sound wave, v_0 is the amplitude.

Hint: Use the complex notation $v_\mu \sim \exp[i(\mathbf{k}\cdot\mathbf{r} - \omega t)]$ and $a_{\mu\nu} \sim \exp[i(\mathbf{k}\cdot\mathbf{r} - \omega t)]$ to solve the inhomogeneous relaxation equation, then determine the real and the imaginary part of $a_{\mu\nu}$.

16.3.7 Heat-Flow Birefringence

Also a heat flux **q** can give rise to birefringence. By analogy to the Maxwell effect
(16.67), the constitutive relation for the *heat-flow birefringence* is

$$\overline{\varepsilon_{\mu\nu}} = -2\,\beta_{\mathrm{q}}\,\overline{\nabla_\mu q_\nu} = 2\,\beta_{\mathrm{q}}\,\lambda\,\overline{\nabla_\mu \nabla_\nu}\,T. \qquad (16.80)$$

Here β_{q} is the *heat-flow birefringence coefficient*. The second equality in (16.80),
involving the second spatial derivative of the temperature field T follows from $q_\nu =$
$-\lambda\nabla_\nu T$ where λ is the heat conductivity. The existence of the effect (16.80) was
predicted [62] and calculated [127] for rarefied molecular gases. First measurements
were presented in [128], see also [17].

In gases, the flow birefringence, just like the viscosity, does not depend on the
density, whereas the heat-flow birefringence is inversely proportional to the density.
Theoretical considerations [129] predict the existence of heat-flow birefringence also
in dense fluids. Experiments in a nematic glass [130] are a manifestation of this effect.

In mixtures, a preferential alignment can also be caused by the gradient of a diffusion flux **j**. The resulting birefringence is referred to as *diffusio-birefringence* [131].

16.3.8 Visco-Elasticity

Elasticity, which is a reversible process, is a typical property of solids. The irreversible
viscous flow behavior is typical for fluids. On a short time scale or for shear rates
varying with high frequencies, however, also fluids show elasticity. The *Maxwell
model* for the symmetric traceless friction pressure tensor $\overline{p_{\mu\nu}}$, viz.

$$\tau_{\mathrm{M}}\,\frac{\partial}{\partial t}\,\overline{p_{\mu\nu}} + \overline{p_{\mu\nu}} = -2\eta\,\overline{\nabla_\mu v_\nu}\,, \qquad \eta = G\tau_{\mathrm{M}}, \qquad (16.81)$$

is a prototype for the description of the *visco-elastic behavior*. Here τ_{M} is the *Maxwell
relaxation time*, η is the shear viscosity and G is the high frequency shear modulus.
Notice that $\overline{\nabla_\mu v_\nu} = \frac{\partial}{\partial t}\overline{u_{\mu\nu}}$. Thus for fast varying processes, where $\tau_{\mathrm{M}}|\frac{\partial}{\partial t}\overline{p_{\mu\nu}}| \gg$
$|\overline{p_{\mu\nu}}|$, (16.81) reduces to $\frac{\partial}{\partial t}\overline{p_{\mu\nu}} = -2G\frac{\partial}{\partial t}\overline{u_{\mu\nu}}$ or $-\overline{p_{\mu\nu}} = \overline{\sigma_{\mu\nu}} = 2G\overline{u_{\mu\nu}}$,
which corresponds to the constitutive law (16.21) for elasticity.

For a periodic velocity gradient proportional to $\exp[-i\omega t]$, the linear equation
(16.81) implies that the friction pressure has the same dependence on the frequency
ω and the time t. In this case (16.81) can be written as $\overline{p_{\mu\nu}} = -2\eta(\omega)\,\overline{\nabla_\mu v_\nu}$ with
the complex frequency dependent viscosity

$$\eta(\omega) = \eta\,(1 - i\omega\,\tau_{\mathrm{M}})^{-1} = \eta'(\omega) + i\eta''(\omega), \qquad (16.82)$$

$$\eta'(\omega) = \eta\,\frac{1}{1 + (\omega\tau_{\mathrm{M}})^2}, \qquad \eta''(\omega) = \eta\,\frac{\omega\tau_{\mathrm{M}}}{1 + (\omega\tau_{\mathrm{M}})^2}.$$

The pertaining complex shear modulus $G(\omega)$ is given by

$$G(\omega) = -i\omega\,\tau_{\mathrm{M}}\,\eta(\omega) = G'(\omega) - iG''(\omega), \qquad (16.83)$$

$$G'(\omega) = G\,\frac{(\omega\tau_{\mathrm{M}})^2}{1 + (\omega\tau_{\mathrm{M}})^2}, \quad G''(\omega) = G\,\frac{\omega\tau_{\mathrm{M}}}{1 + (\omega\tau_{\mathrm{M}})^2}.$$

For low frequencies where $\omega\tau_{\mathrm{M}} \ll 1$, the real part η' of the viscosity approaches the viscosity η, its imaginary part η'', as well as G' and G'' vanish. For high frequencies where $\omega\tau_{\mathrm{M}} \gg 1$, the real part G' of $G(\omega)$ approaches the "high frequency shear modulus" G, whereas G'', as well as η' and η'' become zero.

Maxwell derived the 'Maxwell model' equation from the Boltzmann equation for the velocity distribution function of a gas. In that case, $\overline{p_{\mu\nu}}$ is the kinetic contribution $\overline{p_{\mu\nu}^{\mathrm{kin}}}$ to the symmetric traceless friction pressure tensor, see the first equation of (16.59) with $\tau_{\mathrm{M}}^{-1} = \nu_{\mathrm{p}}$ and $\nu_{\mathrm{pa}} = 0$. Here the Maxwell relaxation time is determined by a Boltzmann collision integral and one has $G = p_0 = nk_{\mathrm{B}}T$.

Multiplication of the kinetic equation (12.120) for the pair correlation function by $r_\nu F_\mu$, and use of (12.107) yields a Maxwell model equation for the potential contribution of the pressure tensor. In this case, τ_{M} is equal to the relaxation time τ introduced in (12.120) and the high frequency shear modulus G is given by the Born-Green expression (16.35).

The Maxwell model equation can also be derived within the framework of irreversible thermodynamics. Taking into account that the entropy density contains a contribution proportional to $G^{-1}\,\overline{p_{\mu\nu}}\,\overline{p_{\mu\nu}}$, where G is the high frequency shear modulus. The ensuing entropy production associated with the second rank tensors is proportional to $G^{-1}\,\overline{p_{\mu\nu}}\,\mathrm{d}\,\overline{p_{\mu\nu}}/\mathrm{d}t$. The *Extended Irreversible Thermodynamics*, cf. [132, 133], takes into account additional contributions to the entropy and consequently to the entropy production, which then contains time derivatives of 'non-conserved' quantities, like of the friction pressure considered here, or of the heat flux. An expression for the extended non-equilibrium entropy, valid for gases, is derived in Exercise 12.3. More general schemes for the treatment of non-equilibrium phenomena are presented in [134, 135].

In molecular fluids and colloidal dispersions containing non-spherical particles, the visco-elastic behavior is associated with the dynamics of the alignment as described by (16.74), cf. [136, 137]. The constitutive equation (16.133) for the friction pressure tensor is equivalent to

$$\overline{p_{\nu\mu}} = -2\,\eta_{\mathrm{iso}}\,\overline{\nabla_\nu v_\mu} + \overline{p_{\nu\mu}}^{\,\mathrm{align}}, \quad \overline{p_{\nu\mu}}^{\,\mathrm{align}} = \sqrt{2}\,\frac{\rho}{m}\,k_{\mathrm{B}}T\,\frac{\tau_{\mathrm{ap}}}{\tau_{\mathrm{a}}}\,a_{\mu\nu}. \qquad (16.84)$$

Here $\overline{p_{\nu\mu}}^{\,\mathrm{align}}$ is the part of the pressure tensor associated with the alignment. The viscosity coefficient η_{iso}, corresponding to a situation, where the alignment vanishes, is smaller than the Newtonian viscosity $\eta = \eta_{\mathrm{New}}$ pertaining to the case where the time derivative of the alignment and effects nonlinear in the shear rate vanish. These viscosity coefficients are given by

$$\eta_{\text{iso}} = \eta_{\text{New}} \left(1 - \frac{\tau_{\text{ap}}^2}{\tau_a \tau_p}\right), \quad \eta_{\text{New}} = \frac{\rho}{m} k_B T \tau_p. \tag{16.85}$$

Similarly, the pressure tensor is also given by

$$\overline{p_{\nu\mu}} = -2\eta_{\text{New}} \overline{\nabla_\nu v_\mu} + \overline{p_{\nu\mu}}^{\text{Gies}}, \tag{16.86}$$

$$\overline{p_{\nu\mu}}^{\text{Gies}} = -\sqrt{2}\frac{\rho}{m} k_B T \tau_{\text{pa}} \left(\frac{\overline{da_{\mu\nu}}}{dt} - 2\overline{\varepsilon_{\mu\lambda\kappa}\omega_\lambda a_{\kappa\nu}}\right).$$

The superscript *Gies* refers to Giesekus [138].

In analogy to (16.82), the real and imaginary parts of the complex viscosity coefficient are now given by

$$\eta'(\omega) = (\eta_{\text{New}} - \eta_{\text{iso}})\frac{1}{1 + (\omega\tau_a)^2} + \eta_{\text{iso}}, \quad \eta''(\omega) = (\eta_{\text{New}} - \eta_{\text{iso}})\frac{\omega\tau_a}{1 + (\omega\tau_a)^2}, \tag{16.87}$$

with

$$\eta_{\text{New}} - \eta_{\text{iso}} = \eta_{\text{New}} \frac{\tau_{\text{ap}}^2}{\tau_a \tau_p}. \tag{16.88}$$

Here $\eta'(\omega)$ approaches the viscosity η_{iso} for high frequencies where $\omega\tau_a \gg 1$ applies. Depending on the type of fluids, the relative viscosity difference $(\eta_{\text{New}} - \eta_{\text{iso}})/\eta_{\text{New}}$ ranges from 10^{-2} to 10^2, or higher.

The expressions (16.82) and (16.87) show, and this is true in general, a fluid can reveal its visco-elastic behavior only, when the frequency ω is not too small compared with the reciprocal relaxation time. Depending on the type of fluid and on the temperature, values for the relaxation time vary over many orders of magnitude. Similarly, a non-linear flow behavior can be observed when the shear rate is not too small compared with the reciprocal relaxation time. For this reason, typical viscoelastic fluids also show nonlinear viscous behavior.

16.3.9 Nonlinear Viscosity

The study of the viscoelastic and nonlinear viscous properties of complex fluids is called *Rheology* [139, 140]. The nonlinear effects of the shear rate on the material properties which are due to shear-induced distortions of the local structure or the shear-induced partial orientation of particles, are also referred to as *rheological properties*.

This type of nonlinearity is to be distinguished from the nonlinear flow effects, e.g. the turbulence, resulting from the convective term $\mathbf{v} \cdot \nabla$ in the local linear momentum balance equation (7.52).

Within a phenomenological description, the nonlinear viscous behavior of a plane Couette flow is characterized by three material coefficients, which depend on the imposed shear rate γ. The first of these coefficients is the *non-Newtonian viscosity* $\eta(\gamma)$ defined via the ratio of the yx-component of the pressure or stress tensor and the shear rate $\gamma = \partial v_x / \partial y$:

$$\sigma_{yx} = -p_{yx} = \eta(\gamma)\, \gamma. \tag{16.89}$$

It is understood that $\eta(\gamma)$ approaches the Newtonian, i.e. shear rate independent, viscosity η_{New} in the limit of small shear rates. When nonlinear effects play no role, it is common practice to use the symbol η instead of η_{New}.

The appropriate ansatz for the friction pressure tensor adapted to the plane Couette symmetry is

$$\overline{p_{\mu\nu}} = 2\, \overline{e_\mu^x e_\nu^y}\, \Pi_+ + (e_\mu^x e_\nu^x - e_\mu^y e_\nu^y)\, \Pi_- + 2\, \overline{e_\mu^z e_\nu^z}\, \Pi_0. \tag{16.90}$$

Viscosity coefficients η_i, with $i = +, -, 0$ are defined by

$$\Pi_+ = -\eta_+ \gamma, \quad \Pi_- = -\eta_- \gamma, \quad \Pi_0 = -\eta_0 \gamma. \tag{16.91}$$

The coefficient η_+ is the non-Newtonian viscosity $\eta(\gamma)$, the coefficients η_- and η_0 characterize the normal pressure differences $p_{xx} - p_{yy} = 2\Pi_-$ and $p_{zz} - \frac{1}{2}(p_{xx} + p_{yy}) = 2\Pi_0$. Equivalently, and this is common practice in the rheological literature, the normal pressure differences $p_{xx} - p_{yy}$ and $p_{yy} - p_{zz}$ are used and referred to as "first" and "second" normal pressure differences. The corresponding stress differences, denoted by N_1 and N_2, are defined by

$$N_1 = \sigma_{xx} - \sigma_{yy} = p_{yy} - p_{xx} = \Psi_1 \gamma^2, \quad N_2 = \sigma_{yy} - \sigma_{zz} = p_{zz} - p_{yy} = \Psi_2 \gamma^2. \tag{16.92}$$

The viscometric functions Ψ_1 and Ψ_2 are related to the viscosity coefficients η_- and η_0 according to

$$\Psi_1 \gamma = 2\eta_-, \quad \Psi_2 \gamma = -2\eta_0 - \eta_-. \tag{16.93}$$

In contradistinction to η_+, which is positive in order to guarantee positive entropy production, the coefficients η_-, η_0, and also Ψ_1, Ψ_2 may have either sign.

The normal pressure differences $p_{xx} - p_{yy}$ and $p_{yy} - p_{zz}$ are zero in the linear flow regime, however, they are non-zero at higher shear rates. In general, the viscometric functions depend on the shear rate. When the nonlinearity of the friction pressure is analytic in γ, i.e. when it can be expressed in terms of a power series in γ, one has $p_{xy} \sim \gamma$ and $p_{xx} - p_{yy} \sim \gamma^2$, $p_{yy} - p_{zz} \sim \gamma^2$, for $\gamma \to 0$. Consequently,

330 Constitutive Relations
16 Constitutive Relations

the shear viscosity and the viscometric functions approach constant values at small shear rates.

A generic model for a non-linear viscous behavior is a Maxwell model, cf. (16.81), with a co-rotational time derivative and an additional deformational contribution, viz.

$$\frac{\partial}{\partial t}\,\overline{p_{\mu\nu}} - 2\,\overline{\varepsilon_{\mu\lambda\kappa}\omega_\lambda\,p_{\kappa\nu}} - 2\kappa\,\overline{\nabla_\mu v_\lambda\,p_{\lambda\nu}} + \tau_M^{-1}\,\overline{p_{\mu\nu}} = -2\,G\,\overline{\nabla_\mu v_\nu}\,. \tag{16.94}$$

The pertaining Newtonian viscosity is $\eta_{\text{New}} = \eta = G\tau_M$. The coefficient κ characterizes the influence of the symmetric traceless part of the velocity gradient on the dynamics of the friction pressure tensor. The case $\kappa = 0$ is referred to as the *co-rotational Maxwell model* or as *Jaumann-Maxwell model*. For a plane Couette flow, this special model yields $\eta_0 = 0$, and η_+, η_- are given by the expressions (16.82) for the real and imaginary parts of the complex viscosity coefficient $\eta(\omega)$, with the frequency ω replaced by the shear rate γ. The resulting decrease of the shear viscosity with increasing shear rate is termed *shear thinning*.

Due to $\eta_0 = 0$, one has $\Psi_2 = -0.5\Psi_1$, in this case. For many polymeric liquids, however, typically $\Psi_2 \approx -0.1\Psi_1$ is observed in the small shear rate limit. As will be pointed out in Sect. 16.4.6, this behavior follows from $\kappa \approx 0.4$. In general, the parameter κ is non-zero, as it can e.g. be inferred from microscopic approaches based on kinetic equations [41, 141]. The phenomenological Maxwell model equation (16.94) with $\kappa \neq 0$ is also referred to as *Johnson-Segalman model* [142, 143], the cases $\kappa = \pm1$ are called *co-deformational* or *convected* Maxwell model.

By analogy, the nonlinear viscous properties associated with the alignment are essentially described by (16.87). Assuming that the dynamics of the alignment is governed by the co-rotational time derivative, η_+ and η_- are given by the expressions for η' and η'' with ω replaced by the shear rate γ, and $\eta_0 = 0$. Again, the nonlinear viscosity shows a shear thinning behavior, however, it approaches a finite value, viz. η_{iso} which is also called *second Newtonian viscosity*.

More general cases with $\kappa \neq 0$ and where terms nonlinear in the alignment and in the friction pressure tensor are included in the dynamic equations, are treated in Sects. 16.4.6 and 17.4.

16.3.10 Vorticity Free Flow

A flow field for which the velocity gradient tensor has no antisymmetric part is referred to as *vorticity free*. Two main types are distinguished: uniaxial and planar biaxial flow fields.

(i) The uniaxial extensional or compressional flow is considered with the special geometry $v_z = 2\varepsilon z$ and $v_x = -\varepsilon x$, $v_y = -\varepsilon y$. Here $\varepsilon = \frac{1}{2}\partial v_z/\partial z = -\partial v_x/\partial x = -\partial v_y/\partial y$ is the extension or compression rate. The symmetry of the flow field is that of the uniaxial squeeze-stretch field as sketched in Fig. 7.3. The velocity gradient tensor $\nabla_\nu v_\mu$ is symmetric traceless and one has

$$\overline{\nabla_\nu v_\mu} = \varepsilon \left[2\, \overline{e_\nu^z e_\mu^z} - (\overline{e_\nu^x e_\mu^x} + \overline{e_\nu^y e_\mu^y}) \right] = 3\varepsilon \, \overline{e_\nu^z e_\mu^z} . \tag{16.95}$$

By analogy to (16.90), here the symmetry adapted ansatz for the friction pressure tensor is $\overline{p_{\mu\nu}} = 2\, \overline{e_\mu^z e_\nu^z}\, \Pi_0$. In the linear flow regime, where one has $\overline{p_{\mu\nu}} = -2\eta_{\mathrm{New}}\, \overline{\nabla_\nu v_\mu}$, with the Newtonian viscosity η_{New}. In this case, the viscosity coefficient η_0, defined by $\Pi_0 = -\eta_0 \varepsilon$, cf. (16.91), is $\eta_0 = 3\eta_{\mathrm{New}}$. This coefficient is also referred to as *extensional viscosity* or *Trouton viscosity*. The nonlinear case, an equation governing the component Π_0 of the pressure tensor, follows from the

the Maxwell model (16.94). Due to $\overline{e_\mu^z e_\lambda^z}\, \overline{e_\lambda^z e_\nu^z} = \tfrac{1}{3}\, \overline{e_\mu^z e_\nu^z}$, the resulting equation for Π_0 is

$$\tau_{\mathrm{M}} \frac{\partial}{\partial t} \Pi_0 - 2\kappa\, \varepsilon \tau_{\mathrm{M}}\, \Pi_0 + \Pi_0 = -3\eta_{\mathrm{New}}\, \varepsilon, \quad \eta_{\mathrm{New}} = G\tau_{\mathrm{M}}. \tag{16.96}$$

For a stationary situation this leads to

$$\Pi_0 = -\eta_0\, \varepsilon, \quad \eta_0 = 3\,\eta_{\mathrm{New}}\, (1 - 2\,\kappa\, \varepsilon \tau_{\mathrm{M}})^{-1}, \tag{16.97}$$

provided that $2\kappa\varepsilon\tau_{\mathrm{M}} < 1$. This inequality plays no role for $\kappa\varepsilon < 0$, where the viscosity is decreasing with an increasing magnitude of the deformation rate. For $\kappa\varepsilon > 0$, (16.97) yields an increasing viscosity and the stationary solution breaks down at the finite deformation rate $\varepsilon_{\mathrm{lim}} = (2\kappa)^{-1}$.

(ii) The planar biaxial extensional or compressional flow is considered with the special geometry $v_x = \varepsilon x$, $v_y = -\varepsilon y$, and $v_z = 0$. Here $\varepsilon = \partial v_x / \partial x = -\partial v_y / \partial y$ is the extension or compression rate. The symmetry of the flow field is that of the biaxial squeeze-stretch field as sketched in Fig. 7.4. Again, the velocity gradient tensor $\nabla_\nu v_\mu$ is symmetric traceless and one has

$$\overline{\nabla_\nu v_\mu} = \varepsilon\, (\overline{e_\nu^x e_\mu^x} - \overline{e_\nu^y e_\mu^y}). \tag{16.98}$$

By analogy to (16.90), here the symmetry adapted ansatz for the friction pressure tensor is $\overline{p_{\mu\nu}} = (\overline{e_\mu^x e_\nu^x} - \overline{e_\mu^y e_\nu^y})\Pi_- + 2\, \overline{e_\mu^z e_\nu^z}\, \Pi_0$. Viscosity coefficients η_- and η_0 are defined by $\Pi_- = -\eta_- \varepsilon$ and $\Pi_0 = -\eta_0 \varepsilon$. In the linear flow regime, where $\overline{p_{\mu\nu}} = -2\eta_{\mathrm{New}}\, \overline{\nabla_\nu v_\mu}$ applies, with the Newtonian viscosity η_{New}, one has $\eta_- = 2\eta_{\mathrm{New}}$ and $\eta_0 = 0$. For the planar biaxial flow, the Maxwell model (16.94), and the use of

$$(\overline{e_\mu^x e_\lambda^x} - \overline{e_\mu^y e_\lambda^y})(\overline{e_\lambda^x e_\nu^x} - \overline{e_\lambda^y e_\nu^y}) = \overline{e_\nu^x e_\mu^x} + \overline{e_\nu^y e_\mu^y} = -\overline{e_\nu^z e_\mu^z}$$

and

$$(e_\mu^x e_\lambda^x - e_\mu^y e_\lambda^y)\,\overline{e_\lambda^z e_\nu^z} = -\frac{1}{3}(e_\mu^x e_\nu^x - e_\mu^y e_\nu^y),$$

lead to two coupled equations for Π_- and Π_0:

$$\tau_M \frac{\partial}{\partial t} \Pi_- + \frac{4}{3}\kappa\,\varepsilon\tau_M\,\Pi_0 + \Pi_- = -2\,\eta_{New}\,\varepsilon, \qquad (16.99)$$

$$\tau_M \frac{\partial}{\partial t} \Pi_0 + \kappa\,\varepsilon\tau_M\,\Pi_- + \Pi_0 = 0.$$

Here the stationary solution is given by $\Pi_- = -\eta_-\varepsilon$, $\Pi_0 = -\eta_0\varepsilon$, with

$$\eta_- = 2\,\eta_{New}\left(1 - \frac{4}{3}(\kappa\varepsilon\tau_M)^2\right)^{-1}, \qquad \eta_0 = -\kappa\,\varepsilon\tau_M\,\eta_-, \qquad (16.100)$$

provided that $4(\kappa\varepsilon\tau_M)^2 < 3$.

16.4 Viscosity and Alignment in Nematics

16.4.1 Well Aligned Nematic Liquid Crystals and Ferro Fluids

A viscous flow with a moderate shear rate does not affect the magnitude of the order parameter of a nematic liquid crystal, cf. Sect. 15.2.1. The direction of the director **n**, however, is influenced by the flow geometry and by external fields. First, the case is considered, where a magnetic field is applied, which is strong enough such that it practically fixes the orientation of the director. On the other hand, it should not be so strong, that it alters the order parameter. The tensor $\overline{\mathbf{nn}}$ determines the anisotropy of the fluid. It is understood that **n** is parallel or anti-parallel to the direction of the applied magnetic field $\mathbf{B} = B\mathbf{h}$, the sign of **n** has no meaning for nematics. The symmetry considerations used for the viscosity coefficients of well aligned nematic liquid crystals also apply to ferro-fluids in the presence a of strong magnetic field. *Ferro-fluids* are colloidal dispersions containing practically spherical particles with permanent or induced magnetic dipoles [110].

Point of departure for the set up of the constitutive relations governing the friction pressure tensor are the expression (16.42) for the entropy production, and the second rank tensor $\overline{\mathbf{nn}}$, or equivalently $\overline{\mathbf{hh}}$ specifying the anisotropy of the unperturbed state. The external field exerts a torque on the system. As a consequence, the antisymmetric part of the pressure tensor is not zero. Compared with the ansatz (16.43), an additional constitutive relation is needed and coupling terms between the symmetric traceless and antisymmetric parts of the pressure tensor occur:

$$\overline{P_{\mu\nu}} = -2\eta \,\overline{\nabla_\mu v_\nu} - 2\tilde{\eta}_1 \,\overline{n_\mu n_\lambda}\,\overline{\nabla_\lambda v_\nu} - 2\tilde{\eta}_3 \,\overline{n_\mu n_\nu}\,\overline{n_\kappa n_\lambda}\,\overline{\nabla_\lambda v_\kappa}$$

$$-2\tilde{\eta}_2\left(-\varepsilon_{\mu\lambda\kappa}\,\omega_\lambda\,\overline{n_\kappa n_\nu}\right) - \zeta\,\overline{n_\mu n_\nu}\,\nabla_\lambda v_\lambda, \qquad (16.101)$$

$$P_\mu = -\gamma_1\left(\omega_\mu - n_\mu n_\nu \omega_\nu\right) + \gamma_2\,\varepsilon_{\mu\nu\lambda}\,\overline{n_\nu n_\kappa}\,\overline{\nabla_\kappa v_\lambda},$$

$$\tilde{p} = -\eta_V \nabla_\lambda v_\lambda - \zeta\,\overline{n_\mu n_\nu}\,\overline{\nabla_\mu v_\nu}.$$

The shear viscosity η, the *twist viscosity* coefficient γ_1 and the bulk viscosity η_V are positive, all other coefficients may have either sign. The coefficient ζ characterizes the coupling between the irreducible tensors of rank 0 and 2, the Onsager symmetry is already taken into account. The coupling between the irreducible tensors of rank 1 and 2 is specified by γ_2 and $\tilde{\eta}_2$. These coefficients obey the Onsager relation

$$2\tilde{\eta}_2 = \gamma_2. \qquad (16.102)$$

The coefficients γ_1 and γ_2 occurring in the equation for the axial vector associated with the antisymmetric part of the pressure tensor are called *Leslie viscosity* coefficients.

The Miesowicz viscosity coefficients η_i, $i = 1, 2, 3$, [144], see also Sect. 16.3.3, are defined for a plane Couette flow with the velocity in the x-direction and its gradient in the y-direction, are inferred from

$$p_{yx} = -\eta_i \frac{\partial v_x}{\partial y}, \qquad (16.103)$$

where the cases $i = 1, 2, 3$ correspond to the direction of the field and thus **n** parallel to the x-, y- and z-axes, respectively. The η_i are related to the viscosities defined in (16.101) by

$$\eta_1 = \eta + \frac{1}{6}\tilde{\eta}_1 + \frac{1}{2}\tilde{\eta}_2 + \frac{1}{4}\gamma_1 + \frac{1}{4}\gamma_2,$$

$$\eta_2 = \eta + \frac{1}{6}\tilde{\eta}_1 - \frac{1}{2}\tilde{\eta}_2 + \frac{1}{4}\gamma_1 - \frac{1}{4}\gamma_2,$$

$$\eta_3 = \eta - \frac{1}{3}\tilde{\eta}_1. \qquad (16.104)$$

Notice that $\eta = (\eta_1 + \eta_2 + \eta_3)/3$ is the average of the η_i. Furthermore, one has

$$\eta_1 - \eta_2 = \tilde{\eta}_2 + \frac{1}{2}\gamma_2 = \gamma_2. \qquad (16.105)$$

The second equality follows from the Onsager symmetry relation (16.102). The experimentally observed difference between η_1 and η_2 is an evidence for the existence of an antisymmetric part of the pressure tensor.

The three Miesowicz coefficients do not involve the viscosity coefficient $\tilde{\eta}_3$. This is different for η_{45} corresponding to the case where $\mathbf{n}$ points along the bisector in the xy-plane. The resulting *Helfrich viscosity coefficient* $\eta_{12} = 4\eta_{45} - 2(\eta_1 - \eta_2)$, cf. (16.53), is given by

$$\eta_{12} = 2\,\tilde{\eta}_3. \tag{16.106}$$

In the theoretical approach of Ericksen and Leslie, the local anisotropy of a nematic liquid crystal is characterized by the director $\mathbf{n}$, which depends on the time t and may also depend on the position $\mathbf{r}$. A frequently used ansatz for the pressure tensor of a nematic is

$$-p_{\nu\mu} = \alpha_1\, n_\nu n_\mu\, n_\kappa n_\lambda\, \overline{\nabla_\lambda v_\kappa} + \alpha_2\, n_\nu N_\mu + \alpha_3\, N_\nu n_\mu \tag{16.107}$$

$$+\alpha_4\, \overline{\nabla_\mu v_\nu} + \alpha_5\, n_\nu n_\lambda\, \overline{\nabla_\lambda v_\mu} + \alpha_6\, \overline{\nabla_\nu v_\lambda}\, n_\lambda n_\mu,$$

where

$$N_\mu = \frac{\partial n_\mu}{\partial t} - \varepsilon_{\mu\lambda\nu}\omega_\lambda n_\nu \tag{16.108}$$

is the co-rotational time derivative of the director. For a constant director field, (16.107) is equivalent to (16.101) when one has $\nabla_\lambda v_\lambda = 0$ and $\zeta = 0$ applies. The *Leslie viscosity coefficients* α_i, $i = 1, 2, .., 6$ are related to the viscosities introduced in (16.101), by

$$\eta = \frac{1}{2}\alpha_4 + \frac{1}{6}(\alpha_5 + \alpha_6), \quad \tilde{\eta}_1 = \frac{1}{2}(\alpha_5 + \alpha_6), \quad \tilde{\eta}_2 = \frac{1}{2}(\alpha_2 + \alpha_3), \quad \tilde{\eta}_3 = \frac{1}{2}\alpha_1, \tag{16.109}$$

$$\gamma_1 = \alpha_3 - \alpha_2, \quad \gamma_2 = \alpha_6 - \alpha_5. \tag{16.110}$$

The symmetry (16.102) corresponds to the *Onsager-Parodi relation* [145]

$$\alpha_2 + \alpha_3 = \alpha_6 - \alpha_5. \tag{16.111}$$

The Miesowicz viscosities of nematic liquid crystals composed of prolate particles obey the inequalities $\eta_2 > \eta_3 > \eta_1$ and consequently $\gamma_2 = \eta_1 - \eta_2$ is negative. For nematics composed of discotic particles, the order of the Miesowicz coefficients is reversed and γ_2 is positive. An affine transformation model considered next gives an instructive insight into the anisotropy of the viscosity.

The connection between the friction pressure tensor involving the various viscosity coefficients of nematics and the alignment tensor is elucidated in Sects. 16.4.4 and 16.4.5. Alternative approaches are, e.g. found in [149, 150].

The rotational viscosity coefficient γ_1 has been measured for many nematics. The full set of viscosity coefficients, viz. γ_1, γ_2 and η_1, η_2, η_3, η_{12} has been determined for a few liquid crystals only, e.g. cf. [151]. For model systems, the anisotropy of the viscous behavior of liquid crystals can also be analyzed by computer simulations. Non-Equilibrium Molecular Dynamics (NEMD) simulations were first performed for perfectly oriented nematics in [147]. Nematics composed of particles interacting via a Gay-Berne potential were studied both in NEMD simulations, cf. [152], and in equilibrium Molecular Dynamics (MD) calculations, e.g. see [153]. In MD simulations, time correlation functions, cf. Sect. 17.1, are determined, the transport coefficients then are obtained with the help of Green-Kubo relation, viz. as an integral over the time. Viscosity coefficients were inferred from the dependence of the flow resistance on the strength of orienting electric or magnetic fields, both in experiments and in NEMD computer simulations [154]. Results from NEMD computations were presented in [155] for the viscosity coefficients γ_1, γ_2 and η_1, η_2, η_3, η_{12}, as functions of the density of the Gay-Berne fluid. For densities approaching the nematic $\leftrightarrow$ smectic A transition, the smallest of the viscosity coefficient, viz. η_1 increases and it overtakes η_3 and η_2. The rod-like particles form disc-like clusters in anticipation of the smectic layers.

16.4.2 Perfectly Oriented Ellipsoidal Particles

With the help of an volume conserving affine transformation, cf. Sect. 5.7, the viscous behavior of a fluid composed of perfectly oriented ellipsoidal particles can be mapped onto that of a fluid of spherical particles [146, 147]. The affine transformation model provides a good qualitative description for the anisotropy of the viscosity [148] of nematics.

Within the framework of this model, the binary interaction potential Φ of the non-spherical particles are expressed in terms of the interaction potentials Φ_{sph} of a reference fluid composed of spherical particles according to

$$\Phi(\mathbf{r}) = \Phi_{\text{sph}}(\mathbf{r}^{A}), \quad r_{\mu}^{A} = A_{\mu\nu}^{1/2} r_{\nu}.$$

The vector $\mathbf{r}$ in real space is linked with $\mathbf{r}^{A}$ via an affine transformation, as given by (5.5.3). This means, the equipotential surfaces are ellipsoids. In simple liquids, the pressure is mainly determined by its potential contribution, see (16.29). For simplicity, the kinetic contribution to the friction pressure tensor is disregarded here. The spatial derivative is transformed according to

$$\nabla_{\mu}^{A} = A_{\mu\nu}^{-1/2} \nabla_{\nu}.$$

The transformation of the pressure tensor $P_{\nu\mu}$, between the affine and the real space is

$$P_{\nu\mu}^{A} = A_{\nu\lambda}^{1/2} P_{\lambda\kappa} A_{\kappa\mu}^{-1/2}.$$

The friction pressure $p_{\nu\mu}^A$, in the affine space, obeys the constitutive law

$$-p_{\nu\mu}^A = 2\eta^A \, \Gamma_{\nu\mu}^A + \eta_V^A \, \nabla_\lambda^A v_\lambda^A \, \delta_{\mu\nu},$$

which is standard for an isotropic fluid with the shear viscosity η^A and the bulk viscosity η_V^A, cf. Sect. 16.3. Notice that $\nabla_\lambda^A v_\lambda^A = \nabla_\lambda v_\lambda$. The symmetric traceless deformation rate tensor $\Gamma_{\nu\mu}^A$ in affine space is related to the velocity gradient in real space by

$$\Gamma_{\mu\nu}^A = \frac{1}{2}\left(A_{\mu\lambda}^{-1/2} A_{\nu\kappa}^{1/2} + A_{\nu\lambda}^{-1/2} A_{\mu\kappa}^{1/2}\right) \nabla_\lambda v_\kappa - \frac{1}{3}\nabla_\lambda v_\lambda \, \delta_{\mu\nu}.$$

The friction pressure tensor, in real space, and for $\nabla \cdot \mathbf{v} = 0$,

$$p_{\nu\mu} = -\eta^A \left(A_{\nu\lambda}^{-1} A_{\mu\kappa} \nabla_\lambda v_\kappa + \nabla_\nu v_\mu\right), \tag{16.112}$$

contains symmetric traceless and antisymmetric parts.

For ellipsoids of revolution, with their symmetry axis parallel to the unit vector $\mathbf{u}$, which is identical with the director $\mathbf{n}$ of the perfectly oriented fluid, the volume conserving transformation is governed by

$$A_{\mu\nu} = Q^{2/3}\left[\delta_{\mu\nu} + (Q^{-2} - 1)n_\mu n_\nu\right], \quad A_{\mu\nu}^{-1} = Q^{-2/3}\left[\delta_{\mu\nu} + (Q^2 - 1)n_\mu n_\nu\right],$$

cf. (5.58). Here $Q = a/b$ is the axes ratio of an ellipsoid with the semi-axes a and $b = c$.

Comparison of the resulting friction pressure in real space with the ansatz for the anisotropic viscosity made in the previous section leads to

$$\eta = \left[1 + \frac{1}{6}(Q - Q^{-1})^2\right]\eta^A, \quad \tilde{\eta}_1 = \frac{1}{2}(Q - Q^{-1})^2\,\eta^A,$$

$$\tilde{\eta}_2 = \frac{1}{2}(Q^{-2} - Q^2)\,\eta^A, \quad \tilde{\eta}_3 = -\frac{1}{2}(Q - Q^{-1})^2\,\eta^A,$$

and

$$\gamma_1 = (Q - Q^{-1})^2\,\eta^A, \quad \gamma_2 = (Q^{-2} - Q^2)\,\eta^A. \tag{16.113}$$

The pertaining Miesowicz and the Helfrich viscosity coefficients are

$$\eta_1 = Q^{-2}\eta^A, \quad \eta_2 = Q^2\eta^A, \quad \eta_3 = \eta^A, \quad \eta_{12} = -(Q - Q^{-1})^2\eta^A. \tag{16.114}$$

The Onsager-Parodi relation $\eta_1 - \eta_2 = \gamma_2$ is fulfilled. Furthermore, one has $\eta_V = \eta_V^A$. Thus all the viscosity coefficients of this perfectly ordered nematic liquid are related to the shear and volume viscosities of the reference liquid with the same density and to the axes ratio Q of the ellipsoids. Clearly. for prolate particles with $Q > 1$, the

Miesowicz coefficients obey the inequalities $\eta_2 > \eta_3 > \eta_1$ and $\gamma_2 < 0$. For oblate particles, i.e. for $Q < 1$, one has $\eta_2 < \eta_3 < \eta_1$ and $\gamma_2 > 0$. An adaption of this model to partially ordered nematics and comparison with experimental data is found in [148]. The alternative approach, where the dynamics of the alignment tensor is taken into account, is discussed in Sects. 16.4.4–16.4.6.

The anisotropy of the heat conductivity and of the diffusion tensor can also be treated by the affine transformation approach. For diffusion, see the following exercise. An amended affine transformation model which takes the partial alignment into account and a comparison with molecular dynamics computer simulations is presented in [156].

16.3 Exercise: Diffusion of Perfectly Oriented Ellipsoids

In the nematic phase, the flux $\mathbf{j}$ of diffusing particles with number density ρ obeys the equation

$$j_\mu = -D_{\mu\nu} \nabla_\nu \rho, \quad D_{\mu\nu} = D_\| n_\mu n_\nu + D_\perp (\delta_{\mu\nu} - n_\mu n_\nu),$$

where $D_\|$ and $D_\perp$ are the diffusion coefficients for the flux parallel and perpendicular to the director $\mathbf{n}$.

Use the volume conserving affine transformation model for uniaxial particles, cf. Sect. 5.7.2, to derive

$$D_\| = Q^{4/3} D_0, \quad D_\perp = Q^{-2/3} D_0$$

for perfectly oriented ellipsoidal particles with axes ratio Q. Here D_0 is the reference diffusion coefficient of spherical particles.

Furthermore, determine the anisotropy ratio $R = (D_\| - D_\perp)/(D_\| + 2D_\perp)$, the average diffusion coefficient $\bar{D} = \frac{1}{3} D_\| + \frac{2}{3} D_\perp$ and the geometric mean $\tilde{D} = D_\|^{1/3} D_\perp^{2/3}$. Discuss the cases $Q > 1$ and $Q < 1$ for prolate and oblate particles.

16.4.3 Free Flow of Nematics, Flow Alignment and Tumbling

The antisymmetric part of the pressure tensor vanishes for a free flow, i.e. when no orienting external field is applied. This implies

$$p_\mu = \varepsilon_{\mu\nu\lambda} n_\nu \left[\gamma_1 \left(\frac{\partial n_\lambda}{\partial t} - \varepsilon_{\lambda\kappa\tau} \omega_\kappa n_\tau \right) + \gamma_2 n_\kappa \overline{\nabla_\kappa v_\lambda} \right] = 0. \qquad (16.115)$$

For a spatially inhomogeneous situation and in the presence of external fields the torque associated with (16.115) is not zero but balanced by the elastic and field-induced torques $\varepsilon_{\mu\nu\lambda} n_\nu (K \Delta n_\lambda + F_{\lambda\kappa} n_\kappa)$ as described by (15.33).

Provided that $|\gamma_2| > \gamma_1$ holds true, a stationary plane Couette flow leads to a *stationary flow alignment* where the director is in the plane spanned by the flow velocity and its gradient. In particular, for the geometry used above, e.g. in Sect. 12.4.6, one has $n_x = \cos \chi$, $n_y = \sin \chi$, $n_z = 0$, then (16.115) implies

$$\gamma_1 \left(\frac{\partial \chi}{\partial t} + \Gamma/2 \right) + (\Gamma/2)\, \gamma_2 \cos 2\chi = 0,$$

where the shear rate $\frac{\partial v_x}{\partial y}$ has been denoted by Γ. In a stationary situation, where $\frac{\partial \chi}{\partial t} = 0$, the *flow alignment angle* χ is determined by

$$\cos 2\chi = -\frac{\gamma_1}{\gamma_2}. \tag{16.116}$$

This result is independent of the shear rate. For (16.116) to be applicable, the *tumbling parameter*

$$\lambda \equiv -\frac{\gamma_2}{\gamma_1} \tag{16.117}$$

has to obey the condition $|\lambda| > 1$. The viscosity for this "free" flow in a flow aligned state is

$$\eta_{\text{free}} = \frac{1}{2}(\eta_1 + \eta_2 - \gamma_1) + \eta_{12}\left(1 - \gamma_1^2/\gamma_2^2\right). \tag{16.118}$$

In many nematics, the flow alignment angle is small, typically around $10°$, such that η_{free} is not much larger than the smallest Miesowicz viscosity η_1. On the other hand, close to the isotropic-nematic transition temperature T_{ni}, the average viscosity $\eta = (\eta_1 + \eta_3 + \eta_3)/3 > \eta_{\text{free}}$ in the nematic phase is approximately equal to the viscosity in the isotropic phase. This explains the surprising result that the viscosity below the phase transition is smaller than the viscosity above T_{ni}.

For $|\lambda| < 1$, no stationary solution of (16.115) is possible, even when the imposed shear rate is constant. Then the director undergoes a tumbling motion, similar to that one described by Jeffrey [161] for an ellipsoid in a streaming fluid. More specifically, the tumbling period is related to the Ericksen-Leslie tumbling parameter λ by $P_J = 4\pi/(\gamma\sqrt{1-\lambda^2})$, for a full rotation of the director.

16.4.4 Fokker-Planck Equation Applied to Flow Alignment

The equation governing the orientational dynamics of liquid crystals, both in the isotropic and nematic phases, can be derived from a generalized Fokker-Planck equation [157–159]. To indicate the physics underlying this approach, the Langevin type equation of a single non-spherical particle, immersed in streaming fluid, is considered first. Let the orientation of the particle be specified by the unit vector $\mathbf{u}$, its angular velocity is written as $\omega + \Omega$, where ω is the vorticity $\frac{1}{2}\nabla \times \mathbf{v}$. Then the time change of $\mathbf{u}$ is given by $\dot{u}_\mu = \varepsilon_{\mu\lambda\nu}\omega_\lambda u_\nu + \varepsilon_{\mu\lambda\nu}\Omega_\lambda u_\nu$, and Ω obeys the equation:

$$\dot{\Omega}_\lambda = -v_r \, \Omega_\lambda + \theta^{-1} T_\lambda^{\text{syst}} + \theta^{-1} T_\lambda^{\text{flct}}.$$

Here $v_r > 0$ is a rotational damping coefficient due to a frictional torque, θ is the relevant moment of inertia, $\mathbf{T}^{\text{syst}}$ and $\mathbf{T}^{\text{flct}}$ denote the systematic and the fluctuating torques acting on a particle. The systematic torque can be derived from a Hamiltonian function H according to $T_\lambda^{\text{syst}} = -\mathscr{L}_\lambda H(\mathbf{u})$, cf. (12.45). It is recalled that

$$\mathscr{L}_\lambda = \varepsilon_{\lambda\kappa\tau} u_\kappa \frac{\partial}{\partial u_\tau}.$$

In the following, the dimensionless function $\mathscr{H} = -\beta H$ is used, where $\beta = (k_B T)^{-1}$, and T is the temperature. Then one has $\beta T_\lambda^{\text{syst}} = \mathscr{L}_\lambda \mathscr{H}(\mathbf{u})$, and the generalized Fokker-Planck equation pertaining to this Langevin equation reads

$$\frac{\partial f(\mathbf{u})}{\partial t} + \omega_\lambda \mathscr{L}_\lambda f(\mathbf{u}) - v_0 \mathscr{L}_\lambda \left(\mathscr{L}_\lambda f(\mathbf{u}) - f(\mathbf{u}) \mathscr{L}_\lambda \mathscr{H}(\mathbf{u}) \right) = 0. \qquad (16.119)$$

The relaxation frequency v_0 is related to the rotational damping coefficient v_r by

$$v_0 = \frac{k_B T}{\theta v_r}. \qquad (16.120)$$

For spherical particles with radius R, one has $v_r \sim R^3$, see the Exercise 10.3, provided that hydrodynamics applies. For non-spherical particles v_r depends on the shape of the particle, but an effective radius R_{eff} can be defined such that $v_r \sim R_{\text{eff}}^3$, and $v_0 \sim R_{\text{eff}}^{-3}$. For vanishing vorticity, the kinetic equation (16.119) is similar to (12.46), where the systematic torque was assumed to be due to external orienting fields. Here, however, the orienting torques are due to the viscous flow and an internal field which, in turn, is caused by the alignment of the surrounding particles. Both torques are of the form $T_\lambda^{\text{syst}} \sim \varepsilon_{\lambda\mu\kappa} u_\mu F_{\kappa\nu} u_\nu$ such that

$$\mathscr{H} = F_{\mu\nu} \phi_{\mu\nu}, \quad \phi_{\mu\nu} = \zeta_2 \, \overline{u_\mu u_\nu}, \quad \zeta_2 = \sqrt{\frac{15}{2}}. \qquad (16.121)$$

The specific expression for the symmetric traceless tensor $F_{\mu\nu}$ is

$$F_{\mu\nu} = (6v_0)^{-1} \mathscr{R} \, \overline{\nabla_\mu v_\nu} + T^{-1} T^* a_{\mu\nu}, \quad a_{\mu\nu} = \langle \phi_{\mu\nu} \rangle. \qquad (16.122)$$

The first term in (16.122) involving the *shape parameter* $\mathscr{R}$ describes the orienting effect of the velocity gradient. When hydrodynamics applies, this parameter is given by

$$\mathscr{R} = \sqrt{\frac{6}{5}} \frac{Q^2 - 1}{Q^2 + 1}, \qquad (16.123)$$

for uniaxial ellipsoids with the axis ratio Q. Values $Q > 1$ and $Q < 1$ pertain to prolate, i.e. rod-like and oblate, i.e. disc-like particles, respectively. One has $\mathscr{R} > 0$ and $\mathscr{R} < 0$ for these cases. The quantity $\mathscr{R}$ vanishes for spherical particles, corresponding to $Q = 1$.

The second term in (16.122) is associated with the internal field proportional to the alignment tensor. The characteristic temperature T^* is linked with the strength of the alignment energy, just as in the Maier-Saupe theory, cf. Sect. 15.2.3. In the absence of a flow, the stationary solution of the kinetic equation (16.119) is

$$f = f_{\text{eq}} \sim \exp[T^{-1}T^* a_{\mu\nu}\,\phi_{\mu\nu}],$$

which is essentially the Maier-Saupe distribution function.

Multiplication of the kinetic equation (16.119) by $\phi_{\mu\nu} = \zeta_2 \, \overline{u_\mu u_\nu}$ and integration over d^2u and use of (16.121) with (16.122) leads to a nonlinear, inhomogeneous equation for $a_{\mu\nu}$, which is, however, not yet a closed equation for the second rank alignment tensor. More specifically, the moment equation for $a_{\mu\nu}$, as inferred from (16.119), is

$$\frac{\partial}{\partial t}a_{\mu\nu} - 2\,\overline{\varepsilon_{\mu\lambda\kappa}\omega_\lambda a_{\kappa\nu}} + \nu_2\,a_{\mu\nu} - \nu_0 \left\langle (\mathscr{L}_\lambda \phi_{\mu\nu})(\mathscr{L}_\lambda \phi_{\alpha\beta}) \right\rangle F_{\alpha\beta} = 0,$$

with $\nu_2 = 6\nu_0$, cf. (12.44). Computation of the expression within the bracket $\langle \ldots \rangle$ of the last term yields

$$4\,\zeta_2^2\,\overline{\varepsilon_{\lambda\kappa\mu}u_\kappa u_\nu}\,\varepsilon_{\lambda\sigma\alpha}u_\sigma u_\beta\,F_{\alpha\beta} = 4\,\zeta_2^2 \left(\overline{u_\nu u_\beta F_{\mu\beta}} - \overline{u_\mu u_\nu}\,u_\alpha u_\beta F_{\alpha\beta} \right).$$

Use of $u_\nu u_\beta = \frac{1}{3}\delta_{\nu\beta} + \overline{u_\alpha u_\beta}$ in the first term on the right hand side and of the relation

$$\overline{u_\mu u_\nu}\,u_\alpha u_\beta\,F_{\alpha\beta} = \frac{2}{15}F_{\mu\nu} + \frac{4}{7}\,\overline{u_\mu u_\kappa}\,F_{\kappa\nu} + \overline{u_\mu u_\nu u_\alpha u_\beta}\,F_{\alpha\beta},$$

cf. (11.58), leads to

$$(\mathscr{L}_\lambda \phi_{\mu\nu})(\mathscr{L}_\lambda \phi_{\alpha\beta}) = 4\,\zeta_2^2 \left(\frac{1}{5}F_{\mu\nu} + \frac{3}{7}\,\overline{u_\mu u_\kappa}\,F_{\kappa\nu} - \overline{u_\mu u_\nu u_\alpha u_\beta}\,F_{\alpha\beta} \right).$$

The orientational average of this expression involves the fourth rank alignment tensor

$$a_{\mu\nu\alpha\beta} = \langle \phi_{\mu\nu\alpha\beta} \rangle, \quad \phi_{\mu\nu\alpha\beta} = \zeta_4 \, \overline{u_\mu u_\nu u_\alpha u_\beta}, \quad \zeta_4 = \frac{3}{4}\sqrt{70},$$

cf. Sect. 12.2.2. Thus the moment equation for the second rank alignment tensor becomes

$$\frac{\partial}{\partial t}a_{\mu\nu} - 2\overline{\varepsilon_{\mu\lambda\kappa}\omega_\lambda a_{\kappa\nu}} + v_2\, a_{\mu\nu}$$

$$-4v_0\,\zeta_2^2\left(\frac{1}{5}F_{\mu\nu} + \frac{3}{7}\zeta_2^{-1}\overline{a_{\mu\kappa}F_{\kappa\nu}} - \zeta_4^{-1}a_{\mu\nu\alpha\beta}F_{\alpha\beta}\right) = 0.$$

Use of the explicit expression (16.122) for the tensor $F_{..}$ leads to

$$\frac{\partial}{\partial t}a_{\mu\nu} - 2\overline{\varepsilon_{\mu\lambda\kappa}\omega_\lambda a_{\kappa\nu}} - 2\kappa\,\overline{\nabla_\mu v_\kappa\, a_{\kappa\nu}} + v_2\left(Aa_{\mu\nu} - \sqrt{6}B\,\overline{a_{\mu\kappa}a_{\kappa\nu}}\right)$$

$$+5v_2\zeta_4^{-1}T^{-1}T^*a_{\mu\nu\alpha\beta}a_{\alpha\beta} = \mathscr{R}\left(\overline{\nabla_\mu v_\nu} - 5\,\zeta_4^{-1}a_{\mu\nu\alpha\beta}\,\overline{\nabla_\alpha v_\beta}\right),$$

$$(16.124)$$

with

$$A = 1 - \frac{T^*}{T}, \qquad B = \frac{\sqrt{5}}{7}\frac{T^*}{T}, \qquad (16.125)$$

and

$$\kappa = \frac{1}{7}\zeta_2\mathscr{R} = \frac{3}{7}\frac{Q^2-1}{Q^2+1}. \qquad (16.126)$$

The second equality in (16.126) pertains to the hydrodynamic expression (16.123) the parameter $\mathscr{R}$. For long, rod-like particles, corresponding to $Q \gg 1$, the quantity κ approaches $3/7 \approx 0.4$.

In the isotropic phase and for small alignment where terms nonlinear in the second rank alignment tensor and the fourth rank alignment tensor can be disregarded, (16.124) reduces to an equation similar to (16.74), viz.

$$\frac{\partial}{\partial t}a_{\mu\nu} - 2\overline{\varepsilon_{\mu\lambda\kappa}\omega_\lambda a_{\kappa\nu}} - 2\kappa\,\overline{\nabla_\mu v_\kappa\, a_{\kappa\nu}} + \tau_a^{-1}Aa_{\mu\nu} = -\tau_a^{-1}\tau_{ap}\sqrt{2}\,\overline{\nabla_\nu v_\mu},$$

$$(16.127)$$

where the relaxation time coefficients are now related to the parameters occurring in the Fokker-Planck approach by

$$\tau_a = v_2^{-1} = (6\,v_0)^{-1}, \quad \sqrt{2}\,\tau_{ap} = -\mathscr{R}\,\tau_a. \qquad (16.128)$$

The comparison of (16.127) with (16.74) shows two additional features. The first is the term involving the parameter κ, which describes an effect of the deformation rate on the alignment. The second is the occurrence of the factor $A = 1 - T^*/T$. This implies the *pre-transitional increase* of the relaxation time $\tau = A^{-1}\tau_a$ and of the flow birefringence $\sim A^{-1}$, when the temperature T approaches the transition temperature T_{ni}, from above. Notice that T_{ni} is slightly larger than the pseudo-critical temperature T^*, cf. Sect. 15.2.2.

In general, however, and in particular in the nematic phase, terms nonlinear in the alignment matter and the fourth rank tensor has to be taken into account. Then an equation is needed for $a_{\mu\nu\alpha\beta}$, in (16.124). Multiplication of the generalized Fokker-Planck equation (16.119) by the fourth order expansion function $\phi_{\mu\nu\alpha\beta}$ and subsequent integration leads to an equation for the fourth rank alignment tensor, that is analogous to that of the second rank tensor. There, however, not only a coupling with the second rank but also with the sixth rank tensor occurs. Clearly, the game may be continued leading to a hierarchy of coupled equations for tensors with rank $\ell = 2, 4, 6, \ldots$. The equation for the second rank tensor is the only one which contains an inhomogeneous term. The tensors of rank $\ell > 2$ relax faster than the second rank tensor, cf. (12.44). In [157], a closure of the set of equations was achieved by disregarding the tensors of rank 6 and higher. When furthermore, the co-rotational time derivative of the fourth rank tensor and terms nonlinear in the deformation rate $\overline{\nabla \mathbf{v}}$ are disregarded, this approximation amounts to putting $a_{\mu\nu\alpha\beta} \sim \overline{a_{\mu\nu}a_{\alpha\beta}}$ with a proportionality coefficient analogous to that one occurring for equilibrium alignment, cf. Sect. 12.2.4. Use of the relation (12.34) for uniaxial alignment in the high temperature approximation, which is equivalent to

$$\zeta_4^{-1}\, a_{\mu\nu\alpha\beta} = \frac{5}{7}\, \zeta_2^{-2}\, \overline{a_{\mu\nu}a_{\alpha\beta}} \,, \tag{16.129}$$

leads to the closed equation governing the second rank alignment tensor

$$\frac{\partial}{\partial t} a_{\mu\nu} - 2\, \overline{\varepsilon_{\mu\lambda\kappa}\omega_\lambda a_{\kappa\nu}} - 2\kappa\, \overline{\nabla_\mu v_\kappa\, a_{\kappa\nu}} + \nu_2 \Phi_{\mu\nu}$$
$$= \mathscr{R}\left(\overline{\nabla_\mu v_\nu} - \frac{10}{21}\, \overline{a_{\mu\nu}a_{\lambda\kappa}}\, \overline{\nabla_\alpha v_\beta} \right). \tag{16.130}$$

Here

$$\Phi_{\mu\nu} = A\, a_{\mu\nu} - B\sqrt{6}\, \overline{a_{\mu\kappa}a_{\kappa\nu}} + C\, a_{\mu\nu}\, a_{\lambda\kappa}a_{\lambda\kappa},$$

is the derivative of the Landau-de Gennes potential, cf. Sect. (15.2.2), where now the coefficients A and B are given by (16.125) and C is found to be

$$C = \frac{12}{49}\left(\frac{T^*}{T} \right)^2. \tag{16.131}$$

Apart from the last term on the right hand side of (16.130), this equation is equivalent to (16.127) where now $\Phi_{\mu\nu}$ appears in the relaxation term instead of $Aa_{\mu\nu}$. The limiting case of a weak alignment in the isotropic phase was already discussed above. The other limiting case corresponds to a weak flow in the nematic phase where the velocity gradient does not alter the uniaxial character of the alignment nor affect the magnitude of the alignment.

The equation for the alignment tensor which underlies a unified theory valid both for the isotropic and the nematic phases of a liquid crystal, can also be derived within the framework of irreversible thermodynamics, see the following section. The flow alignment and also the viscous properties of nematics are treated by this approach. Dynamic phenomena, such as a time dependent and even chaotic response of the alignment to a stationary shear rate are discussed in Sect. 17.3.

Some historical remarks: The application of a Fokker-Planck equation to the flow birefringence in colloidal dispersions was initiated by A. Peterlin and H.A. Stuart in 1939 and reviewed 1943 [160]. They used the torque caused by the flow as derived by Jeffery [161]. The inclusion of a torque associated with the alignment, which allows the treatment of both the isotropic and nematic phases, was first presented by the author [157]. An independent derivation was given later by Doi [158], who considered the application to rod-like polymers, see also [162]. In the literature, both the generalized Fokker-Planck equation and the resulting equation for the second rank alignment tensor are referred to as *Doi-theory* or *Doi-Hess-theory*, see e.g. [163]. Different assumptions were made for the closure of the hierarchy equations. A discussion of the dynamic equations and the underlying physics is also presented in [164–167].

16.4.5 Unified Theory for Isotropic and Nematic Phases

The first unified theory for the flow alignment and the viscous properties of liquid crystals in the isotropic and nematic phases [168], as well as the study of the influence of a shear flow on the phase transition [169] was based on a generalized version of irreversible thermodynamics, where the alignment tensor is treated as an additional macroscopic variable, as in Sect. 16.3.6. As before, the point of departure for a treatment within the framework of irreversible thermodynamics is an expression for the contribution of the alignment to the free energy or the free enthalpy. Now it is assumed that this contribution is proportional to the Landau-de Gennes potential Φ, its time change is proportional to $-\Phi_{\mu\nu}\mathrm{d}a_{\mu\nu}/\mathrm{d}t$, where $\Phi_{\mu\nu}$ is the derivative of Φ with respect to $a_{\mu\nu}$. When the co-rotational time derivative of the alignment is used as in (16.69), the resulting entropy production is similar to the expression (16.70), just with $a_{\mu\nu}(\frac{\delta a_{\mu\nu}}{\delta t})_{\mathrm{irrev}}$ replaced by $\Phi_{\mu\nu}(\frac{\delta a_{\mu\nu}}{\delta t})_{\mathrm{irrev}}$. The ensuing constitutive relations are similar to (16.133), now $a_{\mu\nu}$, in the first of these equations replaced by $\Phi_{\mu\nu}$. As a consequence, the inhomogeneous equation for the alignment tensor analogous to (16.74) contains the nonlinear relaxation term $\tau_{\mathrm{a}}^{-1}\Phi_{\mu\nu}$, instead of $\tau_{\mathrm{a}}^{-1}a_{\mu\nu}$.

Motivated by the first three terms of (16.124), the more general ansatz

$$\frac{\mathrm{d}a_{\mu\nu}}{\mathrm{d}t} - 2\overline{\varepsilon_{\mu\lambda\kappa}\omega_\lambda a_{\kappa\nu}} - 2\kappa\overline{\nabla_\mu v_\kappa\, a_{\kappa\nu}} = \left(\frac{\delta a_{\mu\nu}}{\delta t}\right)_{\mathrm{irrev}}, \tag{16.132}$$

is made. Then the entropy production is given by

$$
-\frac{\rho}{m}T\left(\frac{\delta s}{\delta t}\right)_{\text{irrev}}^{(2)} = \overline{p_{\nu\mu}}\,\overline{\nabla_\nu v_\mu} + \frac{\rho}{m}k_{\text{B}}T\,\Phi_{\mu\nu}\left[\left(\frac{\delta a_{\mu\nu}}{\delta t}\right)_{\text{irrev}} + 2\kappa\,\overline{\nabla_\mu v_\kappa\,a_{\kappa\nu}}\right]
$$

$$
= \left[p_{\nu\mu} + 2\kappa\frac{\rho}{m}k_{\text{B}}T\,\overline{\Phi_{\mu\kappa}a_{\kappa\nu}}\right]\overline{\nabla_\mu v_\nu} + \frac{\rho}{m}k_{\text{B}}T\,\Phi_{\mu\nu}\left(\frac{\delta a_{\mu\nu}}{\delta t}\right)_{\text{irrev}}.
$$

$$(16.133)$$

With $\Phi_{\mu\nu}$ and $(\sqrt{2}\frac{\rho}{m}k_{\text{B}}T)^{-1}(\overline{p_{\nu\mu}} + 2\kappa\frac{\rho}{m}k_{\text{B}}T\,\overline{\Phi_{\mu\kappa}a_{\kappa\nu}})$ chosen as fluxes, and $(\frac{\delta a_{\mu\nu}}{\delta t})_{\text{irrev}}$ and $\sqrt{2}\,\overline{\nabla_\nu v_\mu}$ as forces, as suggested in [171], the constitutive laws for the second rank tensors now are

$$
-\Phi_{\mu\nu} = \tau_a\left(\frac{\delta a_{\mu\nu}}{\delta t}\right)_{\text{irrev}} + \tau_{\text{ap}}\sqrt{2}\,\overline{\nabla_\nu v_\mu}\,,
$$

$$
-\left(\sqrt{2}\frac{\rho}{m}k_{\text{B}}T\right)^{-1}\left(\overline{p_{\nu\mu}} + 2\kappa\frac{\rho}{m}k_{\text{B}}T\,\overline{\Phi_{\mu\kappa}a_{\kappa\nu}}\right) = \tau_{\text{pa}}\left(\frac{\delta a_{\mu\nu}}{\delta t}\right)_{\text{irrev}} + \tau_{\text{p}}\sqrt{2}\,\overline{\nabla_\nu v_\mu}\,.
$$

$$(16.134)$$

As before, the quantities $\tau_{..}$ are relaxation time coefficients where the subscripts a and p refer to "alignment" and "pressure". The non-diagonal coefficients obey the Onsager symmetry relation, cf. (16.72), $\tau_{\text{ap}} = \tau_{\text{pa}}$. Positive entropy production is guaranteed by the inequalities $\tau_a > 0$, $\tau_p > 0$, $\tau_a\tau_p > \tau_{\text{ap}}^2$.

Use of the first of the (16.133) in (16.69) yields the inhomogeneous relaxation equation

$$
\frac{da_{\mu\nu}}{dt} - 2\,\overline{\varepsilon_{\mu\lambda\kappa}\omega_\lambda a_{\kappa\nu}} - 2\kappa\,\overline{\nabla_\mu v_\kappa\,a_{\kappa\nu}} + \tau_a^{-1}\Phi_{\mu\nu} = -\tau_a^{-1}\tau_{\text{ap}}\sqrt{2}\,\overline{\nabla_\nu v_\mu}\,. \quad (16.135)
$$

Apart from the last term on the right hand side of (16.130), the phenomenological equation corresponds to the equation derived from the Fokker-Planck equation, when τ_a^{-1} and $\sqrt{2}\tau_a^{-1}\tau_{\text{ap}}$ are identified with ν_2 and $-\mathcal{R}$, as in (16.128).

The symmetric traceless part of the pressure tensor, as it follows from the constitutive relations, is given by $\overline{p_{\nu\mu}} = -2\eta_{\text{iso}}\,\overline{\nabla_\nu v_\mu} + \overline{p_{\nu\mu}}^{\text{align}}$, with $\eta_{\text{iso}} = \eta_{\text{New}}(1 - \frac{\tau_{\text{ap}}^2}{\tau_a\tau_p})$, $\eta_{\text{New}} = \frac{\rho}{m}k_{\text{B}}T\tau_p$, cf. (16.84) and (16.85), where the friction pressure associated with the alignment is now

$$
\overline{p_{\nu\mu}}^{\text{align}} = \frac{\rho}{m}k_{\text{B}}T\left(\sqrt{2}\frac{\tau_{\text{ap}}}{\tau_a}\Phi_{\mu\nu} - 2\kappa\,\overline{\Phi_{\mu\nu}a_{\kappa\nu}}\right). \quad (16.136)
$$

A remark on the antisymmetric part of the pressure tensor is in order. Prior to putting the average angular velocity w_μ equal to the vorticity ω_μ, the entropy production involving pseudo-vectors is proportional to $p_\mu(w_\mu - \omega_\mu)$, where p_μ is the pseudo vector associated with the antisymmetric part of the pressure tensor, cf. Sect. 16.3.5. This contribution vanishes, when the average angular velocity is equal to the vorticity, as already assumed above. In the presence of an external field, which exerts a torque T_μ, the entropy production contains an additional contribution proportional to $w_\mu(T_\mu + 2k_B T \varepsilon_{\mu\nu\lambda} a_{\nu\kappa} \Phi_{\kappa\lambda})$. The entropy production, however, should not depend explicitly on the vorticity, since $\omega_\mu \neq 0$ can also be achieved by a solid body like rotation. This implies that $T_\mu = -2k_B T \varepsilon_{\mu\nu\lambda} a_{\nu\kappa} \Phi_{\kappa\lambda}$ has to hold true. When furthermore, the relaxation of the internal angular momentum $\mathbf{J}$ is fast, compared with the orientational relaxation, one has effectively $d\mathbf{J}/dt = 0$ and p_μ matches the torque density, viz. $p_\mu = (\rho/m)T_\mu$. Thus the pseudo-vector associated with the antisymmetric part of the pressure is related to the alignment by

$$p_\mu = -2 \frac{\rho}{m} k_B T \, \varepsilon_{\mu\nu\lambda} a_{\nu\kappa} \Phi_{\kappa\lambda}. \tag{16.137}$$

As expected, both the symmetric traceless and the antisymmetric parts of the pressure tensor associated with the alignment vanish in thermal equilibrium where one has $\Phi_{\mu\nu} = 0$.

Multiplication of (16.135) by $\tau_a \varepsilon_{\lambda\kappa\nu} a_{\kappa\mu}$ yields

$$\tau_a M_\lambda + \varepsilon_{\lambda\kappa\nu} a_{\kappa\mu} \Phi_{\mu\nu} = -\varepsilon_{\lambda\kappa\nu} a_{\kappa\mu} \left(\tau_{ap} \sqrt{2} \, \overline{\nabla_\mu v_\nu} - 2\kappa \, \tau_a \, \overline{\nabla_\mu v_\sigma \, a_{\sigma\nu}} \right),$$

with

$$M_\lambda = \varepsilon_{\lambda\kappa\nu} a_{\kappa\mu} \left(\frac{da_{\mu\nu}}{dt} - 2 \overline{\varepsilon_{\mu\alpha\beta} \omega_\alpha a_{\beta\nu}} \right). \tag{16.138}$$

Then (16.137) is equal to

$$p_\lambda = 2\frac{\rho}{m} k_B T M_\lambda + 2\frac{\rho}{m} k_B T \varepsilon_{\lambda\kappa\nu} a_{\kappa\mu} \left(\tau_{ap} \sqrt{2} \, \overline{\nabla_\mu v_\nu} - 2\kappa \, \tau_a \, \overline{\nabla_\mu v_\sigma \, a_{\sigma\nu}} \right). \tag{16.139}$$

16.4.6 Limiting Cases: Isotropic Phase, Weak Flow in the Nematic Phase

For a plane Couette flow, a symmetry adapted ansatz for the alignment and for the pressure tensors can be made in analogy to (16.90). Then each of the tensorial equations reduces to three coupled equations for the relevant 3 components. In detail, these can be inferred from the more general case of all 5 components as presented in Sect. 17.3. Here just some results are stated for the nonlinear viscous behavior in the isotropic phase, where terms nonlinear in the alignment tensor are disregarded.

The non-Newtonian viscosity coefficient η_+, cf., Sect. 16.3.9, is found to be

$$\eta_+ = \eta_{\text{New}} H_+(\Gamma), \quad H_+(\Gamma) = 1 + \frac{\tau_{\text{ap}}^2}{\tau_a \tau_p}\left[\frac{1 + \Gamma^2(1 + \kappa^2/3)}{(1 + \Gamma^2(1 - \kappa^2/3))^2} - 1\right]. \quad (16.140)$$

The dimensionless viscosity coefficient $H_+(\Gamma)$ is a function of the dimensionless shear rate Γ, here defined by

$$\Gamma = \tau\gamma = A^{-1}\tau_a \frac{\partial v_x}{\partial y}. \quad (16.141)$$

Notice that $\frac{\tau_{\text{ap}}^2}{\tau_a \tau_p} = (\eta_{\text{New}} - \eta_{\text{iso}})/\eta_{\text{New}} < 1$, cf. (16.85). For $|\kappa| < 1$, (16.140) describes a shear thinning behavior, for $1 < |\kappa| < \sqrt{3}$, shear thickening, i.e. an increase of the shear viscosity with increasing shear rate results for smaller values of Γ, followed by a shear thinning at higher shear rates. In any case, η_{iso} is approached for $\Gamma \to \infty$. Similarly, the viscosity coefficients η_- and η_0 are given by

$$\eta_- = (\eta_+(\Gamma) - \eta_{\text{iso}})\,\Gamma, \quad \eta_0 = -\kappa\,(\eta_{\text{New}} - \eta_{\text{iso}})\left[1 + \Gamma^2(1 - \kappa^2/3)\right]^{-1}\Gamma. \quad (16.142)$$

Both η_- and η_0 approach 0 for $\Gamma \ll 1$ and $\Gamma \gg 1$. The normal pressure differences and the pertaining viscometric functions, as defined in Sect. 16.3.9, can be inferred from (16.142). In particular, the ratio between the first and the second viscometric function is found to be

$$\frac{-\Psi_2}{\Psi_1} = \frac{1}{2} - \kappa\left[1 + \Gamma^2(1 - \kappa^2/3)\right]^{-1}\left(1 + \Gamma^2(1 + \kappa^2/3)\right). \quad (16.143)$$

The special case $\kappa = 0$, corresponding to a pure co-rotational time derivative of the alignment tensor, implies $\eta_0 = 0$ and the small shear rate limit $\Psi_2/\Psi_1 = -0.5$. The value $\kappa \approx 0.4$, suggested by the Fokker-Planck approach, yields $\Psi_2/\Psi_1 \approx -0.1$, which is typical for many polymeric liquids.

For a weak flow in the nematic phase, the alignment tensor maintains its uniaxial form $a_{\mu\nu} = a_{\mu\nu}^{\text{eq}} = \sqrt{3/2}a_{\text{eq}}\,n_\mu n_\nu$, where the director $\mathbf{n}$, in general, depends on the time and the position. The equilibrium order parameter $a_{\text{eq}} = \sqrt{5}S$, where $S = S_2$ is recalled as the Maier-Saupe order parameter, is assumed not to be affected by the flow. In this case $\mathbf{M}$ reduces to

$$M_\lambda = \frac{3}{2}a_{\text{eq}}^2 \varepsilon_{\lambda\kappa\nu}n_\kappa N_\nu, \quad N_\nu = \frac{dn_\nu}{dt} - \overline{\varepsilon_{\nu\lambda\kappa}\omega_\lambda n_\kappa}. \quad (16.144)$$

The vector $\mathbf{N}$ is the co-rotational time derivative of the director $\mathbf{n}$, cf. (16.108). For $d\mathbf{n}/dt = 0$, as considered in (16.136), the pseudo-vector $\mathbf{M}$ is proportional to the component of the vorticity, which is perpendicular to $\mathbf{n}$, viz. $M_\lambda = -\frac{3}{2}a_{\text{eq}}^2(\omega_\lambda - n_\lambda n_\kappa \omega_\kappa)$. In the weak flow limit, (16.139) reduces to

$$p_\mu = \varepsilon_{\mu\nu\lambda} n_\nu \left(\gamma_1 N_\lambda + \gamma_2 \overline{\nabla_\lambda v_\kappa} \, n_\kappa \right), \tag{16.145}$$

$$\gamma_1 = 3 \frac{\rho}{m} k_B T a_{eq}^2 \tau_a, \quad \gamma_2 = \frac{\rho}{m} k_B T \left(2\sqrt{3} \, a_{eq} \tau_{ap} - \kappa a_{eq}^2 \tau_a \right).$$

In this expression, the Ericksen-Leslie coefficients $\gamma_1 > 0$ and γ_2, already introduced in Sect. 16.4.1, are related to the equilibrium alignment $a_{eq} = \sqrt{5}S$, and to the model parameters occurring in the dynamic equation for the alignment tensor. Notice that the Fokker-Planck approach yields $\tau_{ap} < 0$ for rod-like particles, cf. (16.126), (16.128), and consequently $\gamma_2 < 0$. For disc-like particles, on the other hand, one has $\gamma_2 > 0$. In general, terms of higher power in the order parameter a_{eq} contribute to the Ericksen-Leslie coefficients, when higher rank tensors are taken into account in the solution of the Fokker-Planck equation. The expressions given here contain the leading terms.

In the weak flow approximation, the relation (16.136) for the symmetric traceless part of the friction pressure tensors leads to an expression for $\overline{p_{\mu\nu}}$, as presented in (16.101), now with the viscosity coefficients $\eta, \tilde{\eta}_1, \tilde{\eta}_2, \tilde{\eta}_3$ given by

$$\eta = \frac{\rho}{m} k_B T \left(\tau_p + \frac{1}{6} \kappa^2 a_{eq}^2 \tau_a \right), \quad \tilde{\eta}_1 = -\frac{\rho}{m} k_B T \kappa \, a_{eq} \left(2\sqrt{3} \tau_{ap} + \frac{1}{2} \kappa \, a_{eq} \tau_a \right),$$

$$\tilde{\eta}_2 = \frac{\rho}{m} k_B T \kappa \, a_{eq} \left(\sqrt{3} \tau_{ap} - \frac{1}{2} \kappa \, a_{eq} \tau_a \right), \quad \tilde{\eta}_3 = \frac{1}{2} \frac{\rho}{m} k_B T \kappa^2 a_{eq}^2 \tau_a. \tag{16.146}$$

The Onsager symmetry relation $2\tilde{\eta}_2 = \gamma_2$ is obeyed. Notice that $\kappa = 0$ implies $\tilde{\eta}_1 = \tilde{\eta}_3 = 0$. In lowest order in the alignment, one has $\tilde{\eta}_1 = -\kappa\gamma_2$. This relation can be used to obtain an estimate for the size of the parameter κ from experimental data.

16.4.7 Scaled Variables, Model Parameters

The relaxation term of the inhomogeneous equation (16.135) for the second rank alignment tensor involves the derivative of the Landau-de Gennes potential, which in turn contains the three parameters A, B, C. When the alignment is expressed in units of the nematic order parameter $a_{ni} = 2B/(3C)$, cf. Sect. 15.2.2, the coefficients B and C are replaced by specific numbers. More precisely, the alignment tensor is written as $a_{\mu\nu} = a_{ni} a_{\mu\nu}^*$, the derivative of the potential is expressed as $\Phi_{\mu\nu}(a) = \Phi_{ref} \Phi_{\mu\nu}^*(a^*)$ with the reference value $\Phi_{ref} = a_{ni} 2B^2/(9C) = a_{ni} \delta_{ni} A_0$. As in Sect. 15.2.2, the scaled temperature variable $\vartheta = A(T)/A(T_{ni}) = (1 - T^*/T)(1 - T^*/T_{ni})$ is used. The time is scaled in units of a reference time equal to the relaxation time, at coexistence temperature T_{ni}, viz.

$$t = \tau_{ref} \, t^*, \quad \tau_{ref} = \tau_a (1 - T^*/T_{ni})^{-1} A_0^{-1} = \tau_a \frac{9C}{2B^2} = \tau_a \, a_{ni} \, \Phi_{ref}^{-1}. \tag{16.147}$$

Shear rates, in units of τ_{ref}^{-1} are now denoted by Γ, thus one has $\Gamma = \gamma\,\tau_{\text{ref}}$. Furthermore, the tumbling parameter, cf. Sect. 16.4.3, is written as

$$\lambda_{\text{eq}} = \lambda_K \frac{a_{\text{ni}}}{a_{\text{eq}}} + \frac{1}{3}\kappa, \quad \lambda_K = -\frac{2}{3}\sqrt{3}\,\frac{\tau_{\text{ap}}}{\tau_{\text{a}}}\,a_{\text{ni}}^{-1}. \tag{16.148}$$

The equilibrium order parameter, in the nematic phase, a_{eq} is given by $a_{\text{eq}}/a_{\text{ni}} = \frac{3}{4} + \frac{1}{4}\sqrt{9 - 8\vartheta}$, $\vartheta \le \frac{9}{8}$, see (15.23). Notice that λ_{eq} decreases with increasing order a_{eq}. For small shear rates, λ_{eq} determines the flow alignment angle χ, within the Ericksen-Leslie theory, cf. Sect. 16.4.3, according to $\cos(2\chi) = -\gamma_1/\gamma_2 = 1/\lambda_{\text{eq}}$, provided that $\lambda_{\text{eq}} > 1$. For $\lambda_{\text{eq}} < 1$, no stable flow alignment exists. The actual dynamics following from the alignment tensor theory, as discussed in Sect. 17.3, is more complex than the tumbling motion inferred from the Ericksen-Leslie director approach. The quantity λ_K which is the tumbling parameter at the transition temperature, for $\kappa = 0$, is used as a model parameter in the scaled dynamic equation for the alignment tensor.

In the following, when no danger of confusion exists, the scaled alignment tensor $a_{\mu\nu}^*$ is denoted by the original symbol $a_{\mu\nu}$. Let $\Omega_\lambda = \omega_\lambda \tau_{\text{ref}}$ and $\Gamma_{\mu\nu} = \overline{\partial v_\mu/\partial r_\nu}$ be the dimensionless vorticity and deformation rate tensor. The (16.135) governing the dynamics of the alignment then is equivalent to

$$\frac{da_{\mu\nu}}{dt} - 2\,\overline{\varepsilon_{\mu\lambda\kappa}\Omega_\lambda a_{\kappa\nu}} - 2\kappa\,\overline{\Gamma_{\mu\kappa}a_{\kappa\nu}} + \Phi_{\mu\nu} = \sqrt{\frac{3}{2}}\lambda_K\Gamma_{\mu\nu}, \tag{16.149}$$

where it is understood that $\Phi_{\mu\nu}$ stands for the scaled derivative of the relevant potential, viz.

$$\Phi_{\mu\nu} = \vartheta\,a_{\mu\nu} - 3\sqrt{6}\,\overline{a_{\mu\kappa}a_{\kappa\nu}} + 2a_{\mu\nu}a_{\lambda\kappa}a_{\lambda\kappa}. \tag{16.150}$$

This corresponds to the derivative of a Landau-de Gennes potential with $A = \vartheta$, $B = 3, C = 2$.

A scaled symmetric traceless stress tensor $\Sigma_{\mu\nu}^{\text{al}}$, associated with the alignment is introduced via

$$- \overline{p_{\mu\nu}}^{\text{align}} = \sqrt{2}\,G_{\text{al}}\,\Sigma_{\mu\nu}^{\text{al}}, \quad G_{\text{al}} = \frac{3}{4}\frac{\rho}{m}k_B T\,\lambda_K^2\,\delta_{\text{ni}}\,A_0\,a_{\text{ni}}^2, \tag{16.151}$$

where $\delta_{\text{ni}} = 1 - T^*/T_{\text{ni}}$, and G_{al} is a shear modulus linked with the alignment. For $\overline{p_{\mu\nu}}^{\text{align}}$ see (16.136). The scaled version of this equation corresponds to

$$\Sigma_{\mu\nu}^{\text{al}} = \frac{2}{\sqrt{3}}\lambda_K^{-1}\,\tilde{\Phi}_{\mu\nu}, \quad \tilde{\Phi}_{\mu\nu} = \Phi_{\mu\nu} + \frac{2\kappa}{3\lambda_K}\sqrt{6}\,\overline{a_{\mu\kappa}\Phi_{\kappa\nu}}. \tag{16.152}$$

The Fokker-Planck equation approach implies $2\kappa/(3\lambda_K) = \sqrt{5}\,a_{\text{ni}}/7$.

The model parameters occurring in the scaled equations are the reduced temperature ϑ, the dimensionless shear rate Γ, the tumbling parameter λ_K, and κ. Notice that λ_K serves as a measure for the coupling between the alignment and the flow. In some applications of the equations, as presented in Sect. 17.3, it suffices to treat the special case $\kappa = 0$.

16.4.8 Spatially Inhomogeneous Alignment

In a spatially inhomogeneous situation, the equation governing the dynamics of the alignment tensor contains terms linked with spatial derivatives. There are two sources for terms of this kind: firstly, the divergence of the flux tensor $b_{\lambda\mu\nu} \sim \langle c_\lambda \overline{u_\mu u_\nu} \rangle$, where $\mathbf{c}$ is the peculiar velocity of a particle, and secondly, the terms characterizing the elasticity in the free energy and its derivative with respect to the alignment, see e.g. (15.38). More specifically, the relaxation equation for the alignment tensor is

$$\frac{da_{\mu\nu}}{dt} - 2\,\overline{\varepsilon_{\mu\lambda\kappa}\omega_\lambda a_{\kappa\nu}} - \ldots + \nabla_\lambda b_{\lambda\mu\nu} + \tau_a^{-1}\left(\Phi_{\mu\nu} - \xi_0^2\,\Delta\,a_{\mu\nu}\right) + \ldots = 0,$$

where the dots ... indicate the terms involving the deformation rate tensor $\overline{\nabla_\mu v_\nu}$, as in (16.135). It is understood that $\frac{da_{\mu\nu}}{dt}$ stands for the substantial derivative, i.e. $\frac{da_{\mu\nu}}{dt} = \frac{\partial a_{\mu\nu}}{\partial t} + v_\lambda \nabla_\lambda a_{\mu\nu}$. As before, $\Phi_{\mu\nu}$ is the derivative of the potential Φ, with respect to $a_{\mu\nu}$, e.g. the Landau de Gennes expression (15.13). The term involving $\xi_0^2 \Delta a_{\mu\nu}$ corresponds to the simple case of an isotropic elasticity. The length ξ_0 is linked with the quantities occurring in (15.35) via $\xi_0^2 = \frac{\varepsilon_0}{k_B T}\xi_{ref}^2 \sigma_2$, where ξ_{ref} is the reference length which, in (15.35), was denoted by ξ_0. Equations for the three irreducible parts of the tensor $b_{\lambda\mu\nu}$, which are tensors of ranks 1, 2, 3, can be derived by kinetic theory or by irreversible thermodynamics. When the relaxation times for these three parts are practically equal to a single relaxation time τ_b, the approximation

$$b_{\lambda\mu\nu} = -D_a \nabla_\lambda \left(\Phi_{\mu\nu} - \xi_0^2 \Delta a_{\mu\nu}\right), \quad D_a = \frac{k_B T}{m}\tau_b,$$

is obtained. With the *diffusion length* ℓ_a defined by $\ell_a^2 = D_a \tau_a$, the generalization of (16.135) to a spatially inhomogeneous fluid becomes, [172],

$$\frac{d}{dt}a_{\mu\nu} - 2\,\overline{\varepsilon_{\mu\lambda\kappa}\omega_\lambda a_{\kappa\nu}} - 2\kappa\,\overline{\nabla_\mu v_\kappa\,a_{\kappa\nu}} \tag{16.153}$$

$$+\tau_a^{-1}\left[1 - \ell_a^2 \Delta\right]\left(\Phi_{\mu\nu} - \xi_0^2 \Delta a_{\mu\nu}\right) = -\tau_a^{-1}\tau_{ap}\sqrt{2}\,\overline{\nabla_\nu v_\mu}.$$

Second and fourth order spatial derivatives occur. The second order terms involve diffusional and elastic contributions proportional to ℓ_a^2 and ξ_0^2, respectively. In the isotropic phase, where $\Phi_{\mu\nu} \approx A a_{\mu\nu}$ applies, these contributions are additive. In the nematic phase, and close to equilibrium, one has $\Phi_{\mu\nu} \approx 0$ and the relaxation term reduces to $-\tau_a^{-1} \xi_0^2 [\Delta a_{\mu\nu} - \ell_a^2 \Delta^2 a_{\mu\nu}]$. The fourth order term is proportional to the product $\ell_a^2 \xi_0^2$, in any case.

The part of the friction pressure tensor associated with the alignment is also modified by additional terms involving the spatial derivatives [172]. The solution of a spatial differential equation requires boundary conditions, those appropriate for the (16.153) and some applications are discussed in [173, 174].

Chapter 17
Tensor Dynamics

Abstract This chapter presents examples for dynamical phenomena involving tensors. Firstly, linear tensor equations are considered which provide the basis for the computation of time-correlation functions and of spectral functions describing the frequency dependence, e.g. of scattered radiation. Secondly, nonlinear relaxation phenomena involving the second rank alignment tensor are treated. Basis tensors are introduced which lead to coupled non-linear equations for the relevant components of the tensor. The stability of stationary solutions is analyzed. Thirdly, the effect of an imposed shear flow on the alignment tensor is considered. Depending on the model parameters, stationary as well as periodic and chaotic solutions are obtained. Similar features are found for a nonlinear Maxwell model governing the shear stress tensor.

17.1 Time-Correlation Functions and Spectral Functions

The dynamics of small fluctuations about an equilibrium state, as well as of small macroscopic deviations from equilibrium are described by time-correlation functions. Spectral functions are obtained by a time-Fourier transformation.

17.1.1 Definitions

Let $\psi_i = \psi_i(t)$ with $i = 1, 2, \ldots$ be functions which depend on the time t via their dependence on dynamic variables like the position, the linear momentum or the internal angular momentum of a particle. Appropriately defined averages $\langle \ldots \rangle_0$ of these quantities are assumed to vanish, viz. $\langle \psi_i \rangle_0 = 0$. In general, the ψ_i fluctuate about their average values, their squares averaged are non-zero: $\langle \psi_i^2 \rangle_0 > 0$. The average

$$\mathscr{C}_{ij}(t) = \langle \psi_i(t_0 + t)\psi_j(t_0) \rangle_0, \quad i = 1, 2, \ldots, \quad j = 1, 2, \ldots, \tag{17.1}$$

© Springer International Publishing Switzerland 2015

S. Hess, *Tensors for Physics*, Undergraduate Lecture Notes in Physics,
DOI 10.1007/978-3-319-12787-3_17

defines time-correlation functions. It is assumed that the distribution underlying the average is stationary, e.g. pertaining to an equilibrium state. Thus the correlation function just depends on the time difference, i.e. it is independent of the time t_0. Then $t_0 = 0$ can be chosen. On the other hand, the choice $t_0 = -t$ yields

$$\langle \psi_i(0) \psi_j(-t) \rangle_0 = \langle \psi_j(-t) \psi_i(0) \rangle_0,$$

which implies the symmetry relation

$$\mathscr{C}_{ij}(t) = \mathscr{C}_{ji}(-t). \tag{17.2}$$

Consequently the diagonal correlation functions pertaining to $i = j$ are even functions of the time. The correlation functions with $i = j$ are referred to as *auto-correlation functions*, those with $i \neq j$ as *cross-correlation functions*. In the following, the notation $C_{ij}(t)$ is used for the normalized correlation functions which are defined by

$$C_{ij}(t) = \langle \psi_i(t_0 + t) \psi_j(t_0) \rangle_0 \left\{ \langle \psi_i(t_0) \psi_i(t_0) \rangle_0 \langle \psi_j(t_0) \psi_j(t_0) \rangle_0 \right\}^{-1/2}, \tag{17.3}$$

where $i = 1, 2, \ldots, j = 1, 2, \ldots$, as before. One has $C_{ii}(0) = 1$ for the normalized auto-correlation functions.

Now a non-equilibrium state is considered where $\langle \psi_i \rangle = \langle \psi_i \rangle(t) \neq 0$ applies. When the distortion which caused the deviation from equilibrium, is switched off, the quantity $\langle \psi_i \rangle(t)$ relaxes to 0, in the long time limit. The original derivation of the Onsager symmetry relation was based on the assumption that fluctuations and small macroscopic deviations from thermal equilibrium decay alike [108]. With this argument, time-correlation functions can also be defined via a linear relation between time dependent averages

$$\langle \psi_i \rangle(t) = C_{ij}(t) \langle \psi_j \rangle(0), \quad i = 1, 2, \ldots, \quad j = 1, 2, \ldots \tag{17.4}$$

This allows the calculation of time-correlation functions from linear macroscopic equations.

The averages $\langle \psi_i \rangle$ may depend on the position $\mathbf{r}$ in space $\langle \psi_i \rangle = \langle \psi_i \rangle(t, \mathbf{r})$. The spatial Fourier transform is

$$\langle \psi_i \rangle(t \mid \mathbf{k}) = \int \exp[-i\, \mathbf{k} \cdot \mathbf{r}] \, \langle \psi_i \rangle(t, \mathbf{r}) \, d^3r,$$

where $\mathbf{k}$ is the relevant wave vector. Applications in spectroscopy and light scattering involve wave vector dependent time-correlation functions defined by

$$\langle \psi_i \rangle(t \mid \mathbf{k}) = C_{ij}(t \mid \mathbf{k}) \langle \psi_j \rangle(0 \mid \mathbf{k}), \quad i = 1, 2, \ldots, \quad j = 1, 2, \ldots \tag{17.5}$$

The pertaining *spectral functions* S_{ij} are the Fourier-Laplace transform of the time-correlation functions, viz.

$$S_{ij}(\omega, \mathbf{k}) = \pi^{-1} \text{Re} \int_0^\infty \exp[i\omega t] \, C_{ij}(t \,|\, \mathbf{k}) dt. \qquad (17.6)$$

When the averages $\langle \psi_i \rangle$ and $\langle \psi_j \rangle$ are components of irreducible tensors of ranks ℓ and n, the corresponding time-correlation and spectral functions are tensors of rank $\ell + n$. With $\langle \psi_i \rangle$ and $\langle \psi_j \rangle$ replaced by $A_{\mu_1 \cdots \mu_\ell}$ and $B_{\nu_1 \cdots \nu_n}$, equation (17.5) becomes

$$A_{\mu_1 \cdots \mu_\ell}(t \,|\, \mathbf{k}) = C^{AB}_{\mu_1 \cdots \mu_\ell, \nu_1 \cdots \nu_n}(t \,|\, \mathbf{k}) \, B_{\nu_1 \cdots \nu_n}(0 \,|\, \mathbf{k}), \qquad (17.7)$$

The pertaining spectral function, evaluated according to (17.6), is denoted by

$$S^{AB}_{\mu_1 \cdots \mu_\ell, \nu_1 \cdots \nu_n}(\omega, \mathbf{k}).$$

The symmetry, parity and time reversal consideration discussed in connection with linear constitutive relations, cf. Sect. 16.1, apply to these functions as well. The *depolarized Rayleigh scattering*, to be discussed in the next section, corresponds to a case, where one has $\ell = n = 2$.

Examples for the computation of auto- and cross-correlation functions of the friction pressure tensor and the tensor polarization of a gas of linear molecules, are found in [175]. These correlation functions are linked with the viscosity and the flow birefringence. The influence of a magnetic field also studied there is associated with the Senftleben-Beenakker effect of the viscosity, cf. Sect. 16.3.4.

As originally pointed out by Green and Kubo [101, 176], transport coefficients can be computed as time-integrals of correlation functions. The relevant equations are referred to as *Green-Kubo-* or as *Kubo-relations*. For details of the method e.g. see [48–50]. The Green-Kubo relations imply that material coefficients characterizing non-equilibrium processes can be inferred from fluctuations in an equilibrium state. Instead of performing a time integral, the material coefficients can also be obtained from the dependence of the magnitude of the fluctuations on the length of the time interval, over which the fluctuations are pre-averaged. This has been demonstrated in [177] for the viscosity and the viscoelasticity of a simple fluid.

17.1.2 Depolarized Rayleigh Scattering

Light scattering is caused by fluctuations of the dielectric tensor $\varepsilon_{\mu\nu}$. The *Rayleigh scattering* and the *Brillouin scattering* are associated with the fluctuations of the isotropic part which, in turn, are mainly caused by density fluctuations. In this case, the electric field of the scattered light is parallel to that of the incident light, this is *polarized scattering*. Fluctuations of the anisotropic part $\overline{\varepsilon_{\mu\nu}}$ lead to a scattered

Fig. 17.1 Depolarized
Rayleigh scattering, VH- and
HH-geometries. The *double
arrows* indicate the
directions of the electric field
vectors of the incident and of
the scattered light

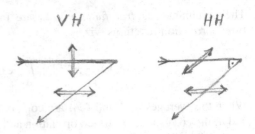

light with a weaker intensity, whose electric field, however, has also a component
perpendicular to that of the incident light. For this reason, the term *depolarized
scattering* is used. The name "Rayleigh" in *depolarized Rayleigh scattering* indicates,
that the frequency of this contribution to the scattered light is centered about the
frequency of the incident light, just as the ordinary Rayleigh scattering. The *rotational
Raman scattering*, where the frequency is shifted, also has a depolarized component.
Let $\mathbf{e}'$ and $\mathbf{e}$ be unit vectors parallel to the electric field vectors of the incident and of
the scattered light. The intensity of the scattered light is proportional to

$$I_{\text{scat}} = e'_\mu e_\nu \, S_{\mu\nu,\lambda\kappa} \, e'_\lambda e_\kappa. \tag{17.8}$$

The spectral function $S_{..}$ depends on $\omega = \omega_1 - \omega_2$ and $\mathbf{k} = \mathbf{k}_1 - \mathbf{k}_2$, where $\omega_1, \mathbf{k}_1$
and $\omega_2, \mathbf{k}_2$ are the frequencies and the wave vectors of the incident and of the scat-
tered light. Depolarized scattering means: $\mathbf{e}$ is perpendicular to $\mathbf{e}'$. Two scattering
geometries, referred to by *VH* and *HH* are sketched in Fig. 17.1. The letters *V* and
H stem from 'vertical' and 'horizontal', with respect to the scattering plane spanned
by $\mathbf{k}_1$ and $\mathbf{k}_2$. The HH-case is for 90° scattering only.

Orientational fluctuations of molecules cause fluctuations of $\overline{\varepsilon_{\mu\nu}}$. Thus the
time-correlation function and consequently the spectral function of the depolarized
Rayleigh scattering can be inferred from relaxation equation of the second rank align-
ment tensor $a_{\mu\nu}$ of liquids, cf. (12.19) or of the tensor polarization $a^{\text{T}}_{\mu\nu}$ in gases of
linear molecules, cf. (13.64).

In the absence of external fields and when the coupling with the friction pressure
tensor is ignored, the (16.59) and (16.74) describe a simple exponential relaxation
for the alignment tensor:

$$\frac{\partial a_{\mu\nu}}{\partial t} + \tau^{-1} a_{\mu\nu} = 0,$$

with a relaxation time τ. This equation implies

$$a_{\mu\nu}(t) = C_{\mu\nu,\lambda\kappa}(t) a_{\lambda\kappa}(0), \quad C_{\mu\nu,\lambda\kappa}(t) = \Delta_{\mu\nu,\lambda\kappa} \, C(t), \quad C(t) = \exp[-t/\tau], \tag{17.9}$$

with an isotropic time-correlation tensor. The resulting spectral function has the
Lorentz line shape

$$S_{\text{Lor}}(\omega) = \pi^{-1} \frac{\tau}{1 + \omega^2 \tau^2} = \pi^{-1} \frac{\nu}{\omega^2 + \nu^2}, \quad \nu = \tau^{-1}. \tag{17.10}$$

The *line width* is determined by the relaxation frequency $\nu = \tau^{-1}$.

The scattered intensity is proportional to

$$I_{\text{scat}} = e'_\mu e_\nu \Delta_{\mu\nu,\lambda\kappa} e'_\lambda e_\kappa S_{\text{Lor}}(\omega) = \overline{e'_\mu e_\nu}\, \overline{e'_\mu e_\nu} S_{\text{Lor}}(\omega)$$

$$= \frac{1}{2} \left(1 + \frac{1}{3}(\mathbf{e}' \cdot \mathbf{e})^2 \right) S_{\text{Lor}}(\omega). \tag{17.11}$$

The depolarized component, with $\mathbf{e}' \cdot \mathbf{e} = 0$, is $\frac{1}{2} S_{\text{Lor}}(\omega)$.

The time-correlation function and the spectral function are no longer isotropic, as in (17.9) when external fields or an ordered structure render the system anisotropic. An instructive example, as treated in [64], is considered next. Application of a magnetic field to a gas of rotating molecules causes a precessional motion of their rotational angular momenta with the frequency ω_B, cf. Sect. 16.3.4. Ignoring the coupling with the friction pressure tensor, the second of the (16.59) reduces to

$$\frac{\partial}{\partial t} a_{\mu\nu} - \omega_B H_{\mu\nu,\mu'\nu'} a_{\mu'\nu'} + \nu\, a_{\mu\nu} = 0,$$

with the relaxation frequency $\nu = \nu_a$. With the help of the projection tensors introduced in Chap. 14 in connection with the rotation of tensors, the solution of this equation is written as

$$a_{\mu\nu}(t) = C_{\mu\nu,\lambda\kappa}(t) a_{\lambda\kappa}(0), \quad C_{\mu\nu,\lambda\kappa}(t) = \exp[-\nu\, t] \sum_{m=-2}^{2} \exp[im\omega_B t]\, \mathscr{P}^{(m)}_{\mu\nu,\lambda\kappa}. \tag{17.12}$$

Now the scattered intensity is proportional to

$$I_{\text{scat}} = \overline{e'_\mu e_\nu}\, \overline{e'_\mu e_\nu} \sum_{m=-2}^{2} W_m\, S_{\text{Lor}}(\omega + m\,\omega_B),$$

$$\overline{e'_\mu e_\nu}\, \overline{e'_\mu e_\nu} W_m = \frac{1}{2} e'_\mu e_\nu \left(\mathscr{P}^{(m)}_{\mu\nu,\lambda\kappa} + \mathscr{P}^{(-m)}_{\mu\nu,\lambda\kappa} \right) e'_\lambda e_\kappa. \tag{17.13}$$

According to the relations presented in Sect. 14.5, the weight coefficients W_m, with the property $\sum_{m=-2}^{2} W_m = 1$, are explicitly given by

$$W_0 = 3(\mathbf{e} \cdot \mathbf{h})^2 (\mathbf{e}' \cdot \mathbf{h})^2, \quad W_1 = W_{-1} = \frac{1}{2} \left[(\mathbf{e} \cdot \mathbf{h})^2 + (\mathbf{e}' \cdot \mathbf{h})^2 \right] - 2(\mathbf{e} \cdot \mathbf{h})^2 (\mathbf{e}' \cdot \mathbf{h})^2,$$

$$W_2 = W_{-2} = \frac{1}{2} - \frac{1}{2} W_0 - W_1. \tag{17.14}$$

The unit vector $\mathbf{h}$ is parallel to the magnetic field. Consider the HH-geometry and put $\mathbf{h}$ perpendicular to both $\mathbf{e}'$ and $\mathbf{e}$. Then one has $W_0 = W_{\pm 1} = 0$ and resulting spectral line is split by the frequency $4|\omega_B|$, provided that the line width ν is not larger than about $|\omega_B|$. Similarly, for the VH-geometry, $W_0 = W_{\pm 2} = 0$ is obtained, when $\mathbf{h}$ is parallel to either $\mathbf{e}'$ or to $\mathbf{e}$. Then the line splitting is $2|\omega_B|$.

17.1.3 Collisional and Diffusional Line Broadening

The examples of time-correlation and spectral functions considered so far do not depend on the wave vector $\mathbf{k}$. For depolarized Rayleigh scattering in gases, this applies when the density n of the gas is large enough, such that $k\ell \ll 1$, where $\ell \sim n^{-1}$ is the mean free path, i.e. the average distance traveled by a molecule, in free flight, between two collisions. Under these conditions, the line width is determined by the collision frequency ν which is proportional to the number density. This type of broadening is called *collisional broadening* or also *pressure broadening*, since the density increases with increasing pressure. In the opposite limiting case, realized at low densities where $k\ell \gg 1$ applies, the line broadening is determined by the *Doppler broadening* where the line shape, reflecting the velocity distribution of the particles, is Gaussian. For intermediate cases, where one has $k\ell \approx 1$, diffusional processes contribute to the line width. This *diffusional broadening* is described by spatial derivatives in the relevant equations.

For a spatially inhomogeneous system, the alignment tensor obeys the equation

$$\frac{\partial a_{\mu\nu}}{\partial t} + \nabla_\lambda b_{\lambda\mu\nu} + \nu\, a_{\mu\nu} = 0, \tag{17.15}$$

where $b_{\lambda\mu\nu} \sim \langle c_\lambda \overline{J_\mu J_\nu} \rangle$ is the flux of the tensor polarization, $\mathbf{c}$ is the velocity of a molecule. Equations for the three irreducible parts of the tensor $b_{\lambda\mu\nu}$, which are tensors of ranks 1, 2, 3, can be derived by kinetic theory. When the collision frequencies for these three parts are practically equal to a single collision frequency ν_b and large compared with ν, the approximation

$$b_{\lambda\mu\nu} = -D_a \nabla_\lambda a_{\mu\nu}, \quad D_a = \frac{k_B T}{m} \nu_b^{-1} \tag{17.16}$$

can be made. Due to collisional changes of the rotational angular momenta, the diffusion coefficient D_a is smaller than a self diffusion coefficient. Insertion of the relation for the flux into the equation for the tensor polarization $a_{\mu\nu}$ leads to

$$\frac{\partial a_{\mu\nu}}{\partial t} - D_a \Delta a_{\mu\nu} + \nu\, a_{\mu\nu} = 0. \tag{17.17}$$

In a spatial Fourier transform of this equation, the Laplacian Δ is replaced by $-k^2$. The resulting time-correlation function is

$$C(t|\mathbf{k}) = \exp\left[-(v + D_a k^2) t\right],$$

and the corresponding spectral function is a Lorentzian with the line width determined by

$$v + D_a k^2, \quad v \sim n, \quad D_a \sim n^{-1}. \tag{17.18}$$

The density dependence of the line width (17.18) shows a minimum at an intermediate density. Such a minimum, referred to as *Dicke narrowing*, is actually observed provided that the collisions change the direction of the velocity of a particle more effectively than its rotational angular momentum. Relation (17.18) does not apply to lower densities where the Doppler broadening takes over [178, 179].

In general, the diffusional broadening is anisotropic in the sense that the **k**-dependent contribution to the line width is different for the VH and HH scattering geometries. The replacement of $D_a k^2 a_{\mu\nu}$ in the spatial Fourier transformer equation (17.17) by

$$D_a \left(k^2 a_{\mu\nu} + \beta \, \overline{k_\mu k_\kappa} \, a_{\kappa\nu} \right),$$

leads to such an effect [180, 181]. The parameter β characterizes the anisotropy of the effective diffusion coefficient.

17.2 Nonlinear Relaxation, Component Notation

In the absence of a flow and of any orienting torque, (16.149) describes a nonlinear relaxation process which can be significantly different from the exponential relaxation following from a linear equation. The symmetric traceless second rank tensor has 5 independent components. A convenient choice of components is introduced next, based on appropriately defined basis tensors.

17.2.1 Second-Rank Basis Tensors

The tensor $a_{\mu\nu}$ is decomposed as

$$a_{\mu\nu} = \sum_{i=0}^{4} a_i T_{\mu\nu}^i, \quad a_i = T_{\lambda\kappa}^i a_{\lambda\kappa}. \tag{17.19}$$

The basis tensors T^i are defined by

$$T^0_{\mu\nu} = \sqrt{\frac{3}{2}}\,\overline{e^z_\mu e^z_\nu}\,, \quad T^1_{\mu\nu} = \frac{1}{2}\sqrt{2}\left(\overline{e^x_\mu e^x_\nu - e^y_\mu e^y_\nu}\right),$$

$$T^2_{\mu\nu} = \sqrt{2}\,\overline{e^x_\mu e^y_\nu}\,, \quad T^3_{\mu\nu} = \sqrt{2}\,\overline{e^x_\mu e^z_\nu}\,, \quad T^4_{\mu\nu} = \sqrt{2}\,\overline{e^y_\mu e^z_\nu}\,, \tag{17.20}$$

where the e^x, e^y, e^z are unit vectors parallel to the coordinate axes. In a principal axes system, just the first two of these tensors occur, cf. Sect. 15.2.1. The first three of these tensors have the symmetry of the plane Couette geometry. In general, all 5 components are needed.

In matrix notation, the basis tensors (17.20) read

$$\sqrt{6}\,T^0 = \begin{pmatrix} -1 & 0 & 0 \\ 0 & -1 & 0 \\ 0 & 0 & 2 \end{pmatrix}, \quad \sqrt{2}\,T^1 = \begin{pmatrix} 1 & 0 & 0 \\ 0 & -1 & 0 \\ 0 & 0 & 0 \end{pmatrix}, \quad \sqrt{2}\,T^2 = \begin{pmatrix} 0 & 1 & 0 \\ 1 & 0 & 0 \\ 0 & 0 & 0 \end{pmatrix},$$

$$\sqrt{2}\,T^3 = \begin{pmatrix} 0 & 0 & 1 \\ 0 & 0 & 0 \\ 1 & 0 & 0 \end{pmatrix}, \quad \sqrt{2}\,T^4 = \begin{pmatrix} 0 & 0 & 0 \\ 0 & 0 & 1 \\ 0 & 1 & 0 \end{pmatrix}. \tag{17.21}$$

The basis tensors obey the ortho-normalization relation

$$T^i_{\mu\nu}\, T^k_{\mu\nu} = \delta_{ik}. \tag{17.22}$$

The square of the alignment tensor is equal to the sum of its squared components, viz.

$$a^2 = a_{\mu\nu} a_{\mu\nu} = \sum_{i=0}^{4} a_i^2. \tag{17.23}$$

Furthermore, as presented in [182], the symmetric traceless part of the product of two of these tensors is explicitly given by

$$\sqrt{6}\,\overline{T^0_{\mu\lambda} T^0_{\lambda\nu}} = T^0_{\mu\nu}, \quad \sqrt{6}\,\overline{T^1_{\mu\lambda} T^1_{\lambda\nu}} = \sqrt{6}\,\overline{T^2_{\mu\lambda} T^2_{\lambda\nu}} = -T^0_{\mu\nu}, \quad \overline{T^1_{\mu\lambda} T^2_{\lambda\nu}} = 0,$$

$$\sqrt{6}\,\overline{T^0_{\mu\lambda} T^1_{\lambda\nu}} = -T^1_{\mu\nu}, \quad \sqrt{6}\,\overline{T^0_{\mu\lambda} T^2_{\lambda\nu}} = -T^2_{\mu\nu}, \quad \sqrt{6}\,\overline{T^0_{\mu\lambda} T^3_{\lambda\nu}} = \frac{1}{2}T^3_{\mu\nu},$$

$$\sqrt{6}\,\overline{T^0_{\mu\lambda} T^4_{\lambda\nu}} = \frac{1}{2}T^4_{\mu\nu}, \quad \sqrt{6}\,\overline{T^1_{\mu\lambda} T^3_{\lambda\nu}} = \sqrt{6}\,\overline{T^2_{\mu\lambda} T^4_{\lambda\nu}} = \frac{1}{2}\sqrt{3}T^3_{\mu\nu},$$

$$\sqrt{6}\,\overline{T^1_{\mu\lambda} T^4_{\lambda\nu}} = -\frac{1}{2}\sqrt{3}T^4_{\mu\nu}, \quad \sqrt{6}\,\overline{T^2_{\mu\lambda} T^3_{\lambda\nu}} = \frac{1}{2}\sqrt{3}T^4_{\mu\nu}, \quad \sqrt{6}\,\overline{T^3_{\mu\lambda} T^4_{\lambda\nu}} = \frac{1}{2}\sqrt{3}T^2_{\mu\nu},$$

$$\sqrt{6}\,\overline{T^3_{\mu\lambda} T^3_{\lambda\nu}} = \frac{1}{2}(T^0_{\mu\nu} + \sqrt{3}T^1_{\mu\nu}), \quad \sqrt{6}\,\overline{T^4_{\mu\lambda} T^4_{\lambda\nu}} = \frac{1}{2}(T^0_{\mu\nu} - \sqrt{3}T^1_{\mu\nu}). \tag{17.24}$$

The scalar constructed from the triple product of these tensors is determined by

$$\sqrt{6}\, T^{i}_{\mu\lambda}\, T^{j}_{\lambda\nu}\, T^{k}_{\nu\mu} \equiv C(i,j,k). \tag{17.25}$$

The coupling coefficient $C(i,j,k)$ is symmetric under the interchange of any two of the labels i,j,k, e.g. $C(i,j,k) = C(i,k,j) = C(k,j,i)$. From (17.24) and the orthogonality relation, one infers: apart from interchanges, the only nonzero coefficients are

$$C(0,0,0) = 1, \quad C(0,1,1) = C(0,2,2) = -1, \quad C(0,3,3) = C(0,4,4) = \frac{1}{2},$$

$$C(1,3,3) = C(2,3,4) = \frac{1}{2}\sqrt{3}, \quad C(1,4,4) = -\frac{1}{2}\sqrt{3}. \tag{17.26}$$

17.1 Exercise: Components of a Uniaxial Alignment

Consider a uniaxial alignment given by $a_{\mu\nu} = \sqrt{3/2}\, a \, \overline{n_{\mu} n_{\nu}}$. Determine the components a_i in terms of the polar coordinates ϑ and φ. Use $n_x = \sin\vartheta\cos\varphi$, $n_y = \sin\vartheta\sin\varphi$, $n_z = \cos\vartheta$.
Consider the special cases $\vartheta = 0, 45, 90°$ and $\cos^2\vartheta = 1/3$.

17.2.2 Third-Order Scalar Invariant and Biaxiality Parameter

The third-order scalar invariant is defined by $I_3 = \sqrt{6}a_{\mu\nu}a_{\nu\kappa}a_{\kappa\mu}$, cf. (15.4). Due to (17.25) with (17.26), $I_3 = \sqrt{6}a_{\mu\nu}a_{\nu\kappa}a_{\kappa\mu}$ is expressed in terms of the components a_i by

$$I_3 = a_0\left[a_0^2 - 3(a_1^2 + a_2^2) + \frac{3}{2}(a_3^2 + a_4^2)\right] + \frac{3}{2}\sqrt{3}a_1(a_3^2 - a_4^2) + 3\sqrt{3}a_2a_3a_4. \tag{17.27}$$

The square of the biaxiality parameter b, cf. Sect. 5.5.2, in particular (15.5), is

$$b^2 = 1 - I_3^2/I_2^3. $$

17.2.3 Component Equations

The relaxation equations for the 5 components, which correspond to (16.149), with (16.150), in the absence of a flow, are

$$\frac{\partial}{\partial t}a_i + \Phi_i = 0, \quad \Phi_i = (\vartheta + 2a^2)\, a_i + Q_i, \quad i = 0,1,2,3,4. \tag{17.28}$$

The quantity $Q_i = -\sqrt{6}\,T^i_{\mu\nu}a_{\nu\lambda}a_{\lambda\mu}$ is explicitly given by

$$Q_0 = -3\,a_0^2 + 3\,(a_1^2 + a_2^2) - \frac{3}{2}\,(a_3^2 + a_4^2), \tag{17.29}$$

$$Q_1 = 6\,a_0 a_1 - \frac{3}{2}\sqrt{3}\,(a_3^2 - a_4^2), \quad Q_2 = 6\,a_0 a_2 - 3\sqrt{3}\,a_3 a_4,$$

$$Q_3 = -3\,a_0 a_3 - 3\sqrt{3}\,(a_1 a_3 + a_2 a_4), \quad Q_4 = -3\,a_0 a_4 + 3\sqrt{3}\,(a_1 a_4 - a_2 a_3).$$

The Φ_i occurring in the relaxation equation (17.28) are the derivatives of the potential function Φ with respect to the components a_i, viz. $\Phi_i = \partial\Phi/\partial a_1$, where

$$\Phi = \frac{1}{2}\vartheta\,a^2 + Q + \frac{1}{2}\,(a^2)^2, \quad Q = -\sqrt{6}\,a_{\mu\nu}a_{\nu\lambda}a_{\lambda\mu}. \tag{17.30}$$

Apart from the sign and a numerical factor, Q is the determinant of the alignment tensor, cf. (5.44). In terms of the a_i, it is given by

$$Q = -a_0^3 + 3\,a_0\left(a_1^2 + a_2^2 - \frac{1}{2}\,a_3^2 - \frac{1}{2}\,a_4^2\right) - \frac{3}{2}\sqrt{3}\,a_1\left(a_3^2 - a_4^2\right) - 3\sqrt{3}\,a_2\,a_3\,a_4. \tag{17.31}$$

Since Q, obviously, is not a function of a^2, the potential Φ is highly anisotropic in the 5-dimensional space of the a_i components.

17.2.4 Stability of Stationary Solutions

Let $a^{st}_{\mu\nu}$ be a stationary solution of the inhomogeneous relaxation equation (16.149) with $\Phi_{\mu\nu}$ given by (16.150). Insertion of $a_{\mu\nu} = a^{st}_{\mu\nu} + \delta a_{\mu\nu}$ into the equation and disregard of terms nonlinear in the small deviation $\delta a_{\mu\nu}$ from the stationary state yields

$$\frac{\partial}{\partial t}\delta a_{\mu\nu} - 2\,\overline{\varepsilon_{\mu\lambda\kappa}\Omega_\lambda \delta a_{\kappa\nu}} - 2\kappa\,\overline{\Gamma_{\mu\kappa}\delta a_{\kappa\nu}} + \Phi_{\mu\nu,\lambda\kappa}\,\delta a_{\lambda\kappa} = 0, \tag{17.32}$$

with the second derivative of the potential, viz.

$$\Phi_{\mu\nu,\lambda\kappa} = \frac{\partial}{\partial a_{\lambda\kappa}}\Phi_{\mu\nu} = (\vartheta + 2\,a^2)\Delta_{\mu\nu,\lambda\kappa} + 4\,a_{\mu\nu}a_{\lambda\kappa} - 6\sqrt{6}\,\Delta^{(2,2,2)}_{\mu\nu,\alpha\beta,\lambda\kappa}\,a_{\alpha\beta}, \tag{17.33}$$

evaluated at the stationary value for the alignment tensor. For the isotropic coupling tensor $\Delta^{(2,2,2)}_{...}$ see (11.36).

In the absence of a flow, one has $a^{st}_{\mu\nu} = a_{eq} T^0_{\mu\nu}$, when the z-direction is put parallel to the director $\mathbf{n}$. The equilibrium value of the order parameter is $a_{eq} = \frac{3}{4} + \frac{3}{4}\sqrt{9 - 8\vartheta}$, $\vartheta < \frac{9}{8}$. The relaxation equation for $\delta a_{\mu\nu}$ then reduces to

$$\frac{\partial}{\partial t}\delta a_{\mu\nu} + (\vartheta + 2a^2_{eq})\,\delta a_{\mu\nu} + 4a^2_{eq}\,T^0_{\mu\nu}\,\delta a_0 - 6\sqrt{6}\,a_{eq}\,\overline{T^0_{\mu\lambda}\,\delta a_{\lambda\nu}} = 0. \quad (17.34)$$

Due to (17.24), the last term in this equation is equivalent to

$$-6\sqrt{6}\,a_{eq}\,\overline{T^0_{\mu\lambda}\,\delta a_{\lambda\nu}}$$

$$= a_{eq}\left(-6T^0_{\mu\nu}\delta a_0 + 6T^1_{\mu\nu}\delta a_1 + 6T^2_{\mu\nu}\delta a_2 - 3T^3_{\mu\nu}\delta a_3 - 3T^4_{\mu\nu}\delta a_4\right).$$

The resulting equations for the i components of the distortion can be written as

$$\frac{\partial}{\partial t}\delta a_i + \nu^{(i)}\,\delta a_i = 0, \quad i = 0, \ldots, 4, \quad (17.35)$$

with dimensionless relaxation frequencies $\nu^{(i)} = \nu^{(i)}(a_{eq})$. The stationary solution is stable against these different distortions δa^i when $\nu^{(i)} > 0$ holds true. The case $\nu^{(i)} = 0$ pertains to a marginal linear stability. Then terms nonlinear in δa_i have to be taken into account.

For a uniaxial distortion where $\delta a_{\mu\nu} = T^0_{\mu\nu}\delta a_0$ applies, one obtains the relaxation frequency

$$\nu^{(0)} = \vartheta - 6a_{eq} + 6a^2_{eq} = 3a_{eq} - 2\vartheta.$$

The last equality follows from the equilibrium condition $\vartheta - 3a_{eq} + 2a^2_{eq} = 0$. At the phase transition temperature T_{ni} one has $\vartheta = 1$, $a_{eq} = 1$ and consequently $\nu^{(0)} = 1$. At lower temperatures $\nu^{(0)}$ becomes larger and the exponential relaxation of a uniaxial distortion is even faster. The highest temperature where a meta-stable nematic phase exists, corresponds to $\vartheta = 9/8$, with $a_{eq} = 3/4$ and consequently $\nu^{(0)} = 0$. This is a marginal stability.

The linear stability analysis for biaxial distortions, in particular the determination of the relaxation frequencies $\nu^{(1)} = \nu^{(2)}$ and $\nu^{(3)} = \nu^{(4)}$, are deferred to the next exercise.

17.2 Exercise: Stability Against Biaxial Distortions

Compute the relaxation frequencies $\nu^{(1)}$ and $\nu^{(3)}$ for biaxial distortions $\delta a_{\mu\nu} = T^1_{\mu\nu}\delta a_1$ and $\delta a_{\mu\nu} = T^3_{\mu\nu}\delta a_3$ from the relevant relations given in Sect. 17.2.4.

Solve the full nonlinear relaxation equation for a_3 with $a_1 = a_2 = a_4 = 0$ and $a_0 = a_{eq}$.

17.3 Alignment Tensor Subjected to a Shear Flow

17.3.1 Dynamic Equations for the Components

In the presence of a shear flow the equation (16.149), viz.

$$\frac{da_{\mu\nu}}{dt} - 2\,\overline{\varepsilon_{\mu\lambda\kappa}\Omega_\lambda a_{\kappa\nu}} - 2\kappa\,\overline{\Gamma_{\mu\kappa}a_{\kappa\nu}} + \Phi_{\mu\nu} = \sqrt{\frac{3}{2}}\,\lambda_K\,\Gamma_{\mu\nu},$$

is governing the dynamics of the alignment tensor. For a plane Couette flow with the velocity in x-direction and its gradient in y-direction and with the imposed shear rate Γ, this equation is equivalent to 5 coupled equations for the a_i:

$$\frac{\partial}{\partial t}a_0 + \frac{1}{3}\sqrt{3}\,\kappa\Gamma a_2 + \Phi_0 = 0, \tag{17.36}$$

$$\frac{\partial}{\partial t}a_1 - \Gamma a_2 + \Phi_1 = 0,$$

$$\frac{\partial}{\partial t}a_2 + \Gamma a_1 + \frac{1}{3}\sqrt{3}\,\kappa\Gamma a_0 + \Phi_2 = \frac{1}{2}\sqrt{3}\,\lambda_K\,\Gamma,$$

$$\frac{\partial}{\partial t}a_3 - \frac{1}{2}\Gamma(1+\kappa)a_4 + \Phi_3 = 0,$$

$$\frac{\partial}{\partial t}a_4 + \frac{1}{2}\Gamma(1-\kappa)a_3 + \Phi_4 = 0,$$

where $\Phi_i = (\vartheta + a^2)a_i + Q_i$, $i = 0, .., 4$. For Q_i see (17.29).

Stationary solutions of these equations correspond to problems discussed in Sects. 16.3.6 and 16.4.6. Another application is the study of the effect of a shear flow on the phase transition isotropic-nematic, as first presented in [169] and independently treated in [170]. The solutions found for equation (17.36), however, are much richer, cf. [183].

17.3.2 Types of Dynamic States

For a stationary imposed shear rate, not only stationary solutions exist. Also periodic and even chaotic behavior is found for the alignment tensor, subjected to a plane Couette flow. In the following, the name *main director* is used for the direction of the principal axis associated with the largest eigenvalue of the tensor. The various types of dynamic states are

• Symmetry adapted states with $a_3 = a_4 = 0$:

A   *Aligning*: stationary in-plane flow alignment with $a_0 < 0$. Furthermore, one may distinguish states A_+ and A_- pertaining to positive and negative values for the flow alignment angle χ. For nematics composed of rod-like particles the first case occurs for small, the latter one for very large shear rates.

T   *Tumbling*: in-plane tumbling of the alignment tensor, the main director is in the flow plane and rotates about the vorticity axis.

W   *Wagging*: in-plane wagging or librational motion of the main director about the flow direction.

L   *Log-rolling*: stationary alignment with $a_1 = a_2 = 0$ and $a_0 > 0$. This out-of-plane solution is instable, in most cases.

• Symmetry breaking states with $a_3 \neq 0$, $a_4 \neq 0$:

SB   *Stationary symmetry breaking states*, which occur in pairs of a_3, a_4 and $-a_3$, $-a_4$.

KT   *Kayaking-tumbling*: the projection of the main director onto the flow plane describes a tumbling motion.

KW   *Kayaking-wagging*: a periodic orbit where the projection of the main director onto the flow plane describes a wagging motion.

C   *Complex*: complicated motion of the alignment tensor. This includes periodic orbits composed of sequences of KT and KW motion with multiple periodicity as well as aperiodic, erratic orbits. The largest Lyapunov exponent for the latter orbits is positive, i.e., these orbits are *chaotic*.

For a given choice of parameters, in general, only a subset of these solutions are found by increasing the shear rate Γ. The T and W states can be distinguished in a plot of a_1 versus a_2. The point $(a_1, a_2) = (0, 0)$ is included in the cycle for tumbling and excluded for wagging. Similarly, in a plot of a_3 versus a_4, the point $(a_3, a_4) = (0, 0)$ is included in the cycle for the KT orbits and excluded for the KW orbits. 'Phase portraits' of this kind are also useful to recognize more complicated periodic and also irregular orbits. Examples for orbits pertaining to *kayaking tumbling*, *kayaking wagging*, and *chaotic* solutions are shown in Figs. 17.2, 17.3 and 17.4. All curves are computed for $\vartheta = 0$, where $a_{eq} = 3/2$, for $\lambda_K = 1.25$, and $\kappa = 0$, the tumbling parameter is $\lambda = 5/6 \approx 0.833$. The initial state has small, but finite values $a_0, .., a_4$.

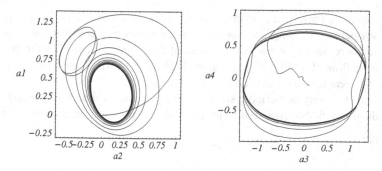

Fig. 17.2 Kayaking tumbling orbits in the 1–2- and 3–4-planes of the alignment

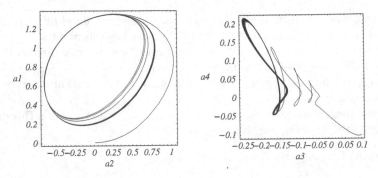

Fig. 17.3 Kayaking wagging orbits in the 1–2- and 3–4-planes of the alignment

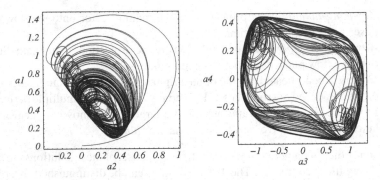

Fig. 17.4 Chaotic orbits in the 1–2- and 3–4-planes of the alignment

The shear rates for the KT and KW solutions are $\Gamma = 2.0$ and 1.75, for the chaotic solution it is 3.75.

The kayaking type of solutions, also referred to as "out of plane solutions", were first discussed in [190]. For a discussion of the complex dynamics of polymeric liquid crystals and of related computer simulation studies see also [165, 191]. Observations of the complex orientational dynamics in solutions of rod-like viruses are reported in [192].

The scenarios for the route to chaos in nonlinear dynamics [194, 195], e.g. transitions via period doubling and via intermittent states do occur for the equations considered here which govern the dynamics of the alignment tensor in the presence of a Couette flow, cf. [183–187]. Equation (17.36) can be supplemented by an equation for the shear rate in order to control the shear stress, cf. [193]. Then it is possible to stabilize stationary or periodic solutions for parameters where a constant shear rate leads to chaotic behavior. For a survey of chaos control in other areas see [196].

17.3.3 Flow Properties

The type of orientational behavior strongly affects the rheological behavior of the fluid, due to the coupling between alignment and flow. The expansion with respect to the basis tensors and the component notation can also be used for the symmetric traceless part of the pressure tensor or the stress tensor. From (16.151) to (16.152) one deduces expressions for the (dimensionless) shear stress σ_{xy}, and the normal stress differences $N_1 = \sigma_{xx} - \sigma_{yy}$ and $N_2 = \sigma_{yy} - \sigma_{zz}$ in terms of the dimensionless tensor components $\Sigma_i \equiv \Sigma_{\mu\nu}^{al} T_{\mu\nu}^i$. These relations are

$$\sigma_{xy} = \eta_{iso}\,\Gamma + \Sigma_2, \quad N_1 = 2\,\Sigma_1, \quad N_2 = -\sqrt{3}\,\Sigma_0 - \Sigma_1. \tag{17.37}$$

Here η_{iso} stands for the scaled second Newtonian viscosity and one has

$$\Sigma_2 = \frac{2}{\sqrt{3}}\lambda_K^{-1}\left[\phi_2 - \tilde{\kappa}\left(a_2\phi_0 + a_0\phi_2 - \frac{\sqrt{3}}{2}(a_4\phi_3 + a_3\phi_4)\right)\right],$$

$$\Sigma_1 = \frac{2}{\sqrt{3}}\lambda_K^{-1}\left[\phi_1 - \tilde{\kappa}\left(a_1\phi_0 + a_0\phi_1 - \frac{\sqrt{3}}{2}(a_3\phi_3 - a_4\phi_4)\right)\right], \tag{17.38}$$

$$\Sigma_0 = \frac{2}{\sqrt{3}}\lambda_K^{-1}\left[\phi_0 - \tilde{\kappa}\left(a_0\phi_0 - a_1\phi_1 - a_2\phi_2 + \frac{1}{2}(a_3\phi_3 + a_4\phi_4)\right)\right],$$

with $\tilde{\kappa} = 2\kappa/(3\lambda_K)$.

Examples for the rheological properties like the shear stress, the non-newtonian viscosity and the normal stress differences as functions of the shear rate for a few selected values of the temperature and for the other model parameters λ_K and κ are e.g. found in [185–188]. *Rheochaos*, a term coined by Cates [197], is found for those parameter ranges, where the dynamics of the alignment tensor is chaotic.

Solutions of the coupled equations for the velocity and the alignment tensor, for a boundary driven plane Couette flow, show pulsed jets in the velocity field, [189].

The coupled dynamics of the alignment and the electric polarization was studied in [97]. An extension of the theory to active materials involving swimmers was introduced in [198], see also [199].

17.4 Nonlinear Maxwell Model

The Maxwell model equation, cf. (16.81) and (16.94), governing the dynamics of the friction pressure tensor contains a linear relaxation term. The model can be extended to include damping terms nonlinear in the pressure tensor [200]. Here the notation follows [201].

17.4.1 Formulation of the Model

Here, the stress tensor rather than the pressure tensor is used. The symmetric traceless part $\overline{\sigma_{\mu\nu}}$ of the stress tensor is written as

$$\overline{\sigma_{\mu\nu}} = \sqrt{2}G_{\rm ref}\,\pi_{\mu\nu} + 2\eta_\infty\Gamma_{\mu\nu}, \qquad (17.39)$$

where $G_{\rm ref}$ is a reference shear modulus, also called Maxwell modulus $G_{\rm M}$ or G. The symmetric traceless tensor $\pi_{\mu\nu}$ is the dimensionless stress tensor, η_∞ is the second Newtonian viscosity and

$$\Gamma_{\mu\nu} = \overline{\nabla_\mu v_\nu}$$

is the symmetric traceless part of the deformation rate tensor. The generalized non-linear Maxwell model is formulated for the dimensionless stress tensor [200]

$$\frac{\partial}{\partial t}\pi_{\mu\nu} - 2\overline{\varepsilon_{\mu\lambda\kappa}\omega_\lambda\pi_{\kappa\nu}} - 2\kappa\,\overline{\Gamma_{\mu\lambda}\pi_{\lambda\nu}} + \tau_0^{-1}(\Phi_{\mu\nu} - \ell_0^2\Delta\pi_{\mu\nu}) = \sqrt{2}\Gamma_{\mu\nu},$$

$$\Phi_{\mu\nu} = \frac{\partial}{\partial\pi_{\mu\nu}}\Phi. \qquad (17.40)$$

The relevant relaxation time is called τ_0 and ℓ_0 is a characteristic length. The tensor $\Phi_{\mu\nu}$ is the derivative of a potential function Φ with respect to $\pi_{\mu\nu}$. The standard Maxwell model with the linear relaxation term corresponds to $\Phi = \frac{1}{2}A\pi_{\mu\nu}\pi_{\mu\nu}$ and $\Phi_{\mu\nu} = A\pi_{\mu\nu}$, with a dimensionless coefficient $A > 0$. In terms of the scalar invariants $I_2 = \pi_{\mu\nu}\pi_{\mu\nu}$ and $I_3 = \sqrt{6}\pi_{\mu\nu}\pi_{\nu\lambda}\pi_{\lambda\mu} = 3\sqrt{6}\det(\pi)$, cf. Sect. 5.5 and (15.4), the ansatz for the potential, up to the sixth power in π, is written as

$$\Phi = \frac{1}{2}AI_2 - \frac{1}{3}BI_3 + \frac{1}{4}CI_2^2 + \frac{1}{5}DI_2\,I_3 + \frac{1}{6}EI_2^3 + \frac{1}{6}FI_3^2. \qquad (17.41)$$

The dimensionless coefficients A, B, C, D, E, F are model parameters. The derivatives of I_2 and I_3 with respect to $\pi_{\mu\nu}$ are $2\pi_{\mu\nu}$ and $3\sqrt{6}\,\overline{\pi_{\mu\lambda}\pi_{\lambda\nu}}$, respectively.

17.4.2 Special Cases

The most widely studied special case is $D = E = F = 0$ with $A = A_0(1 - T_0/T)$ and $A_0, B, C > 0$. Then the potential is analogous to the Landau-de Gennes potential and the generalized nonlinear Maxwell model (17.40) is mathematically equivalent to the dynamic equation for the alignment tensor. For a spatially homogeneous equilibrium situation, a uniaxial stress tensor $\pi_{\mu\nu} = \sqrt{3/2}\pi\,\overline{n_\mu n_\nu}$ is found, where the scalar stress π is $\pi = B/(2C) \pm \sqrt{B^2/(4C^2) - A/C}$, provided that $A < B^*/(4C)$, otherwise one has $\pi = 0$. The case $\pi \neq 0$, in the absence of a flow corresponds to a solid state with a yield stress. At the transition temperature, here called T_c, one has $\pi = \pi_c \equiv \frac{2B}{3C}$, by analogy to the Landau-de Gennes theory for the isotropic-nematic phase transition. Scaled variables can be introduced in analogy to the treatment in Sect. 16.4.7. In particular, $\pi_{\mu\nu}$ is expressed in units of π_c. For convenience, the scaled stress tensor is also denoted by $\pi_{\mu\nu}$. Then $\Phi_{\mu\nu}$ occurring in the Maxwell model equation assumes the form

$$\Phi_{\mu\nu} = A^* \pi_{\mu\nu} - 3\sqrt{6}\, \overline{\pi_{\mu\lambda}\pi_{\lambda\nu}} + 2\pi_{\mu\nu}\,\pi_{\lambda\kappa}\pi_{\lambda\kappa},$$

$$A^* = AA_c^{-1}, \quad A_c = \frac{2B^2}{9C}. \tag{17.42}$$

Furthermore, just as in Sect. 16.4.7, the time is expressed in units of a reference time, here called τ_c, the shear rate and the vorticity are made dimensionless by multiplication with $\tau_c = \tau_0 A_c^{-1}$. A model parameter equivalent to the tumbling parameter of nematics is $\lambda_K = 2/(\sqrt{3}\pi_c)$. The nonlinear Maxwell model equation leads to non-stationary periodic and even chaotic solutions of 'stick-slip' type, when the *stick-slip parameter*, defined by

$$\lambda = 2\left(\sqrt{3}\pi_c\pi_{eq}\right)^{-1} + \kappa/3, \quad \pi_{eq} = \frac{1}{4}\left(3 + \sqrt{9 - 8A^*}\right), \quad A^* < 1.125, \tag{17.43}$$

is less than 1. For $\lambda > 1$, nonlinear flow behavior with shear thinning and shear thickening are found, even for $\kappa = 0$.

An example for the non-steady response of the system to an applied steady shear, of stick-slip type, is presented in Fig. 17.5 for the model parameters $\pi_c = 1.0$, $\kappa = 0.0$, $\eta_\infty = 0.1/A^*$, and for $A^* = 0.25, 0.35, 0.42$, from top to bottom, at the shear rate 3.2. Such a behavior is strikingly similar to that one seen in solid friction processes. Notice that the friction force is proportional to the shear stress. When the plastic

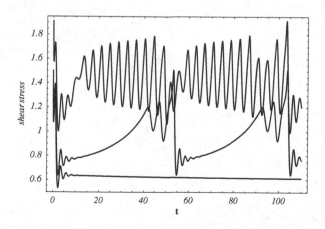

Fig. 17.5 Shear stress versus time for stick-slip motion

flow responsible for the friction occurs in a layer which is approximately constant, a constant velocity corresponds to a constant velocity gradient. The nonlinear dynamics observed here is linked with a *Shilnikov bifurcation* [195].

So far, a plane Couette flow with an imposed the shear rate was considered here. In general, however, the stress tensor has to be inserted into the momentum balance equation and the velocity field has to be solved, in accord with boundary conditions. Calculations for a 3D flow problem show that the generalized nonlinear Maxwell model yields turbulent flow behavior at low Reynolds numbers [202]. Such a behavior is typical for 'elastic turbulence', [204, 205].

The special case of an 'isotropic' potential function which just depends on I_2, but not on I_3, viz. the case $B = D = F = 0$ with $A = A_0(1 - T_0/T)$ and $A_0, E > 0$, $C < 0$, was also treated in [206].

Chapter 18
From 3D to 4D: Lorentz Transformation, Maxwell Equations

Abstract This chapter provides an outlook onto Special Relativity Theory and the four-dimensional formulation of the Maxwell equations of electrodynamics. Co- and contra-variant four-dimensional vectors and tensors are introduced, the Lorentz transformation is discussed, properties of the four-dimensional epsilon tensor are stated, some historical remarks are added. The formulation of the homogeneous Maxwell equations involves the field tensors derived from the four-dimensional electric potential. The inhomogeneous Maxwell equations, which can also be derived from a Lagrange density, contain the four-dimensional flux density as a source term. The transformation behavior of the electromagnetic fields is stated. A discussion of the four-dimensional force density and the Maxwell stress tensor conclude the final chapter. The Maxwell equations in four-dimensional form are closely linked with the Lorentz-invariance of these equations. Similarities and differences between the 3D and 4D formulation are discussed. First the Lorentz transformation as well as four-dimensional vectors and tensors are introduced.

18.1 Lorentz Transformation

18.1.1 Invariance Condition

The Maxwell equations imply that the speed of light c, in vacuum, observed in a coordinate system which moves with a constant velocity $\mathbf{v}$ with respect to the original coordinate system, is the same as in the original system. And this is in accord with experiments. Consequently, the rule for the transformation of coordinates between these two systems must be supplemented by a transformation of the time, as formulated by the Lorentz transformation.

Let $\mathbf{r}$, t be the position vector and the time in the original coordinate system, $\mathbf{r}'$, t' the corresponding variables in the system moving with the constant velocity. The Maxwell equations enforce an invariance condition, viz. the square of the line element or "length" s, viz.

$$s^2 = c^2 t^2 - r^2$$

© Springer International Publishing Switzerland 2015
S. Hess, *Tensors for Physics*, Undergraduate Lecture Notes in Physics,
DOI 10.1007/978-3-319-12787-3_18

is invariant, for two coordinate systems moving with a constant velocity with respect to each other. More specifically, the linear relation between the coordinates and the time

$$\mathbf{r} \rightarrow \mathbf{r}', \quad t \rightarrow t'$$

has to be such that $s^2 = (s')^2$, i.e.

$$c^2 t^2 - \mathbf{r}^2 = c^2 (t')^2 - \mathbf{r}'^2, \tag{18.1}$$

or, with x, y, z instead of r_1, r_2, r_3,

$$c^2 t^2 - (x^2 + y^2 + z^2) = c^2 (t')^2 - (x'^2 + y'^2 + z'^2). \tag{18.2}$$

When the coordinate systems are chosen such that the x- and also the x'-direction is parallel to the constant velocity $\mathbf{v}$, one has $y' = y, z' = z$ and (18.2) reduces to

$$c^2 t^2 - x^2 = c^2 (t')^2 - x'^2.$$

The same relation applies for differences dt and dx between times t and positions x. From

$$c^2 dt^2 - dx^2 = c^2 (dt')^2 - dx'^2 = 0$$

follows

$$\frac{dx}{dt} = \frac{dx'}{dt'} = c,$$

i.e. the speed of light is the same in both coordinate systems.

Four-dimensional vectors, endowed with the appropriate metric, allow to express $c^2 t^2 - \mathbf{r}^2$ as a 4D scalar product. Then (18.1) becomes analogous to the condition that the 3D scalar product is invariant under a rotation of the coordinate system.

18.1.2 4-Vectors

Contra- and co-variant 4-vectors x^i and x_i, with $i = 1, 2, 3, 4$, are introduced by

$$x^i = (r_1, r_2, r_3, ct), \quad x_i = (-r_1, -r_2, -r_3, ct). \tag{18.3}$$

With the Einstein summation convention for the four Roman indices, the scalar product of the 4-vectors is

$$x^i x_i = -(r_1^2 + r_2^2 + r_3^2) + c^2 t^2 = -\mathbf{r}^2 + c^2 t^2. \qquad (18.4)$$

The condition (18.1) for the Lorentz invariance is equivalent to

$$x^i x_i = (x')^i (x')_i. \qquad (18.5)$$

In this notation, the summation index always occurs as a pair of subscript and super-script. Just as for the components of a position vector with respect to a non-orthogonal basis, cf. Sect. 2.2.2, the contra- and co-variant components of the 4-vector are linked with each other by

$$x^i = g^{ik} x_k, \quad x_i = g_{ik} x^k. \qquad (18.6)$$

In matrix notation, the metric tensor is given by

$$g^{ik} = g_{ik} := \begin{pmatrix} -1 & 0 & 0 & 0 \\ 0 & -1 & 0 & 0 \\ 0 & 0 & -1 & 0 \\ 0 & 0 & 0 & 1 \end{pmatrix}. \qquad (18.7)$$

Notice that

$$g^{i\ell} g_{\ell k} = \delta_i^k := \begin{pmatrix} 1 & 0 & 0 & 0 \\ 0 & 1 & 0 & 0 \\ 0 & 0 & 1 & 0 \\ 0 & 0 & 0 & 1 \end{pmatrix}, \qquad (18.8)$$

which is the 4-dimensional unit matrix. Furthermore, one has

$$x^i x_i = g^{ik} x_i x_k = g_{ik} x^i x^k. \qquad (18.9)$$

The parity operator $\mathscr{P}$ replaces $\mathbf{r}$ by $-\mathbf{r}$, the time reversal operator $\mathscr{T}$ replaces t by $-t$, cf. Sects. 2.6.1 and 2.8. Clearly, the combined operation $\mathscr{P}\,\mathscr{T}$ is needed for all components of the 4-vector to reverse sign at once, viz.

$$\mathscr{P}\,\mathscr{T}\, x^i = -x^i. \qquad (18.10)$$

Remarks on notation are in order. Sometimes, ct is treated as the first component and the counting of the four components runs from 0 to 3, viz. the notation $x^0 = ct$, $x^i = r_i$, $i = 1, 2, 3$ is used. Then the metric tensor has the diagonal elements $1, -1, -1 - 1$.

The notation due to Minkowski, where the fourth component of the vector is ict, with the imaginary unit i, avoids the use of a metric tensor. In this case, the square of the pseudo-Euclidian norm of the vector is $x_k x_k = r_1^2 + r_2^2 + r_3^2 - c^2 t^2$.

The formulation of vectors and tensors in 4D-space with a metric tensor is preferred since it is more apt for the generalization from Special to General Relativity.

18.1.3 Lorentz Transformation Matrix

Components of the 4-vector in two coordinate systems which move with a constant velocity **v** with respect to each other are linearly related via the Lorentz transformation matrix L, viz.

$$(x')^i = L^i_k \, x^k, \quad (x')_i = L^k_i \, x_k. \tag{18.11}$$

The condition (18.5) implies

$$L^i_k \, L^n_i = \delta^n_k. \tag{18.12}$$

This relation is analogous to the unitarity condition (2.31) for the 3-dimensional rotation matrix. In (18.12) the summation is over the fore indices. The corresponding relation with a summation over the hind indices also holds true:

$$L^k_i \, L^i_n = \delta^k_n. \tag{18.13}$$

As in the case of the orthogonal transformation discussed in Sect. 2.41 for a rotation in 3D, the reciprocal of the 4D Lorentz transformation matrix L is equal to its transposed matrix $\tilde{L}$, thus $L^{-1} = \tilde{L}$.

18.1.4 A Special Lorentz Transformation

Consider a 'primed' coordinate system which moves with the constant velocity **v** in the 1- or x-direction. With the abbreviations

$$\beta := \frac{v}{c}, \quad \gamma := \frac{1}{\sqrt{1 - \beta^2}}, \tag{18.14}$$

the rule proposed by Lorentz for the interrelation of the components with respect to these coordinate systems are

$$x' = \gamma \, (x - vt) = \gamma \, (x - \beta ct), \quad y' = y, \quad z' = z, \quad ct' = \gamma \, (ct - \beta x). \tag{18.15}$$

Clearly, for $\beta \ll 1$ and consequently $\gamma \approx 1$, the Lorentz transformation rule (18.15) reduces to the corresponding Galilei transformation where $x' = x - vt$ and $t' = t$.

A contra-variant Lorentz vector **a** is transformed according to

$$(a')^1 = \gamma \, (a^1 - \beta a^4), \quad (a')^2 = a^2, \quad (a')^3 = a^3, \quad (a')^4 = \gamma \, (a^4 - \beta a^1). \tag{18.16}$$

The pertaining Lorentz transformation matrix, cf. (18.11), is

$$L_k^i := \begin{pmatrix} \gamma & 0 & 0 & -\beta\gamma \\ 0 & 1 & 0 & 0 \\ 0 & 0 & 1 & 0 \\ -\beta\gamma & 0 & 0 & \gamma \end{pmatrix}. \qquad (18.17)$$

18.1.5 General Lorentz Transformations

The product

$$L_k^i = L_n^i(1)L_k^n(2)$$

of two Lorentz transformations $L_k^i(1)$ and $L_k^i(2)$ is also a Lorentz transformation. A general Lorentz transformation can be expressed as a (multiple) product of special Lorentz transformations. Notice that a rotation of the coordinate system, where $\mathbf{r}^2 = (\mathbf{r}')^2$ and $t = t'$ also obeys the invariance condition (18.1). A 4 by 4 matrix where the first 3 by 3 elements are given by the matrix elements of the orthogonal matrix U pertaining to the 3D rotation, with furthermore, $L_4^4 = 1$ and the other elements in the fourth row and fourth column put equal to zero, is also a Lorentz-transformation.

Thus the general Lorentz-transformation governs the interrelation between the position and the time of two coordinate systems, one of which is rotated and moving with a constant velocity with respect to the other coordinate system.

18.2 Lorentz-Vectors and Lorentz-Tensors

18.2.1 Lorentz-Tensors

The 3D scalars, vectors and tensors, as presented in Sect. 2.5.2, are defined via their transformation behavior under a rotation of the coordinate system. By analogy, the 4D scalars, vectors and tensors needed for special relativity are defined via the behavior of their components under a Lorentz transformation.

A quantity $\mathbf{a}$ with the 4 components a^1, a^2, a^3, a^4 is a *Lorentz vector* when its components in the primed coordinate system are related to those in the original system by the same transformation rule as obeyed by the 4-vector $(\mathbf{r}, ct)$, cf. (18.11), i.e. when

$$(a')^i = L_k^i a^k, \quad (a')_i = L_i^k a_k \qquad (18.18)$$

holds true. A *Lorentz scalar* is a quantity which does not change under a Lorentz transformation. A *Lorentz tensor* of rank ℓ requires a ℓ-fold product of Lorentz matrices for the transformation of its 4 times ℓ components. As an example, the contra-variant components of a second rank tensor T are transformed according to

$$(T')^{ik} = L_n^i\, L_m^k\, T^{nm}. \tag{18.19}$$

Equations of physics, properly formulated in terms of Lorentz tensors of ranks $\ell = 0, 1, 2, \ldots$ are *Lorentz invariant* and consequently are in accord with *Special Relativity*.

18.2.2 Proper Time, 4-Velocity and 4-Acceleration

Let τ be the time in the co-moving coordinate system, i.e. in a system moving with a particle. It is called *proper time* (Eigenzeit). Time differences dt in a space-fixed coordinate system are related to the proper time differences $d\tau$ by

$$dt = \gamma d\tau = \frac{d\tau}{\sqrt{1-\beta^2}}. \tag{18.20}$$

This time-dilation explains the prolonged live time of fast moving π-mesons observed in high altitude radiation.

The proper time

$$d\tau = \sqrt{1-\beta^2} dt \tag{18.21}$$

is a Lorentz scalar, i.e. it is invariant under Lorentz transformations. This is inferred from

$$d\tau^2 = \left(1 - \frac{v^2}{c^2}\right) dt^2 = \frac{1}{c^2}(dt^2 - d\mathbf{r}\cdot d\mathbf{r}) = \frac{1}{c^2} g_{ik} dx^i dx^k.$$

Here $v^2 = d\mathbf{v}\cdot d\mathbf{v}$ and $d\mathbf{r} = \mathbf{v}dt$ were used.

The *4-velocity*, defined by

$$u^i = \frac{dx^i}{d\tau}, \tag{18.22}$$

is a Lorentz vector. Its components are

$$u^1 = \gamma v_1, \quad u^2 = \gamma v_2, \quad u^3 = \gamma v_3, \quad u^4 = \gamma c.$$

Notice that the 3D velocity $\mathbf{v}$ is the derivative of the position vector with respect to t whereas the proper time τ occurs in (18.22).

The norm of the 4-velocity is constant, its square is given by

$$u^i u_i = g_{ik}\, u^i u^k = c^2. \tag{18.23}$$

Thus the 4-velocity, divided by c, is a 4-dimensional unit vector. As a consequence, the *4-acceleration* defined by

$$b^i = \frac{du^i}{d\tau} = \frac{d^2x^i}{d\tau^2},$$ (18.24)

obeys the relation

$$u^i b_i = g_{ik} u^i b^k = 0,$$ (18.25)

i.e. the 4-acceleration is orthogonal to the 4-velocity.

The 4-momentum p^i of a particle with the rest m, is

$$p^i = mu^i.$$ (18.26)

The first three components of p^i are equal to

$$p_\mu = m(v)\, v_\mu,$$

where the effective mass $m(v)$ is defined by

$$m(v) := \gamma m = \frac{m}{\sqrt{1 - v^2/c^2}}.$$ (18.27)

Here, the variable v occurring in γ is the magnitude of the velocity $\mathbf{v}$ of the moving particle, as seen from the rest frame. The fourth component of p^i is equal to the energy E of a free particle moving with speed v, divided by c, viz.

$$p^0 = m(v)\, c = \frac{E}{c},$$

where

$$E := m(v)\, c^2 = \frac{mc^2}{\sqrt{1 - v^2/c^2}}.$$ (18.28)

The pertaining kinetic energy is

$$E_{\text{kin}} = E - mc^2 = mc^2 \left(\frac{1}{\sqrt{1 - v^2/c^2}} - 1 \right).$$

In the limit $v \ll c$, this expression reduces to the non-relativistic limit $E_{\text{kin}} = \frac{1}{2}mv^2$.

18.2.3 Differential Operators, Plane Waves

The 4D generalization of the nabla differential operator ∇ is

$$\partial_i = \frac{\partial}{\partial x^i} := \left(\frac{\partial}{\partial r_1}, \frac{\partial}{\partial r_2}, \frac{\partial}{\partial r_3}, \frac{\partial}{\partial ct} \right). \tag{18.29}$$

Clearly, $\partial_i x^i = 4$ is a scalar, thus ∂_i is a Lorentz-vector.
The second derivative

$$\partial_i \, \partial^i = -\Delta + \frac{\partial^2}{c^2 \, \partial t^2} = -\Box \tag{18.30}$$

is a Lorentz scalar. Here $\Delta = \nabla_\mu \nabla_\mu$ is the 3D Laplace operator, $\Box$ is the d'Alembert operator, cf. (7.62).
A plane wave proportional to

$$\exp\left[-i \left(-k_\nu \, r_\nu + \omega t \right) \right],$$

is a solution of the wave equation $\Box \ldots = 0$, cf. (7.64), provided that the wave vector k_ν and the circular frequency ω obey the *dispersion relation* $k_\nu k_\nu = \omega^2/c^2$, or equivalently, $\omega = kc$, cf. (7.65). Here, k is the magnitude of the wave vector.
The 4-wave vector K^i is defined by

$$K^i := \left(k_1, k_2, k_3, \frac{\omega}{c} \right), \quad K_i := \left(-k_1, -k_2, -k_3, \frac{\omega}{c} \right). \tag{18.31}$$

Thus the phase factor occurring in the expression for the plane wave is equal to the Lorentz scalar

$$K_i x^i = K^i x_i = -k_\nu \, r_\nu + \omega t.$$

Let the function $\Psi = \Psi(\mathbf{r}, t)$ be the plane wave

$$\Psi \sim \exp[-i K_n x^n].$$

Here, the summation index "i" is not used in order to avoid any confusion with the imaginary unit i. Then one has

$$\partial_n \Psi = -i K_n \Psi, \quad \partial^n \Psi = -i K^n \Psi.$$

Consequently, the wave equation

$$\partial_n \partial^n \Psi = -K_n K^n \Psi = -\Box \Psi = 0$$

yields

$$K_n K^n = -k^2 + \omega^2/c^2 = 0. \tag{18.32}$$

This 4D version of the dispersion relation is equivalent to (7.65).

18.2.4 Some Historical Remarks

The appropriate rules for the transformations underlying the Maxwell equations were studied by a number of scientists from about 1890 to 1910. Hendrik Lorentz (1853–1928) published his findings in 1899 and 1904. The distance r travelled by light with speed c during the time t is determined by $r^2 = c^2 t^2$. Then the invariance of the speed of light implies the invariance condition

$$(r/t)^2 = (r'/t')^2, \tag{18.33}$$

which is a special case of (18.1). Lorentz noticed that the more general rule

$$x' = \ell \gamma (x - vt), \quad y' = \ell y, \quad z' = \ell z, \quad t' = \ell \gamma \left(t - \frac{v}{c^2} x \right), \tag{18.34}$$

with an yet unspecified scale function $\ell = \ell(v^2)$ guarantees the validity of (18.33). It is understood that the velocity has the components $v_x = v$ and $v_y = v_z = 0$. In 1905, Henri Poincaré pointed out that the scale function should be $\ell = 1$, then (18.34) reduces to the special transformation (18.15). Poincaré coined the term 'Lorentz-transformation' for this transformation rule. In the same year, Albert Einstein gave an alternative derivation of the transformation rules and elucidated their meaning. In particular, he postulated that the transformation rule should also apply for the motion of particles, not just for the propagation of light. He also introduced a scale function, similar to ℓ, and presented arguments for $\ell = 1$. In 1905, Einstein did not refer to the work of Lorentz and Poincaré, later he also used the term Lorentz transformation.

Woldemar Voigt, who introduced in 1898 the word and the notion tensor in the sense we still use it nowadays, had already noticed in 1887: the Maxwell equations and the invariance of the speed of light require the invariance condition (18.5). For the case where the primed coordinate system moves in x-direction, as in Sect. 18.1.4, Voigt proposed the transformation rule

$$x' = x - vt, \quad y' = \gamma^{-1} y, \quad z' = \gamma^{-1} z, \quad ct' = ct - \frac{v}{c} x. \tag{18.35}$$

This corresponds to the more general transformation (18.34) with the choice $\ell = \gamma^{-1}$ for the scale function. It was by analogy to a volume conserving elastic deformation of a solid body, which shrinks in two directions, when it is stretched in one, which led Voigt to assume also a change of the y- and z-components. This is in contradistinction to the simpler assumption made by Lorentz and Einstein: the y- and z-components are not affected when the motion is in x-direction.

The time t' shown by a clock moving with the primed system, as seen from the original system at the position $x = vt$, is $t' = \ell\gamma t(1-v^2/c^2) = \ell\gamma^{-1}$. The time delay, expressed by t'/t is $\ell\gamma^{-1}$ and consequently equal to $\sqrt{1 - v^2/c^2}$ for the Lorentz-transformation (18.15) and $1 - v^2/c^2$, for the Voigt-transformation. Experiments on time delay confirm the validity of the Poincaré-Einstein choice $\ell = 1$ and thus the Lorentz-transformation (18.15).

18.1 Exercise: Doppler Effect

Let ω_0 be the circular frequency of the electromagnetic radiation in a system which moves with velocity $\mathbf{v} = v\mathbf{e}^x$ with respect to the observer, who records the frequency ω. Determine the Doppler-shift $\delta\omega = \omega_0 - \omega$ for the two cases, where the wave vector of the radiation is parallel (longitudinal effect) and perpendicular (transverse effect) to the velocity, respectively.

Hint: use the Lorentz transformation rule (18.16) for the components of the 4-wave vector K^i, cf. (18.31). Furthermore, identify ω' with ω_0 and use $k_1 \equiv k_x$.

18.3 The 4D-Epsilon Tensor

18.3.1 Levi-Civita Tensor

In 4D, the totally antisymmetric isotropic tensor, which is analogous to the 3D epsilon tensor, is a tensor of rank 4. Here 'isotropic' means, the tensor is form invariant under a Lorentz transformation, just as the unit tensor and the metric tensor. The antisymmetric 4D-epsilon tensor is also called *Levi-Civita tensor*. By analogy to (4.1), it is defined according to

$$\varepsilon^{k\ell mn} = -\varepsilon_{k\ell mn} := \begin{vmatrix} \delta_{1k} & \delta_{1\ell} & \delta_{1m} & \delta_{1n} \\ \delta_{2k} & \delta_{2\ell} & \delta_{2m} & \delta_{2n} \\ \delta_{3k} & \delta_{3\ell} & \delta_{3m} & \delta_{3n} \\ \delta_{4k} & \delta_{4\ell} & \delta_{4m} & \delta_{4n} \end{vmatrix}. \tag{18.36}$$

This implies

$$\varepsilon^{i\ell mn} = \begin{cases} 1, & k,\ell,m,n = \text{even permutation of } 1234 \\ -1, & k,\ell,m,n = \text{odd permutation of } 1234 \\ 0, & k,\ell,m,n = \text{else,} \end{cases} \tag{18.37}$$

e.g. one has $\varepsilon^{1234} = 1$ and $\varepsilon^{2134} = -1$.

18.3.2 Products of Two Epsilon Tensors

Formulas for the product of two 4D epsilon tensors, similar to those presented for the 3D tensor in Sect. 4.1.2, follow from the definition (18.36). The product of two epsilon-tensors is a tensor of rank 8 which can be expressed in terms of fourfold products of the unit second rank tensor. Contractions yield tensors of ranks 6, 4, 2, 0, in analogy the formulas valid for the 3D epsilon tensor, cf. Sect. 4.1.2. In particular, the first contraction of the product is

$$\varepsilon_{k\ell mn}\,\varepsilon^{k'\ell'm'n} = - \begin{vmatrix} \delta_k^{k'} & \delta_k^{\ell'} & \delta_k^{m'} \\ \delta_\ell^{k'} & \delta_\ell^{\ell'} & \delta_\ell^{m'} \\ \delta_m^{k'} & \delta_m^{\ell'} & \delta_m^{m'} \end{vmatrix}. \tag{18.38}$$

In many applications, the two-fold contracted version of the product of two epsilon-tensors is needed. For $m = m'$, (18.38) reduces to

$$\varepsilon_{k\ell mn}\,\varepsilon^{k'\ell'mn} = -2 \begin{vmatrix} \delta_k^{k'} & \delta_k^{\ell'} \\ \delta_\ell^{k'} & \delta_\ell^{\ell'} \end{vmatrix} = -2\,(\delta_k^{k'}\delta_\ell^{\ell'} - \delta_k^{\ell'}\delta_\ell^{k'}). \tag{18.39}$$

The further contraction of (18.39), with $\ell = \ell'$, yields

$$\varepsilon_{k\ell mn}\,\varepsilon^{k'\ell mn} = -6\,\delta_k^{k'}. \tag{18.40}$$

The total contraction of two epsilon tensors is equal to -24, viz.:

$$\varepsilon_{k\ell mn}\,\varepsilon^{k\ell mn} = -24. \tag{18.41}$$

This numerical value $24 = 4!$ is equal to the number of non-zero elements of the epsilon tensor in 4D.

18.3.3 Dual Tensor, Determinant

In 3D, the antisymmetric part of a second rank tensor has 3 independent components which can be related to a vector. In 4D, the antisymmetric part of a second rank tensor has 6 independent components. Here a similar *duality relation* exists, which, however, links an antisymmetric second rank tensor with another second rank tensor referred to as its *dual tensor*. More specifically, let A be an antisymmetric tensor with $A_{k\ell} = -A_{\ell k}$, then its dual $\tilde{A}$ is defied by

$$\tilde{A}^{k\ell} = \frac{1}{2}\,\varepsilon^{k\ell mn}\,A_{mn}. \tag{18.42}$$

This implies, e.g. $\tilde{A}^{12} = (A_{34} - A_{43})/2 = A_{34}$, $\tilde{A}^{13} = -(A_{24} - A_{42})/2 = -A_{24}$, and $\tilde{A}^{14} = -(A_{23} - A_{32})/2 = -A_{23}$.

As an example, consider a special antisymmetric tensor associated with two vectors **a** and **b** according to

$$A_{ik} = \begin{pmatrix} 0 & b_3 & -b_2 & a_1 \\ -b_3 & 0 & b_1 & a_2 \\ b_2 & -b_1 & 0 & a_3 \\ -a_1 & -a_2 & -a_3 & 0 \end{pmatrix}. \tag{18.43}$$

The components of the contra-variant tensor have just the opposite sign in the fourth row and column, viz.

$$A^{ik} = \begin{pmatrix} 0 & b_3 & -b_2 & -a_1 \\ -b_3 & 0 & b_1 & -a_2 \\ b_2 & -b_1 & 0 & -a_3 \\ a_1 & a_2 & a_3 & 0 \end{pmatrix}. \tag{18.44}$$

In the dual tensor the role of the a- and b-components are interchanged, in particular

$$\tilde{A}^{ik} = \begin{pmatrix} 0 & a_3 & -a_2 & -b_1 \\ -a_3 & 0 & a_1 & -b_2 \\ a_2 & -a_1 & 0 & -b_3 \\ b_1 & b_2 & b_3 & 0 \end{pmatrix}. \tag{18.45}$$

The double contracted product of the tensor with its dual is a Lorentz scalar. For the special case (18.43) and (18.45), the result is

$$\tilde{A}^{ik} A_{ik} = -(a_1 b_1 + a_2 b_2 + a_3 b_3) = -\mathbf{a} \cdot \mathbf{b}. \tag{18.46}$$

For comparison, the product of the contra-variant tensor with its co-variant version is, in this special case,

$$A^{ik} A_{ik} = 2 (\mathbf{b} \cdot \mathbf{b} - \mathbf{a} \cdot \mathbf{a}). \tag{18.47}$$

The determinant $\det(A)$ of the tensor **A** is equal to

$$\det(A) = (\tilde{A}^{ik} A_{ik})^2 = (\mathbf{a} \cdot \mathbf{b})^2. \tag{18.48}$$

The determinant is also determined by a fourfold product of $\mathscr{A}$ according to

$$\varepsilon^{k'\ell'm'n'} A_{kk'} A_{\ell\ell'} A_{mm'} A_{nn'} = - \det(A) \, \varepsilon_{k\ell mn}, \tag{18.49}$$

or, equivalently

$$\det(A) = \frac{1}{24} \varepsilon^{k\ell mn} \varepsilon^{k'\ell'm'n'} A_{kk'} A_{\ell\ell'} A_{mm'} A_{nn'}. \tag{18.50}$$

18.4 Maxwell Equations in 4D-Formulation

18.4.1 Electric Flux Density and Continuity Equation

The *4-flux density* J^i is defined by

$$J^i := (j_1, j_2, j_3, c\rho), \tag{18.51}$$

where **j** is the 3D electric flux density and ρ is the charge density. The continuity equation, cf. (7.59),

$$\frac{\partial \rho}{\partial t} + \nabla_\mu j_\mu = 0,$$

is equivalent to

$$\partial_i J^i = 0. \tag{18.52}$$

The differential operator ∂_i is a Lorentz vector, thus the 4-flux density is also a Lorentz vector.

18.4.2 Electric 4-Potential and Lorentz Scaling

In terms of the 3D vector potential **A** and the scalar potential ϕ, the electrodynamic 4-potential is defined by

$$\Phi^i := \left(A_1, A_2, A_3, \frac{\phi}{c}\right), \quad \Phi_i := \left(-A_1, -A_2, -A_3, \frac{\phi}{c}\right). \tag{18.53}$$

The Lorentz scaling, cf. (7.67),

$$\frac{\partial}{\partial t}\phi + \nabla_\lambda A_\lambda = 0,$$

corresponds to

$$\partial_i \Phi^i = 0. \tag{18.54}$$

Clearly, the 4-potential is a Lorentz vector.

18.4.3 Field Tensor Derived from the 4-Potential

In the 3D formulation of electrodynamics, the **B**-field and the **E**-field are related to the vector and scalar potential functions by

$$\mathbf{B} = \nabla \times \mathbf{A}, \quad \mathbf{E} = -\nabla\phi - \frac{\partial \mathbf{A}}{\partial t}.$$

The first components, e.g. of these equations are

$$B_1 = \frac{\partial A_3}{\partial r_2} - \frac{\partial A_2}{\partial r_3} = \frac{\partial \Phi_2}{\partial x^3} - \frac{\partial \Phi_3}{\partial x^2}, \quad E_1 = -\frac{\partial \phi}{\partial r_1} - \frac{\partial A_1}{\partial t} = c\left(\frac{\partial \Phi_1}{\partial x^4} - \frac{\partial \Phi_4}{\partial x^1}\right).$$

The equations for the other components can be inferred by analogy. All these equations are combined by introducing the second rank field tensor F:

$$F_{ik} := \frac{\partial \Phi_i}{\partial x^k} - \frac{\partial \Phi_k}{\partial x^i} = \partial_k \Phi_i - \partial_i \Phi_k. \tag{18.55}$$

Its contra-variant version is

$$F^{ik} := \partial^k \Phi^i - \partial^i \Phi^k.$$

The field tensor is antisymmetric:

$$F_{ik} = -F_{ki}. \tag{18.56}$$

In matrix notation, the field tensor is related to the components of the magnetic and electric fields by

$$F_{ik} := \begin{pmatrix} 0 & B_3 & -B_2 & \frac{1}{c} E_1 \\ -B_3 & 0 & B_1 & \frac{1}{c} E_2 \\ B_2 & -B_1 & 0 & \frac{1}{c} E_3 \\ -\frac{1}{c} E_1 & -\frac{1}{c} E_2 & -\frac{1}{c} E_3 & 0 \end{pmatrix}. \tag{18.57}$$

Notice, the top-left 3×3 part of this antisymmetric 4×4 matrix is just the magnetic field tensor introduced in Sect. 7.5.5. Thus (18.57) can be regarded as the 4-dimensional extension of (7.70) made such that the components of **E** are also incorporated. This works because an antisymmetric tensor, in 4D, has 6 components, just like **B** and **E** together.

The matrix for the contra-variant tensor F^{ik} is given by an expression analogous to (18.57) where the terms involving the **E**-field have the opposite sign.

18.4.4 The Homogeneous Maxwell Equations

From the definition $F_{ik} = \partial_k \Phi_i - \partial_i \Phi_k$, cf. (18.55) follows

$$\partial_n F_{ik} + \partial_i F_{kn} + \partial_k F_{ni} = 0. \tag{18.58}$$

The left hand side of this equation is identical to zero unless all three indices (n, i, k) are different. The case $(1, 2, 3)$ corresponds to $\nabla_\mu B_\mu = 0$, the cases $(2, 3, 4)$, $(3, 1, 4)$, $(1, 2, 4)$ are equivalent to the induction law

$$\varepsilon_{\mu\nu\lambda} \nabla_\nu E_\lambda = -\frac{\partial B_\mu}{\partial t},$$

cf. (7.57). Thus (18.58) is the 4D formulation of the homogeneous Maxwell equations, which are a consequence of the field tensor being given in terms of the 4-potential by (18.55).

The dual field tensor, cf. (18.42), is

$$\tilde{F}^{ik} = \frac{1}{2}\varepsilon^{ikmn} F_{mn} = \varepsilon^{ikmn} \partial_m \Phi_n. \tag{18.59}$$

One has

$$\partial_k \tilde{F}^{ik} = \varepsilon^{ikmn} \partial_k \partial_m \Phi_n = 0,$$

since $\partial_k \partial_m$ is symmetric under the interchange of k and m, while the epsilon-tensor is antisymmetric. Thus the homogeneous Maxwell equations (18.58) are equivalent to

$$\partial_k \tilde{F}^{ik} = 0. \tag{18.60}$$

18.4.5 The Inhomogeneous Maxwell Equations

By analogy to (18.57), the four-dimensional H-tensor is defined by

$$H_{ik} := \begin{pmatrix} 0 & H_3 & -H_2 & c\,D_1 \\ -H_3 & 0 & H_1 & c\,D_2 \\ H_2 & -H_1 & 0 & c\,D_3 \\ -c\,D_1 & -c\,D_2 & -c\,D_3 & 0 \end{pmatrix}. \tag{18.61}$$

The field tensor H^{ik} has the same form as H_{ik}, just with the opposite sign of the terms involving D_1, D_2, D_3.

The inhomogeneous Maxwell equations (7.56), viz.

$$\nabla_\mu D_\mu = \rho, \quad \varepsilon_{\mu\nu\lambda} \nabla_\nu H_\lambda = j_\mu + \frac{\partial}{\partial t} D_\mu,$$

are equivalent to

$$\partial_k H^{ik} = J^i. \tag{18.62}$$

To complete the set of Maxwell equations, constitutive relations are needed which link the field tensors H_{ik} and F_{ik}. In vacuum, the simple linear relation

$$H_{ik} = \frac{1}{\mu_0} F_{ik} \tag{18.63}$$

applies. Here $\mu_0 = 4\pi\,10^{-7}$As/Vm is the magnetic induction constant of the vacuum, As/Vm stands for the SI-units Ampere seconds/Volt meter.

18.4.6 Inhomogeneous Wave Equation

For currents and fields in vacuum, where (18.63) applies, the inhomogeneous Maxwell equations, with (18.55) lead to

$$J^i = \partial_k H^{ik} = \frac{1}{\mu_0} \partial_k F^{ik} = \partial_k (\partial^k \Phi^i - \partial^i \Phi^k).$$

Due to $\partial_k \Phi^k = 0$, cf. (18.54) and with the d'Alembert operator $\Box$, cf. (7.62) and (18.30), the inhomogeneous wave equation reads

$$\Box \Phi^i = -\mu_0 J^i. \tag{18.64}$$

18.4.7 Transformation Behavior of the Electromagnetic Fields

The field tensor F^{ik} is a Lorentz tensor which transforms according to (18.19). For the special case where the primed coordinate system moves with the constant velocity $v = v_1 = \beta c$, with respect to the original coordinate system, the resulting transformed tensor is

$$(F')^{ik} := \gamma \begin{pmatrix} 0 & F^{12} + \beta F^{24} & F^{13} + \beta F^{34} & \gamma^{-1} F^{14} \\ -(F^{12} + \beta F^{24}) & 0 & \gamma^{-1} F^{23} & F^{24} + \beta F^{12} \\ -(F^{13} + \beta F^{34}) & -\gamma^{-1} F^{23} & 0 & F^{34} + \beta F^{13} \\ -\gamma^{-1} F^{14} & -(F^{24} + \beta F^{12}) & -(F^{34} + \beta F^{13}) & 0 \end{pmatrix}. \tag{18.65}$$

As before, the abbreviation $\gamma = (1 - \beta^2)^{-1/2}$ is used. In terms of the pertaining components of the **E**- and **B**-fields, (18.65) corresponds to

$$E_1' = E_1, \quad E_2' = \gamma \, (E_2 - vB_3), \quad E_3' = \gamma \, (E_3 + vB_2), \tag{18.66}$$

and

$$B_1' = B_1, \quad B_2' = \gamma \, (B_2 + vE_3/c^2), \quad B_3' = \gamma \, (B_3 - vE_2/c^2). \tag{18.67}$$

The components of the electromagnetic fields perpendicular to the direction of the relative velocity v are modified and do depend on v. In vector notation, the relations (18.66) and (18.67) correspond to

$$\mathbf{E}' = \gamma \, (\mathbf{E} + \mathbf{v} \times \mathbf{B}), \quad \mathbf{B}' = \gamma \, (\mathbf{B} - \mathbf{v} \times \mathbf{E}/c^2).$$

Notice that

$$\mathbf{E}' \cdot \mathbf{B}' = \mathbf{E} \cdot \mathbf{B}, \quad \mathbf{E}' \cdot \mathbf{E}' - c^2 \mathbf{B}' \cdot \mathbf{B}' = \mathbf{E} \cdot \mathbf{E} - c^2 \mathbf{B} \cdot \mathbf{B}. \tag{18.68}$$

Due to (18.47) and (18.48), these transformation properties of the fields are associated with the *scalar invariants of the field tensor*, viz.

$$F_{ik} F^{ik} = 2 \, (B^2 - E^2/c^2), \quad \det(F^{ik}) = (\mathbf{E} \cdot \mathbf{B})^2/c^2. \tag{18.69}$$

Relations analogous to (18.65) and (18.66), (18.67) apply for the field tensor H^{ik} and for the field vectors **D** and **H**.

18.4.8 Lagrange Density and Variational Principle

The Maxwell equations can be derived from a variational principle involving a Lagrange density depending on the relevant scalar invariants. For electric charges and currents in vacuum, where $F^{ik} = \mu_0 H^{ik}$ applies, the Lagrange density $\mathscr{L}$ is defined by

$$\mathscr{L} = - \left(J^i \, \Phi_i + \frac{1}{4\mu_0} \, F_{ik} F^{ik} \right). \tag{18.70}$$

Its 4D 'action' integral is denoted by

$$\mathscr{S} := \int \mathscr{L} \, d^4 x. \tag{18.71}$$

The variational principle states: the action integral $\mathscr{S}$ is extremal under a variation $\delta \Phi$ of the 4-potential Φ such that

$$\delta\mathscr{S} := \int \delta\mathscr{L}\, d^4 x = 0. \tag{18.72}$$

It is understood that $\delta\Phi$ is zero at the 'surface' of the 4D integration range. Use of the variational principle (18.72) leads to

$$\left(J^{\mathrm{i}} - (\mu_0)^{-1}\, \partial_{\mathrm{n}}\, F^{\mathrm{in}}\right) \delta\Phi_{\mathrm{i}} = 0,$$

and consequently

$$J^{\mathrm{i}} = (\mu_0)^{-1}\, \partial_{\mathrm{n}}\, F^{\mathrm{in}}. \tag{18.73}$$

This is the relation (18.62) for the special case where $F^{\mathrm{ik}} = \mu_0 H^{\mathrm{ik}}$ applies, i.e. for charges, currents and fields in vacuum. The derivation of (18.73) from (18.72) is deferred to the Exercise (18.2).

The scalars $J^{\mathrm{i}}\Phi_{\mathrm{i}}$ and $F_{\mathrm{ik}} F^{\mathrm{ik}}$ occurring in the Lagrange density (18.70) are invariant under the parity operation $\mathscr{P}$ and the time reversal $\mathscr{T}$, despite the fact that the quantities J^{i}, Φ_{i} and F_{ik} have a well defined symmetry only under the combined operation $\mathscr{P}\mathscr{T}$.

The Lagrange density (18.70) leading to the inhomogeneous Maxwell equations involves just the first one of the scalar invariants associated with the field tensor, cf. (18.69). Inclusion of the second scalar invariant, as suggested by Born and Infeld [207], leads to extended Maxwell equations with terms nonlinear in the **E** and **B** fields, even in vacuum. The resulting electrostatic potential of an electron located at $r = 0$, no longer diverges for $r \to 0$.

18.2 Exercise: Derivation of the Inhomogeneous Maxwell Equations
Derive the inhomogeneous Maxwell equations for fields in vacuum from the variational principle (18.72) with the Lagrange density (18.70).

18.5 Force Density and Stress Tensor

18.5.1 4D Force Density

The 4D force density f^{i} or f_{i} is defined by

$$f^{\mathrm{i}} = J_{\mathrm{k}}\, F^{\mathrm{ki}}, \quad f_{\mathrm{i}} = J^{\mathrm{k}}\, F_{\mathrm{ki}}. \tag{18.74}$$

The first three components of f^{i} correspond to the 3D force density **k** with

$$k_{\mu} = \rho(E_{\mu} + \varepsilon_{\mu\nu\lambda}\, v_{\nu}\, B_{\lambda}) = \rho\, E_{\mu} + \varepsilon_{\mu\nu\lambda}\, j_{\nu}\, B_{\lambda},$$

where ρ is the charge density and the 3D flux density is $\mathbf{j} = \rho\mathbf{v}$. The fourth component of f^{i} is the power density associated with **k**, more specifically

$$f^4 = j_\nu\, E_\nu/c = \rho\, v_\nu\, E_\nu/c = v_\nu\, k_\nu/c. \qquad (18.75)$$

By analogy to the 3D description, where the force density is expressed as a spatial derivative of the Maxwell stress tensor $T_{\mu\nu}$ according to $k_\mu = \nabla_\nu T_{\mu\nu}$, cf. Sect. 8.5.4, the force density f^i is related to the 4D stress tensor T^{ki} by

$$f^i = \partial_k\, T^{ki}. \qquad (18.76)$$

18.5.2 Maxwell Stress Tensor

The explicit expression for the stress tensor in terms of the field tensor is obtained from (18.74) with the help of the Maxwell equations (18.58), (18.62). The derivation is deferred to the Exercise 18.3. For a linear medium, the result is

$$T^{ki} = g_{\ell m}\, F^{mi}\, H^{\ell k} - \frac{1}{4}\, g^{ik}\, F_{\ell m}\, H^{\ell m}. \qquad (18.77)$$

For a comparison with the 3D tensor $T_{\mu\nu}$ notice that g^{ik}, in (18.77) plays the role of $\delta_{\mu\nu}$ in (8.120) and (8.121), furthermore one has

$$\frac{1}{4}\, F_{\ell m}\, H^{\ell m} = \frac{1}{2}(\mathbf{E}\cdot\mathbf{D} - \mathbf{B}\cdot\mathbf{H}).$$

The top-left 3 by 3 part of the 4D tensor, i.e. elements in the which do not involve the component 4, are equal to the components of the 3D Maxwell stress tensor (8.122). The components T^{14}, T^{24}, T^{34} are linked with the Poynting vector $\mathbf{S} = \mathbf{E}\times\mathbf{H}$, cf. (8.110), while the components T^{41}, T^{42}, T^{43} are proportional to the density of the linear momentum of the electromagnetic fields, cf. (8.125), viz.

$$\mathbf{j_S} = \mathbf{D}\times\mathbf{B}. \qquad (18.78)$$

The T^{44} component is essentially the energy density $u = \frac{1}{2}(\mathbf{E}\cdot\mathbf{D} + \mathbf{B}\cdot\mathbf{H})$. In matrix notation, one has

$$T^{ik} = \begin{pmatrix} T_{11}^{\mathrm{Max}} & T_{12}^{\mathrm{Max}} & T_{13}^{\mathrm{Max}} & -\frac{1}{c}\,S_1 \\ T_{21}^{\mathrm{Max}} & T_{22}^{\mathrm{Max}} & T_{23}^{\mathrm{Max}} & -\frac{1}{c}\,S_2 \\ T_{31}^{\mathrm{Max}} & T_{32}^{\mathrm{Max}} & T_{33}^{\mathrm{Max}} & -\frac{1}{c}\,S_3 \\ -c\,j_{S1} & -c\,j_{S2} & -c\,j_{S3} & -u \end{pmatrix}. \qquad (18.79)$$

Here the components of the 3D Maxwell tensor are denoted by T^{Max}. For electromagnetic fields in vacuum and for linear isotropic media one has $T^{ik} = T^{ki}$. In general, however, the stress tensor T^{ik} is not symmetric.

18.3 Exercise: Derivation of the 4D Stress Tensor

18.4 Exercise: Flatlanders Invent the Third Dimension and Formulate their Maxwell Equations

The flatlanders of Exercise 7.3 noticed, they can introduce contra- and co-variant vectors

$$x^i = (r_1, r_2, ct), \quad x_i = (-r_1, -r_2, ct).$$

With the Einstein summation convention for the three Roman indices, the scalar product of their 3-vectors is

$$x^i x_i = -(r_1^2 + r_2^2) + c^2 t^2 = -\mathbf{r}^2 + c^2 t^2.$$

They use the differential operator

$$\partial_i = \left(\frac{\partial}{\partial r_1}, \frac{\partial}{\partial r_2}, \frac{\partial}{\partial ct} \right),$$

and form the 3-vectors

$$J^I = (j_1, j_2, c\rho), \quad \Phi^i = (A_1, A_2, \phi/c)$$

from their current and charge densities and their vector and scalar potentials.

How are the relations

$$B = \partial_1 A_2 - \partial_2 A_1, \quad E_i = -\partial_i \phi - \partial A / \partial t, \quad i = 1, 2,$$

which are equivalent to the homogeneous Maxwell equations, cast into the pertaining three-dimensional form? Introduce a three-by-three field tensor F_{ij} and formulate their homogeneous Maxwell equations.

How about the inhomogeneous Maxwell equations in flatland?

Appendix
Exercises: Answers and Solutions

Exercise Chapter 1

1.1 Complex Numbers as 2D Vectors (p. 6)

Convince yourself that the complex numbers $z = x + iy$ are elements of a vector space, i.e. that they obey the rules (1.1)–(1.6). Make a sketch to demonstrate that $z_1 + z_2 = z_2 + z_1$, with $z_1 = 3 + 4i$ and $z_2 = 4 + 3i$, in accord with the vector addition in 2D.

Exercise Chapter 2

2.1 Exercise: Compute Scalar Product for given Vectors (p. 14)

Compute the length, the scalar products and the angles betwen the three vectors $\mathbf{a}$, $\mathbf{b}$, $\mathbf{c}$ which have the components $\{1, 0, 0\}$, $\{1, 1, 0\}$, and $\{1, 1, 1\}$.

Hint: *to visualize the directions of the vectors, make a sketch of a cube and draw them there!*

The scalar products of the vectors with themselves are: $\mathbf{a} \cdot \mathbf{a} = 1$, $\mathbf{b} \cdot \mathbf{b} = 2$, $\mathbf{c} \cdot \mathbf{c} = 3$, and consequently

$$a = |\mathbf{a}| = 1, \quad b = |\mathbf{b}| = \sqrt{2}, \quad c = |\mathbf{c}| = \sqrt{3}.$$

The mutual scalar products are

$$\mathbf{a} \cdot \mathbf{b} = 1, \quad \mathbf{a} \cdot \mathbf{c} = 1, \quad \mathbf{b} \cdot \mathbf{c} = 2.$$

The cosine of the angle φ between these vectors are

$$\frac{1}{\sqrt{2}}, \quad \frac{1}{\sqrt{3}}, \quad \frac{2}{\sqrt{6}},$$

© Springer International Publishing Switzerland 2015
S. Hess, *Tensors for Physics*, Undergraduate Lecture Notes in Physics,
DOI 10.1007/978-3-319-12787-3

respectively. The corresponding angles are exactly 45° for the angle between **a** and **b**, and ≈70.5° and ≈35.3°, for the other two angles.

Exercises Chapter 3

3.1 Symmetric and Antisymmetric Parts of a Dyadic in Matrix Notation (p. 38)
Write the symmetric traceless and the antisymmetric parts of the dyadic tensor $A_{\mu\nu} = a_\mu b_\nu$ in matrix form for the vectors **a** : {1, 0, 0} *and* **b** : {0, 1, 0}. *Compute the norm squared of the symmetric and the antisymmetric parts and compare with $A_{\mu\nu}A_{\mu\nu}$ and $A_{\mu\nu}A_{\nu\mu}$.*
In matrix notation, the tensor $A_{\mu\nu}$ is equal to

$$A = \begin{pmatrix} 0 & 1 & 0 \\ 0 & 0 & 0 \\ 0 & 0 & 0 \end{pmatrix}. \tag{A.1}$$

The trace of this matrix is zero. So its symmetric part coincides with its symmetric traceless part

$$\overline{A} = \frac{1}{2}\begin{pmatrix} 0 & 1 & 0 \\ 1 & 0 & 0 \\ 0 & 0 & 0 \end{pmatrix}. \tag{A.2}$$

The antisymmetric part of this tensor is

$$A^{\text{asy}} = \frac{1}{2}\begin{pmatrix} 0 & 1 & 0 \\ -1 & 0 & 0 \\ 0 & 0 & 0 \end{pmatrix}. \tag{A.3}$$

The tensor product $\overline{A} \cdot \overline{A}$ yields

$$\overline{A_{\mu\lambda}}\,\overline{A_{\lambda\nu}} = \frac{1}{4}\begin{pmatrix} 1 & 0 & 0 \\ 0 & 1 & 0 \\ 0 & 0 & 0 \end{pmatrix}, \tag{A.4}$$

and consequently

$$\overline{A_{\mu\lambda}}\,\overline{A_{\lambda\mu}} = \frac{1}{2}.$$

Similarly, the product of the antisymmetric part with its transposed, viz.

$$A^{\text{asy}}_{\mu\lambda} A^{\text{asy}}_{\nu\lambda}$$

yields the same matrix as in (A.4). Thus one has also

$$A^{asy}_{\mu\lambda} A^{asy}_{\mu\lambda} = \frac{1}{2}.$$

Due to

$$A_{\mu\nu} A_{\mu\nu} = 1, \quad A_{\mu\nu} A_{\nu\mu} = 0,$$

this is in accord with (3.10), viz.

$$A_{\mu\nu} B_{\nu\mu} = \frac{1}{3} A_{\lambda\lambda} B_{\kappa\kappa} + A^{asy}_{\mu\nu} B^{asy}_{\nu\mu} + \overline{A_{\mu\nu}} \; \overline{B_{\nu\mu}},$$

with $B_{\nu\mu} = A_{\mu\nu}$.

3.2 Symmetric Traceless Dyadics in Matrix Notation (p. 39)

(i) *Write the symmetric traceless parts of the dyadic tensor $C_{\mu\nu} = C_{\mu\nu}(\alpha) = 2a_\mu b_\nu$ in matrix form for the vectors $\mathbf{a} = \mathbf{a}(\alpha) : \{c, -s, 0\}$ and $\mathbf{b} = \mathbf{b}(\alpha) : \{s, c, 0\}$, where c and s are the abbreviations $c = \cos\alpha$ and $s = \sin\alpha$. Discuss the special cases $\alpha = 0$ and $\alpha = \pi/4$.*

The desired tensor is

$$C_{\mu\nu} = \begin{pmatrix} 2cs & c^2 & 0 \\ -s^2 & -2cs & 0 \\ 0 & 0 & 0 \end{pmatrix}, \tag{A.5}$$

and consequently, due to $2cs = \sin 2\alpha$, $c^2 - s^2 = \cos 2\alpha$, one obtains

$$\overline{C_{\mu\nu}}(\alpha) = \begin{pmatrix} \sin 2\alpha & \cos 2\alpha & 0 \\ \cos 2\alpha & -\sin 2\alpha & 0 \\ 0 & 0 & 0 \end{pmatrix}. \tag{A.6}$$

For $\alpha = 0$ and $\alpha = \pi/4$, this tensor reduces to

$$\begin{pmatrix} 0 & 1 & 0 \\ 1 & 0 & 0 \\ 0 & 0 & 0 \end{pmatrix}, \quad \begin{pmatrix} 1 & 0 & 0 \\ 0 & -1 & 0 \\ 0 & 0 & 0 \end{pmatrix}, \tag{A.7}$$

respectively. The diagonal expression follows from the first of these tensors when the Cartesian components of the vectors and tensors are with respect to a coordinate system rotated by 45°.

(ii) *Compute the product $B_{\mu\nu}(\alpha) = \overline{C_{\mu\lambda}}(0) \overline{C_{\lambda\nu}}(\alpha)$, determine the trace and the symmetric traceless part of this product. Determine the angle α, for wich one has $B_{\mu\mu} = 0$.*

The result is

$$B_{\mu v}(\alpha) = \begin{pmatrix} \cos 2\alpha & -\sin 2\alpha & 0 \\ \sin 2\alpha & \cos 2\alpha & 0 \\ 0 & 0 & 0 \end{pmatrix}. \tag{A.8}$$

Consequently. one has

$$\overline{B_{\mu v}}\,(\alpha) = \frac{1}{3}\cos 2\alpha \begin{pmatrix} 1 & 0 & 0 \\ 0 & 1 & 0 \\ 0 & 0 & -2 \end{pmatrix}, \tag{A.9}$$

and $B_{\mu\mu} = 2\cos 2\alpha$. Thus one has $B_{\mu\mu} = 0$ for $\alpha = \pi/4$, or 45°. For this angle, the two tensors (A.7) are 'orthogonal' in the sense that the trace of their product vanishes.

3.3 Angular Momentum in Terms of Spherical Components (p. 43)
Compute the z-component of the angular momentum in terms of the spherical components.

For a particle with mass m, the z-component of the angular momentum is $L_z = m(x\dot{y} - y\dot{x})$, in cartesian coordinates. In polar coordinates, cf. Sect. 2.1.4, one has $x = r\sin\vartheta\cos\varphi$, $y = r\sin\vartheta\sin\varphi$, $z = r\cos\vartheta$. The time change of x and y is

$$\dot{x} = \dot{r}\,r^{-1}x + \dot{\vartheta}r\cos\vartheta\cos\varphi - \dot{\varphi}r\sin\vartheta\sin\varphi,$$

$$\dot{y} = \dot{r}\,r^{-1}y + \dot{\vartheta}r\cos\vartheta\sin\varphi + \dot{\varphi}r\sin\vartheta\cos\varphi.$$

In the calculation of L_z, the terms involving $\dot{r}$ and $\dot{\vartheta}$ cancel, the remaining terms add up to

$$L_z = mr^2\dot{\varphi}.$$

3.4 Torque Acting on an Anisotropic Harmonic Oscillator (p. 44)
Determine the torque for the force

$$\mathbf{F} = -k\mathbf{r} \cdot \mathbf{ee} - (\mathbf{r} - \mathbf{r} \cdot \mathbf{ee}),$$

where the parameter k and unit vector $\mathbf{e}$ are constant. Which component of the angular momentum is constant, even for $k \neq 1$?

The torque is $\mathbf{T} = \mathbf{r} \times \mathbf{F} = -(k-1)(\mathbf{r}\cdot\mathbf{e})\mathbf{r} \times \mathbf{e}$. Clearly, the torque vanishes for $k = 1$. For $k \neq 1$, the torque still is zero, when $\mathbf{r} \times \mathbf{e} = 0$ or $(\mathbf{r} \cdot \mathbf{e}) = 0$ hold true. The first case corresponds the one-dimensional motion along a line parallel to $\mathbf{e}$ which passes through the origin $\mathbf{r} = 0$. This is a one-dimensional harmonic oscillator. The second case is a motion in the plane perpendicular to $\mathbf{e}$. This corresponds to an isotropic two-dimensional harmonic oscillator.

3.5 Velocity of a Particle Moving on a Screw Curve (p. 46)

Hint: Use $\alpha = \omega t$ for the parameter occurring in the screw curve (3.48), ω is a frequency.

Differentiation with respect to the time t yields the velocity

$$\mathbf{v} = \rho\omega\left[-\mathbf{e}\sin(\omega t) + \mathbf{u}\cos(\omega t)\right] + \chi\frac{\omega}{2\pi}\mathbf{e}\times\mathbf{u},$$

where it is assumed that not only the orthogonal unit vectors $\mathbf{e}$ and $\mathbf{u}$, but also the radius ρ and the pitch parameter χ are constant.

Exercise Chapter 4

4.1 2D Dual Relation in Complex Notation (p. 54)

Let the two 2D vectors (x_1, y_1) and (x_2, y_2) be expressed in terms of the complex numbers $z_1 = x_1 + iy_1$ and $z_2 = x_2 + iy_2$. Write the dual relation corresponding to (4.25) in terms of the complex numbers z_1 and z_2. How about the scalar product of these 2D vectors?

Hint: the complex conjugate of $z = x + iy$ is $z^* = x - iy$.

The product $z_1^* z_2$ is $x_1 x_2 + y_1 y_2 + i(x_1 y_2 - x_2 y_1)$, thus the dual scalar is

$$x_1 y_2 - x_2 y_1 = \frac{1}{2i}(z_1^* z_2 - z_1 z_2^*).$$

Similarly, the scalar product of the two vectors is

$$x_1 y_1 + x_2 y_2 = \frac{1}{2}(z_1^* z_2 + z_1 z_2^*).$$

In other words, the scalar product is the real part and the dual scalar is the imaginary part of $z_1^* z_2$.

Exercises Chapter 5

5.1 Show that the Moment of Inertia Tensors for Regular Tetrahedra and Octahedra are Isotropic (p. 62)

Hint: Use the coordinates $(1, 1, 1)$, $(-1, -1, 1)$, $(1, -1, -1)$, $(-1, 1, -1)$ for the four corners of the tetrahedron and $(1, 0, 0)$, $(-1, 0, 0)$, $(0, 1, 0)$, $(0, -1, 0)$, $(0, 0, 1)$, $(0, 0, -1)$, for the six of the octahedron.

4

94Appendix: Exercises…

(i) Tetrahedron: the position vectors of the corners are

$$\mathbf{u}^1 = \mathbf{e}^x + \mathbf{e}^y + \mathbf{e}^z, \quad \mathbf{u}^2 = -\mathbf{e}^x - \mathbf{e}^y + \mathbf{e}^z,$$
$$\mathbf{u}^3 = \mathbf{e}^x - \mathbf{e}^y - \mathbf{e}^z, \quad \mathbf{u}^4 = -\mathbf{e}^x + \mathbf{e}^y - \mathbf{e}^z.$$

In the products $u^i_\mu u^i_\nu$ the mixed terms involving $e^x_\mu e^y_\nu$, $e^x_\mu e^z_\nu$, $e^y_\mu e^z_\nu$ have the signs $(+, +, +)$, $(+, -, -)$, $(-, -, +)$, $(-, +, -)$ for $i = 1, 2, 3, 4$, respectively. The sum of these mixed terms vanishes and one finds

$$\sum_{i=1}^4 u^i_\mu u^i_\nu = 4(e^x_\mu e^x_\nu + e^y_\mu e^y_\nu + e^z_\mu e^z_\nu) = 4\,\delta_{\mu\nu},$$

thus the moment of inertia tensor

$$\Theta_{\mu\nu} = 8\,m\,\delta_{\mu\nu}$$

is isotropic.

(ii) Octahedron: here the sum $\sum_{i=1}^6 u^i_\mu u^i_\nu$ yields $2(e^x_\mu e^x_\nu + e^y_\mu e^y_\nu + e^z_\mu e^z_\nu)$ and consequently

$$\Theta_{\mu\nu} = 4\,m\,\delta_{\mu\nu}.$$

5.2 Verify the Relation (5.51) for the Triple Product of a Symmetric Traceless Tensor (p. 72)

Hint: use the matrix notation

$$\begin{pmatrix} a & 0 & 0 \\ 0 & b & 0 \\ 0 & 0 & c \end{pmatrix},$$

with $c = -(a + b)$, for the symmetric traceless tensor in its principal axis system. Compute the expressions on both sides of (5.51) and compare.

In matrix notation, the left hand side of

$$\overline{\mathbf{a} \cdot \mathbf{a} \cdot \mathbf{a}} = \frac{1}{2}\,\mathbf{a}(\mathbf{a} : \mathbf{a})$$

is

$$\begin{pmatrix} 2a^3/3 - b^3/3 - c^3/3 & 0 & 0 \\ 0 & 2b^3/3 - c^3/3 - c^3/3 & 0 \\ 0 & 0 & 2c^3/3 - a^3/3 - b^3/3 \end{pmatrix}. \quad (A.10)$$

Due to $c^3 = -(a^3 + 3a^2b + 3ab^2 + b^3)$, the diagonal elements are equal to $a^3 + a^2b + ab^2 = a(a^2 + ab + b^2)$, $b^3 + a^2b + ab^2 = b(a^2 + ab + b^2)$, and $-a^3 - b^3 - 2a^2b - 2ab^2$. On the other hand, the 11-element of $\mathbf{a}(\mathbf{a} : \mathbf{a})$ is equal to $(a^2 + b^2 + c^2)a = 2a^3 + 2b^3 + 2a^2b = 2a(a^2 + ab + b^2)$. Similarly, one finds for the 22-element $2b(a^2 + ab + b^2)$. The 33-element is $(a^2 + b^2 + c^2)c = -2(a+b)(a^2 + ab + b^2) = -2(a^3 + b^3 + 2a^2b + 2ab^2)$. Comparison of the diagonal matrix elements shows the validity of the relation (5.51).

Exercises Chapter 7

7.1 Divergence, Rotation and the Symmetric Traceless Part of the Gradient Tensor for the Vector Fields iv to vi of Sect. 7.2.1 (p. 90)
(iv) *Uniaxial Squeeze-stretch Field*

$$v_\mu = 3\, e_\mu\, e_\nu r_\nu - r_\mu.$$

The gradient is $\nabla_\nu v_\mu = 3 e_\mu e_\nu - \delta_{\mu\nu}$. One finds

$$\nabla \cdot \mathbf{v} = 0, \quad \nabla \times \mathbf{v} = 0, \quad \overline{\nabla_\nu v_\mu} = 3\, \overline{e_\nu e_\mu}.$$

(v) *Planar Squeeze-stretch Field*

$$v_\mu = e_\mu\, u_\nu r_\nu + u_\mu\, e_\nu r_\nu,$$

where $\mathbf{e}$ and $\mathbf{u}$ are two orthogonal unit vectors, $\mathbf{e} \cdot \mathbf{u} = 0$. Since here $\nabla_\nu v_\mu = e_\mu u_\nu + e_\nu u_\mu$,

$$\nabla \cdot \mathbf{v} = 0, \quad \nabla \times \mathbf{v} = 0, \quad \overline{\nabla_\nu v_\mu} = 2\, \overline{e_\nu u_\mu},$$

is found. When the coordinate axes are rotated by $45°$, this vector field reads

$$v_\mu = e_\mu\, e_\nu r_\nu - u_\mu\, u_\nu r_\nu.$$

Now one finds

$$\nabla \cdot \mathbf{v} = 0, \quad \nabla \times \mathbf{v} = 0, \quad \overline{\nabla_\nu v_\mu} = \overline{e_\nu e_\mu} - \overline{u_\nu u_\mu}.$$

(vi) *Solid-like Rotation or Vorticity Field*. A circular flow with a constant angular velocity $\mathbf{w}$:

$$v_\mu = \varepsilon_{\mu\kappa\lambda} w_\kappa r_\lambda.$$

Here one has $\nabla_\nu v_\mu = \varepsilon_{\mu\kappa\nu} w_\kappa$, and consequently

$$\nabla \cdot \mathbf{v} = 0, \quad (\nabla \times \mathbf{v})_\lambda = \varepsilon_{\lambda\nu\mu} \varepsilon_{\mu\kappa\nu} w_\kappa = 2 w_\lambda, \quad \overline{\nabla_\nu v_\mu} = 0.$$

7.2 Test Solutions of the Wave Equation (p. 102)

Proof that both the ansatz (7.63) *and the plane wave* (7.64) *obey the wave equation. Furthermore, show that the* **E**-*field is perpendicular to the wave vector, and that the* **B**-*field is perpendicular to both.*
(i) *Ansatz* (7.63). From (7.63), i.e. from

$$E_\mu = E_\mu^{(0)} f(\xi), \quad \xi = \widehat{k}_\nu r_\nu - ct,$$

follows

$$\nabla_\nu E_\mu = (\nabla_\nu \xi) E_\mu^{(0)} f(\xi)' = \widehat{k}_\nu E_\mu^{(0)} f(\xi)',$$

where the prime indicates the derivative with respect to ξ. Clearly, $\nabla_\nu E_\nu = 0$ implies $\widehat{k}_\nu E_\nu^{(0)} = 0$, the **E**-field is perpendicular to its direction of propagation. The second spatial derivative of the field yields

$$\nabla_\nu \nabla_\nu E_\mu \equiv \Delta E_\mu = E_\mu^{(0)} f(\xi)''.$$

Similarly, the time derivative of the ansatz for the the **E**-field is given by

$$\frac{\partial}{\partial t} E_\mu = \left(\frac{\partial}{\partial t} \xi\right) E_\mu^{(0)} f(\xi)' = -c E_\mu^{(0)} f(\xi)',$$

and the second time derivative is

$$\frac{\partial^2}{\partial t^2} E_\mu = c^2 E_\mu^{(0)} f(\xi)''.$$

Thus the wave equation

$$\Delta \mathbf{E} - \frac{1}{c^2} \frac{\partial^2}{\partial t^2} \mathbf{E} = 0,$$

cf. (7.60), is obeyed.
For the **B**-field, the ansatz

$$B_\mu = B_\mu^{(0)} f(\xi), \quad \xi = \widehat{k}_\nu r_\nu - ct,$$

is made. The Maxwell equation $-\partial B_\mu/\partial t = \varepsilon_{\mu\nu\lambda}\nabla_\nu E_\lambda$ leads to

$$c B_\mu^{(0)} = \varepsilon_{\mu\nu\lambda}\,\widehat{k}_\nu\, E_\lambda^{(0)}.$$

Thus the magnetic field is perpendicular to both the wave vector and the electric field.

(ii) *Plane Wave* (7.64). From the plane wave ansatz

$$E_\mu = E_\mu^{(0)} \exp[i\, k_\nu\, r_\nu - i\,\omega\, t]$$

follows, by analogy to the calculations above,

$$\nabla_\nu E_\mu = i\, k_\nu\, E_\mu.$$

Again $\nabla_\nu E_\nu = 0$, corresponding to $k_\nu E_\nu = 0$, implies that the **E**-field is perpendicular to the wave vector **k**. The second spatial derivative leads to

$$\Delta E_\mu = -k^2\, E_\mu.$$

The first and second time derivatives of the field are

$$\frac{\partial}{\partial t}E_\mu = -i\,\omega\,E_\mu, \qquad \frac{\partial^2}{\partial t^2}E_\mu = -\omega^2\,E_\mu.$$

Thus the wave equation (7.60) imposes the condition

$$k^2 = \omega^2/c^2,$$

which proofs the dispersion relation (7.65).

Here $-\partial B_\mu/\partial t = \varepsilon_{\mu\nu\lambda}\nabla_\nu E_\lambda$ leads to

$$\omega\, B_\mu = \varepsilon_{\mu\nu\lambda}\, k_\nu\, E_\lambda.$$

As expected, also for plane waves, the magnetic field is perpendicular to both the wave vector and the electric field.

7.3 Electromagnetic Waves in Flatland? (p. 105)

In flatland, one has just 2 dimensions. Cartesian components are denoted by Latin letters $i, j, \ldots, i = 1, 2, j = 1, 2$, etc. The summation convention is used. In vacuum, and for zero charges and currents, the adapted Maxwell equations are

$$\nabla_i E_i = 0, \quad -\nabla_i H_{ij} = \varepsilon_0\frac{\partial}{\partial t}E_j, \quad \nabla_i E_j - \nabla_j E_i = -\mu_0\frac{\partial}{\partial t}H_{ij}.$$

In 2D, there is no equation corresponding to $\nabla_\lambda B_\lambda = 0$. It is not defined and not needed in 2D. The magnetic field tensors have only one independent component.

Differentiation of the second of these equations with respect to t, insertion of the time change of the magnetic field tensor as given by the third equation and use of the first one leads to the wave equation

$$\nabla_i \nabla_i E_j = \varepsilon_0 \mu_0 \frac{\partial^2}{\partial t^2} E_j.$$

Again, the speed c of the radiation is determined by $c^2 = (\varepsilon_0 \mu_0)^{-1}$. The electric field is perpendicular to the direction of the propagation, even in 2D.

How about 1D? Obviously, the Maxwell equations loose their meaning in a true one-dimensional world, thus electromagnetic waves do not exist in 1D. On the other hand, longitudinal sound waves still can propagate in 1D.

7.4 Radial and Angular Parts of the Nabla Operator, Compare (7.81) with (7.78) (p. 106)

Due to the definition (7.80), the term $\varepsilon_{\mu\nu\lambda} \widehat{r_\nu} \mathscr{L}_\lambda$ in (7.81) is equal to

$$\varepsilon_{\mu\nu\lambda} \widehat{r_\nu} \, \varepsilon_{\lambda\alpha\beta} \widehat{r_\alpha} \frac{\partial}{\partial \widehat{r_\beta}} = (\delta_{\mu\alpha}\delta_{\nu\beta} - \delta_{\mu\beta}\delta_{\nu\alpha}) \widehat{r_\nu} \widehat{r_\alpha} \frac{\partial}{\partial \widehat{r_\beta}} = \widehat{r_\mu} \widehat{r_\nu} \frac{\partial}{\partial \widehat{r_\nu}} - \frac{\partial}{\partial \widehat{r_\mu}}.$$

Here, the relation (4.10) was used for the product of the two epsilon-tensors. The last term can be written as $\frac{\partial}{\partial \widehat{r_\mu}} = \delta_{\mu\nu} \frac{\partial}{\partial \widehat{r_\nu}}$ and hence

$$\varepsilon_{\mu\nu\lambda} \widehat{r_\nu} \mathscr{L}_\lambda = \varepsilon_{\mu\nu\lambda} \widehat{r_\nu} \, \varepsilon_{\lambda\alpha\beta} \widehat{r_\alpha} \frac{\partial}{\partial \widehat{r_\beta}} = (\widehat{r_\mu}\widehat{r_\nu} - \delta_{\mu\nu}) \frac{\partial}{\partial \widehat{r_\nu}}.$$

Now (7.81) is seen to be equal to (7.78).

7.5 Prove the Relations (7.82) and (7.83) for the Angular Nabla Operator (p. 106)

From the definition (7.80) of the differential operator $\mathscr{L}$ follows

$$\mathscr{L}_\mu \mathscr{L}_\nu = \varepsilon_{\mu\alpha\beta} r_\alpha \nabla_\beta \varepsilon_{\nu\kappa\tau} r_\kappa \nabla_\tau = \varepsilon_{\mu\alpha\beta} \varepsilon_{\nu\kappa\tau} \delta_{\beta\kappa} \nabla_\tau + \varepsilon_{\mu\alpha\beta} \varepsilon_{\nu\kappa\tau} r_\alpha r_\kappa \nabla_\beta \nabla_\tau$$
$$= \varepsilon_{\mu\alpha\beta} \varepsilon_{\nu\beta\tau} r_\alpha \nabla_\tau + \varepsilon_{\mu\alpha\beta} \varepsilon_{\nu\kappa\tau} r_\alpha r_\kappa \nabla_\beta \nabla_\tau.$$

Due to $\varepsilon_{\mu\alpha\beta} \varepsilon_{\nu\beta\tau} = \varepsilon_{\mu\alpha\beta} \varepsilon_{\tau\nu\beta} = \delta_{\mu\tau}\delta_{\nu\alpha} - \delta_{\mu\nu}\delta_{\alpha\tau}$, cf. (4.10), one obtains

$$\mathscr{L}_\mu \mathscr{L}_\nu = r_\nu \nabla_\mu - \delta_{\mu\nu} r_\alpha \nabla_\alpha + \varepsilon_{\mu\alpha\beta} \varepsilon_{\nu\kappa\tau} r_\alpha r_\kappa \nabla_\beta \nabla_\tau.$$

The second term on the right hand side is obviously symmetric under the exchange of μ and ν. The same applies to the third term. To see this, notice that the simultaneous interchanges $\alpha \leftrightarrow \kappa$ and $\beta \leftrightarrow \tau$ corresponds to an interchange $\mu \leftrightarrow \nu$. Thus one has

$$\varepsilon_{\lambda\mu\nu} \mathscr{L}_\mu \mathscr{L}_\nu = \varepsilon_{\lambda\mu\nu} r_\nu \nabla_\mu = -\mathscr{L}_\lambda,$$

which is the relation (7.83), and

$$\mathscr{L}_\mu \mathscr{L}_\nu - \mathscr{L}_\nu \mathscr{L}_\mu = r_\nu \nabla_\mu - r_\mu \nabla_\nu.$$

The right hand side of the last equation can be written as $-\varepsilon_{\mu\nu\lambda}\mathscr{L}_\lambda$, since one has $-\varepsilon_{\mu\nu\lambda}\mathscr{L}_\lambda = -\varepsilon_{\mu\nu\lambda}\varepsilon_{\lambda\alpha\beta}r_\alpha \nabla_\beta = -(r_\mu \nabla_\nu - r_\nu \nabla_\mu)$. This proves the commutation relation (7.82).

7.6 Determine the Radial Part of the Laplace Operator in D Dimensions (p. 108)

Let $f = f(r)$ a function of the magnitude $r = |\mathbf{r}|$. It does not depend on the direction of $\mathbf{r}$, it has no angular dependence. Thus only the radial part Δ_r of the Laplace operator $\Delta \equiv \nabla_\mu \nabla_\mu$ gives a contribution, when Δ is applied on $f(r)$. Due to $\nabla_\mu \nabla_\mu f = \nabla_\mu(\nabla_\mu f)$ and $\nabla_\mu f = \frac{df}{dr}\nabla_\mu r = \frac{df}{dr}r^{-1}r_\mu$, one obtains $\nabla_\mu \nabla_\mu f = r_\mu \nabla_\mu(r^{-1}\frac{df}{dr}) + r^{-1}\frac{df}{dr}\nabla_\mu r_\mu$. Notice that $r_\mu \nabla_\mu$ is r times the spatial derivative in radial direction, here equal to $r\frac{d}{dr}$. On account of $\nabla_\mu r_\mu = D$, for D dimensions, one finds $\nabla_\mu \nabla_\mu f = \frac{d^2f}{dr^2} + (D-1)r^{-1}\frac{df}{dr}$. For the case where the Laplace operator is applied to a function which also depends on the direction of $\mathbf{r}$, the derivative $\frac{d}{dr}$ with respect to r is replaced by the partial derivative $\frac{\partial}{\partial r}$, in the previous expression. Thus the radial part of the Laplace operator is inferred to be

$$\Delta_r = \frac{\partial^2}{\partial r^2} + (D-1)\,r^{-1}\frac{\partial}{\partial r} = r^{-(D-1)}\frac{\partial}{\partial r}\left(r^{(D-1)}\frac{\partial}{\partial r}\right).$$

Prove $\Delta r^{(2-D)} = 0$

Application of the expression given above yields $\Delta r^n = n(D+n-2)r^{(n-2)}$, where the exponent n is a real number. Apart from the trivial solution $n = 0$, the requirement $\Delta r^n = 0$ implies $n = 2 - D$. Thus r^{-1} is a solution of the Laplace equation, for $D = 3$. One finds r^{-2} for $D = 4$.

The case $D = 2$ has to be considered separately. Here $\ln(r/r_{\text{ref}})$ is a solution of the Laplace equation, please check it. The quantity r_{ref} is a reference length introduced such that the argument of the logarithm ln is dimensionless.

Exercises Chapter 8

8.1 Compute Path Integrals Along a Closed Curve for Three Vector Fields (p. 116)

The differential $d\mathbf{r}$ needed for the integration is recalled to be $dx\{1, 0, 0\}$, for the straight line C_1, and $\rho d\varphi\{-\sin\varphi, \cos\varphi, 0\}$ for the semi-circle C_2. The start and end points are $x = -\rho$ and $x = \rho$, for C_1. For C_2 one has $\varphi = 0$ and $\varphi = \pi$, see Fig. 8.3.

(i) *Homogeneous Field*, where $\mathbf{v} = \mathbf{e} = $ const., with $\mathbf{e}$ parallel to the x-axis.
The expectation is $\oint \mathbf{v} \cdot d\mathbf{r} = \mathscr{I} = 0$ since the vector field is the gradient of a scalar potential. The explicit calculation of the line integrals $\mathscr{I}_1$ and $\mathscr{I}_2$ along the curves C_1 and C_2 yields

$$\mathscr{I}_1 = \int_{C_1} \mathbf{v} \cdot d\mathbf{r} = \int_{-\rho}^{\rho} dx = 2\rho,$$

$$\mathscr{I}_2 = \int_{C_2} \mathbf{v} \cdot d\mathbf{r} = -\rho \int_0^\pi \sin\varphi d\varphi = \rho \cos\varphi|_0^\pi = -2\rho.$$

Thus $\mathscr{I} = \mathscr{I}_1 + \mathscr{I}_2 = 0$, is found, as expected.
(ii) *Radial Field*, where $\mathbf{v} = \mathbf{r}$.
Again $\oint \mathbf{v} \cdot d\mathbf{r} = \mathscr{I} = 0$ is expected since the vector field possesses a scalar potential function, viz.: $\Phi = (1/2)r^2$. Here the integration along C_1 yields

$$\mathscr{I}_1 = \int_{C_1} \mathbf{r} \cdot d\mathbf{r} = \int_{-\rho}^{\rho} x dx = (1/2) x^2 |_{-\rho}^{\rho} = 0.$$

The integral $\mathscr{I}_2$, performed along C_2, also gives zero since one has $\mathbf{r} \cdot d\mathbf{r} = 0$ on the semi-circle. Thus again, the explicit calculation confirms the expectation $\mathscr{I} = \oint \mathbf{v} \cdot d\mathbf{r} = 0$.
(iii) *Solid-like Rotation or Vorticity Field*, where $\mathbf{v} = \mathbf{w} \times \mathbf{r}$, with the constant axial vector $\mathbf{w}$ parallel to the z-axis.
In this case, the curl $\nabla \times \mathbf{v}$ is not zero and no scalar potential exists. So $\oint \mathbf{v} \cdot d\mathbf{r} = \mathscr{I} \neq 0$ is expected.

For the explicit calculation of the integrals, the scalar product $\mathbf{v} \cdot d\mathbf{r}$ is needed. With $w_z \equiv w$, the vector field $\mathbf{v}$ has the components $\{-wy, wx, 0\}$ and consequently $\mathbf{v} \cdot d\mathbf{r} = w(-ydx + xdy)$. For the integral $\mathscr{I}_1$ this implies $\mathscr{I}_1 = -w \int_{-\rho}^{\rho} ydx = 0$, since $y = 0$ along the line C_1. The integration along the semi-circle yields

$$\mathscr{I}_2 = \int_{C_2} \mathbf{v} \cdot d\mathbf{r} = w\rho^2 \int_0^\pi (\sin^2\varphi + \cos^2\varphi)d\varphi = w\rho^2 \int_0^\pi d\varphi = \pi w \rho^2.$$

Thus the non-zero result $\mathscr{I} = \mathscr{I}_1 + \mathscr{I}_2 = \pi w \rho^2$ is obtained here for $\oint \mathbf{v} \cdot d\mathbf{r}$.

8.2 Surface Integrals Over a Hemisphere (p. 124)

Consider a hemisphere with radius R and its center at the origin. The unit vector pointing from the center to the North pole is $\mathbf{u}$. Surface integrals $\mathscr{S}_{\mu\nu} = \int v_\nu ds_\mu$ are to be computed over the hemisphere.
(i) *Homogeneous Vector Field* $v_\nu = v\widehat{v}_\nu = $ const. By symmetry, the integral is proportional to $u_\mu \widehat{v}_\nu$. The ansatz

$$\mathscr{S}_{\mu\nu} = c_1 u_\mu \widehat{v}_\nu$$

is made, with a scalar coefficient c_1. Multiplication of this equation by $u_\mu \widehat{v_\nu}$ leads to

$$c_1 = u_\mu \widehat{v_\nu} \mathscr{S}_{\mu\nu} = v \int u_\mu ds_\mu.$$

For the hemisphere, located on the x–y-plane, one has $u_\mu ds_\mu = R^2 \cos\theta \sin\theta d\theta d\varphi$. With $\zeta = \cos\theta$, where θ is the angle between $\mathbf{u}$ and the vector $\widehat{\mathbf{r}}$, the integral for c_1 is

$$c_1 = v 2\pi R^2 \int_0^1 \zeta d\zeta = v\pi R^2.$$

The flux $\mathscr{S} = \mathscr{S}_{\mu\mu}$ is found to be

$$\mathscr{S} = \mathbf{v} \cdot \mathbf{u}\,\pi R^2,$$

which, as expected, is equal to the flux through the circular base of the hemisphere. (ii) *Radial Field* $v_\nu = r_\nu$. Here the ansatz

$$\mathscr{S}_{\mu\nu} = c_2 u_\mu u_\nu$$

is made. Multiplication of this equation by $u_\mu u_\nu$ yields an expression for the scalar coefficient c_2, viz.

$$c_2 = \int u_\nu r_\nu u_\mu ds_\mu = 2\pi R^3 \int_0^1 \zeta^2 d\zeta = \frac{2}{3}\pi R^3.$$

The resulting flux is just the area of the hemisphere.

8.3 Verify the Stokes Law for a Vorticity Field (p. 127)
The vector field is given by $\mathbf{v} = \mathbf{w} \times \mathbf{r}$ *with* $\mathbf{w} = $ const. *The curl of the field is* $\nabla \times \mathbf{v} = 2\mathbf{w}$, *cf. Exercise* 7.1.
Since the surface element $d\mathbf{s}$ is parallel to $\mathbf{w}$, one has

$$\mathscr{S} \equiv \int (\nabla \times \mathbf{v}) \cdot d\mathbf{s} = 2w \int \widehat{\mathbf{w}} \cdot d\mathbf{s},$$

where w is the magnitude of $\mathbf{w}$. Using the planar polar coordinates ρ and φ yields

$$\mathscr{S} = 2w \int_0^R d\rho \int_0^{2\pi} d\varphi \rho = 2\pi R^2 w.$$

The line integral to be compared with is $\oint \mathbf{v} \cdot d\mathbf{r}$. Now $\mathbf{v}$ and $d\mathbf{r}$ are parallel to each other and $\mathbf{v} \cdot d\mathbf{r} = Rw R d\varphi$. This leads to

$$\mathscr{I} \equiv \oint \mathbf{v} \cdot d\mathbf{r} = R^2 w \int_0^{2\pi} d\varphi = 2\pi R^2 w.$$

The equality $\mathscr{S} = \mathscr{I}$ is in accord with the Stokes law.

8.4 Moment of Inertia Tensor of a Half-Sphere (p. 136)

A half-sphere with radius a and constant mass density is considered. The orientation is specified by the unit vector $\mathbf{u}$, pointing from the center of the sphere to the center of mass of the half-sphere. For the present geometry, the moment of inertia tensor is uniaxial and of the form

$$\Theta_\parallel u_\mu u_\nu + \Theta_\perp (\delta_{\mu\nu} - u_\mu u_\nu).$$

The z-axis is chosen parallel to $\mathbf{u}$, just as in examples discussed in Sects. 8.3.2 and 8.3.3. First, moments of inertia are computed with respect to the geometric center of the sphere. The pertaining moment of inertia $\Theta_\parallel^{\text{eff}}$ is

$$\Theta_\parallel^{\text{eff}} = \rho_0 2\pi 2 \int_0^a r^4 dr \int_0^1 \zeta^2 d\zeta = \frac{4\pi}{15} \rho_0 a^5 = \frac{2}{5} M a^2,$$

where $M = (2\pi/3)\rho_0 a^3$ is the mass of the half-sphere. As before, $\zeta = \cos\theta$ is used. The moment $\Theta_\perp^{\text{eff}}$ is inferred from the mean moment of inertia $\bar{\Theta}^{\text{eff}}$, via $\Theta_\perp^{\text{eff}} = (3\bar{\Theta}^{\text{eff}} - \Theta_\parallel^{\text{eff}})/2$. Equation (8.69) yields

$$3\bar{\Theta}^{\text{eff}} = 2 \int_V \rho(\mathbf{r}) r^2 d^3 r = \frac{4\pi}{5} \rho_0 a^5 = \frac{6}{5} M a^2.$$

This implies $\Theta_\perp^{\text{eff}} = \Theta_\parallel^{\text{eff}}$, i.e. the moment of inertia tensor is isotropic, when evaluated with respect to the geometric center. According to the law of Steiner (8.66), the moment of inertia tensor with respect to the center of mass is

$$\Theta_{\mu\nu} \equiv \Theta_{\mu\nu}^{\text{cm}} = \Theta_{\mu\nu}^{\text{eff}} - M(R^2 \delta_{\mu\nu} - R_\mu R_\nu) = \frac{2}{5} M a^2 \delta_{\mu\nu} M R^2 (\delta_{\mu\nu} - u_\mu u_\nu),$$

with $R = \frac{3}{8}a$ is the distance between the center of mass and the geometric center, cf. (8.63). Thus one finds

$$\Theta_\parallel = \frac{2}{5} M a^2, \quad \Theta_\perp = \Theta_\parallel - \frac{9}{64} M a^2 = \frac{83}{128} \Theta_\parallel.$$

As expected, $\Theta_\perp$ is smaller than $\Theta_\parallel$.

Exercise Chapter 9

9.1 Verify the Required Properties of the Third and Fourth Rank Irreducible Tensors (9.5) and (9.6) (p. 157)

The required symmetry of the third and fourth rank tensors $\overline{a_\mu a_\nu a_\lambda}$ and $\overline{a_\mu a_\nu a_\lambda a_\kappa}$, as given explicitly by (9.5) and (9.6) is seen by inspection, note that $a_\mu a_\nu = a_\nu a_\mu$ and $\delta_{\mu\nu} = \delta_{\nu\mu}$.

Setting $\lambda = \nu$, in (9.5), leads to

$$\overline{a_\mu a_\nu a_\nu} = a_\mu a^2 - \frac{1}{5} a^2 (a_\mu 3 + a_\nu \delta_{\mu\nu} + a_\nu \delta_{\mu\nu}) = 0.$$

Here $\delta_{\nu\nu} = 3$ and $a_\nu \delta_{\mu\nu} = a_\mu$ are used.

Likewise, putting $\kappa = \lambda$ in (9.6), yields

$$\overline{a_\mu a_\nu a_\lambda a_\kappa} = a_\mu a_\nu a^2$$

$$- \frac{1}{7} a^2 (a_\mu a_\nu 3 + a_\mu a_\lambda \delta_{\nu\lambda} + a_\mu a_\lambda \delta_{\nu\lambda} + a_\nu a_\lambda \delta_{\mu\lambda} + a_\nu a_\lambda \delta_{\mu\lambda} + a^2 \delta_{\mu\nu})$$

$$+ \frac{1}{35} a^4 (\delta_{\mu\nu} 3 + \delta_{\mu\lambda} \delta_{\nu\lambda} + \delta_{\mu\lambda} \delta_{\nu\lambda})$$

$$= a_\mu a_\nu a^2 \left(1 - \frac{1}{7}7\right) - \frac{1}{7} a^4 \delta_{\mu\nu} + \frac{1}{35} a^4 \delta_{\mu\nu} 5 = 0,$$

where, e.g. $a_\lambda \delta_{\mu\lambda} = a_\mu$ and $\delta_{\mu\lambda} \delta_{\nu\lambda} = \delta_{\mu\nu}$ have been used.

Exercises Chapter 10

10.1 Prove the Product Rule (10.13) for the Laplace Operator (p. 166)
The proof of (10.13), viz.

$$\Delta(g(r) X_{\mu_1\mu_2\cdots\mu_\ell}) = (g'' - 2\ell r^{-1} g') X_{\mu_1\mu_2\cdots\mu_\ell}$$

starts from the general relation

$$\Delta(fg) = f \Delta g + 2 (\nabla_\kappa f)(\nabla_\kappa g) + g \Delta f,$$

for any two functions f and g. Now choose for f the ℓ-th descending multipole tensor $X_{\mu_1\mu_2\cdots\mu_\ell}$ and assume that the scalar g depends on $r = |\mathbf{r}|$ only. Since $\Delta X_{...} = 0$, one obtains

$$\Delta(g(r) X_{\mu_1\mu_2\cdots\mu_\ell}) = X_{\mu_1\mu_2\cdots\mu_\ell} \Delta g + 2 g' r^{-1} r_\kappa \nabla_\kappa X_{\mu_1\mu_2\cdots\mu_\ell}.$$

Due to $r_\kappa \nabla_\kappa = r\partial/\partial r$ and $X_{\mu_1\mu_2\cdots\mu_\ell} \sim r^{-(\ell+1)}$, the second term in the preceding equation is equal to

$$-2g'r^{-1}\left((\ell+1)X_{\dots}\right).$$

Use of $\Delta g = g'' + 2r^{-1}g'$ then leads to (10.13). The condition $\ell \geq 1$ is obvious for the validity of the present considerations. When the function $r^{(-1)}$ stands for X in the case $\ell = 0$, the value $\ell = 0$ is also included in (10.13).

10.2 Multipole Potentials in D Dimensional Space (p. 166)

In D dimensions, $r^{(2-D)}$ is the radially symmetric solution of the Laplace equation, cf. exercise 7.6, for $D \geq 3$. By analogy with (10.2), D dimensional multipole potential tensors are defined by

$$X^{(D)}_{\mu_1\mu_2\cdots\mu_\ell} \equiv (-1)^\ell \frac{\partial^\ell}{\partial r_{\mu_1}\partial r_{\mu_2}\cdots\partial r_{\mu_\ell}} r^{(2-D)} = (-1)^\ell \nabla_{\mu_1}\nabla_{\mu_2}\cdots\nabla_{\mu_\ell} r^{(2-D)},$$

(A.11)

where now ∇ is the in D dimensional Nabla operator. For $D = 2$, $r^{(2-D)}$ is replaced by $-\ln r$. Compute the first and second multipole potentials, for $D \geq 3$ and for $D = 2$.

The first descending multipole is the D dimensional vector

$$X^{(D)}_\mu = (D-2)r^{(1-D)}\widehat{r_\mu} = (D-2)r^{-D}r_\mu, \quad D \geq 3, \quad X^{(2)}_\mu = r^{-1}\widehat{r_\mu} = r^{-2}r_\mu.$$

The resulting second multipole potential $X^{(D)}_{\mu\nu} = -\nabla_\nu X^{(D)}_\mu$, $D \geq 2$, is

$$X^{(D)}_{\mu\nu} = (D-2)r^{-(D+2)}(D\,r_\mu r_\nu - \delta_{\mu\nu}r^2), \quad D \geq 3, \quad X^{(2)}_\mu = r^{-4}(2\,r_\mu r_\nu - \delta_{\mu\nu}r^2).$$

It is understood that $\delta_{\mu\nu}$ is the D dimensional unit tensor with $\delta_{\mu\mu} = D$. Thus the tensors $X^{(D)}_{\mu\nu}$ are traceless.

10.3 Compute the Torque on a Rotating Sphere (p. 181)

A sphere rotating with the angular velocity Ω experiences a friction torque

$$T_\mu = -8\pi\,\eta\,R^3\,\Omega_\mu.$$

To derive this result from the creeping flow equation, considerations similar to those used for the Stokes force, should be made.

The distortion of the pressure and the flow velocity should be linear in Ω and the respective expressions should have the appropriate parity. The only scalar available for p is proportional to $X_\nu \Omega_\nu$. This term, however has the wrong parity behavior, thus one has $p = 0$, in this case. The possible vectors are proportional to Ω_μ, $X_{\mu\nu}\Omega_\nu$ and $\varepsilon_{\mu\lambda\nu}\Omega_\lambda X_\nu$. The first two of these expressions have the wrong parity. The only polar vector is the ansatz $v_\kappa = c(r)\varepsilon_{\kappa\lambda\nu}\Omega_\lambda X_\nu$, with a coefficient c. For the present problem, the creeping flow equation reduces to $\Delta v_\kappa = 0$. Thus $c = \text{const.}$ is a solution, which has still to be specified by the boundary conditions.

The fluid is assumed to be at rest, far away from the sphere. This is obeyed by the ansatz with $c = \text{const.}$ A no-slip boundary condition at the surface of the sphere means $v_\kappa = R\varepsilon_{\kappa\lambda\nu}\Omega_\lambda r_\nu$, at $r = 1$. Due to $X_\nu = r^{-3}r_\nu$, the solution for the flow velocity is

$$v_\kappa = R\,\varepsilon_{\kappa\lambda\nu}\,\Omega_\lambda\,X_\nu. \tag{A.12}$$

According to (8.98), for the present problem, the torque is given by

$$T_\mu = -\varepsilon_{\mu\alpha\beta}R^2 \oint r_\alpha\,\widehat{r_\tau}\,p_{\tau\beta}\mathrm{d}^2\widehat{r} = \eta\,\varepsilon_{\mu\alpha\beta}R^3 \oint \widehat{r_\alpha}\,\widehat{r_\tau}\,(\nabla_\tau v_\beta + \nabla_\beta v_\tau)\mathrm{d}^2\widehat{r}.$$

The first term in the integrand involves

$$\widehat{r_\tau}\,\nabla_\tau v_\beta = R^{-1}\frac{\partial}{\partial r}v_\beta = -2R^{-1}\,v_\beta.$$

The expression $\widehat{r_\tau}\nabla_\beta v_\tau$, occurring in the second term of the integrand, is equal to

$$-\widehat{r_\tau}\,\varepsilon_{\tau\lambda\nu}\,\Omega_\lambda\,X_{\nu\beta} = r^{-3}\,\widehat{r_\tau}\,\varepsilon_{\tau\lambda\beta}\,\Omega_\lambda = -R^{-1}\,v_\beta.$$

Thus one obtains

$$T_\mu = -3\eta\varepsilon_{\mu\alpha\beta}\varepsilon_{\beta\lambda\tau}\Omega_\lambda R^3 \oint \widehat{r_\alpha}\widehat{r_\tau}\mathrm{d}^2\widehat{r}.$$

The surface integral $\oint \widehat{r_\alpha}\widehat{r_\tau}\mathrm{d}^2\widehat{r} = (4\pi/3)\delta_{\alpha\tau}$ leads to

$$T_\mu = -4\pi\eta R^3\varepsilon_{\mu\alpha\beta}\varepsilon_{\beta\lambda\alpha}\Omega_\lambda = -8\pi\eta R^3\Omega_\mu,$$

which is the expression for the friction torque mentioned above. The minus sign and $\eta > 0$ imply that the rotational motion is damped. Time reversal invariance is broken, as typical for irreversible processes.

Exercises Chapter 11

11.1 Contraction Rules for Delta-Tensors (p. 185)
Verify (11.6) for $\ell = 2$.
The contraction rule for the ℓ-th Δ^ℓ-tensor is

$$\Delta^{(\ell)}_{\mu_1\mu_2\cdots\mu_{\ell-1}\lambda,\mu_1'\mu_2'\cdots\mu_{\ell-1}'\lambda} = \frac{2\ell+1}{2\ell-1}\Delta'^{(\ell-1)}_{\mu_1\mu_2\cdots\mu_{\ell-1},\mu_1'\mu_2'\cdots\mu_{\ell-1}'}.$$

For $\ell = 2$, it reduces to $\Delta_{\mu\lambda,\nu\lambda} = \frac{5}{3}\delta_{\mu\nu}$.

From the definition of the Δ-tensor follows

$$\Delta_{\mu\lambda,\nu\lambda} = \frac{1}{2}(\delta_{\mu\nu}\delta_{\lambda\lambda} + \delta_{\mu\lambda}\delta_{\nu\lambda}) - \frac{1}{3}\delta_{\mu\lambda}\delta_{\nu\lambda} = \left(2 - \frac{1}{3}\right)\delta_{\mu\nu} = \frac{5}{3}\delta_{\mu\nu},$$

as expected.

11.2 Determine $\Delta^{(3)}_{\mu\nu\lambda,\mu'\nu'\lambda'}$ (p. 186)

Hint: compute $\Delta^{(3)}_{\mu\nu\lambda,\mu'\nu'\lambda'}$, *in terms of triple products of δ-tensors, from* (11.15) *for* $\ell = 3$.

Use of (11.15) for $\ell = 3$ with (10.6) yields

$$\Delta^{(3)}_{\mu\nu\lambda,\mu'\nu'\lambda'} = \frac{1}{6}\nabla_\mu\nabla_\nu\nabla_\lambda \overline{r_{\mu'}r_{\nu'}r_{\lambda'}}$$

$$= \frac{1}{6}\nabla_\mu\nabla_\nu\nabla_\lambda\left[r_{\mu'}r_{\nu'}r_{\lambda'} - \frac{1}{5}r^2(r_{\mu'}\delta_{\nu'\lambda'} + r_{\nu'}\delta_{\mu'\lambda'} + r_{\lambda'}\delta_{\mu'\nu'})\right].$$

Successive application of the differential operators $\nabla_\mu\nabla_\nu\nabla_\lambda$ on the first term after the second equality sign of the equation above yields

$$\nabla_\mu\nabla_\nu\nabla_\lambda(r_{\mu'}r_{\nu'}r_{\lambda'}) = \nabla_\mu\nabla_\nu(r_{\mu'}r_{\nu'}\delta_{\lambda\lambda'} + r_{\mu'}r_{\lambda'}\delta_{\lambda\nu'} + r_{\lambda'}r_{\nu'}\delta_{\lambda\mu'})$$

$$= \nabla_\mu\left[(r_{\mu'}\delta_{\nu\nu'} + r_{\nu'}\delta_{\nu\mu'})\delta_{\lambda\lambda'}\right.$$

$$\left.+(r_{\mu'}\delta_{\nu\lambda'} + r_{\lambda'}\delta_{\nu\mu'})\delta_{\lambda\nu'} + (r_{\lambda'}\delta_{\nu\nu'} + r_{\nu'}\delta_{\nu\lambda'})\delta_{\lambda\mu'}\right]$$

$$= (\delta_{\mu\mu'}\delta_{\nu\nu'} + \delta_{\mu\nu'}\delta_{\nu\mu'})\delta_{\lambda\lambda'}$$

$$+(\delta_{\mu\mu'}\delta_{\nu\lambda'} + \delta_{\mu\lambda'}\delta_{\nu\mu'})\delta_{\lambda\nu'} + (\delta_{\mu\lambda'}\delta_{\nu\nu'} + \delta_{\mu\nu'}\delta_{\nu\lambda'})\delta_{\lambda\mu'}.$$

Likewise, the application of the same differential operators on $r^2 r_{\mu'}$ occurring above as factor of $\delta_{\nu'\lambda'}$ yields

$$\nabla_\mu\nabla_\nu\nabla_\lambda(r^2 r_{\mu'}) = \nabla_\mu\nabla_\nu(2r_\lambda r_{\mu'} + r^2\delta_{\lambda\mu'}) = 2\nabla_\mu(r_{\mu'}\delta_{\nu\lambda} + r_\lambda\delta_{\nu\mu'} + r_\nu\delta_{\lambda\mu'})$$

$$= 2(\delta_{\mu\mu'}\delta_{\nu\lambda} + \delta_{\mu\lambda}\delta_{\nu\mu'} + \delta_{\mu\nu}\delta_{\lambda\mu'}).$$

The derivatives of $r^2 r_{\nu'}$ and $r^2 r_{\lambda'}$ which are factors of $\delta_{\mu'\lambda'}$ and $\delta_{\mu'\lambda'}$ are given by analogous expressions where just μ' is replaced by ν' and λ'. All terms for $\Delta^{(3)}$ put together, with the numerical factors, leads to

$$\Delta^{(3)}_{\mu\nu\lambda,\mu'\nu'\lambda'} = \frac{1}{6}\big[(\delta_{\mu\mu'}\delta_{\nu\nu'} + \delta_{\mu\nu'}\delta_{\nu\mu'})\delta_{\lambda\lambda'} + (\delta_{\mu\mu'}\delta_{\nu\lambda'} + \delta_{\mu\lambda'}\delta_{\nu\mu'})\delta_{\lambda\nu'}$$

$$+ (\delta_{\mu\lambda'}\delta_{\nu\nu'} + \delta_{\mu\nu'}\delta_{\nu\lambda'})\delta_{\lambda\mu'}\big]$$

$$- \frac{1}{15}\big[(\delta_{\nu\lambda}\delta_{\mu\mu'} + \delta_{\mu\lambda}\delta_{\nu\mu'} + \delta_{\mu\nu}\delta_{\lambda\mu'})\delta_{\nu'\lambda'}$$

$$+ (\delta_{\nu\lambda}\delta_{\mu\nu'} + \delta_{\mu\lambda}\delta_{\nu\nu'} + \delta_{\mu\nu}\delta_{\lambda\nu'})\delta_{\mu'\lambda'}$$

$$+ (\delta_{\nu\lambda}\delta_{\mu\lambda'} + \delta_{\mu\lambda}\delta_{\nu\lambda'} + \delta_{\mu\nu}\delta_{\lambda\lambda'})\delta_{\mu'\nu'}\big]. \tag{A.13}$$

Notice that all δ-tensors in the bracket [...] behind the factor $\frac{1}{6}$ in (A.13) contain one primed and one unprimed subscript. On the other hand, the triple products of δ-tensors behind the factor $\frac{1}{15}$ contain one unit tensor with two primed subscripts, one with two unprimed ones, and one with mixed subscripts.

The contraction $\lambda' = \lambda$ yields $\Delta^{(3)}_{\mu\nu\lambda,\mu'\nu'\lambda} = \frac{7}{5}\Delta^{(2)}_{\mu\nu,\mu'\nu'}$. This is in accord with (11.6).

Exercises Chapter 12

12.1 Verify the Numerical Factor in (12.7) for the Integral over a Triple Product of Tensors (p. 202)

Hint: Put $\nu = \lambda$, $\kappa = \sigma$, $\tau = \mu$ and use the relevant formulae given in Sect. 11.4. The recommended contraction, on the left hand side of (12.7) involves

$$\overline{\widehat{r_\mu r_\nu}}\;\overline{\widehat{r_\nu r_\kappa}}\;\overline{\widehat{r_\kappa r_\mu}} = \frac{1}{3}\overline{\widehat{r_\mu r_\kappa}}\;\overline{\widehat{r_\kappa r_\mu}} = \frac{2}{9},$$

and the subsequent integration still yields $\frac{2}{9}$.

On the other hand, due to $\Delta_{\mu\nu,\nu\kappa,\kappa\mu} = \frac{35}{12}$, cf. (11.38), the right hand side becomes $\frac{8}{105}\frac{35}{12} = \frac{2}{9}$, just as expected.

12.2 Prove that the Fokker-Planck Equation Implies an Increase of the Orientational Entropy with Increasing Time (p. 212)

Hint: The time change of an orientational average is $d\langle\psi\rangle/dt = \int \partial(\psi f)/\partial t d^2u$. The time change of $f\ln(f/f_0)$ is $\ln(f/f_0)\partial f/\partial t + ff^{-1}\partial f/\partial t$. Thus the time change of the orientational entropy $s_a = -k_B \int f\ln(f/f_0)d^2u$ is

$$\frac{d}{dt}s_a = -k_B \int \ln(f/f_0)\frac{\partial}{\partial t}f d^2u = -k_B\nu_0 \int \ln(f/f_0)\,\mathcal{L}_\mu\mathcal{L}_\mu f d^2u,$$

where $\int \frac{\partial}{\partial t}f d^2u = 0$ and (12.41) have been used. Here $\mathcal{L}_\mu = \varepsilon_{\mu\nu\lambda}u_\nu\frac{\partial}{\partial u_\lambda}$ is the relevant differential operator. An integration by part leads to the expression

$$\frac{d}{dt}s_a = k_B\,\nu_0 \int f^{-1}(\mathcal{L}_\mu f)\,\mathcal{L}_\mu f d^2u,$$

where the integrand is positive, furthermore $v_0 > 0$, and thus

$$\frac{d}{dt} s_a \geq 0$$

holds true.

12.3 Second Order Contributions of the Kinetic Heat Flux and Friction Pressure Tensor to the Entropy (p. 221)

The 'non-equilibrium' entropy, per particle, associated with the velocity distribution function $f = f_M(1 + \Phi)$, is given by $s = -k_B \langle \ln(f/f_M) \rangle = -k_B \langle (1 + \Phi) \ln(1 + \Phi) \rangle_M$, where f_M is the local Maxwell distribution and Φ is the deviation of f from f_M. By analogy with (12.39), the contribution up to second order in rhe deviation is $s = -k_B \frac{1}{2} \langle \Phi^2 \rangle_M$.

Determine the second order contributions to the entropy associated with heat flux and the symmetric traceless pressure tensor.

The quantity Φ to be used here is, cf. (12.96),

$$\Phi = \langle \phi_\mu \rangle \, \phi_\mu + \langle \phi_{\mu\nu} \rangle \, \phi_{\mu\nu},$$

where ϕ_μ and $\phi_{\mu\nu}$ are the expansion tensors pertaining to the kinetic contributions to the heat flux and the friction pressure tensor, see (12.93) and (12.94). Due to the orthogonality of the expansion tensors, the expression for the entropy becomes $s = -k_B \frac{1}{2} (\langle \phi_\mu \rangle \langle \phi_\mu \rangle + \langle \phi_{\mu\nu} \rangle \langle \phi_{\mu\nu} \rangle)$. The dimensionless moments are related to the heat flux vector q_μ^{kin} and the pressure tensor $\overline{p_{\mu\nu}^{kin}}$ via (12.93) and (12.94). In terms of these quantities, the desired contribution to the entropy is

$$s = -k_B \frac{1}{2} (nk_B T)^{-2} \left[\frac{2}{5} \frac{m}{k_B T} q_\mu^{kin} q_\mu^{kin} + \frac{1}{2} \overline{p_{\mu\nu}^{kin}} \, \overline{p_{\mu\nu}^{kin}} \right].$$

Compared with thermal equilibrium, the non-equilibrium state has a smaller entropy and thus a higher order.

12.4 Pair Correlation Distorted by a Plane Couette Flow (p. 230)

Compute the functions g_+, g_- and g_0 in first and second order in the shear rate γ, in steady state, from the plane Couette version of the kinetic equation (12.120)

$$\gamma y \frac{\partial}{\partial x} \delta g + \tau^{-1} \delta g = -\gamma y \frac{\partial}{\partial x} g_{eq}.$$

Hint: Use $y^2 = (x^2 + y^2)/2 - (x^2 - y^2)/2$ and $x^2 + y^2 = r^2 - z^2 = 2r^2/3 - (z^2 - r^2/3)$, furthermore decompose $x^2 y^2 = e_\mu^x e_\nu^x e_\lambda^y e_\kappa^y r_\mu r_\nu r_\lambda r_\kappa$ into its parts associated with tensors of ranks $\ell = 0, 2, 4$ with the help of (9.6). Compare g_- with g_+.

The first order result is

$$\delta g^{(1)} = -\gamma \tau \, x y \, r^{-1} g'_{eq},$$

where the prime indicates the differentiation with respect to r. This corresponds to

$$g_+ = -\gamma \tau r g'_{eq},$$

see (12.128).

The second order contribution is given by

$$\delta g^{(2)} = -\gamma \tau y \frac{\partial}{\partial x} \delta g^{(1)} = (\gamma \tau)^2 \left[x^2 y^2 r^{-1} (r^{-1} g'_{eq})' + y^2 r^{-1} g'_{eq} \right].$$

With the help of (9.6),

$$x^2 y^2 = e^x_\mu e^x_\nu e^y_\lambda e^y_\kappa \overline{r_\mu r_\nu r_\lambda r_\kappa} + \frac{1}{7} r^2 (x^2 + y^2) - \frac{1}{35} r^4$$

is obtained, where the first term involves irreducible tensors of rank 4. The resulting second rank contributions proportional to $x^2 - y^2$ and $z^2 - r^2/3$, cf. (12.128), are

$$g_- = -(\gamma \tau)^2 r g'_{eq},$$

$$g_0 = -(\gamma \tau)^2 \left[\frac{1}{2} r g'_{eq} + \frac{1}{7} r^3 (r^{-1} g'_{eq})' \right] = -(\gamma \tau)^2 \frac{1}{7} \left[\frac{5}{2} r g'_{eq} + r^2 g''_{eq} \right],$$

and one has $g_- = \gamma \tau g_+$.

12.5 Compute the Vector and Tensor Polarization for a $\ell = 1$ State (p. 237)

Hint: use the wave function (12.141) with $e_\mu = e^x_\mu$ and $e_\mu = (e^x_\mu + i e^y_\mu)/\sqrt{2}$ for the linear and circular polarized cases. For the angular momentum operator and its properties see Sect. 7.6.2. Furthermore, notice, the term polarization is used here with two distinct, although related meanings. In connection with electric field, "polarization" indicates the direction of the field. In connection with angular momenta and spins, this term refers to their average orientation.

Application of the angular momentum operator $L_\mu = -i\mathcal{L}_\mu$ on the wave function $\Phi_1 = e_\nu \phi_\nu$ with $\phi_\nu = \sqrt{3} \hat{r}_\nu$ yields

$$L_\mu \Phi_1 = -i \varepsilon_{\mu\kappa\tau} \hat{r}_\kappa \frac{\partial}{\partial \hat{r}_\tau} \phi_\lambda e_\lambda = -i \varepsilon_{\mu\kappa\lambda} \phi_\kappa e_\lambda.$$

Multiplication by Φ_1^* and subsequent integration over $d^2\hat{r}$ leads to

$$\langle L_\mu \rangle = -i \varepsilon_{\mu\kappa\lambda} e^*_\kappa e_\lambda.$$

Clearly, one has $\langle L_\mu \rangle = 0$ when $e^*_\kappa = e_\kappa$, as in the case of linear polarization. On the other hand, for the circular polarization, the unit vector is given by $e_\nu = (e^x_\lambda + i e^y_\lambda)/\sqrt{2}$ and $e^*_\kappa = (e^x_\kappa - i e^y_\kappa)/\sqrt{2}$. Thus one has $\langle L_\mu \rangle = -i\varepsilon_{\mu\kappa\lambda}(-i e^y_\kappa e^x_\lambda + i e^x_\kappa e^y_\lambda)/2 = \varepsilon_{\mu\kappa\lambda} e^x_\kappa e^y_\lambda = e^z_\mu$.

For circular polarized light with its electric field vector parallel to $\mathbf{e} = \mathbf{e}^x \pm i\mathbf{e}^y$, the vector polarization is

$$\langle \mathbf{L} \rangle = \pm \mathbf{e}^z.$$

Application of L_ν on $L_\mu \Phi_1 = -i\varepsilon_{\mu\kappa\lambda}\phi_\kappa e_\lambda$ yields $L_\nu L_\mu \Phi_1 = -\varepsilon_{\nu\alpha\beta}\widehat{r}_\alpha \frac{\partial}{\partial r_\beta}\varepsilon_{\mu\kappa\lambda}\phi_\kappa$
$e_\lambda = -\varepsilon_{\nu\alpha\beta}\varepsilon_{\mu\beta\lambda}\phi_\alpha e_\lambda = \delta_{\nu\mu}\phi_\lambda e_\lambda - \phi_\mu e_\nu$. Now multiplication by Φ_1^* and subsequent integration over $d^2\widehat{r}$ leads to

$$\langle L_\nu L_\mu \rangle = \delta_{\nu\mu} e_\lambda^* e_\lambda - e_\mu^* e_\nu.$$

Notice that $e_\lambda^* e_\lambda = 1$ and $\langle L_\nu L_\nu \rangle = 2$, in accord with $\ell(\ell+1)$ for $\ell = 1$. The symmetric traceless part of $\langle L_\nu L_\mu \rangle$ is the tensor polarization

$$\langle \overline{L_\mu L_\nu} \rangle = -\overline{e_\mu^* e_\nu}.$$

Thus for the linear polarization $e_\nu = e_\nu^x$, and for the circular polarization one finds

$$\langle \overline{L_\mu L_\nu} \rangle = -\overline{e_\mu^x e_\nu^x},$$

and

$$\langle \overline{L_\mu L_\nu} \rangle = -\frac{1}{2}\left(\overline{e_\mu^x e_\nu^x} + \overline{e_\mu^y e_\nu^y}\right) = \frac{1}{2}\overline{e_\mu^z e_\nu^z}.$$

Exercises Chapter 13

13.1 Verify the Normalization for the Spin 1 Matrices (p. 241)
Compute explicitly $s_x^2 + s_y^2 + s_z^2$ for the spin matrices (13.6) in order to check the normalization relation (13.4), viz. $\mathbf{s}\cdot\mathbf{s} = s(s+1)\mathbf{1}$.

Spin 1. Matrix multiplication yields

$$s_x s_x = \frac{1}{2}\begin{pmatrix}1&0&1\\0&2&0\\1&0&1\end{pmatrix}, \quad s_y s_y = \frac{1}{2}\begin{pmatrix}1&0&-1\\0&2&0\\-1&0&1\end{pmatrix}, \quad s_z s_z = \begin{pmatrix}1&0&0\\0&0&0\\0&0&1\end{pmatrix}, \quad (A.14)$$

thus

$$s_x^2 + s_y^2 + s_z^2 = 2\begin{pmatrix}1&0&0\\0&1&0\\0&0&1\end{pmatrix}, \tag{A.15}$$

in accord with (13.4), for $s = 1$.

13.2 Verify a Relation Peculiar for Spin $1/2$ (p. 243)

For spin $s = 1/2$, the peculiar relation

$$s_\mu s_\nu = \frac{i}{2}\varepsilon_{\mu\nu\lambda}\, s_\lambda + \frac{1}{4}\delta_{\mu\nu}$$

holds true. To prove it, start from $\overline{s_\mu s_\nu} = 0$, *for* $s = 1/2$, *and use the commutation relation.*

For spin $s = 1/2$, one has, by definition,

$$2\,\overline{s_\mu s_\nu} = s_\mu s_\nu + s_\nu s_\mu - \frac{1}{2}\delta_{\mu\nu}.$$

The general commutation relation (13.1) implies $s_\nu s_\mu = s_\mu s_\nu - i\varepsilon_{\mu\mu\lambda}s_\lambda$. Then, from $\overline{s_\mu s_\nu} = 0$, the desired relation for spin $1/2$ is obtained.

Exercise Chapter 14

14.1 Scalar Product of two Rotated Vectors (p. 262)

Let $\tilde{a}_\mu = R_{\mu\nu}(\varphi)a_\nu$ *and* $\tilde{b}_\mu = R_{\mu\kappa}(\varphi)a_\kappa$ *be the cartesian components of the vectors* **a** *and* **b** *which have been rotated by the same angle φ about the same axis. Prove that the scalar products* $\tilde{\mathbf{a}} \cdot \tilde{\mathbf{b}}$ *is equal to* $\mathbf{a} \cdot \mathbf{b}$.

Application of (14.15) leads to

$$\tilde{a}_\mu \tilde{b}_\mu = \sum_{m=-1}^{1}\sum_{m'=-1}^{1} \exp[i\,m\,\varphi]\exp[i\,m'\,\varphi]P_{\mu\nu}^{(m)}P_{\mu\kappa}^{(m')}a_\nu b_\kappa.$$

Due to $P_{\mu\nu}^{(m)} = P_{\nu\mu}^{(-m)}$, cf. (14.9), and with the help of the orthogonality relation $P_{\nu\mu}^{(-m)}P_{\mu\kappa}^{(m')} = \delta^{-mm'}P_{\nu\kappa}^{(m')}$ the exponential functions cancel each other. The completeness relation $\sum_m P_{\nu\kappa}^{(m)} = \delta_{\nu\kappa}$, cf. (14.10), then implies $\tilde{a}_\mu \tilde{b}_\mu = a_\nu b_\nu$, as expected.

Exercises Chapter 15

15.1 Derivation of the Landau-de Gennes Potential (p. 282)

In general, the free energy F is related to the internal energy U and the entropy S by $F = U - TS$. Thus the contributions to these thermodynamic functions which are associated with the alignment obey the relation

$$\mathscr{F}_a = \mathscr{U}_a - T\mathscr{S}_a.$$

Assume that the relevant internal energy is equal to

$$\mathscr{U}_a = -N\frac{1}{2}\varepsilon\, a_{\mu\nu}a_{\mu\nu},$$

where $\varepsilon > 0$ is a characteristic energy, per particle, associated with the alignment. It is related to the temperature T^* by $k_B T^* = \varepsilon/A_0$. Furthermore, approximate the entropy by the single particle contribution

$$\mathscr{S}_a = -N\,k_B\,\langle\ln(f/f_0)\rangle_0,$$

cf. Sect. 12.2.6 where the entropy per particle s_a was considered, Notice that $\mathscr{S}_a = N s_a$. Use $f = f_0(1 + a_{\mu\nu}\phi_{\mu\nu})$ and (12.39) to compute the entropy and consequently the free energy up to fourth order in the alignment tensor. Compare with the expression (15.12) to infer A_0, B, C. Finally, use these values to calculate a_{ni} and $\delta = (T_{ni} - T^*)/T_{ni}$, cf. (15.17) and (15.18).

Due to (12.39), the alignment entropy per particle is given by

$$s_a = -k_B\left(\frac{1}{2}\langle\Phi^2\rangle_0 - \frac{1}{6}\langle\Phi^3\rangle_0 + \frac{1}{12}\langle\Phi^4\rangle_0 \pm \ldots\right),$$

where here $\Phi = a_{\mu\nu}\phi_{\mu\nu}$ is used. The normalization of $\phi_{\mu\nu}$ implies that the second order contribution is $\frac{1}{2}a_{\mu\nu}a_{\mu\nu}$. The third order term involves the triple product $a_{\mu\nu}a_{\mu'\nu'}a_{\mu''\nu''}\langle\phi_{\mu\nu}\phi_{\mu'\nu'}\phi_{\mu''\nu''}\rangle_0$. Due to the integral relation (12.7), this expression becomes $\frac{2}{7}\sqrt{30}a_{\mu\nu}a_{\nu\kappa}a_{\kappa\mu}$ and the third order contribution is $-\frac{1}{21}\sqrt{30}a_{\mu\nu}a_{\nu\kappa}a_{\kappa\mu}$.

The fourth order term involves the quadruple product

$$a_{\mu_1\nu_1}a_{\mu_2\nu_2}a_{\mu_3\nu_3}a_{\mu_4\nu_4}\langle\phi_{\mu_1\nu_1}\phi_{\mu_2\nu_2'}\phi_{\mu_3\nu_3}\phi_{\mu_4\nu_4}\rangle_0.$$

Due to the integral relation (12.9) and the use of (12.10), this expression becomes $\frac{15}{7}(a_{\mu\nu}a_{\nu\mu})^2$ and the fourth order contribution is $\frac{5}{28}(a_{\mu\nu}a_{\nu\mu})^2$.

As a consequence, the Landau-de Gennes Φ^{LdG}, defined via $F_a = U_a - T S_a = N k_B T \Phi^{LdG}$ is given by

$$\Phi^{LdG} = \frac{1}{2}A a_{\mu\nu}a_{\mu\nu} - \frac{1}{3}\sqrt{6}B a_{\mu\nu}a_{\nu\kappa}a_{\kappa\mu} + \frac{1}{4}C(a_{\mu\nu}a_{\mu\nu})^2,$$

with

$$A = A_0(1 - T^*/T), \quad A_0 = 1, \quad k_B T^* = \varepsilon, \quad B = \frac{\sqrt{5}}{7}, \quad C = \frac{5}{7}.$$

These specific values, based on the single particle contribution only for the entropy, yield for the order parameter $a_{ni} = 2B/(3C)$ at the transition temperature T_{ni} the value $\frac{2}{15}\sqrt{5}$, corresponding to a Maier-Saupe order parameter of $\frac{2}{15} \approx 0.133$. The quantity $\delta = (T_{ni} - T^*)/T_{ni}$, as given in (15.18), assumes the value $\frac{2}{63} \approx 0.032$.

15.2 Compute the Cubic Order Parameter $\langle H_4 \rangle$ for Systems with Simple Cubic, bcc and fcc Symmetry (p. 294)

Hint: The coordinates of one the nearest neighbors, in first coordination shells, are $(1, 0, 0)$ for simple cubic, $(1, 1, 1)/\sqrt{3}$ for bcc and $(1, 1, 0)/\sqrt{2}$ for fcc. Use symmetry arguments!

By definition, one has $H_4 = u_1^4 + u_2^4 + u_3^4 - \frac{3}{5}$, where the u_i are the Cartesian components of one of the nearest neighbors. Due to the symmetry of the first coordination shell in these cubic systems, all nearest neighbors give the same contribution in the evaluation of the average $\langle H_4 \rangle$, thus it suffices to consider one of them. The result then is

$$\langle H_4 \rangle = \frac{2}{5}, \quad -\frac{4}{15}, \quad -\frac{1}{10},$$

for simple cubic, bcc, and fcc symmetry, respectively.

15.3 Renormalization of Landau-de Gennes Coefficients (p. 296)

Consider the special case where $\Phi^b = \frac{1}{2} A_b b_{\mu\nu} b_{\mu\nu}$, for simplicity put $A_b = 1$. Determine $b_{\mu\nu}$ from $\frac{\partial \Phi}{\partial b_{\mu\nu}} = 0$ with the help of the second equation of (15.52). Insert this expression into the first equation of (15.52) to obtain a derivative of a Landau-de Gennes potential with coefficients A, B, C which differ from the original coefficients A_a, B_a, C_a due to the coupling between the tensors.

Hint: use relation (5.51) for a, viz. $\overline{a_{\mu\kappa} a_{\kappa\lambda} a_{\lambda\nu}} = \frac{1}{2} a_{\mu\nu} a_{\lambda\kappa} a_{\lambda\kappa}$.

With the assumptions made here, $\frac{\partial \Phi}{\partial b_{\mu\nu}} = 0$, used for the second equation of (15.52), implies

$$b_{\mu\nu} = - \left(c_1 a_{\mu\nu} + c_2 \sqrt{6} \, \overline{a_{\mu\kappa} a_{\kappa\nu}} \right).$$

Insertion into the first equation of (15.52) yields

$$\frac{\partial \Phi}{\partial a_{\mu\nu}} = \Phi_{\mu\nu}^a - \left(c_1^2 a_{\mu\nu} + 3\sqrt{6} c_1 c_2 \, \overline{a_{\mu\lambda} a_{\lambda\nu}} + 12 c_2^2 \, \overline{a_{\mu\kappa} \, \overline{a_{\kappa\lambda} a_{\lambda\nu}}} \right).$$

Due to

$$\overline{a_{\kappa\lambda} a_{\lambda\nu}} = a_{\kappa\lambda} a_{\lambda\nu} - \frac{1}{3} \delta_{\kappa\nu} a_{\alpha\beta} a_{\alpha\beta},$$

the triple product of the tensors in the last term is

$$\overline{a_{\mu\kappa} \, \overline{a_{\kappa\lambda} a_{\lambda\nu}}} = a_{\mu\kappa} a_{\kappa\lambda} a_{\lambda\nu} - \frac{1}{3} a_{\mu\nu} a_{\alpha\beta} a_{\alpha\beta}$$

$$= \left(\frac{1}{2} - \frac{1}{3} \right) a_{\mu\nu} a_{\alpha\beta} a_{\alpha\beta} = \frac{1}{6} a_{\mu\nu} a_{\lambda\kappa} a_{\lambda\kappa}.$$

Since $\Phi_{\mu\nu}^{a} = A_{a}a_{\mu\nu} - \sqrt{6}B_{a}\,\overline{a_{\mu\lambda}a_{\lambda\nu}} + C_{a}a_{\mu\nu}a_{\lambda\kappa}a_{\lambda\kappa}$,

$$\frac{\partial\Phi}{\partial a_{\mu\nu}} = Aa_{\mu\nu} - \sqrt{6}B\,\overline{a_{\mu\lambda}a_{\lambda\nu}} + Ca_{\mu\nu}a_{\lambda\kappa}a_{\lambda\kappa}$$

is obtained, where

$$A = A_{a} - c_{1}^{2}, \quad B = B_{a} + 3c_{1}c_{2}, \quad C = C_{a} - 2c_{1}^{2},$$

in accord with (15.53).

15.4 Flexo-electric Coefficients (p. 297)

Start from equation (15.56) for the vector d_{μ}, use $a_{\mu\nu} = \sqrt{\frac{3}{2}}a_{eq}\,\overline{n_{\mu}n_{\nu}}$ and $P_{\mu} = P^{ref}d_{\mu}$ in order to derive an expression of the form (15.57) and express the flexo-electric coefficients e_{1} and e_{3} to c_{1}, c_{2} and $a_{eq} = \sqrt{5}S$, where S is the Maier-Saupe order parameter. Furthermore, compute the contribution to the electric polarization which is proportional to the spatial derivative of $a_{eq} = \sqrt{5}S$.

Hint: treat the components of **P** *parallel and perpendicular to* **n** *separately.*
With the recommended ansatz for $a_{\mu\nu}$, the right hand side of (15.56) becomes

$$c_{1}\nabla_{\nu}a_{\nu\mu} = \sqrt{\frac{3}{2}}c_{1}\left[a_{eq}n_{\mu}\nabla_{\nu}n_{\nu} + a_{eq}n_{\nu}\nabla_{\nu}n_{\mu} + \overline{n_{\mu}n_{\nu}}\,\nabla_{\nu}a_{eq}\right].$$

Likewise, the left hand side of (15.56) becomes

$$\left(\delta_{\mu\nu} + \sqrt{\frac{3}{2}}c_{2}a_{eq}\,\overline{n_{\mu}n_{\nu}}\right)d_{\nu} = \left(1 + \sqrt{\frac{2}{3}}c_{2}a_{eq}\right)d_{\mu}^{\parallel} + \left(1 - \sqrt{\frac{1}{6}}c_{2}a_{eq}\right)d_{\mu}^{\perp},$$

where $d_{\mu}^{\parallel} = n_{\mu}n_{\nu}d_{\nu}$ and $d_{\mu}^{\perp} = d_{\mu} - d_{\mu}^{\parallel}$ are the components parallel and perpendicular to the director. The solution of (15.56) for **d** is

$$d_{\mu}^{\parallel} = \left(1 + \sqrt{\frac{2}{3}}c_{2}a_{eq}\right)^{-1}\sqrt{\frac{3}{2}}c_{1}n_{\mu}\left(a_{eq}\nabla_{\nu}n_{\nu} + \frac{2}{3}n_{\nu}\nabla_{\nu}a_{eq}\right),$$

$$d_{\mu}^{\perp} = \left(1 - \sqrt{\frac{1}{6}}c_{2}a_{eq}\right)^{-1}\sqrt{\frac{3}{2}}c_{1}\left(a_{eq}n_{\nu}\nabla_{\nu}n_{\mu} - \frac{1}{3}\nabla_{\mu}^{\perp}a_{eq}\right).$$

The resulting flexo-electric coefficients are

$$e_1 = P^{\text{ref}} \left(1 + \sqrt{\frac{2}{3}} c_2 a_{\text{eq}} \right)^{-1} \sqrt{\frac{3}{2}} c_1 a_{\text{eq}},$$

$$e_3 = P^{\text{ref}} \left(1 - \sqrt{\frac{1}{6}} c_2 a_{\text{eq}} \right)^{-1} \sqrt{\frac{3}{2}} c_1 a_{\text{eq}}.$$

For $c_2 = 0$ one has $e_1 = e_3$.

Exercises Chapter 16

16.1 Nonlinear Electric Susceptibility in a Polar Material (p. 302)
In a medium without hysteresis, the electric polarization **P** *can be expanded in powers of the electric field* **E**, *cf.* (2.59), *thus*

$$P_\mu = \varepsilon_0 \left(\chi^{(1)}_{\mu\nu} E_\nu + \chi^{(2)}_{\mu\nu\lambda} E_\nu E_\lambda + \dots \right).$$

The second rank tensor $\chi^{(1)}_{\mu\nu} \equiv \chi_{\mu\nu}$ *characterizes the linear susceptibility. The third rank tensor* $\chi^{(2)}_{\mu\nu\lambda}$ *describes the next higher order contributions to* **P**. *Consider a material whose isotropy is broken by a polar unit vector* **d**. *Formulate the expressions for these tensors which are in accord with the symmetry and with parity conservation. Treat the cases* **E** *parallel and perpendicular to* **d**.

To conserve parity, even rank tensors must contain even powers of **d**, the third rank tensor must be an odd function of **d**. The ansatz which fulfills the required properties is

$$\chi^{(1)}_{\mu\nu} = \chi^{10} \delta_{\mu\nu} + \chi^{12} \overline{d_\mu d_\nu},$$

$$\chi^{(2)}_{\mu\nu\lambda} = \chi^{20} d_\mu \delta_{\nu\lambda} + \chi^{21} \frac{1}{2} (\delta_{\mu\nu} d_\lambda + \delta_{\mu\lambda} d_\nu) + \chi^{23} \overline{d_\mu d_\nu d_\lambda},$$

with scalar coefficients $\chi^{10}, \dots, \chi^{23}$. The resulting expression for $\mathbf{P} = \mathbf{P}^{(1)} + \mathbf{P}^{(2)} + \dots$ is

$$P^{(1)}_\mu = \chi^{10} E_\mu + \chi^{12} \overline{d_\mu d_\nu} \, E_\nu,$$

$$P^{(2)}_\mu = \chi^{20} d_\mu E^2 + \chi^{21} E_\mu d_\nu E_\nu + \chi^{23} \overline{d_\mu d_\nu d_\lambda} \, E_\nu E_\lambda.$$

For $\mathbf{E} \parallel \mathbf{d}$ one has $\mathbf{P}^{(1)} = (\chi^{10} + \frac{2}{3} \chi^{12}) \mathbf{E}$ and $\mathbf{P}^{(2)} = (\chi^{20} + \chi^{21} + \frac{2}{5} \chi^{23}) E^2 \mathbf{d}$. Similarly, the result for $\mathbf{E} \perp \mathbf{d}$ is $\mathbf{P}^{(1)} = (\chi^{10} - \frac{1}{3} \chi^{12}) \mathbf{E}$ and $\mathbf{P}^{(2)} = (\chi^{20} - \frac{1}{5} \chi^{23}) E^2 \mathbf{d}$. Here

$$\overline{d_\mu d_\nu d_\lambda}\, E_\nu E_\lambda = d_\mu (\mathbf{d} \cdot \mathbf{E})^2 - \frac{1}{5}\left[d_\mu E^2 + 2E_\mu (\mathbf{d} \cdot \mathbf{E}) \right]$$

was used, cf. (9.5).

16.2 Acoustic Birefringence (p. 326)

Sound waves cause an alignment of non-spherical particles in fluids. The ensuing birefringence is called acoustic birefringence. *Use (16.74) to compute the sound-induced alignment tensor for the velocity field* $\mathbf{v} = v_0 k^{-1}\mathbf{k}\cos(\mathbf{k}\cdot\mathbf{r} - \omega t)$ *where* $\mathbf{k}$ *and* ω *are the wave vector and the frequency of the sound wave,* v_0 *is the amplitude. Hint: Use the complex notation* $v_\mu \sim \exp[i(\mathbf{k}\cdot\mathbf{r} - \omega t)]$ *and* $a_{\mu\nu} = \tilde{a}_{\mu\nu}\exp[i(\mathbf{k}\cdot\mathbf{r} - \omega t)]$ *to solve the inhomogeneous relaxation equation. Discuss the meaning of the real and imaginary parts of* $\tilde{a}_{\mu\nu}$.

With the recommended ansatz, (16.74) yields

$$(-i\omega\tau_a + 1)\tilde{a}_{\mu\nu} = -i\sqrt{2}\tau_{ap}v_0 k^{-1}\overline{k_\mu k_\nu}\,.$$

Consequently, the alignment is determined by

$$\tilde{a}_{\mu\nu} = -i\sqrt{2}\tau_{ap}v_0 k\,\widehat{\overline{k_\mu k_\nu}}\,\mathscr{A}(\omega),$$

where

$$\mathscr{A}(\omega) = \frac{1}{1 - i\omega\tau_a} = \frac{1 + i\omega\tau_a}{1 + (\omega\tau_a)^2}.$$

The real part of $\mathscr{A}(\omega)$, corresponding to the imaginary part of $\tilde{a}_{\mu\nu}$ describes the part of the alignment which is in phase with the velocity gradient, i.e. it is proportional to $\sin(\mathbf{k}\cdot\mathbf{r} - \omega t)$. The imaginary part of $\mathscr{A}(\omega)$ is linked with the part of the alignment which is in phase with the velocity, i.e. it is proportional to $\cos(\mathbf{k}\cdot\mathbf{r} - \omega t)$.

16.3 Diffusion of Perfectly Oriented Ellipsoids (p. 337)

In the nematic phase, the flux $\mathbf{j}$ *of diffusing particles with number density* ρ *obeys the equation*

$$j_\mu = -D_{\mu\nu}\nabla_\nu\rho, \quad D_{\mu\nu} = D_\parallel n_\mu n_\nu + D_\perp(\delta_{\mu\nu} - n_\mu n_\nu),$$

where $D_\parallel$ and $D_\perp$ are the diffusion coefficients for the flux parallel and perpendicular to the director $\mathbf{n}$.

Use the volume conserving affine transformation model for uniaxial particles, cf. Sect. 5.7.2, to derive

$$D_\parallel = Q^{4/3}D_0, \quad D_\perp = Q^{-2/3}D_0$$

for perfectly oriented ellipsoidal particles with axes ratio Q. *Here* D_0 *is the reference diffusion coefficient of spherical particles.*

Furthermore, determine the anisotropy ratio $R = (D_\parallel - D_\perp)/(D_\parallel + 2D_\perp)$, *the average diffusion coefficient* $\bar{D} = \frac{1}{3}D_\parallel + \frac{2}{3}D_\perp$ *and the geometric mean* $\tilde{D} = D_\parallel^{1/3} D_\perp^{2/3}$. *Discuss the cases* $Q > 1$ *and* $Q < 1$ *for prolate and oblate particles.*

The flux **j** transforms like the position vector, viz. $j_\mu = A_{\mu\nu}^{-1/2} j_\nu^A$. The spatial gradient in real space is linked with the nabla operator in affine space according to $\nabla_\nu = A_{\nu\kappa}^{1/2} \nabla_\kappa^A$.

From the diffusion law in the affine space, $j_\lambda^A = -D_{\lambda\kappa}^A \nabla_\nu^A \rho$, where $D_{\lambda\kappa}^A$ is the diffusion tensor in that space, follows

$$ j_\mu = A_{\mu\lambda}^{-1/2} j_\lambda^A = -A_{\mu\lambda}^{-1/2} D_{\lambda\kappa}^A \nabla_\nu^A \rho = -A_{\mu\lambda}^{-1/2} D_{\lambda\kappa}^A A_{\kappa\nu}^{-1/2} \nabla_\nu \rho. $$

Comparison with the diffusion law in real space and $D_{\lambda\kappa}^A = D_0 \delta_{\lambda\kappa}$ implies

$$ D_{\mu\nu} = A_{\mu\lambda}^{-1/2} D_{\lambda\kappa}^A A_{\kappa\nu}^{-1/2} = A_{\mu\nu}^{-1} D_0. $$

For uniaxial particles with their symmetry axis parallel to the director, one has

$$ A_{\mu\nu}^{-1} = Q^{-2/3} \left[\delta_{\mu\nu} + (Q^2 - 1)n_\mu n_\nu \right], $$

cf. (5.58). Here $Q = a/b$ is the axes ratio of an ellipsoid with the semi-axes a and $b = c$. Thus the above mentioned result $D_\parallel = Q^{4/3} D_0$, $D_\perp = Q^{-2/3} D_0$ is obtained. This implies

$$ R = (D_\parallel - D_\perp)/(D_\parallel + 2D_\perp) = (Q^2 - 1)/(Q^2 + 2), $$

and

$$ \bar{D} = \frac{1}{3}D_\parallel + \frac{2}{3}D_\perp = \frac{1}{3}Q^{-2/3}(Q^2 + 2)D_0, $$

for the ansotropy ratio R and for the average diffusion coefficient $\bar{D}$. The geometric mean diffusion coefficient $\tilde{D} = D_\parallel^{1/3} D_\perp^{2/3} = D_0$ is independent of the axes ratio Q.

As expected intuitively, prolate particles with $Q > 1$ diffuse faster in the direction parallel to **n**, since $D_\parallel > D_\perp$, in this case. For prolate particles, pertaining to $Q < 1$, one has $D_\parallel < D_\perp$. There the diffusion along **n** is slower compared with the diffusion in a direction perpendicular to **n**. The anisotropy ratio R is positive for prolate and negative for oblate particles.

Exercises Chapter 17

17.1 Components of a Uniaxial Alignment (p. 359)

Consider a uniaxial alignment given by $a_{\mu\nu} = \sqrt{3/2}a\,\overline{n_\mu n_\nu}$. Determine the components a_i in terms of the polar coordinates ϑ and φ. Use $n_x = \sin\vartheta\cos\varphi$, $n_y = \sin\vartheta\sin\varphi$, $n_z = \cos\vartheta$. Consider the special cases $\vartheta = 0, 45, 90°$ and $\cos^2\vartheta = 1/3$.

By definition, the components are $a_i = T^i_{\mu\nu}a_{\mu\nu}$. Thus, due to (17.20), one finds

$$a_0 = a\frac{3}{2}(\cos^2\vartheta - \frac{1}{3}), \quad a_1 = \frac{1}{2}\sqrt{3}a\sin^2\vartheta\cos(2\varphi), \quad a_2 = \frac{1}{2}\sqrt{3}a\sin^2\vartheta\sin(2\varphi),$$

$$a_3 = \frac{1}{2}\sqrt{3}a\sin(2\vartheta)\cos\varphi, \quad a_4 = \frac{1}{2}\sqrt{3}a\sin(2\vartheta)\sin\varphi.$$

For $\vartheta = 0$, one has $a_0 = a$, $a_1 = a_2 = a_3 = a_4 = 0$, as expected. The result for $\vartheta = 90°$ is $a_0 = -a/2$, $a_1 = \frac{1}{2}\sqrt{3}a\cos(2\varphi)$, $a_2 = \frac{1}{2}\sqrt{3}a\sin(2\varphi)$, $a_3 = a_4 = 0$.

For $\vartheta = 45°$, one has $\cos^2\vartheta = \sin^2\vartheta = 1/2$, and $a_0 = a/4$, $a_1 = \frac{1}{4}\sqrt{3}a\cos(2\varphi)$, $a_2 = \frac{1}{4}\sqrt{3}a\sin(2\varphi)$, $a_3 = \frac{1}{2}\sqrt{3}a\cos\varphi$, $a_4 = \frac{1}{2}\sqrt{3}a\sin\varphi$.

For the case $\cos^2\vartheta = 1/3$, corresponding to $\vartheta \approx 55°$, the components are $a_0 = 0$, $a_1 = \frac{1}{3}\sqrt{3}a\cos(2\varphi)$, $a_2 = \frac{1}{3}\sqrt{3}a\sin(2\varphi)$, $a_3 = \frac{1}{3}\sqrt{6}a\cos\varphi$, $a_4 = \frac{1}{3}\sqrt{6}a\sin\varphi$.

17.2 Stability against Biaxial Distortions (p. 362)

Compute the relaxation frequencies $\nu^{(1)}$ and $\nu^{(3)}$ for biaxial distortions $\delta a_{\mu\nu} = T^1_{\mu\nu}\delta a_1$ and $\delta a_{\mu\nu} = T^3_{\mu\nu}\delta a_3$ from the relevant relations given in Sect. 17.2.4.

Solve the full nonlinear relaxation equation for a_3 with $a_1 = a_2 = a_4 = 0$ and $a_0 = a_{eq}$.

The expressions given in Sect. 17.2.4 yield

$$\nu^{(1)} = \vartheta + 6a_{eq} + 2a_{eq}^2 = 9a_{eq}.$$

For all temperatures $\vartheta < 9/8$, this relaxation frequency is positive and the stationary nematic state is stable against biaxial distortions of this type described by δa_1, the same applies for δa_2. On the other hand, one finds

$$\nu^{(3)} = \vartheta - 3a_{eq} + 2a_{eq}^2 = 0.$$

The last equality follows from the equilibrium condition. Thus the nematic state has just marginal stability against biaxial distortions of the type δa_3 and also δa_4. In this case the nonlinear equation for a_3 has to be studied. From (17.28) with (17.29) follows

$$\frac{da_3}{dt} + 2a_3^3 = 0.$$

Separation of variables yields $a_3^{-2}(t) - a_3^{-2}(t_0) = 4(t - t_0)$ and, with $t_0 = 0$,

$$a_3(t) = \frac{a_3(0)}{\sqrt{1 + 4a_3(0)^2 t}}.$$

Still, the distortion relaxes to zero, but it is a non-exponential decay with $a_3(t) \sim t^{-1/2}$ for long times.

Exercises Chapter 18

18.1 Doppler Effect (p. 378)
Let ω_0 be the circular frequency of the electromagnetic radiation in a system which moves with velocity $\mathbf{v} = v\mathbf{e}^x$ with respect to the observer, who records the frequency ω. Determine the Doppler-shift $\delta\omega = \omega_0 - \omega$ for the two cases where the wave vector of the radiation is parallel (longitudinal effect) and perpendicular (transverse effect) to the velocity, respectively.

The Lorentz transformation rule (18.16) implies

$$(k')_x = \gamma(k_x - \beta\omega), \quad (k')_y = k_y, \quad (k')_z = k_z, \quad \omega' = \gamma(\omega - vk_x). \quad \text{(A.16)}$$

When $\mathbf{k}$ is parallel or anti-parallel to the velocity, one has $k_x = \pm\omega/c$ and consequently

$$\omega = \omega_0\sqrt{1 - \beta^2}/(1 \mp \beta).$$

The longitudinal relative Doppler shift is

$$(\omega - \omega_0)/\omega_0 = \sqrt{1 - \beta^2}/(1 \mp \beta) - 1 \approx \pm\beta + \dots, \quad \text{(A.17)}$$

where ... stands for terms of second and higher order in $\beta = v/c$. The $\pm$ signs indicate: the frequency is enhanced when the light source is approaching the observer, when it is moving away, the frequency is lowered.

When the velocity is perpendicular to the direction of observation, one has $k_x = 0$. Then the transverse relative Doppler shift is

$$(\omega - \omega_0)/\omega_0 = \sqrt{1 - \beta^2} \approx \frac{1}{2}\beta^2 + \dots, \quad \text{(A.18)}$$

where now ... stands for terms of fourth and higher order in β.

18.2 Derivation of the Inhomogeneous Maxwell Equations from the Lagrange Density (p. 386)

Point of departure is the variational principle (18.72), viz.

$$\delta \mathscr{S} = \int \delta \mathscr{L} \mathrm{d}^4 x = 0,$$

with the Lagrange density (18.70). It is understood that the variation of $\mathscr{L}$ is brought about by a variation $\delta \Phi$ of the 4-potential. Thus

$$\delta \mathscr{L} = - \left(J^i \delta \Phi_i + \frac{1}{2\mu_0} F^{ik} \delta F_{ik} \right),$$

with

$$\delta F_{ik} = \partial_k \delta \Phi_i - \partial_i \delta \Phi_k,$$

cf. (18.55). Due to $F^{ik} = -F^{ki}$, one has

$$F^{ik} \delta F_{ik} = 2 F^{ik} \partial_k \delta \Phi_i.$$

Integration by parts and $\delta \Phi_i = 0$ at the surface of the 4D integration range, leads to

$$\delta \mathscr{S} = - \int \delta \left(J^i - \frac{1}{\mu_0} \partial_k F^{ik} \right) \delta \Phi_i \mathrm{d}^4 x = 0. \qquad (A.19)$$

The 4D integration volume is arbitrary. Thus the integrand has to vanish and one obtains (18.73), viz.

$$J^i = (\mu_0)^{-1} \partial_k F^{ik}.$$

This relation corresponds to the inhomogeneous Maxwell equation (18.62) for fields in vacuum, where $F^{ik} = \mu_0 H^{ik}$ holds true.

18.3 Derive the 4D Stress Tensor (p. 388)

The force density $f^i = J_k F^{ki}$ cf. (18.74), can be rewritten with help of the inhomogeneous Maxwell equation (18.62). Where necessary, co- and contra-variant components are interchanged with the help of the metric tensor. The first few steps of the calculation are

$$f^i = J_k F^{ki} = g_{k\ell} J^\ell F^{ki} = g_{k\ell} \partial_m H^{\ell m} F^{ki} = g_{k\ell} \partial_m H^{\ell m} g^{kk'} g^{ii'} F_{k'i'}.$$

Due to $g_{k\ell} g^{kk'} = \delta_{\ell k'}$ and with the renaming $i \to k$ of the summation indices, one finds

$$f^i = g^{ik} F_{\ell k} \partial_m H^{\ell m} = \partial_m (g^{ik} F_{\ell k} H^{\ell m}) - g^{ik} H^{\ell m} \partial_m F_{\ell k}.$$

The second term is related to the derivative of the energy density u. This is seen as follows. Renaming the summation indices $\ell m \to m\ell$ and using $H^{\ell m} = -H^{m\ell}$, $F_{mk} = -F_{km}$ leads to $H^{\ell m}\partial_m F_{\ell k} = H^{\ell m}\partial_\ell F_{km}$ and consequently

$$H^{\ell m}\partial_m F_{\ell k} = \frac{1}{2}H^{\ell m}(\partial_m F_{\ell k} + \partial_\ell F_{km}).$$

Due to the homogeneous Maxwell equation (18.58), this expression is equal to

$$H^{\ell m}\partial_m F_{\ell k} = -\frac{1}{2}H^{\ell m}\partial_k F_{m\ell} = \frac{1}{2}H^{\ell m}\partial_k F_{\ell m}.$$

For the special case of a linear medium where $H^{\ell m} \sim F_{\ell m}$ applies, one has

$$H^{\ell m}\partial_m F_{\ell k} = \frac{1}{4}\partial_k(H^{\ell m}F_{\ell m}),$$

and this then leads to the expression (18.77) for the 4D stress tensor.

18.4 Flatlanders Invent the Third Dimension and Formulate their Maxwell Equations (p. 388)

The flatlanders of Exercise 7.3 noticed, they can introduce contra- and contra-variant vectors

$$x^i = (r_1, r_2, ct), \quad x_i = (-r_1, -r_2, ct).$$

With the Einstein summation convention for the three Roman indices, the scalar product of their 3-vectors is

$$x^i x_i = -(r_1^2 + r_2^2) + c^2 t^2 = -\mathbf{r}^2 + c^2 t^2.$$

They use the differential operator

$$\partial_i = \left(\frac{\partial}{\partial r_1}, \frac{\partial}{\partial r_2}, \frac{\partial}{\partial ct}\right),$$

and form the 3-vectors

$$J^I = (j_1, j_2, c\rho), \quad \Phi^i = (A_1, A_2, \phi/c)$$

from their current and charge densities and their vector and scalar potentials.
  How are the relations

$$B = \partial_1 A_2 - \partial_2 A_1, \quad E_i = -\partial_i\phi - \partial A/\partial t, \quad i = 1, 2,$$

which are equivalent to the homogeneous Maxwell equations, cast into the pertaining three-dimensional form? Introduce a three-by-three field tensor F_{ij} and formulate their homogeneous Maxwell equations. How about the inhomogeneous Maxwell equations in flatland?

The antisymmetric field tensor is defined by

$$F_{ik} := \partial_k \Phi_i - \partial_i \Phi_k, \quad i, k = 1, 2, 3.$$

In matrix notation, it reads

$$F_{ik} := \begin{pmatrix} 0 & B & \frac{1}{c}E_1 \\ -B & 0 & \frac{1}{c}E_2 \\ -\frac{1}{c}E_1 & -\frac{1}{c}E_2 & 0 \end{pmatrix}.$$

From the definition of the field tensor follows

$$\partial_1 F_{23} + \partial_2 F_{31} + \partial_3 F_{12} = 0.$$

This Jacobi identity corresponds to

$$\partial_1 E_2 - \partial_2 E_1 = -\frac{\partial B}{\partial t},$$

The two-dimensional **D**-field and the H-field tensor are combined in the tensor

$$H_{ik} := \begin{pmatrix} 0 & H & c\,D_1 \\ -H & 0 & c\,D_2 \\ -c\,D_1 & -c\,D_2 & 0 \end{pmatrix}. \tag{A.20}$$

The inhomogeneous Maxwell equations are equivalent to

$$\partial_k H^{ik} = J^i.$$

The case $i = 3$ corresponds to

$$\partial_1 D_1 + \partial_2 D_2 = \rho.$$

The cases $i = 1$ and $i = 2$ are

$$\partial_2 H = j_1 + \partial D_1/\partial t, \quad -\partial_1 H = j_2 + \partial D_2/\partial t.$$

Also in flatland, constitutive relations are needed to close the Maxwell equations. In vacuum and in a linear medium, the linear relation $H_{ik} \sim F_{ik}$ applies. This corresponds to $D_i \sim E_i$, $i = 1, 2$ and $H \sim B$.

References

1. W. Voigt, Elemente der Krystallphysik. Die fundamentalen Eigenschaften der Krystalle in elementarer Darstellung (Leipzig, 1898); Lehrbuch der Kristallphysik (Mit Ausschluss der Kristalloptik) (Leipzig, 1910); reproduced, with an addition in 1928, ed. M.v. Laue; reprinted by Teubner (Stuttgart, 1966)
2. P. Curie, Sur la symétrie dans les phénomènes physiques. J. Phys. **3**(3), 393–416 (1894)
3. G. Ricci Curbastro, Résumé de quelque travaux sur les systèmes variable de fonctions associés a une forme différentielle quadratique. Bulletin des Sciences Mathématiques **2**(16), 167–189 (1892); G. Ricci, T. Levi-Civita, Méthodes de calcul différentiel absolu et leurs applications. Mathematische Annalen (Springer) **54**, 125–201 (1900)
4. A. Einstein, Die Grundlage der allgemeinen Relativitätstheorie. Ann. Phys. **49**, 769–822 (1916)
5. C.W. Misner, K.S. Thorne, J.A. Wheeler, *Gravitation* (Freeman W.H., San Francisco, 1973)
6. R. Wald, *General Relativity* (University of Chicago Press, Chicago, 1984)
7. H. Stefani, *General Relativity. Translation of Allgemeinen Relativitätstheorie* (Cambridge University Press, Cambridge, 1990)
8. H. Jeffreys, *Cartesian Tensors* (Cambridge University Press, Cambridge, 1931)
9. L. Brillouin, *Les Tenseur en Méquanique et en Elasticité* (Masson, Paris, 1948)
10. A. Duschek, A. Hochrainer, *Tensorrechnung in Analytischer Darstellung*, vols. I-III (Springer, New York, 1955–1961)
11. G. Temple, *Cartesian Tensors* (Methuen, London, 1960)
12. S. Chapman, T.G. Cowling, *The Mathematical Theory of Non-uniform Gases* (Cambridge University Press, London, 1939)
13. L. Waldmann, in *Transporterscheinungen in Gasen von mittlerem Druck*, vol. XII, ed. by S. Flügge. Handbuch der Physik (Springer, Berlin, 1958), pp. 295–514
14. S. Hess, *Vektor- und Tensor-Rechnung* (Palm und Enke, Erlangen, 1980)
15. S. Hess, W. Köhler, *Formeln zur Tensor-Rechnung* (Palm und Enke, Erlangen, 1980)
16. J.R. Coope et al., Irreducible Cartesian Tensors I, II, III. J. Chem. Phys. **43**, 2269 (1965); J. Math. Phys. **11**, 1003; 1591 (1970)
17. F.R.W. McCourt, J.J.M. Beenakker, W.E. Köhler, I. Kuščer, *Nonequilibrium Phenomena in Polyatomic Gases*, vols. 1, 2 (Clarendon, Oxford, 1990)
18. J.W. Gibbs, *Elements of Vector Analysis* (Dover Publications, New York, 1881)
19. H. Ehrentraut, W. Muschik, On symmetric irreducible tensors in d-dimensions. ARI **51**, 149–159 (1998)
20. L. Waldmann, E. Trübenbacher, Formale kinetische Theorie von Gasgemischen aus anregbaren Moleülen. Z. Naturforsch. **17a**, 363–376 (1962)

© Springer International Publishing Switzerland 2015   423
S. Hess, *Tensors for Physics*, Undergraduate Lecture Notes in Physics,
DOI 10.1007/978-3-319-12787-3

21. S. Hess, L. Waldmann, Kinetic theory for a dilute gas of particles with spin. Z. Naturforsch. **21a**, 1529–1546 (1966)
22. S. Hess, On light scattering by polyatomic fluids. Z. Naturforsch. **24a**, 1675–1687 (1969)
23. A. Kilian, S. Hess, Derivation and application of an algorithm for the numerical calculation of the local orientation of nematic liquid crystals. Z. Naturforsch. **44a**, 693–703 (1989); T. Gruhn, S. Hess, Monte Carlo simulation of the director field of a nematic liquid crystal with three elastic coefficients. Z. Naturforsch. **51a**, 1–9 (1996)
24. A.M. Sonnet, S. Hess, in *Alignment Tensor Versus Director Description in Nematic Liquid Crystals*, ed. by O. Lavrentovich, P. Pasini, C. Zannoni, S. Zunmer. Defects in Liquid Crystals: Computer Simulations Theory and Experiments (Kluwer Academic Publishing, Dordrecht 2001), pp. 17–33
25. S. Hess, Decay of BCC-structure and of bond-orientational order in a fluid. Physica **127A**, 509–528 (1984)
26. F.C. von der Lage, H.A. Bethe, A method for obtaining electronic eigenfunctions and eigenvalues in solids with an application to sodium. Phys. Rev. **71**, 612 (1947)
27. A. Hüller, J.W. Kane, Electrostatic model for the rotational potential in ammonium halides. J. Chem. Phys. **61**, 3599 (1974)
28. N. Herdegen, S. Hess, Influence of a shear flow on the phase transition liquid-solid. Physica **138A**, 382–403 (1986)
29. J. Happel, H. Brenner, *Low Reynolds Number Hydrodynamics* (Noordhoff, Leiden, 1973)
30. C.W.J. Beenakker, W. van Saarloos, P. Mazur, Many-sphere hydrodynamic interactions III. The influence of a plane wall. Physica **127A**, 451–472 (1984)
31. J.F. Brady, G. Bossis, Stokesian dynamics. Ann. Rev. FluidMech. **20**, 111–157 (1988)
32. J.K.G. Dhont, *An Introduction to Dynamics of Colloids* (Elsevier Science, Amsterdam, 1996)
33. S. Reddig, H. Stark, Nonlinear dynamics of spherical particles in Poiseuille flow under creeping-flow condition. J. Chem. Phys. **138**, 234902 (2013)
34. C. Gray, K.E. Gubbins, *Theory of Molecular Fluids* (Oxford University Press, Oxford, 1984)
35. T. Erdmann, M. Kröger, S. Hess, Phase behavior and structure of Janus fluids. Phys. Rev. E **67**, 041209 (2003)
36. S. Hess, B. Su, Pressure and isotropic-nematic transition temperature for model liquid crystals. Z. Naturforsch. **54a**, 559–569 (1999); H. Steuer, S. Hess, M. Schoen, Pressure, alignment and phase behavior of a simple model liquid crystal. A Monte Carlo simulation study. Physica A **328**, 322–334 (2003)
37. J.G. Gay, B.J. Berne, Modification of the overlap potential to mimic a linear sits-site potential. J. Chem. Phys. **74**, 3316 (1981); G.R. Luckhurst, P.S.J. Simmonds, Computer simulation studies of anisotropic systems - XXI parametrization of the Gay-Berne potential for model mesogens. Mol. Phys. **80**, 233–252 (1992)
38. S. Hess, Flow velocity and effective viscosity of a fluid containing rigid cylindrical inclusions. Z. Naturforsch. **60a**, 401–407 (2005)
39. M. Ellero, M. Kröger, S. Hess, Viscoelastic flows studied by smoothed particle hydrodynamics. J. Nonnewton. Fluid Mech. **105**, 35–51 (2002)
40. S. Hess, M. Faubel, in *Gase und Molekularstrahlen*, vol. 5, ed. by K. Kleinermanns. Bergmann-Schaefer, Lehrbuch der Experimentalphysik (de Gruyter, Berlin, New York, 2006), p. 1–134
41. W. Loose, S. Hess, Velocity distribution function of a streaming gas via nonequilibrium molecular dynamics. Phys. Rev. Lett. **58**, 2443 (1987); Nonequilibrium velocity distribution function of gases: kinetic theory and molecular dynamics. Phys. Rev. A **37**, 2099 (1988)
42. S. Hess, L. Waldmann, Kinetic theory for a dilute gas of particles with spin, III. The influence of collinear static and oscillating magnetic fields on the viscosity. Z. Naturforsch. **26a**, 1057–1071 (1971)
43. A. Fokker, Die mittlere Energie rotierender elektrischer Dipole im Strahlungsfeld. Ann. Physik **43**, 810–820 (1914)
44. Max Planck, *Über einen Satz der statistischen Dynamik und eine Erweiterung der Quantentheorie* (Sitzungsberichte der Preuss. Akademie der Wissenschaften, Berlin, 1917), pp. 324–341

45. H. Grad, in *Principles of the Kinetic Theory of Gases*, vol. XII, ed. by S. Flügge. Handbuch der Physik (Springer, Berlin, 1958) pp. 205–294
46. S. Hess, L.J.H. Hermans, Evidence for Maxwell's thermal pressure via light-induced velocity selective heating or cooling. Phys. Rev. A **45**, 829–832 (1992)
47. S. Hess, M. Malek Mansour, Temperature profile of a dilute gas undergoing a plane Poiseuille flow. Physica A **272**, 481–496 (1999)
48. J.P. Hansen, I.R. McDonald, *Theory of Simple Liquids*, 2nd edn. (Academic Press, London, 1986)
49. M.P. Allen, D.J. Tildesley, *Computer Simulation of Liquids* (Clarendon, Oxford, 1987)
50. D.J. Evans, G.P. Morriss, *Statistical Mechanics of Nonequilibrium Liquids* (Academic Press, London, 1990); 2nd edn. (Cambridge University Press, 2008)
51. P.A. Egelstaff, *An Introduction to the Liquid State* (Clarendon, Oxford, 1992); 2nd edn. (1994)
52. S. Hess, M. Kröger, P. Fischer, in *Einfache und Disperse Flüssigkeiten*, vol. 5, ed. by K. Kleinermanns. Bergmann-Schaefer, Lehrbuch der Experimentalphysik (de Gruyter, Berlin, New York, 2006) pp. 385–467
53. J.G. Kirkwood, Statistical mechanical theory of transport processes. J. Chem. Phys. **14**, 180–201 (1946); 15, 72–76 (1947)
54. S. Hess, in *Similarities and Differences in the Nonlinear Flow Behavior of Simple and of Molecular Liquids*, ed. by H.J.M. Hanley. Nonlinear Flow Behavior (North-Holland, Amsterdam, 1983) (Physica 118A, 79–104 (1983))
55. S. Hess, Shear-flow induced distortion of the pair correlation function. Phys. Rev. A **22**, 2844 (1980); S. Hess, H.J.M. Hanley, Stokes-Maxwell relations for the distorted fluid microstructure. Phys. Lett. **98A**, 35 (1983); J.F. Schwarzl, S. Hess, On the shear-flow-induced distortion of the structure of a fluid: Applications of a simple kinetic equation. Phys. Rev. A **33**, 4277 (1986); H.J.M. Hanley, J.C. Rainwater, S. Hess, Shear-induced angular dependence of the liquid pair-correlation function. Phys. Rev. A **36**, 1795–1802 (1987)
56. D. el Masri, T. Vissers, S. Badaire, J.C.P. Stiefelhagen, H.R. Vutukuri, P.H. Helfferich, T. Zhang, W.K. Kegel, A. Imhof, A. van Blaaderen, A qualitative confocal microscopy study on a range of colloidal processes by simulating microgravity conditions through slow rotations. Soft Matter **8**, 6979–6990 (2012); D. Derks, Y.L. Wu, A. van Blaaderen, A. Imhof, Dynamics of colloidal crystals in shear flow. Soft Matter **5**, 1060–1065 (2009)
57. W.G. Hoover, Molecular Dynamics (Springer, Berlin, 1986); Computational Statistical Mechanics (Elsevier, Amsterdam 1991); G. Ciccotti, C. Pierleoni, J.-P. Ryckaert, Theoretical foundation and rheological application of nonequilibrium molecular dynamics, in ed. by M. Mareschal, B.L. Holian. Microscopic Simulation of Complex Hydrodynamic Phenomena (Plenum, New York, 1992) p. 25–46
58. D.J. Evans, H.J.M. Hanley, S. Hess, Non-newtonian phenomena in simple fluids. Phys. Today **37**, 26 (1984); S. Hess, M. Kröger, W. Loose, C. Pereira Borgmeyer, R. Schramek, H. Voigt, T. Weider, in *Simple and Complex Fluids Under Shear*, vol. 49, ed. by K. Binder, G. Ciccotti. Monte Carlo and Molecular Dynamics of Condensed Matter Systems. IPS Conference Proceedings (Bologna, 1996) p. 825–841
59. O. Hess, W. Loose, T. Weider, S. Hess, Shear-induced anisotropy of the structure of dense fluids. Physica B **156/157**, 505 (1989)
60. N.A. Clark, B.J. Ackerson, Observation of the coupling of concentration fluctuations to steady-state shear flow. Phys. Rev. Lett. **44**, 1005 (1980)
61. H.M. Laun, R. Bung, S. Hess, W. Loose, O. Hess, K. Hahn, E. Hädicke, R. Hingmann, P. Lindner, Rheological and small angle neutron scattering investigation of shear-induced structures of concentrated polymer dispersions submitted to plane Poiseuille and Couette flow. J. Rheol. **36**, 742–787 (1992)
62. S. Hess, Flow birefringence of polyatomic gases. Phys. Lett. **30A**, 239–240 (1969)
63. S. Hess, Kinetic theory for the broadening of the depolarized Rayleigh line. Z. Naturforsch. **24a**, 1852–1853 (1969)
64. S. Hess, *Depolarisierte Rayleigh-Streuung und Strömungsdoppelbrechung in Gasen* (Springer Tracts in Modern Physics; 54) (Springer, Berlin, Heidelberg, New York, 1970), pp. 136–176

65. A.R. Edmonds, Angular momentum in Quantum Mechanics (Princeton University Press); Drehimpulse in der Quantenmechanik (BI Mannheim, 1964)
66. A. Sommerfeld, Vorlesungen über Theoretische Physik, Bd. 6, Partielle Differentialgleichungen in der Physik (Harri Deutsch, 1992)
67. P.G. de Gennes, *The Physics of Liquid Crystals* (Clarendon Press, Oxford, 1974); P.G. de Gennes, J. Prost, *The Physics of Liquid Crystals* (Clarendon, Oxford, 1993)
68. G. Hauck, G. Heppke, Flüssigkristalle, ed. by K. Kleinermanns. Bergmann-Schaefer, Lehrbuch der Experimentalphysik, vol. 5 (de Gruyter, Berlin, New York, 2006), pp. 649–710
69. D. Demus, in *Phase Types, Structures and Chemistry of Liquid Crystals*, vol. 3, ed. by H. Stegemeyer. Topics in Physical Chemistry (Steinkopff Darmstadt, Springer, New York, 1994), pp. 1–50
70. H. Kelker, R. Hatz, *Handbook of Liquid Crystals* (Verlag Chemie, Weinheim, 1980)
71. P. Oswald, P. Pieranski, *Nematic and Cholesteric Liquid Crystals: Concepts and Physical Properties Illustrated by Experiments* (Taylor and Francis, Boca Raton, 2005)
72. T.J. Sluckin, D.A. Dunmur, H. Stegemeyer, *Crystals That Flow - Classic Papers from the History of Liquid Crystals, Liquid Crystals Series* (Taylor and Francis, London, 2004)
73. G.G. Fuller, *Optical Rheometry of Complex Fluids* (Oxford University Press, New York, 1995)
74. G. Hess, *Anisotrope Fluide - was ist das? Forschung Aktuell*, vol. 8 (TU Berlin, Berlin, 1991), pp. 2–5
75. H.J. Coles, M.N. Pivnenko, Liquid crystal blue phases with a wide temperature range. Nature **436**, 997–1000 (2005); F. Castles, S.M. Morris, E.M. Terentjev, H.J. Coles, Thermodynamically stable blue phases. Phys. Rev. Let. **104**, 157801 (2010)
76. P. Poulin, H. Stark, T.C. Lubensky, D.A. Weitz, Novel colloidal interactions in anisotropic fluids. Science **275**, 1770–1773 (1997)
77. M. Ravnik, G.P. Alexander, J.M. Yeomans, S. Zumer, Three-dimensional colloidal crystals in liquid crystalline blue phases. PANS **108**, 5188–5192 (2011)
78. R.M. Hornreich, S. Shtrikman, C. Sommers, Photonic band gaps in body-centered-cubic structures. Phys. Rev. B **49**, 10914–10917 (1994)
79. R.R. Netz, S. Hess, Static and dynamic properties of ferroelectric liquid crystals in the vicinity of a first-order SmA-SmC* phase transition. Z. Naturforsch. **47a**, 536–542 (1992)
80. W. Maier, A. Saupe, Eine einfache molekular-statistische Theorie der nematischen kristallinflüssigen Phase, Teil I. Z. Naturforsch. **13a**, 564–566 (1958); Teil II. Z. Naturforsch. **14a**, 882–889 (1959); Teil III. Z. Naturforsch. **53a**, 287–292 (1960)
81. L. Onsager, The effects of shape on the interaction of colloidal particles. Ann. N.Y. Acad. Sci. **51**, 627 (1949)
82. C.W. Oseen, Theory of liquid crystals. Trans. Faraday Soc. **29**, 883–899 (1933); H. Zocher, The effect of a magnetic field on the nematic state. Trans. Faraday Soc. **29**, 945–957 (1933); F.C. Frank, On the theory of liquid crystals, Discuss. Faraday Soc. **25**, 19–28 (1958)
83. I. Pardowitz, S. Hess, Elasticity coefficients of nematic liquid crystals. J. Chem. Phys. **76**, 1485–1489 (1982); Molecular foundation of a unified theory for the isotropic and nematic or cholesteric phases of liquid crystals. Physica **121A**, 107–121 (1983)
84. M.A. Osipov, S. Hess, The elastic constants of nematic and nematic discotic liquid crystals with perfect local orientational order. Mol. Phys. **78**, 1191–1201 (1993)
85. H. Steuer, S. Hess, Direct computation of the twist elastic coefficient of a nematic liquid crystal via Monte Carlo simulation. Phys. Rev. Lett. **94**, 027802 (2005)
86. S. Heidenreich, P. Ilg, S. Hess, Robustness of the periodic and chaotic orientational behavior of tumbling nematic liquid crystals. Phys. Rev. E **73**, 06710–1/11 (2006)
87. N. Schopohl, T.J. Sluckin, Defect core structure in nematic liquid crystals. Phys. Rev. Lett. **59**, 2582–2584 (1987)
88. A. Sonnet, A. Kilian, S. Hess, Alignment tensor versus director: description of defects in nematic liquid crystals. Phys. Rev. E **52**, 718 (1995)
89. L. Longa, H.-R. Trebin, Structure of the elastic free energy for chiral nematic liquid crystals. Phys. Rev. A **39**, 2160 (1989)

90. S. Hess, Dynamic Ginzburg-Landau theory for the liquid-solid phase transition. Z. Natur-forsch. **35a**, 69–74 (1980)
91. A.C. Mitus, A.Z. Patashinski, The theory of crystal ordering. Phys. Lett. A **87**, 179 (1982); D.K. Nelson, J. Toner, Bond orientational order, dislocation loops, and melting of solids and smetic-A liquid crystals. Phys. Rev. B **24**, 363 (1981)
92. S. Hess, On the shock front thickness in water and other molecular liquids, Z. Naturforsch. **52a**, 213–219 (1997)
93. S. Hess, Complex fluid behavior: coupling of the shear stress with order parameter tensors of ranks two and three in nematic liquid crystals and in tetradic fluids. Physica A **314**, 310–319 (2002)
94. H.R. Brand, P.E. Cladis, H. Pleiner, Symmetry and defects in the C_M phase of polymeric liquid crystals. Macromolecules **25**, 7223 (1992)
95. C. Pujolle-Robic, L. Noirez, Observation of shear-induced nematic-isotropic transition in side-chain liquid crystal polymers. Nature **409**, 167–171 (2001)
96. P. Ilg, S. Hess, Two-alignment tensor theory for the dynamics of side chain liquid-crystalline polymers in planar shear flow. J. Non-Newton. Fluid Mech. **134**(1–3), 2–7 (2006)
97. S. Grandner, S. Heidenreich, P. Ilg, S.H.L. Klapp, S. Hess, Dynamic electric polarization of nematic liquid crystals subjected to a shear flow. Phys. Rev. E **75**, 040701(R) (2007); G. Grandner, S. Heidenreich, S. Hess, S.H.L. Klapp, Polar nano-rods under shear: from equilibrium to chaos. Eur. Phys. J. E **24**, 353–365 (2007); S. Heidenreich, S. Hess, S.H.L. Klapp, Shear-induced dynamic polarization and mesoscopic structure in suspensions of polar nanorods. Phys. Rev. Lett. **102**, 028301 (2009)
98. H.R. Brand, H. Pleiner, Flexoelectric effects in cholesteric liquid crystals. Mol. Cryst. Liq. Cryst. **292**, 141 (1997)
99. H. Pleiner, H.R. Brand, Low symmetry tetrahedral nematic liquid crystal phases: ambidextrous chirality and ambidextrous helicity. Eur. Phys. J. E **37**, 11 (2014)
100. M. Born, Thermodynamics of crystals and melting. J. Chem. Phys. **7**, 591–601 (1939); H.S. Green, The Molecular Theory of Fluids (North Holland, Amsterdam, 1952)
101. M.S. Green, Markoff random processes and the statistical mechanics of time-dependent phenomena. II. Irreversible processes in fluids. J. Chem. Phys. **22**, 398–413 (1954)
102. D.R. Squire, A.C. Holt, W.G. Hoover, Isothermal elastic constants for Argon. Theory and Monte Carlo calculations. Physica **42**, 388–397 (1969); W.G. Hoover, A.C. Holt, D.R. Squire, Adiabatic elastic constants for Argon. Theory and Monte Carlo calculations. Physica **44**, 437–443 (1969)
103. S. Hess, M. Kröger, W.G. Hoover, Shear modulus of fluids and solids. Physica A **239**, 449–466 (1997)
104. M.S. Daw, M.J. Baskes, Semiempirical, quantum mechanical calculations of Hydrogen embrittlement in metals. Phys. Rev. Lett. **50**, 1285 (1983); Embedded-atom method: derivation and application to impurities, surfaces and other defects in metals. Phys. Rev. B **29**, 6443 (1984)
105. R.A. Johnson, Analytic nearest neighbor model for fcc metals. Phys. Rev. B **37**, 3924, 6121 (1988); Alloy models with the embedded-atom method. Phys. Rev. B **39**, 12554 (1989)
106. B.I. Holian, A.F. Voter, N.J. Wagner, R.J. Ravelo, S.P. Chen, W.G. Hoover, C.G. Hoover, J.F. Hammerberg, T.D. Dontje, Effects of pairwise versus many-body forces on high-stress plastic deformations. Phys. Rev. A **43**, 2655 (1991)
107. I. Stankovic, S. Hess, M. Kröger, Structural changes and viscoplastic behavior of a generic embedded-atom model metal in steady shear flow. Phys. Rev. E **69**, 021509 (2004)
108. S.R. de Groot, P. Mazur, *Non-eqilibrium Thermodynamics* (North-Holland, Amsterdam, 1962)
109. L. Onsager, Reciprocal relations in irreversible processes I. Phys. Rev. **37**, 405–426 (1931)
110. S. Odenbach, Magnetoviscous Effects in Ferrofluids (Lecture Notes in Physics; 71) (Springer, Berlin, Heidelberg, New York, 2002); P. Ilg, S. Odenbach, in *Ferrofluid Structure and Rheology*, ed. by S. Odenbach. Colloidal Magnetic Fluids: Basics, Development and Applications of Ferrofluids (Lecture Notes in Physics; 763) (Springer, Berlin, Heidelberg, New York, 2009)
111. E. Blums, A. Cebers, M.M. Maiorov, *Magnetic Fluids* (de Gruyter, Berlin, 1997)

112. P. Ilg, M. Kröger, S. Hess, A.Y. Zubarev, Dynamics of colloidal suspensions of ferromagnetic particles in plane Couette flow: comparison of approximate solutions with Brownian dynamics simulations. Phys. Rev. E **67**, 061401 (2003); M. Kröger, P. Ilg, S. Hess, Magnetoviscous model fluids. J. Phys. Condens. Matter **15**, S1403–S1423 (2003)

113. H. Senftleben, Magnetische Beeinflussung des Wärmeleitvermögens paramagnetischer Gase. Phys. Z. **31**, 822, 961 (1930); H. Engelhardt, H. Sack, Beeinflussung der inneren Reibung von O_2 durch ein Magnetfeld. Phys. Z. **33**, 724 (1933); M. Trautz, E. Fröschel, Notiz zur Beeinflussung der inneren Reibung von O_2 durch ein Magnetfeld. Phys. Z. **33**, 947 (1933)

114. J.J.M. Beenakker, G. Scoles, H.F.P. Knaap, R.M. Jonkman, The influence of a magnetic field on the transport properties of diatomic molecules in the gaseous state. Phys. Lett. **2**, 5 (1962)

115. L. Waldmann, Die Boltzmann-Gleichung für Gase aus rotierenden Molekülen, Z. Naturforsch. **12a**, 660 (1957); Die Boltzmann-Gleichung für Gase aus Spin-Teilchen. Z. Naturforsch. **13a**, 609 (1958): R.F. Snider, Quantum-mechanical modified Boltzmann equation for degenerate internal states. J. Chem. Phys. **32**, 1051 (1960)

116. S. Hess, Verallgemeinerte Boltzmann-Gleichung für mehratomige Gase, Z. Naturforsch. **22a**, 1871–1889 (1967)

117. S.J. Barnett, Magnetization by rotation. Phys. Rev. **6**, 239–270 (1915)

118. R. Cerf, J. Chim. Phys. **68**, 479 (1969)

119. C. Aust, S. Hess, M. Kröger, Rotation and deformation of a finitely extendable flexible polymer molecule in a steady shear flow. Macromolecules **35**, 8621–8630 (2002)

120. S. Hess, Construction and test of thermostats and twirlers for molecular rotations. Z. Naturforsch. **58a**, 377–391 (2003)

121. S. Hess, G.P. Morriss, in *Rotation and Deformation of Polymer Molecules in Solution Subjected to a Shear Flow*, ed. by P. Pasini, C. Zannoni, S. Zumer. Computer Simulations Bridging Liquid Crystals and Polymers (Kluwer, Dordrecht, 2005), pp. 269–294

122. J.C. Maxwell, On double refraction in a viscous fluid in motion. Proc. Roy. Soc. London (A) **22**, 46 (1873); Pogg. Ann. Physik **151**, 151 (1874)

123. H. Janeschitz-Kriegel, *Polymer Melt Rheology and Flow Birefringence* (Springer, Berlin, 1983)

124. F. Baas, Phys. Lett. **36A**, 107 (1971); F. Baas, P. Oudeman, H.F.P. Knaap, J.J.M. Beenakker, Flow birefringence in gases of linear and symmetric top molecules. Physica **88A**, 1 (1977)

125. G.R. Boyer, B.F. Lamouroux, B.S. Prade, Air-flow birefringence measurements. J. Opt. Soc. Ann. **65**, 1319 (1975)

126. S. Hess, The effect reciprocal to flow birefringence in gases. Z. Naturforsch. **28a**, 1531–1532 (1973)

127. S. Hess, Heat-flow birefringence. Z. Naturforsch. **28a**, 861–868 (1973)

128. F. Baas, J.N. Breunese, H.F.P. Knaap, J.J.M. Beenakker, Heat-flow birefringence in gaseous O_2. Physica **88A**, 44 (1977)

129. D. Baalss, S. Hess, Heat flow birefringence in liquids and liquid crystals. Z. Naturforsch. **40a**, 3 (1985)

130. R. Elschner, R. Macdonald, H.J. Eichler, S. Hess, A.M. Sonnet, Molecular reorientation of a nematic glass by laser-induced heat flow. Phys. Rev. E **60**, 1792–1798 (1999)

131. S. Hess, Diffusio birefringence in colloidal suspensions and macromolecular liquids. Phys. Lett. **45A**, 77–78 (1973); Birefringence caused by the diffusion of macromolecules or colloidal particles. Physica **74**, 277–293 (1974)

132. D. Jou, J. Casas Vazquez, G. Lebon, *Extended Irreversible Thermodynamics* (Springer, Berlin, Heidelberg, New York, 1993); G. Lebon, D. Jou, J. Casas-Vzquez, *Understanding Non-equilibrium Thermodynamics: Foundations, Applications, Frontiers* (Springer, Berlin, Heidelberg, New York, 2008)

133. W. Muschik, Survey of some branches of thermodynamics. J. Non-Equilib. Thermodyn. **33**, 165–198 (2008)

134. A.N. Beris, *Thermodynamics of Flowing Systems with Internal Microstructure* (Oxford University Press, Oxford, 1994)

135. H.C. Öttinger, *Beyond Equilibrium Thermodynamics* (Wiley, Hoboken, 2005)

136. S. Hess, Viscoelasticity associated with molecular alignment. Z. Naturforsch. **35a**, 915–919 (1980)

137. H. Thurn, M. Löbl, H. Hoffmann, Viscoelastic detergent solutions. a quantitative comparison between theory and experiment. J. Phys. Chem. **89**, 517–522 (1985)

138. H. Giesekus, Constitutive equations for polymer fluids based on the concept of configuration-dependent molecular mobility: a generalized mean-configuration model. J. Non-Newtonian Fluid Mech. **17**, 349–372 (1985); 43. (1985); Flow phenomena in viscoelastic fluids and their explanation using statistical methods. J. Non-Equilib Thermodyn. **11**, 157–174 (1986)

139. M. Reiner, *Twelve Lectures on Theoretical Rheology* (North Holland, 1949); *Rheologie* (Carl Hanser Verlag, Leipzig, München, 1968)

140. P. Coussot, *Rheophysics, Matter in All Its States* (Springer, 2014)

141. S. Hess, Non-newtonian viscosity and normal pressure differences of simple fluids. Phys. Rev. A **25**, 614–616 (1982)

142. M.W. Johnson, D. Segalman, A model for viscoelastic fluid behavior which allows non-affine deformation. J. Non-newt. Fluid Mech. **2**, 255–270 (1977)

143. O. Rodulescu, P. Olmsted, Matched asymptotic solutions for the steady banded flow of the Johnson-Segalman model in various geometries. Nonnewton. Fluid Mech. **91**, 143–162 (2000); P.D. Olmsted, O. Radulescu, C.Y.D. Lu, The Johnson-Segalman model with a diffusion term: a mechanism for stress selection. J. Rheol. **44**, 257–275 (2000); H.J. Wilson, S.M. Fielding, Linear instability of planar shear banded flow of both diffusive and non-diffusive Johnson-Segalman fluids. Nonnewton. Fluid Mech. **138**, 181–196 (2006)

144. M. Miesowicz, The three coefficients of viscosity of anisotropic liquids. Nature **158**, 27–27 (1946)

145. O. Parodi, Stress tensor for a nematic fluid crystal. J. Phys. (Paris) **31**, 581 (1970)

146. W. Helfrich, Torques in sheared nematic liquid crystals: a simple model in terms of the theory of dense fluids. J. Chem. Phys. **53**, 2267 (1970)

147. D. Baalss, S. Hess, Nonequilibrium molecular dynamics studies on the anisotropic viscosity of perfectly aligned nematic liquid crystals. Phys. Rev. Lett. **57**, 86 (1986); Viscosity coefficients of oriented nematic and nematic discotic liquid crystals; Affine transformation model. Z. Naturforsch. **43a**, 662–670 (1988)

148. H. Ehrentraut, S. Hess, On the viscosity of partially aligned nematic and nematic discotic liquid crystals. Phys. Rev. E **51**, 2203 (1995); S. Blenk, H. Ehrentraut, S. Hess, W. Muschik, Viscosity coefficients of partially aligned nematic liquid crystals. ZAMM **7**, 235 (1994)

149. M.A. Osipov, E.M. Terentjev, Rotational diffusion and rheological properties of liquid crystals. Z. Naturforsch. **44a**, 785–792 (1989)

150. A.M. Sonnet, P.L. Maffettone, E.G. Virga, Continuum theory for nematic liquid crystals with tensorial order. J. Nonnewton. Fluid Mech. **119**, 51–59 (2004)

151. H. Kneppe, F. Schneider, N.K. Sharma, Ber. Bunsenges. Phys. Chem. **85**, 784 (1981); H.-H. Graf, H. Kneppe, F. Schneider, MolPhys. **77**, 521 (1992)

152. S. Sarman, D.J. Evans, J. Chem. Phys. **99**, 9021 (1993). S. Sarman, J. Chem. Phys. **101**, 480 (1994)

153. S. Cozzini, L.F. Rull, G. Ciccotti, G.V. Paolini, Intrinsic frame transport for a model of nematic liquid crystal. Physica A **240**, 173–187 (1997)

154. A. Eich, B.A. Wolf, L. Bennett, S. Hess, Electro- and magneto-rheology of nematic liquid crystals - experiment and non-equilibrium molecular dynamics (NEMD) computer simulation. J. Chem. Phys. **113**, 3829–3838 (2000)

155. L. Bennett, S. Hess, Nonequilibrium-molecular dynamics investigation of the presmectic behavior of the viscosity of a nematic liquid crystal. Phys. Rev. E **60**, 5561–5567 (1999)

156. S. Hess, D. Frenkel, M.P. Allen, On the anisotropy of diffusion in nematic liquid crystals: test of a modified affine transformation model via molecular dynamics. Mol. Phys. **74**, 765–774 (1991)

157. S. Hess, Fokker-Planck equation approach to flow alignment in liquid crystals. Z. Naturforsch. **31a**, 1034–1037 (1976)

158. M. Doi, Rheological properties of rodlike polymers in isotropic and liquid crystalline phases. Ferroelectrics **30**, 247 (1980); Molecular dynamics and rheological properties of concentrated solutions of rodlike polymers in isotropic liquids and liquid crystals. J. Polym. Sci. Polym. Phys. **19**, 229 (1981)

159. M. Doi, S.F. Edwards, *The Theory of Polymer Dynamics* (Clarendon, Oxford, 1986)

160. A. Peterlin, H.A. Stuart, *Doppelbrechung, insbesondere künstliche Doppelbrechung*, vol. 8, ed. by A. Eucken, K.L. Wolf. Hand- und Jahrbuch der chem. Physik, I B (Leipzig, 1943)

161. G.B. Jeffery, The motion of ellipsoidal particles immersed in a viscous fluid. Proc. Roy. Soc. Lond. (A) **102**, 161 (1922)

162. G. Marrucci, P.L. Maffettone, A description of the liquid crystalline phase of rodlike polymers at high shear rates. Macromolecules **22**, 4076–4082 (1989)

163. M. Gregory Forest, Q. Wang, R. Zhou, The flow-phase diagram of Doi-Hess theory for sheared nematic polymers II: finite shear rates. Rheol. Acta **44**, 80–93 (2004)

164. G. Marrucci, N. Grizzuti, Rheology of liquid-crystalline polymers. Theory and experiments. Makromolekulare Chemie, Macromolecular Symposia **48–49**, 181–188 (1991)

165. R.G. Larson, *The Structure and Rheology of Complex Fluids* (Oxford University Press, Oxford, 1999)

166. M. Kröger, Simple models for complex nonequilibrium fluids. Phys. Rep. **390**, 453–551 (2004)

167. M. Kröger, *Models for Polymeric and Anisotropic Liquids* (Lecture Notes in Physics; 675) (Springer, Berlin, Heidelberg, New York, 2005)

168. S. Hess, Irreversible thermodynamics of non-equilibrium alignment phenomena in molecular liquids and liquid crystals, I. Derivation of nonlinear constitutive laws, relaxation of the alignment, phase transition. Z. Naturforsch. **30a**, 728–738 (1975). II. Viscous flow and flow alignment in the isotropic (stable and metastable) and nematic phase. Z. Naturforsch. **30a**, 1224–1232 (1975)

169. S. Hess, Pre- and post-transitional behavior of the flow alignment and flow-induced phase transition in liquid crystals. Z. Naturforsch. **31a**, 1507–1513 (1976)

170. P.D. Olmsted, P. Goldbart, Theory of the non-equilibrium phase transition for nematic liquid crystals under shear flow. Phys. Rev. A **41**, 4588 (1990); Nematogenic fluids under shear flow: state selection, coexistence, phase transitions, and critical behavior. Phys. Rev. A **46**, 4966–4993 (1992)

171. C. Pereira Borgmeyer, S. Hess, Unified description of the flow alignment and viscosity in the isotropic and nematic phases of liquid crystals. J. Non-Equilib. Thermodyn **20**, 359–384 (1995)

172. S. Hess, I. Pardowitz, On the unified theory for nonequilibrium phenomena in the isotropic and nematic phases of a liquid crystal; spatially inhomogeneous alignment. Z. Naturforsch. **36a**, 554–558 (1981)

173. S. Hess, H.-M. Koo, Boundary effects on the flow-induced orientational anisotropy and on the flow properties of a molecular liquid. J. Non-Equilib. Thermodyn. **14**, 159 (1989)

174. S. Heidenreich, P. Ilg, S. Hess, Boundary conditions for fluids with internal orientational degree of freedom: apparent slip velocity associated with the molecular alignment. Phys. Rev. E **75**, 066302 (2007)

175. A.G.S. Pierre, W.E. Köhler, S. Hess, Time-correlation functions for gases of linear molecules in a magnetic field. Z. Naturforsch. **27a**, 721–732 (1972)

176. R. Kubo, Statistical-mechanical theory of irreversible processes. I. general theory and simple applications to magnetic and conduction problems. J. Phys. Soc. Jpn. **12**, 570–586 (1957)

177. S. Hess, D. Evans, Computation of the viscosity of a liquid from time averages of stress fluctuations. Phys. Rev. E **64**, 011207 (2001); S. Hess, M. Kröger, D.J. Evans, Crossover between short- and long-time behavior of stress fluctuations and viscoelasticity of liquids. Phys. Rev. E **67**, 042201 (2003)

178. S. Hess, Kinetic theory of spectral line shapes - the transition from Doppler broadening to collisional broadening. Physica **61**, 80–94 (1972)

179. S. Hess, A. Mörtel, Doppler broadened spectral functions and the pertaining time correlation functions for a gas in non-equilibrium. Z. Naturforsch. **32a**, 1239–1244 (1977)
180. B.K. Gupta, S. Hess, A.D. May, Anisotropy in the Dicke narrowing of rotational raman lines - a new measure of the non-sphericity of intermolecular forces. Can. J. Phys. **50**, 778–782 (1972)
181. S. Hess, R. Müller, On the depolarized Rayleigh scattering from macromolecular and colloidal solutions - anisotropy of the diffusional broadening. Opt. Com. **10**, 172–174 (1974)
182. P. Kaiser, W. Wiese, S. Hess, Stability and instability of an uniaxial alignment against biaxial distortions in the isotropic and nematic phases of liquid crystals. J. Non-Equilib. Thermodyn **17**, 153–169 (1992)
183. G. Rienäcker, Orientational dynamics of nematic liquid crystals in a shear flow, Thesis TU Berlin, 2000 (Shaker Verlag, Aachen, 2000)
184. M. Grosso, R. Keunings, S. Crescitelli, P.L. Maffettone, Phys. Rev. Lett. **86**, 3184 (2001)
185. G. Rienäcker, M. Kröger, S. Hess, Phys. Rev. E **66**, 040702(R) (2002). Physica A **315**, 537 (2002)
186. S. Hess, M. Kröger, Regular and chaotic orientational and rheological behaviour of liquid crystals. J. Phys. Condens. Matter **16**, S3835–S3859 (2004)
187. S. Hess, M. Kröger, in *Regular and Chaotic Rheological Behavior of Tumbling Polymeric Liquid Crystals*, ed. by P. Pasini, C. Zannoni, S. Zumer. Computer Simulations Bridging Liquid Crystals and Polymers (Kluwer, Dordrecht, 2005), pp. 295–334
188. D.A. Strehober, H. Engel, S.H.L. Klapp, Oscillatory motion of sheared nanorods beyond the nematic phase. Phys. Rev. E **88**, 012505 (2013)
189. M.G. Forest, S. Heidenreich, S. Hess, X. Yang, R. Zhou, Robustness of pulsating jet-like layers in sheared nano-rod dispersions. J. Nonnewton. Fluid Mech. **155**, 130–145 (2008)
190. R.G. Larson, H.C. Öttinger, Effect of molecular elasticity on out-of-plane orientations in shearing flows of liquid-crystalline polymers. Macromolecules **24**, 6270–6282 (1991)
191. Y.-G. Tao, W.K. den Otter, W.J. Briels, Kayaking and wagging of liquid crystals under shear: comparing director and mesogen motions. Europhys. Lett. **86**, 56005 (2009)
192. M.P. Lettinga, H. Wang, J.K.G. Dhont, Flow behaviour of colloidal rodlike viruses in the nematic phase. Langmuir **21**, 8048–8057 (2005)
193. S.H.L. Klapp, S. Hess, Shear-stress-controlled dynamics of nematic complex fluids. Phys. Rev. E **81**, 051711 (2010)
194. H.G. Schuster, W. Just, *Deterministic Chaos: an introduction*, 4th edn. (Wiley VHC, Weinheim, 2005)
195. L.P. Shilnikov, A. Shilnikov, D. Turaev, and L. Chua, Methods of Qualitative Theory in Nonlinear Dynamics, Part I. World Sci. (1998); Part II. World Sci. (2001)
196. E. Schoell, H.G. Schuster (eds.), *Chaos Control*, 2nd edn. (Wiley VHC, Weinheim, 2007)
197. M.E. Cates, D.A. Head, A. Ajdari, Rheological chaos in a scalar shear-thickening model. Phys. Rev. E **66**, 025202 (2002); A. Aradian, M.E. Cates, Instability and spatiotemporal rheochaos in a shear-thickening fluid model. Europhys. Lett. **70**, 397–403 (2005)
198. Y. Hatwalne, S. Ramaswamy, M. Rao, R.A. Simha, Rheology of active particle systems. Phys. Rev. Lett. **92**, 118101 (2004)
199. S. Heidenreich, S. Hess, S.H.L. Klapp, Nonlinear rheology of active particle suspensions: Insight from an analytical approach. Phys. Rev. E **83**, 011907 (2011); (2009)
200. O. Hess, S. Hess, Nonlinear fluid behavior: from shear thinning to shear thickening. Physica A **207**, 517 (1994)
201. O. Hess, C. Goddard, S. Hess, From shear-thickenning and periodic flow behavior to rheochaos in nonlinear Maxwell model fluids. Physica A **366**, 31–54 (2006); C. Goddard, O. Hess, A. Balanov, S. Hess, Shear-induced chaos in nonlinear Maxwell-model. Phys. Rev. E **77**, 026311 (2008)
202. C. Goddard, Rheological Chaos and Elastic Turbulence in a Generalized Maxwell Model for Viscoelastic Fluid Flow. Thesis, University of Surrey, Guildford, England, 2008
203. C. Goddard, O. Hess, S. Hess, Low Reynolds number turbulence in nonlinear Maxwell model fluids. Phys. Rev. E **81**, 036310 (2010)

204. A. Groisman, V. Steinberg, Elastic turbulence in a polymer solution flow. Nature **405**, 53–55 (2000)
205. B.A. Schiamberg, L.T. Shereda, H. Hu, R.G. Larson, Transitional pathway to elastic turbulence in torsional, parallel-plate flow of a polymer solution. J. Fluid Mech. **554**, 191–216 (2006)
206. S. Hess, B. Arlt, S. Heidenreich, P. Ilg, C. Goddard, O. Hess, Flow properties inferred from generalized Maxwell models. Z. Naturforsch. **64a**, 81–95 (2009)
207. M. Born, Modified field equations with a finite radius of the electron. Nature **132**, 282 (1933); On the quantum theory of the electromagnetic field. Proc. Roy. Soc. A **143**, 410 (1934); M. Born, L. Infeld, Electromagnetic mass. Nature **132**, 970 (1933); Foundations of the new field theory. Nature **132**, 1004 (1933); Proc. Roy. Soc. A 1 **44**, 425 (1934)

Index

Symbols
□-tensor, 186, 193
Δ-tensors, 184
$\Delta^{(\ell)}$-tensors, 184
4-acceleration, 375
4-momentum, 375
4-velocity, 374
4-wave vector, 376
4D-epsilon tensor, 378

A
Actio equal reactio, 81
Active rotation of a tensor, 259
Affine transformation, 17, 335
Aligning, 362
Alignment tensor, 95, 203
Alignment tensor elasticity, 288
Angular momentum, 43
Angular momentum balance, 151
Angular momentum commutation relations, 107
Angular momentum conservation, 320
Angular velocity, 51, 89
Anisotropic fluids, 274
Anisotropic part, 57
Antisymmetric part, 33, 40
Antisymmetric part of the pressure tensor, 344
Antisymmetric pressure, 320
Antisymmetric tensor, 50
Antisymmetric traceless part, 34
Ascending multipole potentials, 163
Auto-correlation functions, 352
Axes ratio, 336
Azimuthal component, 30

B
Banana phases, 276, 294
Barnett effect, 321
Basis tensors, 358
Bend deformations, 285
Bessel functions, 233
Biaxial distortions, 361
Biaxial extensional or compressional flow, 331
Biaxial nematics, 274
Biaxial tensor, 58
Biaxiality parameter, 70, 359
Bilinear form, 65
Biot-Savart relation, 103
Birefringence, 63, 204, 253
Blue phase liquid crystals, 290
Blue phases, 275
Boltzmann equation, 318, 327
Bond orientational order, 291
Born-Green expression, 310
Boundary conditions, 177, 368
Brillouin scattering, 353
Brownian particles, 217
Bulk modulus, 306
Bulk viscosity, 333

C
Cartesian components, 11
Cartesian coordinate system, 11
Cartesian tensors of rank, 23
Cauchy relation, 312
Center of mass, 132
Central force, 44, 83
Chaotic, 363
Charge density, 173, 381
Cholesteric, 274

© Springer International Publishing Switzerland 2015
S. Hess, *Tensors for Physics*, Undergraduate Lecture Notes in Physics,
DOI 10.1007/978-3-319-12787-3

Cholesteric liquid crystal, 274, 287
Cholesterics, 290
Circular frequency, 102
Clebsch-Gordan tensors, 191
Clesch-Gordan coefficients, 256
Closed curve, 112
Co-rotational Maxwell model, 330
Co-rotational time derivative, 330
Collision frequency, 319
Collision integrals, 319
Collisional broadening, 356
Colloidal dispersions, 332
Commutation relation, 106, 239
Complex viscosity coefficients, 315
Component equations, 359
Component notation, 13
Conductivity coefficients, 268
Conductivity tensor, 267
Configurational canonical average, 309
Configurational partition integral, 309
Confocal microscopy, 229
Conservation of mass, 97, 140
Constitutive laws, 344
Constitutive relations, 299
Continuity equation, 97, 100, 140, 381
Contra- and co-variant components, 15
Contraction, 157
Contraction number, 307
Contraction of tensors, 35
Convected Maxwell model, 330
Convective transport, 97
Conventional classification of vector fields,
       94
Cotton-Mouton effect, 208
Couette flow, 315
Couette flow geometry, 229
Coulomb energy, 145
Coulomb force, 139
Coupling tensors, 191
Creeping flow approximation, 178
Cross-correlation functions, 352
Cross effect, 325
Cross product, 41
Cubatics, 291
Cubic crystals, 161, 292, 308
Cubic harmonic, 162
Cubic order parameter, 291
Cubic symmetry, 161, 231, 307
Curie Principle, 300
Curl, 89
Curve integral, 112
Curve integral of a vector field, 114
Cylinder coordinates, 130

Cylinder mantle, 118, 122
Cylindrical geometry, 78, 82

D
d'Alembert operator, 101
Decomposition, 75
Deformation, 304
Deformation rate, 227
Deformation tensor, 304, 305
Depolarized Rayleigh light scattering, 253
Depolarized Rayleigh scattering, 354
Depolarized scattering, 354
Descending multipole potentials, 163
Determinant, 47, 380
Deviatoric part, 89
Diagonal operators, 252
Diamagnetic gases, 318
Dicke narrowing, 357
Dielectric permeability, 98
Dielectric tensor, 63, 145, 148, 255, 302
Differential change, 79
Differential operator, 189
Diffusional broadening, 356
Diffusion coefficient, 218
Diffusion length, 349
Diffusion tensor, 337
Dipolar orientation, 206
Dipolar symmetry, 171
Dipole moment, 169, 171
Dipole potential, 164
Dipole–quadrupole interactions, 177
Dipole transition matrix elements, 236
Director elasticity, 285
Disc-like particles, 347
Discotic nematic, 274
Dispersion relation, 102
Divergence, 89
Divergence-free, 91
Doi-Hess-theory, 343
Doi-theory, 343
Doppler broadening, 356
Double refraction, 63
Double twist structure, 275
Dual relation, 40
Dual tensor, 379
Dyad, 37
Dyadic, 37
Dyadic tensor, 37
Dynamic states, 362
Dynamics of the alignment tensor, 362

E
Effective mass, 375
Effective shear viscosity, 315
Eigenvalues, 56
Einstein summation convention, 370
Elastic behavior, 284
Elastic modulus tensor, 310
Elastic properties, 304
Elastic turbulence, 368
Electric and magnetic torques, 64
Electric dipole transitions, 236
Electric displacement field, 63, 98
Electric field, 63, 98
Electric polarization, 99, 174, 204, 301
Electric quadupole transitions, 237
Electrodynamic 4-potential, 381
Electrodynamics, 98, 99, 138
Electromagnetic potential functions, 102
Electromagnetic waves, 101
Electro-optic Kerr effect, 209
Electroscalar potential, 102
Electrostatic energy, 174
Electrostatic force density, 146
Electrostatic potential, 168, 170
Electrostatics, 94, 145
Electrostatic stress tensor, 146
Ellipsoid, 66
Ellipsoidal particles, 335
Ellipsoids of revolution, 336
Embedded atom method, 312
Energetic coupling, 294
Energy balance, 147
Energy density, 146
Energy flux density, 147, 151
Energy principle, 302
Entropy production, 302, 313, 320, 323, 344
Epsilon-tensor, 47
Equal potential surfaces, 78
Equation of motion, 80
Equilibrium average, 214
Ericksen-Leslie coefficients, 347
Expansion, 225
Expansion coefficients, 203, 220
Expansion functions, 215
Extended irreversiblet thermodynamics, 327
Extensional viscosity, 331

F
Faraday induction, 101, 129
Ferro-fluids, 314, 332
Field, 77
Field tensor, 382

Field-induced orientation, 205
Flexo-electric effect, 296
Flow alignment, 337, 343
Flow alignment angle, 338
Flow birefringence, 322
Fluctuations, 351
Flux density, 140, 381
Flux of a vector field, 123
Fokker-Planck equation, 210
Fokker-Planck relaxation operator, 217
Force, 79, 175
Force balance, 144
Force density, 143, 386
Four-dimensional vectors, 370
Four-field formulation, 98
Fourier-Laplace transform, 353
Fourth rank projection tensors, 36, 263
Fourth rank rotation tensor, 264
Frank elasticity coefficients, 285
Frank-Oseen elasticity, 285
Free currents, 99
Free energy, 309
Free flow of nematics, 337
Frequency dependent viscosity, 326
Frictional torque, 339
Friction force, 178
Friction pressure tensor, 177, 220

G
Galilei transformation, 372
Gases of rotating molecules, 324
Gauss law, 100, 139
Gauss theorem, 136
Generalized cross product, 187
Generalized Fokker-Planck equation, 338
Generalized Gauss theorem, 136
Generalized Legendre polynomial, 196
Generalized Stokes law, 124
Geometric interpretation, 65
Gibbs relation, 292
Gradient, 79
Graphical representation, 95
Green-Kubo relation, 335
Gyromagnetic factor, 318

H
Hall-effect, 268
Hamilton-Cayley, 241, 260
Hamilton-Cayley theorem, 71
Head-tail symmetry, 274, 277
Hear deformations, 305

Heat conductivity, 337
Heat flux vector, 220
Heat-flow birefringence, 326
Heisenberg picture, 244
Helfrich viscosity, 316
Helfrich viscosity coefficient, 334
Helical axis, 274
Hermitian operator, 107
Hexadecapole moment, 172
High frequency shear modulus, 310
High temperature approximation, 206
High temperature expansion, 205
Homgeneous field, 84, 89
Homogeneous Maxwell equations, 98, 383
Homogeneous vector field, 92, 115
Hooke's law, 305
Hydrostatic pressure, 97
Hysteresis-free medium, 148

I
Incompressible flow, 141
Infinitesimal rotation, 262
Inhomogeneous Maxwell equations, 98, 384
Integrability condition, 91
Integration by parts, 140
Interaction potential, 196
Intermittent states, 364
Internal angular momentum, 98, 320
Internal field, 340
Internal force density, 142
Internal rotational degree of freedom, 98
Invariance condition, 369
Inverse transformation, 18
Irreducible, 55
Irreducible tensor, 155, 186
Irreversible processes, 301
Irreversible thermodynamics, 303
Isotropic fluid, 314
Isotropic linear medium, 149
Isotropic part, 34
Isotropic phase transition, 279
Isotropic system, 306
Isotropic tensor, 56, 183

J
Janus spheres, 197
Jaumann-Maxwell model, 330
Johnson-Segalman model, 330

K
Kayaking tumbling, 363

Kayaking wagging, 363
Kerr effect, 208
Kinetic energy, 108, 375
Kinetic energy operator, 108
Kinetic equation, 216, 227
Kirkwood-Smoluchowski equation, 228
Kronecker symbol, 15

L
Lagrange density, 385
Landau-de Gennes potential, 343
Landau-de Gennes theory, 279
Laplace equation, 93, 163
Laplace fields, 93
Legendre polynomial, 157, 172
Leslie viscosity coefficients, 333, 334
Levi-Civita tensor, 47, 378
Line integral, 111, 113
Linear mapping, 25, 66
Linear medium, 146
Linear molecules, 60
Linear momentum density, 97, 142
Linear momentum of the electromagnetic
    field, 151
Linear momentum operator, 107
Linear relation, 24, 27
Linear rotator, 250
Linear transformations, 16
Linearly increasing field, 85, 89
Liquid crystals, 273
Local momentum conservation equation,
    143
Log-rolling, 363
Lorentz field approximation, 204
Lorentz force, 45, 314
Lorentz invariance, 104, 371
Lorentz invariant, 374
Lorentz scalar, 373
Lorentz scaling, 381
Lorentz tensor, 373
Lorentz transformation, 372
Lorentz vector, 373
Lyotropic liquid crystal, 274

M
Magnetic field, 98, 127
Magnetic field tensors, 103
Magnetic induction, 98
Magnetic moment, 241
Magnetic permeability, 148
Magnetic quantum numbers, 241
Magnetic susceptibility, 98

Magnetic vector potential, 102
Magnetization, 99
Maier-Saupe distribution function, 340
Maier-Saupe mean field theory, 283
Maier-Saupe order parameter, 279
Main director, 362
Mapping, 73
Material coefficients, 299
Maxwell coefficient, 322
Maxwell distribution, 213
Maxwell effect, 322
Maxwell model, 326
Maxwell relaxation time, 228, 326
Maxwell stress tensor, 387
Maxwell's thermal pressure, 221
Mean free path, 356
Miesowicz viscosities, 316
Miesowicz viscosity coefficients, 333
Model parameters, 349
Molecular polarizability tensor, 63
Moment equation, 217, 341
Moment of inertia, 52
Moment of inertia tensor, 52, 60, 134
Moments of the distribution function, 203
Momentum balance, 151
Momentum conservation equation, 97
Momentum flux density, 151
Monopole function, 166
Multipole moment tensors, 171
Multipole–multipole interaction, 176
Multipole potential, 163, 165
Multipole potential tensors, 164

N
Nabla operator, 79, 105
Navier-Stokes equations, 98
Nematic, 274
Nematic liquid crystal, 273, 332
Nematic phase transition, 279
Newton, 80
Newtonian viscosity, 331
Non-diagonal tensor operators, 255
Non-equilibrium alignment, 303
Non-Equilibrium Molecular Dynamics
    (NEMD), 229, 322, 335
Non-Newtonian viscosity, 329
Non-Newtonian viscosity coefficient, 346
Non-orthogonal basis, 15
Non-spherical particles, 327
Nonlinear dynamics, 364
Nonlinear Maxwell model, 365
Nonlinear viscosity, 328

Normal pressure differences, 316, 329
Normal pressure gradient, 317

O
Octahedron, 172
Octupole moment, 170, 171
Octupole potential, 164
Oersted law, 101
Ohm's law, 267
Onsager relation, 333
Onsager symmetry relation, 303, 323
Onsager-Casimir symmetry relation, 303
Onsager-Parodi relation, 334
Orbital angular momentum, 43, 97
Orbital angular momentum operator, 190
Order parameter tensor, 203, 277
Orientational average, 200
Orientational distribution function, 202
Orientational entropy, 209
Orientational fluctuations, 354
Ortho-normalization relation, 358
Orthogonal basis, 14
Orthogonal matrix, 20
Orthogonal transformation, 19
Orthogonality relation, 19
Out of plane solutions, 364

P
Pair-correlation function, 222
Paramagnetic gases, 318
Parameter representation, 112
Parameter representation of surfaces, 117
Parity, 25, 300
Parity operation, 25
Path integral, 112
Pauli matrices, 239
Peculiar velocity, 218
Period doubling, 364
Permanent dipoles, 174
Phase transition, 343
Planar biaxial, 69
Planar geometry, 78, 82
Planar squeeze-stretch field, 86
Plane, 118, 121
Plane Couette symmetry, 329
Poiseuille flow, 316
Poisson equation, 91, 165
Polar coordinates, 14
Polarizability, 173
Polarized scattering, 353
Pole–dipole interaction energies, 176

Pole–pole interaction energies, 176
Pole–quadrupole interaction energies, 176
Polymer coils, 321
Polymeric liquids, 346
Position vector, 11
Potential energy, 223
Potential of a vector field, 114
Potentials, 78
Power density, 386
Poynting vector, 147
Pre-transitional increase, 341
Precession frequency, 319
Pressure broadening, 356
Pressure tensor, 97, 142, 224
Principal axes representation, 56
Principal values, 56
Principle of Archimedes, 143
Projection operator, 241
Projection tensor, 36, 260, 268
Proper rotation, 21
Proper time, 374

Q
Quadruple product, 72
Quadrupole moment, 169, 171
Quadrupole potential, 164
Quadrupole–quadrupole interactions, 177
Quantization axis, 161
Quantum mechanical angular momentum
    operator, 106

R
Radial and angular parts, 105
Radial and cylindrical fields, 85, 90
Radial component, 30
Radius of gyration tensor, 62
Rayleigh expansion, 233
Rayleigh scattering, 353
Reciprocal effect, 325
Reciprocal relations, 303
Recursion relation, 194
Reduced mass, 81
Regular tetrahedra, 293
Relaxation coefficients, 211
Relaxation time, 211
Relaxation time approximation, 228
Rheochaos, 365
Rheological behavior, 365
Rheological properties, 328
Rheology, 328
Richtungs-Quantelung, 241
Rod-like particles, 347

Rotated coordinate system, 24
Rotation, 89
Rotation axis, 134
Rotation tensor, 260
Rotational damping, 339
Rotational eigenstates, 251
Rotational quantum numbers, 253
Rotational Raman scattering, 256, 354
Rotational velocity, 320
Route to chaos, 364

S
Scalar fields, 78, 113
Scalar invariants, 69, 195
Scalar product, 13
Scaled variables, 347
Scattering wave vector, 224
Schrödinger equation, 234
Screw curve, 45
Second law of thermodynamics, 301
Second rank alignment tensor, 277, 341
Selection rules, 236
Senftleben-Beenakker effect, 318
Shape parameter, 339
Shear flow, 343, 362
Shear-flow induced distortion, 227
Shear modulus, 306
Shear rate tensor, 227
Shear thickening, 346
Shear thinning, 330, 346
Shear viscosity, 98, 177, 333
Shear viscosity tensor, 313
Shilnikov bifurcation, 368
Simple shear field, 87
Simple shear flow, 87, 90
Smectic, 274
Smectic A, 276
Smectic B, 276
Smectic C, 276
Smectic liquid crystals, 276
Solid body, 304
Solid-like rotation, 87
Solid-like rotational flow, 92
Sonine polynomials, 215
Source free, 89
Spatial Fourier transform, 352
Spatially inhomogeneous alignment, 349
Special relativity, 374
Spectral functions, 353
Speed of light, 370
Spherical Bessel functions, 233
Spherical components, 158, 262

Spherical coordinates, 130
Spherical geometry, 79
Spherical harmonic, 160, 165
Spherical symmetry, 83
Spherical tensor operators, 255
Spherical top molecules, 61
Spherical unit vectors, 158
Spin, 239
Spin averages, 247
Spin density, 98
Spin density matrix, 247
Spin density operator, 247
Spin matrices, 240
Spin operator, 239
Spin particles, 321
Spin tensors, 246
Spin traces, 245
Splay deformations, 285
Spontaneous birefringence, 275
Stability, 360
Static structure factor, 224
Stick-slip parameter, 367
Stokes force, 178
Stokes law, 124
Strain tensor, 305
Streaming double refraction, 322
Stress differences, 329
Stress tensor, 151
Substantial time derivative, 141
Summation convention, 12
Surface area, 67
Surface integrals, 120
Surface of a sphere, 119, 122
Susceptibility tensors, 27
Symmetric and antisymmetric parts, 33
Symmetric dyadic tensor, 59
Symmetric part, 33
Symmetric tensor, 55
Symmetric top molecules, 61
Symmetric traceless, 55
Symmetric traceless part, 34
Symmetric traceless tensor, 155
Symmetry, 300
Symmetry adapted ansatz, 324, 331
Symmetry adapted states, 362
Symmetry breaking, 300
Symmetry breaking states, 363
Symmetry relation, 352

T
Tangential component, 30
Tensor divergence, 96

Tensor polarization, 249
Tensor polarizations, 237
Tensor product, 192
Tetradics, 290
Tetrahedral symmetry, 291
Thermal equilibrium, 97
Thermodynamic fluxes, 303
Thermodynamic forces, 303
Thermotropic liquid crystal, 274
Third-order scalar invariant, 359
Thirteen moments approximation, 221
Time-correlation functions, 352
Time derivatives, 28
Time reversal, 30
Time reversal behavior, 301
Torque, 43, 176
Torque density, 152
Torque on a rotating solid body, 144
Trace of the tensor, 34
Trajectory, 29
Transformation behavior, 384
Transformation matrix, 22
Translation, 16
Transport-relaxation equations, 318
Transverse pressure gradient, 317
Transverse viscosity, 317
Transverse viscosity coefficients, 315
Transverse wave, 101
Triple product, 72
Trouton viscosity, 331
Tumbling, 363
Tumbling parameter, 338, 348
Turbulence, 329
Twist deformations, 285
Twist viscosity coefficient, 333
Two-particle density, 222
Typical for hydrodynamics, 178

U
Uniaxial ellipsoid, 73
Uniaxial extensional or compressional flow,
    330
Uniaxial non-spherical particles, 196
Uniaxial squeeze-stretch field, 86
Uniaxial tensor, 57
Unified theory, 343
Unit vector, 13

V
Van der Waals interaction, 177
Variational principle, 286, 385
Vector field, 84, 114

...tion, 249
...ntial, 91
...roduct, 41
...city, 29
velocity distribution function, 212
Visco-elasticity, 326
Viscometric functions, 329, 346
Viscous behavior, 313
Viscous fluid, 178
Viscous properties, 343
Voigt elasticity coefficients, 306
Volume, 67
Volume integrals, 129
Volume viscosity, 98, 313
Vortex, 89

Vorticity field, 87
Vorticity free flow, 330

W
Wagging, 363
Waldmann-Snider equation, 318
Wave equation, 101, 384
Wave mechanics, 107
Wave vector, 102, 352

Y
Yield stress, 367
Young elastic modulus, 307